Physics for Entertainment

By
Yakov Perelman.

The Inventions Researches and Writings of Nikola Tesla With Special Reference To His Work In Polyphase Currents And High Potential Lighting

By
Thomas Commerford Martin

Yakov Perelman's
Physics For Entertainment.
By
Yakov Perelman

Physics for Entertainment

By Yakov Perelman

Published in 1913, a best-seller in the 1930s and long out of print, *Physics for Entertainment* was translated from Russian into many languages and influenced science students around the world. Among them was Grigori Yakovlevich Perelman, the Russian mathematician (unrelated to the author), who solved the Poincaré conjecture, and who was awarded and rejected the Fields Medal. Grigori's father, an electrical engineer, gave him Physics for Entertainment to encourage his son's interest in mathematics. In the foreword, the book's author describes the contents as "conundrums, brain-teasers, entertaining anecdotes, and unexpected comparisons," adding, "I have quoted extensively from Jules Verne, H. G. Wells, Mark Twain and other writers, because, besides providing entertainment, the fantastic experiments these writers describe may well serve as instructive illustrations at physics classes." The book's topics included how to jump from a moving car, and why, "according to the law of buoyancy, we would never drown in the Dead Sea." Ideas from this book are still used by science teachers today. Yakov Isidorovich Perelman died in the siege of Leningrad in 1942.

CONTENTS

Chapter Three

ATMOSPHERIC RESISTANCE

Chapter Four

ROTATION. "PERPETUAL MOTION" MACHINES

Chapter Five

PROPERTIES OF LIQUIDS AND GASES

Chapter Six

HEAT

Chapter Seven

LIGHT

Chapter Eight

REFLECTION AND REFRACTION

Chapter Nine

VISION

Chapter Ten

SOUND AND HEARING

FROM THE AUTHOR'S FOREWORD
TO THE 13th EDITION

The aim of this book is not so much to give you some fresh knowledge, as to help you "learn what you already know". In other words, my idea is to brush up and liven your basic knowledge of physics, and to teach you how to apply it in various ways. To achieve this purpose conundrums, brain-teasers, entertaining anecdotes and stories, amusing experiments, paradoxes and unexpected comparisons—all dealing with physics and based on our everyday world and sci-fic—are afforded. Believing sci-fic most appropriate in a book of this kind, I have quoted extensively from Jules Verne, H. G. Wells, Mark Twain and other writers, because, besides providing entertainment, the fantastic experiments these writers describe may well serve as instructive illustrations at physics classes.

I have tried my best both to arouse interest and to amuse, as I believe that the greater the interest one shows, the closer the heed one pays and the easier it is to grasp the meaning—thus making for better knowledge.

However, I have dared to defy the customary methods employed in writing books of this nature. Hence, you will find very little in the way of parlour tricks or spectacular experiments. My purpose is different, being mainly to make you think along scientific lines from the angle of physics, and amass associations with the variety of things from everyday life. I have tried in rewriting the original copy to follow the principle that was formulated by Lenin thus: "The popular writer leads his reader towards profound thoughts, towards profound study, proceeding from simple and generally known facts; with the aid of simple argu-

9

ments or striking examples he shows the main *conclusions* to be drawn from those facts and arouses in the mind of the thinking reader ever newer questions. The popular writer does not presuppose a reader that does not think, that cannot or does not wish to think; on the contrary, he assumes in the undeveloped reader a serious intention to use his head and *aids* him in his serious and difficult work, leads him, helps him over his first steps, and *teaches* him to go forward independently. (*Collected Works*, Vol. 5, p. 311, Moscow 1961.)

Since so much interest has been shown in the history of this book, let me give you a few salient points of its "biography".

Physics for Entertainment first appeared a quarter of a century ago, being the author's first-born in his present large family of several score of such books. So far, this book—which is in two parts—has been published in Russian in a total print of 200,000 copies. Considering that many are to be found on the shelves of public libraries, where each copy reaches dozens of readers, I daresay that millions have read it. I have received letters from readers in the furthermost corners of the Soviet Union.

A Ukrainian translation was published in 1925, and German and Yiddish translations in 1931. A condensed German translation was published in Germany. Excerpts from the book have been printed in French—in Switzerland and Belgium—and also in Hebrew—in Palestine.

Its popularity, which attests to the keen public interest displayed in physics, has obliged me to pay particular note to its standard, which explains the many changes and additions in reprints. In all the 25 years it has been in existence the book has undergone constant revision, its latest edition having barely half of the maiden copy and practically not a single illustration from the first edition.

Some have asked me to refrain from revision, not to be compelled "to buy the new revised edition for the sake of a dozen or so new pages". Scarcely can such considerations absolve me of my obligation constantly to improve this book in every way. After all *Physics for Entertainment* is not a work of fiction. It is a book on science—be it even popular science—and the subject taken, physics, is enriched even in its fundamentals with every day. This must necessarily be taken into consideration.

On the other hand, I have been reproached more than once for failing to deal in this book with questions such as the latest achievements in radio engineering, nuclear fission, modern theories and the like. This springs from a misunderstanding. This book has a definite purpose; it is the task of other books to deal with the points mentioned.

Physics for Entertainment has, besides its second part, some other associated books of mine. One, *Physics at Every Step*, is intended for the unprepared layman who has still not embarked upon a systematic study of physics. The other two are, on the contrary, for people who have gone through a secondary school course in physics. These are *Mechanics for Entertainment* and *Do You Know Your Physics?*, the last being the sequel, as it were, to this book.

1936   *Y. Perelman*

CHAPTER ONE

SPEED AND VELOCITY. COMPOSITION
OF MOTIONS

HOW FAST DO WE MOVE?

A good athlete can run 1.5 km in about 3 min 50 sec—the 1958 world record was 3 min 36.8 sec. Any ordinary person usually does, when walking, about 1.5 metres a second. Reducing the athlete's rate to a common denominator, we see that he covers seven metres every second. These speeds are not absolutely comparable though. Walking, you can keep on for hours on end at the rate of 5 km. p.h. But the runner will keep up his speed for only a short while. On quick march, infantry move at a speed which is but a third of the athlete's, doing 2 m/sec, or 7 odd km.p.h. But they can cover a much greater distance.

I daresay you would find it of interest to compare your normal walking pace with the "speed" of the proverbially slow snail or tortoise. The snail well lives up to its reputation, doing 1.5 mm/sec, or 5.4 metres p.h.—exactly one thousand times less than your rate. The other classically slow animal, the tortoise, is not very much faster, doing usually 70 metres p.h.

Nimble compared to the snail and the tortoise, you would find yourself greatly outraced when comparing your own motion with other motions—even not very fast ones—that we see all around us. True, you will easily outpace the current of most rivers in the plains and be a pretty good second to a moderate wind. But you will successfully vie with a fly, which does 5 m/sec, only if you don skis. You won't overtake a hare or a hunting dog even when riding a fast horse and you can rival the eagle only aboard a plane.

Still the machines man has invented make him second to none for speed. Some time ago a passenger hydrofoil ship, capable of 60-70 km. p.h., was launched in the U.S.S.R. (*Fig. 1*). On land you can move faster

Fig. 1. Fast passenger hydrofoil ship

than on water by riding trains or motor cars—which can do up to 200 km. p.h. and more (*Fig. 2*). Modern aircraft greatly exceed even these speeds. Many Soviet air routes are serviced by the large TU-104

Fig. 2. New Soviet ZIL-111 motor car

(*Fig. 3*) and TU-114 jet liners, which do about 800 km. p.h. It was not so long ago that aircraft designers sought to overcome the "sound barrier", to attain speeds faster than that of sound, which is 330 m/sec,

14

or 1,200 km. p.h. Today this has been achieved. We have some small but very fast supersonic jet aircraft that can do as much as 2,000 km.p.h.

There are man-made vehicles that can work up still greater speeds. The initial launching speed of the first Soviet sputnik was about

Fig. 3. TU-104 jet airliner

8 km/sec. Later Soviet space rockets exceeded the so-called "escape" velocity, which is 11.2 km/sec at ground level.

The following table gives some interesting speed data.

A snail	1.5 mm/sec	or	5.4 metres p.h.
A tortoise	20 "	or	70 "
A fish	1 m/sec	or	3.5 km. p.h
A pedestrian	1.4 "	or	5 "
Cavalry, pacing	1.7 "	or	6 "
" trotting	3.5 "	or	12.6 "
A fly	5 "	or	18 "
A skier	5 "	or	18 "
Cavalry, galloping	8.5 "	or	30 "
A hydrofoil ship	16 "	or	58 "
A hare	18 "	or	65 "
An eagle	24 "	or	86 "
A hunting dog	25 "	or	90 "
A train	28 "	or	100 "
A ZIL-111 passenger car	50 "	or	170 "
A racing car (record)	174 "	or	633 "
A TU-104 jet airliner	220 "	or	800 "
Sound in air	330 "	or	1,200 "
Supersonic jet aircraft	550 "	or	2,000 "
The earth's orbital velocity	30,000 "	or	108,000 "

RACING AGAINST TIME

Could one leave Vladivostok by air at 8 a.m. and land in Moscow at 8 a.m. on the same day?

I'm not talking through my hat. We can really do that. The answer lies in the 9-hour difference in Vladivostok and Moscow zonal times. If our plane covers the distance between the two cities in these 9 hours, it will land in Moscow at the very same time at which it took off from Vladivostok. Considering that the distance is roughly 9,000 kilometres, we must fly at a speed of 9,000:9=1,000 km. p.h., which is quite possible today.

To "outrace the Sun" (or rather the earth) in Arctic latitudes, one can go much more slowly. Above Novaya Zemlya, on the 77th parallel, a plane doing about 450 km. p.h. would cover as much as a definite point on the surface of the globe would cover in an identical space of time in the process of the earth's axial rotation. If you were flying in such a plane you would see the sun suspended in immobility. It would never set, provided, of course, that your plane was moving in the proper direction.

It is still easier to "outrace the Moon" in its revolution around the earth. It takes the moon 29 times longer to spin round the earth than it takes the earth to complete one rotation (we are comparing, naturally, the so-called "angular", and not linear, velocities). So any ordinary steamer making 15-18 knots could "outrace the Moon" even in the moderate latitudes.

Mark Twain mentions this in his *Innocents Abroad*. When sailing across the Atlantic, from New York to the Azores "... we had balmy summer weather, and nights that were even finer than the days. We had the phenomenon of a full moon located just in the same spot in the heavens at the same hour every night. The reason for this singular conduct on the part of the moon did not occur to us at first, but it did afterward when we reflected that we were gaining about twenty minutes every day, because we were going east so fast—we gained just enough every day to keep along with the moon."

16

THE THOUSANDTH OF A SECOND

For us humans, the thousandth of a second is nothing from the angle of time. Time intervals of this order have only started to crop up in some of our practical work. When people used to reckon the time according to the sun's position in the sky, or to the length of a shadow (*Fig. 4*), they paid no heed to minutes, considering them even unworthy

Fig. 4. How to reckon the time according to the position of the sun (left), and by the length of a shadow (right)

of measurement. The tenor of life in ancient times was so unhurried that the timepieces of the day—the sun-dials, sand-glasses and the like—had no special divisions for minutes (*Fig. 5*). The minute hand first appeared only in the early 18th century, while the second sweep came into use a mere 150 years ago.

But back to our thousandth of a second. What do you think could happen in this space of time? Very much, indeed! True, an ordinary train would cover only some 3 cm. But sound would already fly 33 cm and a plane half a metre. In its orbital movement around the sun, the earth would travel 30 metres. Light would cover the great distance of 300 km. The minute organisms around us wouldn't think the thousandth of a second so negligible an amount of time—if they could think of course. For insects it is quite a tangible interval. In the space of a second a mosquito flaps its wings 500 to 600 times. Consequently in the space of a thousandth of a second, it would manage either to raise its wings or lower them.

We can't move our limbs as fast as insects. The fastest thing we can do is to blink our eyelids. This takes place so quickly that we fail even to notice the transient obscurement of our field of vision. Few know, though, that this movement, "in the twinkling of an eye"—which has

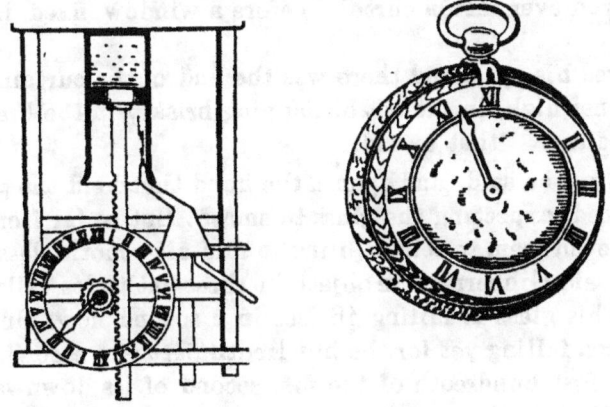

Fig. 5. An ancient water clock (left) and an old pocket-watch (right). Note that neither has the minute hand

become synonymous for incredible rapidity—is quite slow if measured in thousandths of a second. A full "twinkling of an eye" averages—as exact measurement has disclosed—two-fifths of a second, which gives us 400 thousandths of a second. This process can be divided into the following stages: firstly, the dropping of the eyelid which takes 75-90 thousandths of a second; secondly, the closed eyelid in a state of rest, which takes up 130-170 thousandths; and, thirdly, the raising of the eyelid, which takes about 170 thousandths.

As you see, this one "twinkling of an eye" is quite a considerable time interval, during which the eyelid even manages to take a rest. If we could photograph mentally impressions lasting the thousandth of a second, we would catch in the "twinkling of an eye" two smooth motions of the eyelid, separated by a period during which the eyelid would be at rest.

Generally speaking, the ability to do such a thing would completely transform the picture we get of the world around us and we would see the odd and curious things that H. G. Wells described in his *New Accelerator*. This story relates of a man who drank a queer mixture which caused him to see rapid motions as a series of separate static phenomena. Here are a few extracts.

"'Have you ever seen a curtain before a window fixed in that way before?'

"I followed his eyes, and there was the end of the curtain, frozen, as it were, corner high, in the act of flapping briskly in the breeze.

"'No,' said I, 'that's odd.'

"'And here,' he said, and opened the hand that held the glass. Naturally I winced, expecting the glass to smash. But so far from smashing it did not even seem to stir; it hung in mid-air—motionless. 'Roughly speaking,' said Gibberne, 'an object in these latitudes falls 16 feet in a second. This glass is falling 16 feet in a second now. Only you see, it hasn't been falling yet for the hundredth part of a second. [Note also that in the first hundredth of the first second of its downward flight a body, the glass in this case, covers not the hundredth part of the distance, but the 10,000th part (according to the formula $S = 1/2\ gt^2$). This is only 0.5 mm and in the first thousandth of the second it would be only 0.01 mm.]

"'That gives you some idea of the pace of my Accelerator.' And he waved his hand round and round, over and under the slowly sinking glass.

"Finally he took it by the bottom, pulled it down and placed it very carefully on the table. 'Eh?' he said to me, and laughed....

"I looked out of the window. An immovable cyclist, head down and with a frozen puff of dust behind his driving-wheel, scorched to overtake a galloping *char-à-banc* that did not stir....

"We went out by his gate into the road, and there we made a minute examination of the statuesque passing traffic. The top of the wheels

2*

and some of the legs of the horses of this *char-à-banc*, the end of the whip lash and the lower jaw of the conductor—who was just beginning to yawn—were perceptibly in motion, but all the rest of the lumbering conveyance seemed still. And quite noiseless except for a faint rattling that came from one man's throat! And as parts of this frozen edifice there were a driver, you know, and a conductor, and eleven people!...

"A purple-faced little gentleman was frozen in the midst of a violent struggle to refold his newspaper against the wind; there were many evidences that all these people in their sluggish way were exposed to a considerable breeze, a breeze that had no existence so far as our sensations went....

"All that I had said, and thought, and done since the stuff had begun to work in my veins had happened, so far as those people, so far as the world in general went, in the twinkling of an eye...."

Would you like to know the shortest stretch of time that scientists can measure today? Whereas at the beginning of this century it was only the 10,000th of a second, today the physicist can measure the 100,000 millionth of a second; this is about as many times less than a second as a second is less than 3,000 years!

THE SLOW-MOTION CAMERA

When H. G. Wells was writing his story, scarcely could he have ever thought he would see anything of the like. However he did live to see the pictures he had once imagined, thanks to what has been called the slow-motion camera. Instead of 24 shots a second—as ordinary motion-picture cameras do—this camera makes many times more. When a film shot in this way is projected onto the screen with the usual speed of 24 frames a second, you see things taking place much more slowly than normally—high jumps, for instance, seem unusually smooth. The more complex types of slow-motion cameras will almost simula H. G. Wells's world of fantasy.

WHEN WE MOVE ROUND THE SUN FASTER

Paris newspapers once carried an ad offering a cheap and pleasant way of travelling for the price of 25 centimes. Several simpletons mailed this sum. Each received a letter of the following content:

"Sir, rest at peace in bed and remember that the earth turns. At the 49th parallel—that of Paris—you travel more than 25,000 km a day. Should you want a nice view, draw your curtain aside and admire the starry sky."

The man who sent these letters was found and tried for fraud. The story goes that after quietly listening to the verdict and paying the fine demanded, the culprit struck a theatrical pose and solemnly declared, repeating Galileo's famous words: "It turns."

He was right, to some extent, after all, every inhabitant of the globe "travels" not only as the earth rotates. He is transported with still greater speed as the earth revolves around the sun. *Every second* this planet of ours, with us and everything else on it, moves 30 km in space, turning meanwhile on its axis. And thereby hangs a question not devoid of interest: When do we move around the sun faster? In the daytime or at night?

A bit of a puzzler, isn't it? After all, it's always day on one side of the earth and night on the other. But don't dismiss my question as senseless. Note that I'm asking you not when the earth itself moves faster, but when we, who live on the earth, move faster in the heavens. And that is another pair of shoes.

In the solar system we make two motions; we revolve around the sun and simultaneously turn on the earth's axis. The two motions add, but with different results, depending whether we are on the daylit side or on the nightbound one.

Fig. 6 shows you that at midnight the speed of rotation is *added* to that of the earth's translation, while at noon it is, on the contrary, *subtracted* from the latter. Consequently, *at midnight we move faster in the solar system than at noon.* Since any point on the equator travels about half a kilometre a second, the difference there between midnight and midday speeds comes to as much as a whole kilometre a second.

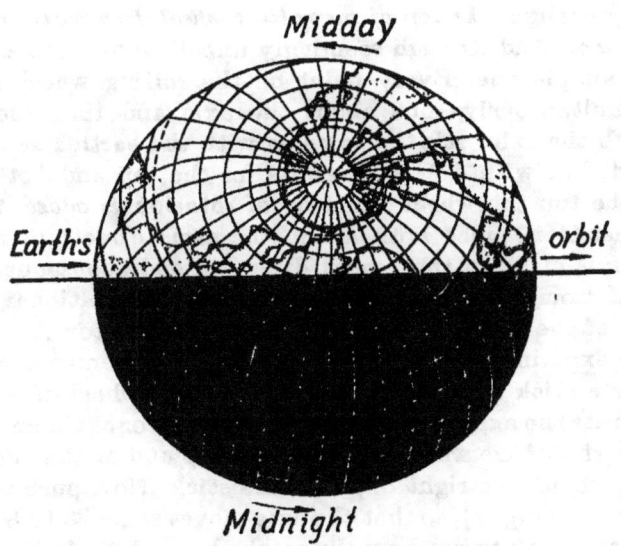

Fig. 6. On the dark side we move around the sun faster
than on the sunlit side

Any of you who are good at geometry will easily reckon that for Leningrad, which is on the 60th parallel, this difference is only half as much. At 12 p.m. Leningraders travel in the solar system half a kilometre more a second than they would do at 12 a.m.

THE CART-WHEEL RIDDLE

Attach a strip of coloured paper to the side of the rim of a cart-wheel or bicycle tire, and watch to see what happens when the cart, or bicycle, moves. If you are observant enough, you will see that near the ground the strip of paper appears rather distinctly, while on top it flashes by so rapidly that you can hardly spot it.

Doesn't it seem that the top of the wheel is moving faster than the bottom? And when you look at the upper and lower spokes of the moving wheel of a carriage, wouldn't you think the same? Indeed, the upper spokes seem to merge into one solid body, whereas the lower spokes can be made out quite distinctly.

Incredibly enough, *the top of the rolling wheel does really move faster than the bottom.* And, though seemingly unbelievable, the explanation is a pretty simple one. Every point on the rolling wheel makes *two* motions simultaneously—one about the axle and the other forward together with the axle. It's the same as with the earth itself. The two motions add, but with different results for the top and bottom of the wheel. At the top the wheel's motion of rotation *is added* to its motion of translation, since both are in the same direction. At the bottom rotation is made in the *reverse* direction and, consequently, must be *subtracted* from translation. That is why the stationary observer sees the top of the wheel moving faster than the bottom.

A simple experiment which can be done at convenience proves this point. Drive a stick into the ground next to the wheel of a stationary vehicle opposite the axle. Then take a piece of coal or chalk and make two marks on the rim of the wheel—at the very top and at the very bottom. Your marks should be right opposite the stick. Now push the vehicle a bit to the right (*Fig. 7*), so that the axle moves some 20 to 30 cm away from the stick. Look to see how the marks have shifted. You will find that the upper mark *A* has shifted much further away than the lower one *B* which is almost where it was before.

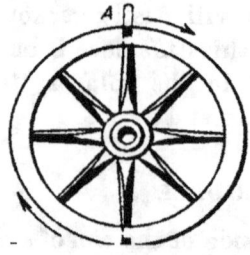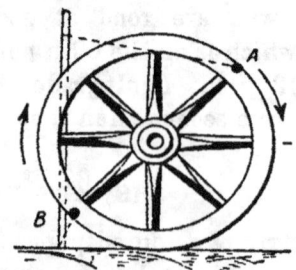

Fig. 7. A comparison between the distances away from the stick of points *A* and *B* on a rolling wheel (right) shows that the wheel's upper segment moves faster than its lower part

THE WHEEL'S SLOWEST PART

As we have seen, not all parts of a rolling cart-wheel move with the same speed. Which part is slowest? That which touches the ground. Strictly speaking, at the moment of contact, this part is absolutely stationary. This refers only to a rolling wheel. For the one that spins round a fixed axis, this is not so. In the case of a flywheel, for instance, all its parts move with the same speed.

BRAIN-TEASER

Here is another, just as ticklish, problem. Could a train going from Leningrad to Moscow have any points which, in relation to the railroad track, would be moving in the opposite direction? It could, we find. All the train wheels have such points every moment. They are at the bottom of the protruding rim of the wheel (the bead). When the train goes forward, these points move backward. The following experiment, which you can easily do yourself, will show you how this happens. Attach a match to a coin with some plasticine so that the match protrudes in the plane of the radius, as shown in *Fig. 8*. Set the coin together with the match in a vertical position on the edge of a flat ruler and hold it with your thumb at its point of contact—*C*. Then roll it to and fro. You will see that points *F*, *E* and *D* of the jutting part of the match

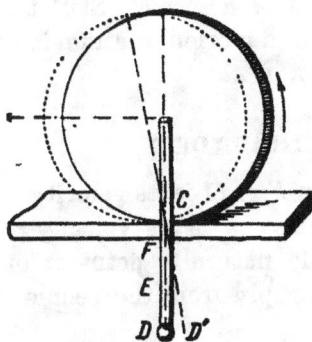

Fig. 8. When the coin is rolled leftwards, points *F*, *E* and *D* of the jutting part of the match move backwards

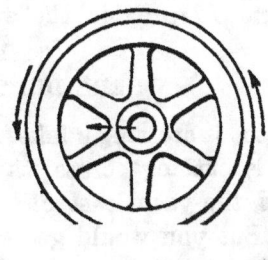

Fig. 9. When the train wheel rolls leftwards the lower part of its rim rolls the other way

Fig. 10. Top: the curve (a cycloid) described by every point on the rim of a rolling cart-wheel. Bottom: the curve described by every point on the rim of a train wheel

move not forwards but backwards. The further point *D*—the end of the match—is from the edge of the coin, the more noticeable backward motion is (point *D* shifts to *D'*).

The points on the bead of the train wheel move similarly. So when I tell you now that there are points in a train that move not *forward* but *backward*, this should no longer surprise you. True, this backward motion lasts only the negligible fraction of a second. Still there is, despite all our habitual notions, a backward motion in a moving train. *Figs. 9* and *10* provide the explanation.

WHERE DID THE YACHT CAST OFF?

A rowboat is crossing a lake. Arrow *a* in *Fig. 11* is its velocity vector. A yacht is cutting across its course; arrow *b* is its velocity vector. Where did the yacht cast off? You would naturally point at once to point *M*. But you would get a different reply from the people in the dinghy. Why?

They don't see the yacht moving at right angles to their own course, because they don't realise that they are moving themselves. They think

Fig. 11. The yacht is cutting across the rowboat's course. Arrows *a* and *b* designate the velocities. What will the people in the dinghy see?

they're stationary, while everything around is moving with their own speed but in the opposite direction. From their point of view the yacht is moving not only in the direction of the arrow *b* but also in the direction of the dotted line *a*—opposite to their own direction (*Fig. 12*). The two motions of the yacht—the real one and the seeming one—are resolved according to the rule of the parallelogram. The result is that the people in the rowboat think the yacht to be moving along the diagonal of the parallelogram *ab;* that is also why they think the yacht cast off not at point *M*, but at point *N*, way in front of the rowboat (*Fig. 12*).

Travelling together with the earth in its orbital path, we also plot the position of the stars wrongly—just as the people in the dinghy did when asked where the yacht cast off from. We see the stars displaced slightly forward in the direction of the earth's orbital motion. Of course, the earth's speed is negligible compared with that of light (10,000

26

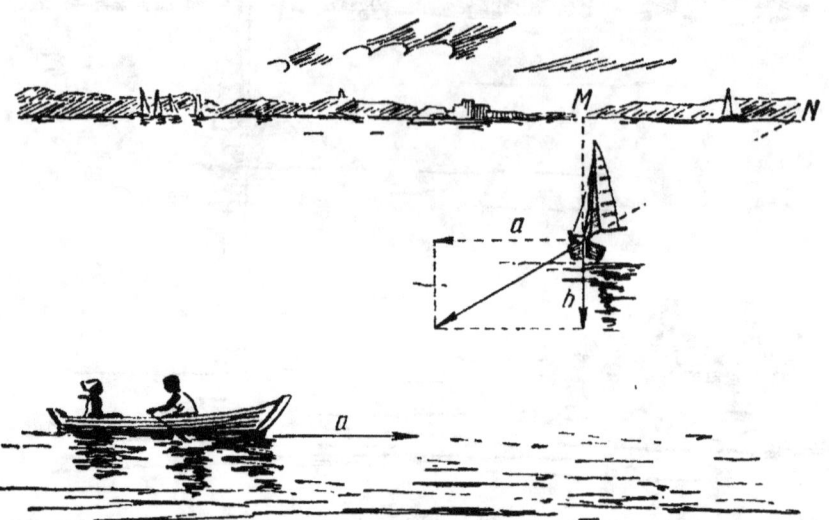

Fig. 12. The people in the dinghy think the yacht to be coming towards them slantwise—from point *N*

times less) and, consequently, this stellar displacement, known as aberration of light, is insignificant. However, we can detect it with the aid of astronomical instruments.

Did you like the yacht problem? Then answer another two questions related to the same problem. Firstly, give the direction in which the yachtsmen think the dinghy is moving. Secondly, say where the yachtsmen think the dinghy is heading. To answer, you must construct a parallelogram of velocities on the vector *a* (*Fig. 12*), whose diagonal will indicate that from the yachtsmen's point of view the dinghy seems to be moving slantwise, as if heading for the shore.

CHAPTER TWO

GRAVITY AND WEIGHT. LEVERS. PRESSURE

TRY TO STAND UP!

You'd think I was joking if I told you that you wouldn't be able to get up from a chair—provided you sat on it in a certain way, even though you wouldn't be strapped down to it. Very well, let's have a go. Sit down on a chair in the same way the boy in *Fig. 13* is sitting. Sit upright and *don't shove your feet under the chair*. Now try to get up without moving your feet or bending forward. You can't, however hard you try. You'll never stand up until you push your feet under the chair or lean forwards.

Before I explain, let me tell you about the equilibrium of bodies in general, and of the human body in particular. A thing will not topple only when the perpendicular from its centre of gravity goes through its base. The leaning cylinder in *Fig. 14* is bound to fall. If, on the other hand, the perpendicular from its centre of gravity fell through its base, it wouldn't topple over. The famous leaning towers of Pisa and Bologna, or the leaning campanile in Arkhangelsk (*Fig. 15*), don't fall, despite their tilt, for the same reason. The perpendiculars from their centres of gravity do not lie outside their bases. Another reason is that their foundations are sunk deep in the ground.

Fig. 13. It's impossible to get up

You won't fall only when the perpendicular from your centre of gravity lies within the area bound by the outer edge of your feet (*Fig. 16*). That is why it is so hard to stand on one leg and still harder to balance on a tight-rope. Our "base" is very small and the perpendicular from the centre of gravity may easily come to lie outside its limits. Have you noticed the odd gait of an "old sea dog"? He spends most of his life aboard a pitching ship where the perpendicular from the centre of gravity of his body may come to fall outside his "base" any moment. That accustoms him to walk on deck so that his feet are set wide apart and take in as large a space as

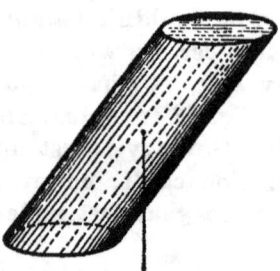

Fig. 14. The cylinder must topple as the perpendicular from its centre of gravity lies outside its base

Fig. 15. Arkhangelsk leaning campanile. A reproduction from an old photograph

possible, which saves him from falling. Naturally, he'll waddle in the same habitual fashion on hard ground as well.

Another instance—of an opposite nature this time. This is when the effort to keep one's balance results in a beautiful pose. Porters who carry loads on their heads are well-built—a point, I presume, you have noticed. You may have also seen exquisite statues of women holding jars on their heads. It is because they carry a load on their heads that these people have to hold their heads and bodies upright. If they were to lean in any direction, this would shift the perpendicular from the centre of gravity higher than usual, because of the head-load, outside the base and unbalance them.

Back now to the problem I set you at the beginning of the chapter. The sitting boy's centre of gravity is inside the body near the spine— about 20 centimetres above the level of his navel. Drop a perpendicular from this point. It will pass through the chair behind the feet. You already know that for the man to stand up it should go through the area *taken up by the feet*. Consequently, when we get up we must either bend forward to shift the centre of gravity, or shove our feet beneath the chair to place our "base" below the centre of gravity. That is what we usually do when getting up from a chair. If we are not allowed to do this, we'll never be able to stand up—as you have already gathered from your own experience.

Fig. 16. When one stands, the perpendicular from the centre of gravity passes through the area bound by the soles of one's feet

WALKING AND RUNNING

The things you do thousands of times a day, and day after day all your life, ought to be things you have a very good idea about, oughtn't they? Yes, you will say. But that is far from so. Take walking and running, for instance. Could anything be more familiar? But I wonder how many of you have a clear picture of what we really do when we walk and run, or of the difference between the two. Let's see what a physiologist has to say about walking and running. I'm sure most of you will find his description startlingly novel. (The passage is from Prof. Paul Bert, *Lectures on Zoology*. The illustrations are my own.)

"Suppose a person is standing on one leg, the right leg, for instance. Suppose further that he is lifting his heel, meanwhile bending forwards. [When walking or running a person exerts on the ground, when pushing his foot away from it, a pressure of some 20kg in addition to his weight. Hence a person exerts a greater pressure on the ground when he is moving than when standing.—*Y. P.*] In such a position the

30

perpendicular from the centre of gravity will naturally be outside the base and the person is bound to fall forwards. Scarcely has he started doing this than he quickly throws forward his left leg, which was suspended thus far, to put it down on the ground in front of the perpendicular from the centre of gravity. The perpendicular thus comes to drop through the area bound by the lines linking the points of

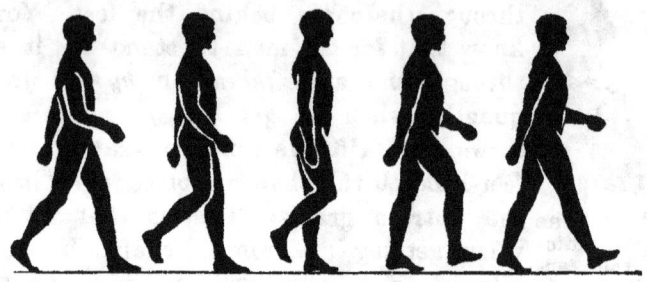

Fig. 17. How one walks. The series of positions in walking

support of both feet. Balance is thus restored; the person has taken a step forward.

"He may remain in this rather tiring position, but should he wish to continue forward, he will lean still further forward, shift the perpendicular from the centre of gravity outside the base, and again throw his leg—the right one this time—forwards when about to fall. He thus

B

Fig. 18. A graph showing how one's feet move when walking. Line *A* is the left foot and line *B* is the right foot. The straight sections show when the foot is on the ground, and the curves—when the foot is in the air. In the time-interval *a* both feet are on the ground; in the time-interval *b*, foot *A* is in the air and foot *B* still on the ground; in the timeinterval *c* both feet are again on the ground. The faster one walks, the shorter the time-intervals *a* and *c* get (compare with the "running" graph in *Fig. 20*).

takes another step forward. And so on and so forth. Consequently, walking is just *a series of forward fallings*, punctually forestalled by throwing the leg left behind into a supporting position.

Fig. 19. How one runs. The series of positions in running, showing moments when both feet are in the air

"Let's try to get to the root of the matter. Suppose the first step has already been made. At this particular moment the right foot is still on the ground and the left foot is already touching it. If the step is not **very** short the right heel should be lifted, because it is this rising heel that enables one to bend forward and change one's balance. It is the heel of the left foot that touches the ground first. When next the entire

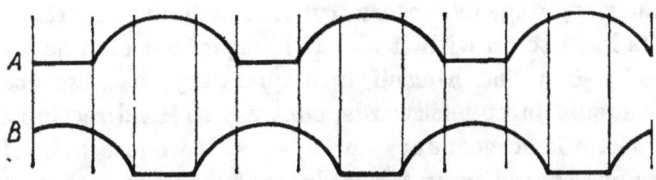

Fig. 20. A graph showing how one's feet move when running (compare with *Fig. 18*). There are time-intervals (*b*, *d* and *f*) when both feet are in the air. This is the difference between running and walking

sole stands on the ground, the right foot is lifted completely and no longer touches the ground. Meanwhile the left leg, which is slightly bent at the knee, is straightened by a contraction of the femoral triceps to become for an instant vertical. This enables the half-bent right

32

leg to move forward without touching the ground. Following the body's movement the heel of the right foot comes to touch the ground in time for the next step forwards. The left leg, which at this moment has only the toes of the foot touching the ground and which is about to rise, goes through a similar series of motions.

"Running differs from walking in that the foot on the ground is energetically straightened by a sudden contraction of its muscles to throw the body forwards so that the latter is *completely off the ground for a very short interval of time*. Then the body again falls to come to rest on the other leg, which quickly moves forward while the body is still in the air. Thus, running consists of a series of *hops* from one foot to the other."

As for the energy a person expends in walking along a horizontal pavement it is not at all nil as some might think. With every step made, the centre of gravity of a walker's body is lifted by a few centimetres. A reckoning shows that the work spent in walking along a horizontal path is about a fifteenth of that required to raise the walker's body to a height equivalent to the distance covered.

HOW TO JUMP FROM A MOVING CAR

Most will surely say that one must jump forward, in the direction in which the car is going, in conformity with the law of inertia. But what does inertia have to do with it all? I'll wager that anyone you ask this question will soon find himself in a quandary, because according to inertia one should jump backwards, contrary to the direction of motion. Actually inertia is of secondary importance. If we lose sight of the main reason why one should jump forwards—one that has nothing to do with inertia—we will indeed come to think that we must jump backwards and not forwards.

Suppose you have to jump off a moving car. What happens? When you jump, your body has, at the moment you let go, the same velocity as the car itself—by inertia—and tends to move forwards. By jumping forwards, far from diminishing this velocity, we, on the contrary, increase it. Then shouldn't we jump *backwards*—since in that case the velocity thus imparted would be *subtracted* from the velocity our body possesses by inertia, and hence, on touching the ground, our body would have less of a toppling impetus?

But, when one jumps from a moving carriage, one always jumps forwards in the direction of its movement. That is indeed the best way, a time-honoured one, and I strongly warn you against trying to test the awkwardness of jumping backwards.

We seem to have a contradiction, don't we? Now whether we jump forwards or backwards we risk falling, since our bodies are still moving when our feet touch the ground and come to a halt. (See "When Is a Horizontal Line Not Horizontal?" from the third chapter of *Mechanics for Entertainment* for another explanation.) When jumping forwards, the speed with which our bodies move is even greater than when jumping backwards, as I have already noted. But it is much *safer* to jump forwards than backwards, because then we mechanically throw a leg forwards or even run a few steps, to steady ourselves. We do this *without thinking*; it's just like walking. After all, according to mechanics, walking, as was noted before, is nothing but a *series of forward fallings of our body, guarded against by the throwing out of a leg.* Since we don't have this guarding movement of the leg when falling *backwards* the danger is much greater. Then even if we do fall forwards we can soften the impact with our hands, which we can't do if we fall on our backs.

As you see, it is safer to jump forwards, not so much because of inertia, but because of ourselves. This rule is plainly inapplicable to *one's belongings*, for instance. A bottle thrown from a moving car forwards stands more chances of crashing when it hits the ground than if thrown backwards. So if you have to jump from a moving car and have some luggage with you, first chuck out the luggage *backwards* and then jump *forwards* yourself. Old hands like tramcar conductors and ticket inspectors often jump off stepping *backwards but with their backs turned to the direction in which they jump.* This gives them a double advantage: firstly they reduce the velocity that the body acquires by inertia, and, secondly, guard themselves against falling on their backs, as they jump with their faces forward, in the direction where they are most likely to fall.

CATCHING A BULLET

The following curious incident was reported during the First World War. One French pilot, while flying at an altitude of two kilometres, saw what he took to be a fly near his face. Trapping it with his hands, he was flabbergasted to find that he had caught a German bullet! How like the tall stories told by Baron Munchausen of legendary fame, who claimed he had caught cannon balls with bare hands! But there is nothing incredible in the bullet-catching story.

A bullet does not fly everlastingly with its initial velocity of 800-900 m/sec. Air resistance causes it to slow down gradually to a mere 40 m/sec towards the end of its journey. Since aircraft fly with a similar speed, we can easily have a situation when bullet and plane will be flying with the same speed, in which case the bullet, in its relation to the plane and its pilot, will be stationary or barely moving. The pilot can easily catch it with his hand, especially if gloved, because a bullet heats up considerably while whizzing through the air.

MELON AS BOMB

We have seen that in certain circumstances a bullet can lose its "sting". But there are instances when a gently thrown "peaceful" object has a destructive impact. During the Leningrad-Tiflis motor run in 1924, Caucasian peasants tossed melons, apples, and the like at the racing cars to express their admiration. However, these innocuous gifts made terrible dents and seriously injured the motorists. This happened because the car's velocity added to that of the tossed melons or apples, transforming them into dangerous projectiles. A ten-gramme bullet possesses the same energy of motion as a 4kg melon thrown at a car doing 120 km.p.h. Of course, the impact of a melon is not the same as the bullet's since melons, after all, are squashy.

When we have super-fast planes doing about 3,000 km.p.h.— a bullet's approximate velocity—their pilots may chance to encounter what we have just described. Everything in the way of a super-fast aircraft will ram into it. Machine-gun fire or just a chance handful of bullets dropped from another plane will have the same effect; these bullets will strike the aircraft with the same impact as if fired from a machine gun. Since the relative velocities in both cases are the same—the plane and bullet meet with a speed of about 800 m/sec—the destruction done when they collide is the same as well . On the contrary, bullets fired from behind at a plane moving with the same speed are harmless, as we have already seen.

Fig. 21. Water-melons tossed at a fast-moving car are as dangerous as bombs

In 1935 engine driver Borshchov prevented a railway disaster by cleverly taking advantage of the fact that objects moving in the same direction at practically the same speed come into contact without knocking each other to pieces. He was driving a train between Yelnikov and Olshanka, in Southern Russia. Another train was puffing along in front. The driver of this train couldn't work up enough steam to make the grade. He uncoupled his engine and several waggons and set off for the nearest station, leaving a string of 36 waggons behind. But as he did not place brake-shoes to block their wheels, these waggons started to roll back down the grade. They gathered up a speed of some 15 km. p.h. and a collision seemed imminent. Luckily enough, Borshchov had his wits about him and was able to figure out at once what to do. He braked his own train and also started a backward manoeuvre, gradually working up the same speed of 15 km.p.h. This enabled him to bring the 36 waggons to rest against his own engine, without causing any damage.

Finally this same principle is applied in a device making it easier for us to write in a moving train. You all know that this is hard to do because of the jolts when the train passes over the rail joints. They do not act simultaneously on both paper and pen. So our task is to

Fig. 22. Contraption for writing in a moving train

contrive something that would make the jolts act simultaneously on both. In this case they would be in a state of rest with respect to each other.

Fig. 22 shows one such device. The right wrist is strapped to the smaller board *a* which slides up and down in the slots in board *b*, which, in turn, slides to and fro along the grooves of the writing board placed on the train compartment table. This arrangement provides plenty of "elbow-room" for writing and at the same time causes each jolt to act simultaneously on both paper and pen, or rather the hand holding the pen. This makes the process as simple as writing on an ordinary table at home. The only unpleasant thing about it is that since the jolts again do not act simultaneously on both wrist and head, you get a jerky picture of what you're writing.

HOW TO WEIGH YOURSELF

You will get your correct weight only if you stand on the scales without moving. As soon as you bend down, the scales show less. Why? When you bend, the muscles that do this also pull up the lower half of your body and thus diminish the pressure it exerts on the scales. On the contrary, when you straighten up, your muscles push the upper and lower halves of the body away from each other; in this case the scales will register a greater weight since the lower half of your body exerts a greater pressure on the scales.

You will change your weight-readings—provided the scales are sensitive enough—even by lifting an arm. This motion already slightly increases your body's seeming weight. The muscles you use to lift your arm up have the shoulder as their fulcrum and, consequently, push it together with the body down, increasing the pressure exerted on the scales. When you stop lifting your arm you start using another, opposite set of muscles; they pull the shoulder up, trying to bring it closer to the end of the arm; this reduces the weight of your body, or rather its pressure on the scales. On the contrary, when you lower your arm you reduce the weight of your body, to increase it when you stop lowering it. In brief, by using your muscles you can increase or reduce your weight, meaning of course the pressure your body exerts on the scales.

WHERE ARE THINGS HEAVIER?

The earth's pull diminishes the higher up we go. If we could lift a kilogramme weight 6,400 km up, to twice the earth's radius away from its centre, the force of gravity would grow $2^2=4$ times weaker, in which case a spring balance would register only 250 grammes instead of 1,000. According to the law of gravity the earth attracts bodies as if its entire mass were concentrated in the centre; the force of this attraction diminishes inversely to the square of the distance away. In our particular instance, we lifted the kilogramme weight twice the distance away from the centre of the earth; hence attraction grew $2^2=4$ times weaker. If we set the weight at a distance of 12,800 km away from the surface of the earth—three times the earth's radius—the force of attraction would grow $3^2=9$ times weaker, in which case our kilogramme weight would register only 111 grammes on a spring balance.

You might conclude that the deeper down in the earth we were to put our one-kilogramme weight, the greater the force of attraction would grow and the more it should weigh. However, you would be mistaken. The weight of a body does not increase; on the contrary, it diminishes.

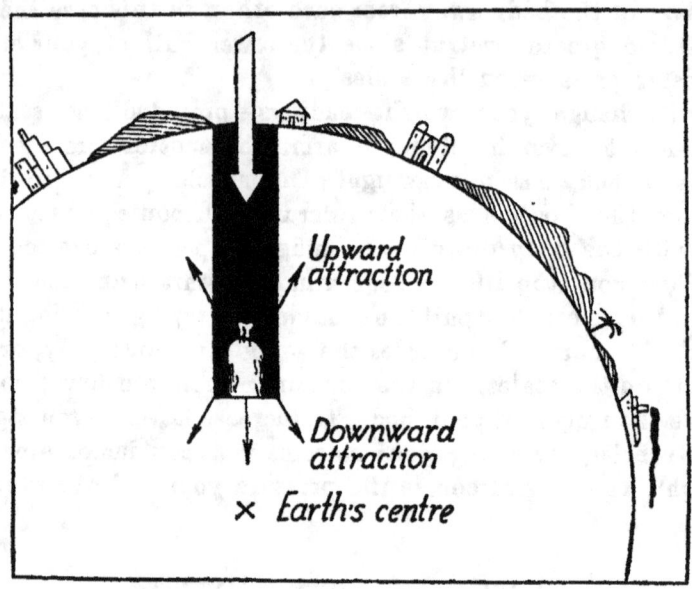

Fig. 23. Gravitational pull lessens the closer we get to the middle of the Earth

This is because now the earth's attracting forces no longer act just on one side of the body but all around it. *Fig. 23* shows you the weight in a well; it is pulled down by the forces below it and simultaneously up by the forces above it. It is really only the pull of that spherical part of the earth, the radius of which is equal to the distance from the centre of the earth to the body, that is of importance. Consequently, the deeper down we go, the less a body should weigh. At the centre of the earth it should weigh nothing, as here it is attracted by equal forces on all sides.

To sum up: a body weighs most at the earth's surface; its weight diminishes whether it is lifted up from the earth's surface or interred (this would stand, naturally, only if the earth were homogeneous in density throughout). Actually, the closer to its centre, the greater the earth's density; at first the force of gravity grows to some distance down; only then does it start to diminish.

HOW MUCH DOES A FALLING BODY WEIGH?

Have you noticed that odd sensation you experience when you *start* to go down in a lift? You feel abnormally light; if you were falling into a bottomless abyss you would feel the same. This sensation is caused by weightlessness. At the very first moment when the lift-cabin floor has already started to go down but you yourself have still not acquired its velocity, your body exerts scarcely any pressure at all on the floor, and, consequently, *weighs* very little. An instant later this queer sensation is gone. Now your body seeks to fall faster than the smoothly running lift; it exerts a pressure on the cabin floor, reacquiring its full weight.

Tie a weight to the hook of a spring balance and observe the pointer as you quickly lower the balance together with the weight. For convenience's sake insert a small piece of cork in the slot and observe how it moves. The pointer will fail to register the full weight; it will be much less! If the balance were falling freely and you would be able to watch its pointer meanwhile, you would see it register a zero weight.

The heaviest object will lose all its weight when falling. The reason is simple. "Weight" is the force with which a body pulls at something holding it up or presses down on something supporting it. A *falling* body cannot pull the balance spring as it is falling together with it. A falling body does not pull at anything or press down on anything. Hence, to ask how much something weighs when falling is the same as to ask how much it weighs when it does not weigh.

Galileo, the father of mechanics, wrote way back in the 17th century in his *Mathematical Proofs Concerning Two Fields of a New Science:* "We feel a load on our back when we try to prevent it from dropping. But if we were to drop as fast as the load does, how could it press upon and burden us? This would be the same as to try to transfix with a spear [without letting go of it—*Y. P.*] somebody running ahead of us as fast as we are running ourselves."

The following simple experiment well illustrates this point. Place a nutcracker on one of the scale pans, with one arm on the pan and the

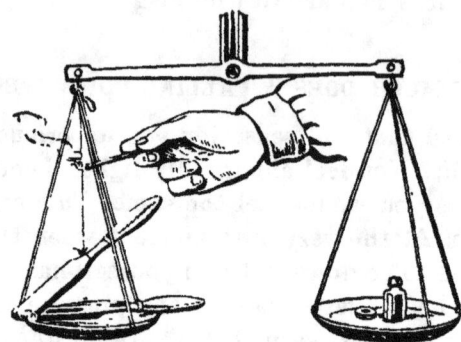

Fig. 24. Falling bodies are weightless

other tied by a piece of thread to the hook of the scale arm (*Fig. 24*). Add weights to the other pan to balance the nutcracker. Apply a lighted match to the thread. The thread will burn through and the suspended nutcracker arm will fall onto the pan. Will the pan holding the nutcracker dip? Will it rise? Or will it remain in equilibrium? Since you know by now that a falling body weighs nothing, you should be able to give the correct answer. The pan will rise for a moment. Indeed, though joined to the lower arm the nutcracker's upper arm nevertheless exerts less of a pressure on the pan when falling than when stationary. For a moment the nutcracker's weight diminishes, and thus the pan holding it rises.

FROM EARTH TO MOON

The years between 1865 and 1870 saw the publication in France of Jules Verne's *From the Earth to the Moon*, in which he set forth a fantastic scheme to shoot at the Moon an enormous projectile with people inside. His description seemed so credible that most of you who have read this book have probably hazarded whether this really could be done. Well, let's discuss it. (Today, after Sputnik and Lunik, we know that it is rockets, not cannon projectiles, that will be used for space travel. However, since a rocket flies after its last engine burns out, in accord with the same laws of ballistics, don't think Perelman is behind the times.)

Let's see at first whether we can fire a shell from a gun—at least theoretically—so that it never falls back to earth again. Theory tells us that it's possible. Indeed, why does a shell fired horizontally eventually fall back on earth again? Because the earth attracts it, curving its trajectory. Instead of keeping up a straight course, it curves towards the ground and is, therefore, bound to hit it sooner or later. The earth's surface is also curved, but the shell's trajectory is bent still more. However, if we made the shell follow a trajectory curved in exactly the same way as the earth's surface it would never fall back on earth again. Instead, it would trace an orbit concentric with the earth's circumference, becoming its satellite, a baby moon.

But how are we to make the shell follow such a trajectory? All we must do is to impart a sufficient initial velocity. Look at *Fig. 25* which depicts a cross-section of part of the earth. A cannon is mounted on the hilltop at point *A*. A shell fired horizontally from it would reach point *B* a second later—if not for the earth's gravitational pull. Instead, it reaches point *C* five metres lower than *B*. Five metres is the distance any freely falling body travels (in a void) in the first second—due to earth's surface gravitational pull. If, after it drops these five metres, our shell is at exactly the same distance away from the ground as it was when fired at point *A*, it means that the shell is

Fig. 25. How to reckon a projectile's "escape" velocity

following a trajectory curved concentrically to the earth's circumference.

All that remains is to reckon the distance AB (*Fig. 25*), or, in other words, the distance the shell travels horizontally in the space of a second, which will tell us the speed we need. In the triangle AOB, the side OA is the earth's radius (roughly 6,370,000 m); $OC=OA$ and $BC=5$m; h nce OB is 6,370,005 m. Applying Pythagoras's theorem we get:

$$(AB)^2 = (6,370,005)^2 - (6,370,000)^2.$$

We resolve this equation to find AB equal to roughly 8 km.

So, if there were no drag a shell shot horizontally with a muzzle velocity of 8 km/sec *would never fall back to earth again*; it would be an everlasting baby moon.

Now suppose we imparted to our shell a still greater initial velocity. Where would it fly then? Scientists dealing with celestial mechanics have proved that velocities of 8, 9 and even 10 km/sec give a trajectory shaped like an ellipse which would be the more elongated the greater the initial speed is. When the velocity reaches 11.2 km/sec, the shell will describe not an ellipse but a non-locked curve, a parabola, and fly away from the earth never to return (*Fig. 26*). So, *theoretically* it is quite possible to fly to the Moon inside a cannon ball, provided its muzzle speed is big enough. This, however, is a problem that may

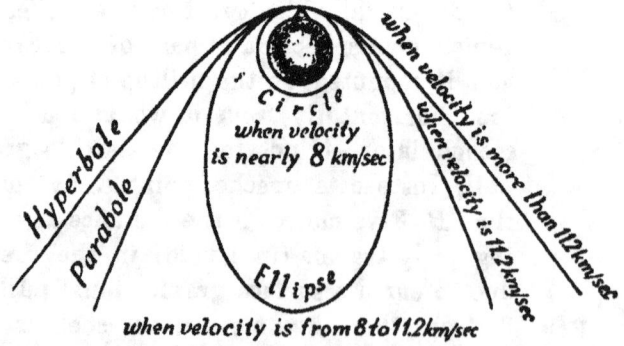

Fig. 26. When a projectile is fired with a starting velocity of 8 km/sec and more

present some quite specific difficulties. Let me refer you, for greater detail, to Book Two of *Physics for Entertainment* and also to *Interplanetary Travel*—another book of mine. (In the foregoing we dismissed the drag which in real life would exceedingly complicate the attainment of such great velocities and perhaps render the task absolutely impossible.)

FLYING TO THE MOON: JULES VERNE VS. THE TRUTH

Any of you who have read *From the Earth to the Moon* most likely remembers the interesting passage describing the projectile's intersection of the boundary where the Moon matches the Earth in attraction. Wondrous things happened. All the objects inside the projectile became weightless; the travellers themselves began to float in the air.

There is nothing wrong in all this. What Jules Verne did lose sight of was that this happens not only at the point the novelist gave. It happens before and after as well—in fact, *as soon as free flight begins.*

It seems incredible, doesn't it? I'm sure though that soon you will be surprised not to have noticed this signal omission before. Let's turn to Jules Verne for an example. You haven't forgotten how the space travellers ejected the dead dog and how surprised they were to see it continue to trail behind the projectile instead of falling back to earth. Jules Verne described and explained this correctly. In a void all bodies fall with the same speed, with gravity imparting an identical acceleration to each. So, owing to gravity, both the projectile and the dead dog should have acquired the same falling velocity (an identical acceleration). Rather should we say that due to gravity their starting velocities diminished in the same measure. Consequently, both should whizz along with the same velocity; that is why after its ejection the dead dog kept on trailing along in the projectile's wake.

Jules Verne's omission was: if the dead dog did not fall back to earth again *after the ejection*, why should it fall when *inside* the projectile? The same forces act in both cases! The dead dog suspended in mid-air inside the projectile should remain in that state as its speed *is* absolutely the same as the projectile's; hence it is in a state of rest *in respect to the projectile.*

44

What goes for the dead dog also goes for the travellers and all objects, in general, inside the projectile, as they all fly along the trajectory with the same speed as the projectile and should not fall, even though having nothing to stand, sit, or lie on. One could take a chair, turn it upside down and lift it to the ceiling; it won't fall "down", because it will go on travelling together with the ceiling. One could sit on this chair also upside down and not fall either. What, after all, could make him fall? If he did fall or float down, this would mean that the projectile's speed would be greater than that of the man on the chair; otherwise the chair wouldn't float or fall. But this is impossible since we know that everything inside the projectile has the same acceleration as the projectile itself. This was what Jules Verne failed to take into account. He thought everything inside the projectile would continue to press down on its floor when it was in space. He forgot that a weight presses down on what supports it only because this support is stationary. But if both object and its support hurtle with the same velocity in space they simply can't press down on each other.

So, as soon as the projectile began to fly further on by its own momentum, its travellers became completely weightless and could float inside it, just as everything else could, too. That alone would have immediately told the travellers whether they were hurtling through space or still inside the cannon. Jules Verne, however, says that in the first half hour after the projectile was shot into space they couldn't guess whether they were moving or not, however hard they tried.

"'Nicholl, are we moving?'

"Nicholl and Barbicane looked at each other; they had not yet troubled themselves about the projectile.

"'Well, are we really moving?' repeated Michel Ardan.

"'Or quietly resting on the soil of Florida?' asked Nicholl.

"'Or at the bottom of the Gulf of Mexico?' added Michel Ardan."

These are doubts a steamboat passenger may entertain; they are absolutely out of the question for a space traveller, because he can't help noticing his complete loss of weight, which the steamboat passenger naturally retains.

Jules Verne's projectile must certainly be a very queer place, a tiny world of its own, where things are weightless and float and stay where they are, where objects retain their equilibrium wherever they are placed, where even water won't pour out of an inclined bottle. A pity Jules Verne slipped up, when this offers such a delightful opportunity for fantasy to run riot! (If this problem interests you, we could refer you to the appropriate chapter in A. Sternfeld's *Artificial Earth Satellites.*)

FAULTY SCALES CAN GIVE RIGHT WEIGHT

What is more important to get the right weight—scales or weights? Don't think both identically important. You can get the right weight even on faulty scales as long as you have the right weights. Of the several methods used, we shall deal with two.

One was suggested by the great Russian chemist Dmitry Mendeleyev. You begin by placing anything handy on one of the pans. Make sure that it is heavier than the object you want to weigh. Balance it with weights on the other pan. Then place what you want to weigh on the pan holding the weights and remove the necessary number of weights to bring to balance again. Tote up the weights removed to get the weight of what you wanted to weigh. This is called "the constant load method" and is particularly convenient when several objects need to be weighed in succession. The initial load is used to weigh everything you have to weigh.

Another method, called the "Borda method" after the scientist who proposed it, is as follows:

Place the object you want to weigh on one of the pans. Then pour sand or shot into the other pan till the scales balance. Remove your object from the pan—but don't touch the sand or shot in the other pan!—and place weights in the emptied pan till the scales balance again. Tote up these weights to find how much your object weighs. This is also called "replacement weighing".

This simple method can also be used for a one-pan spring balance, provided of course you have correct weights. In this case you don't need either sand or shot. Just put your object on the pan and note the reading. Then remove the object and place in the pan as many weights as needed to get the same reading. Their combined weight will give the weight of the object they replace.

How much can you lift with one arm? Let's say it's ten kilogrammes. Does this amount qualify your arm's muscle-power? Oh, no. Your biceps is much stronger. *Fig. 27* shows how this muscle works. It is attached close to the fulcrum of the lever that the bone of your forearm represents. The load you are lifting acts on the other end of this live lever. The distance between the load and the fulcrum, that is, the joint, is almost eight times more than that between the end of the biceps and the fulcrum. This means that if you are lifting a load of 10 kg your biceps is exerting eight times as much power, and, consequently, could lift 80 kg.

It would be no exaggeration to say that everybody is much stronger than he is, or rather that one's muscles are much more powerful than what we can really do with them. Is this an expedient arrangement? Not at all, you might think at first glance. We seem to have totally unrewarded loss. Recall, however, an old "golden rule" of mechanics: whatever you lose in power you gain in displacement. Here you gain in speed; your arm moves eight times faster than its muscles do. The muscular arrangement in animals enables them to move extremities quickly, which is more important than strength in the struggle to survive. Otherwise, we would move around at literally a snail's pace.

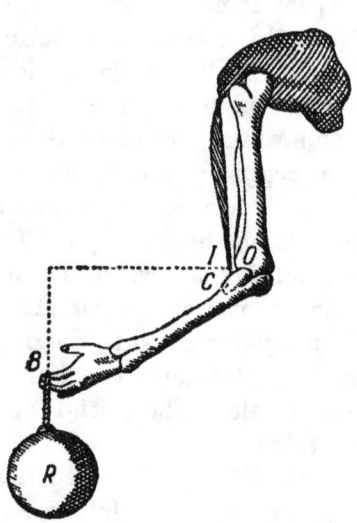

Fig. 27. Forearm *C* acts as a lever. The force acts on point *I*; the fulcrum is at point *O* and the load *R* is being lifted from point *B*. *BO* is roughly eight times longer than *IO*. (This drawing is from an ancient book called *Concerning the Motions of Animals* by the 17th-century Florentine scholar Borelli who was the first to apply the laws of mechanics to physiology.)

WHY DO SHARP THINGS PRICK?

Have you ever wondered why a needle so easily pierces things? Why is it so easy to drive a needle through a piece of cloth or cardboard and so hard to do the same thing with a blunt nail? After all, doesn't the same force act in both cases? The force is the same, but the *pressure* isn't. In the case of the needle the entire force is concentrated on its point; in the case of the nail the same amount of force is distributed over the larger area of the blunt end. So, though we exert the same force, the needle gives a much greater pressure than the blunt nail.

You all know that a twenty-toothed harrow loosens the soil more deeply than a sixty-toothed one of the same weight. Why? Because the *load on each tooth* of the first harrow is more than on each tooth of the second.

When we speak of pressure, we must always take into consideration, besides force, also the area upon which this force acts. When we are told that a worker is paid a hundred rubles, we don't know whether this is much or little, because we don't know whether this is for a whole year or for just one month.

Similarly does the action of a force depend on whether it is distributed over a square centimetre or concentrated on the hundredth of a square millimetre. Skis easily take us across fresh snow; without them we fall through. Why? On skis the weight of your body is distributed over a much greater area. Supposing the surface of our skis is 20 times more than the surface of our soles, on skis we would exert on the snow a· pressure which is only a twentieth of the pressure we exert when we have no skis on. As we have noticed, fresh snow will bear you when you are on skis, but will treacherously let you down when you're without them.

For the same reason horses used in marshlands are shod in a special fashion giving them a wider supporting area and lessening the pressure exerted per square centimetre. For the same reason people take the same precautions when they want to cross a bog or thin ice, often crawling to distribute their weight over a greater area.

Finally, tanks and caterpillar tractors don't get stuck in loose ground, though they are very heavy, again because their weight is distributed over a rather great supporting area. An eight-ton tractor exerts a pressure of only 600 grammes per square centimetre. There are caterpillars which exert a pressure of only 160 gr/cm² despite a two-ton load, which makes for the easy crossing of peatbogs and sand-beaches. Here it is a large supporting area which gives the advantage, whereas in the case of the needle it is the other way round.

This all shows that a sharpened edge pierces things only because it has a very minute area for the force to act upon. That is why a sharp knife cuts better than a blunt one: the force is concentrated on a smaller area of the knife edge. To sum up: sharp objects prick and cut well, because much pressure is concentrated on their points and edges.

COMFORTABLE BED ... OF ROCK

Why is it pleasanter to sit on a chair than on a flat-topped stool though both are of wood? Why is it pleasant to lie in a hammock though the pieces of rope that go to make it are by no means soft?

I suppose you've already guessed why. The stool-top is flat; when you sit on it, you press down with your entire weight on a small area. Chairs, on the other hand, usually have a concave seat; in this case you press down on a much greater area, over which your weight is distributed. To every unit of surface you have a smaller weight, smaller pressure.

The trick, as you see, is to distribute pressure more evenly. On a soft bed we make depressions that conform to the uneven shape of our bodies. Pressure is distributed rather evenly, with only a few grammes per square centimetre. No wonder we find it so pleasant.

The following reckoning well illustrates the difference. An adult person has a body surface of about 2m², or 20,000 cm². In bed roughly a quarter of it—0.5 m², or 5.000 cm²—supports him. Presuming that he weighs about 60 kg, or 60,000 gr, this would mean that we have a pressure of only 12 gr/cm². On bare boards he would have a supporting area of only some 100 cm². There are fewer points of contact. This means a pressure per sq. cm. of half a kilogramme instead of a dozen grammes. Quite a noticeable difference, isn't it? And one feels it at once.

But even the *hardest* of beds would be as soft as eiderdown, provided the weight of your body were distributed all over it. Suppose you left the imprint of your body in wet clay. When it hardens—drying clay shrinks by some five to ten per cent, but we shall discount this—you could lie in it again and think yourself in a featherbed. Though you would be lying on what is practically rock, it would feel soft, because your weight would be distributed over a much greater area of support.

CHAPTER THREE

ATMOSPHERIC RESISTANCE

BULLET AND AIR

Every schoolboy knows that the air impedes a bullet in its flight. Few, however, know what a great impediment it is. Most think such a "caressing" environment as the air—which is something we usually never feel—could not really get in the way of a fast-flying rifle bullet.

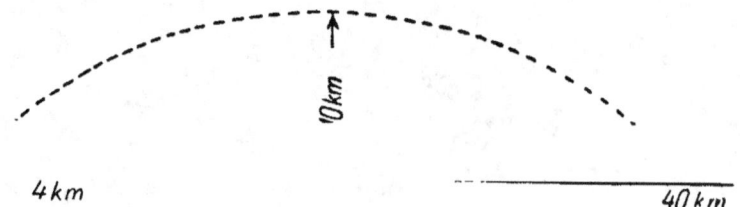

Fig. 28. Flight of a bullet in the air and in a vacuum. The big arc is the trajectory described when there is no atmosphere. The tiny, left-hand arc is the real trajectory

However, one good glance at *Fig. 28* will already make you realise that the air places quite a serious obstacle in the bullet's way. The large curve on the diagram designates the trajectory the bullet would describe were there no air. In this case, after flying out of a rifle tilted at 45°, and with an initial velocity of 620 m/sec, the bullet would describe a vast arc ten kilometres high and fly almost 40 km. But actually our bullet flies only 4 km, describing the tiny arc which is scarcely noticeable side by side with the first one. That is what the resistance of the air, the air drag, does!

BIG BERTHA

The Germans were the first—in 1918, towards the close of the First World War, when French and British aircraft had put a stop to German air raids—to practise long-range artillery bombardment from a distance of 100 kilometres and more.

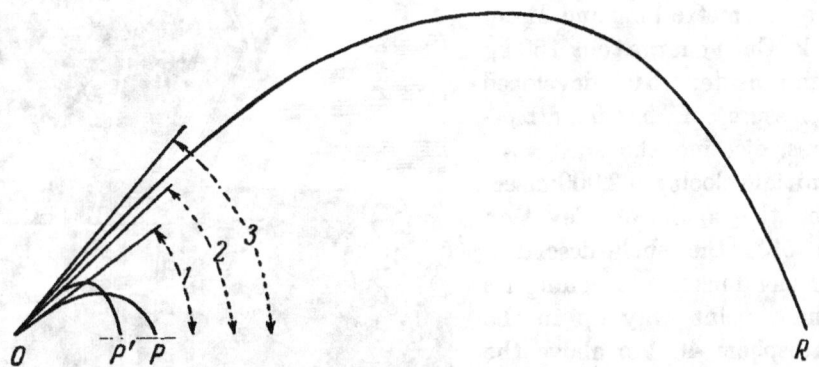

Fig. 29. The range changes when the mouth of a long-distance gun is tilted at different angles. In the case of angle *1*, the projectile strikes *P*, and in the case of angle *2*, *P'*, but in the case of angle *3*, it flies much farther as it goes through the rarefied stratosphere

It was by chance that German gunners hit upon their absolutely novel method for shelling the French capital, which was then at least 110 km away from the front lines. Firing shells from a big cannon tilted up at a wide angle, they unexpectedly discovered that they could make them fly 40 km instead of 20. When a shell is fired steeply upwards with a great initial velocity, it reaches a high-altitude, rarefied atmospheric strata, where the air drag is rather weak. Here it flies for quite a distance, before veering steeply to fall back to earth again. *Fig. 29* illustrates the great difference in trajectory at different angles of the gun barrel. This became the basic principle of the long-range gun that the Germans designed to bombard Paris from 115 km away. Such a gun was made—Big Bertha—and it fired more than 300 shells at Paris throughout the summer of 1918.

It was learned later that Big Bertha consisted of a tremendous steel tube 34 metres long and 1 metre thick. The breech walls were 40 cm thick. The gun itself weighed 750 tons. Its 120 kg shells were one metre long and 21 cm thick. Each charge took 150 kg of gunpowder which developed a pressure of 5,000 atmospheres, ejecting the shell with an initial velocity of 2,000m/sec. Since the angle of elevation was 52°, the shell described a tremendous arc, reaching its highest point way up in the stratosphere 40 km above the ground. It took the shell only 3.5 minutes to reach Paris, 115 km away; two minutes were spent in the stratosphere.

Fig. 30. Big Bertha

Big Bertha was the first long-range gun in history, the progenitor of modern long-range artillery.

Let me note that the greater the initial velocity of a bullet or shell, the more resistance the air puts up, increasing, moreover, in proportion to the square, cube, etc., of the velocity, depending on its amount.

WHY DOES A KITE FLY?

Do you know why a kite soars when pulled forward by the twine? If you do, you will also be able to understand why airplanes fly and maple seeds float. You'll even be able to fathom to some extent the causes of the boomerang's very odd behaviour. Because all these things are related. The very same air which is so great an impediment to a bullet or a shell enables the light maple seed to float and even heavy airliners to fly.

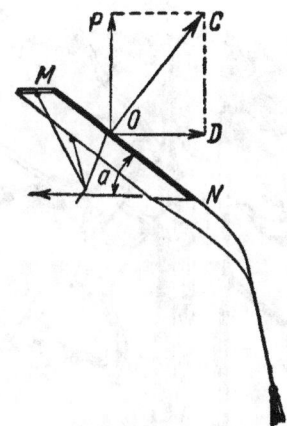

Fig. 31. The forces that make a kite fly

If you don't know why a kite flies, the simple drawing in *Fig. 31* will provide the explanation. Let line *MN* designate the kite's cross-section. When you let the kite go and pull at the cord, the kite, because of its heavy tail, moves at an angle to the ground. Let the kite move from right to left and *a* be the angle at which the plane of the kite is inclined to the horizon. We shall now proceed to examine the forces that act on the kite. The air, of course, should obstruct its movement and exert some pressure on it, designated on *Fig. 31* by the vector *OC*. Since the air always presses perpendicular to the plane, *OC* is at right angles to *MN*. The force *OC* may be resolved into two forces by constructing what is called a *parallelogram of forces*. This gives us the two forces *OD* and *OP*. Of these two, the force *OD* pushes the kite back, thus reducing its initial velocity. The other force, *OP*, pulls the kite up, reducing its weight. When this force is big enough it overcomes the weight of the kite and lifts it. That is why the kite goes *up* when you pull it *forwards*.

The airplane is also a kite really, with the difference that its forward motion, which makes it go up, is imparted not by our pulling at it but by the propeller or jet engine. This is, of course, a very crude explanation. There are other factors that cause an airplane to rise. They are explained in Book Two of *Physics for Entertainment* under the heading "Waves and Whirlwinds".

LIVE GLIDERS

As you see aircraft are not made like birds, as one usually thinks, but rather like flying squirrels or.flying fish, which, by the way, employ their flying mechanism not to fly up but merely to take rather big leaps—or what a flier would call "glides". In their case, the force *OP*

(*Fig. 31*) is too small to offset their weight; it merely reduces their weight, enabling them to make very big jumps from some high point (*Fig. 32*). A flying squirrel can jump 20-30 m from the top of one tree to the lower branches of another. In the East Indies and in Ceylon a much larger species of flying squirrel is found. This is the kaguan, a flying lemur, which is about the size of our house cat and which has a wing spread of about half a metre, enabling it to leap some 50 m, despite its great weight. As for the phalangers that inhabit the Sunda Isles and the Philippines, they can jump as far as 70 m.

Fig. 32. Flying squirrels jump from 20 to 30 m

BALLOONING SEEDS

Plants also often employ a gliding mechanism to propagate. Many seeds have either a parachuting tuft or hairy appendages (the pappus), as in dandelions, cotton balls, and "goat's beards", or "wings", as in conifers, maples, white birches, elms, lindens, many kinds of umbelliferae, etc.

In Kerner von Marilaum's well-known *Plant Life*, we find the following relevant passage:

"On windless sunny days a host of seeds and fruits are lifted high up by vertical air currents. However, after dusk they usually float down a short cry away. It is important for seeds to fly, not so much to cover a wide area as to inhabit cracks in terraces and cliffs, which they would never reach in any other way. Meanwhile, horizontal air currents may carry these hovering seeds and fruits rather far.

"The seeds of some plants retain their wings and parachutes only while they fly. Thistle seeds quietly float until they encounter an

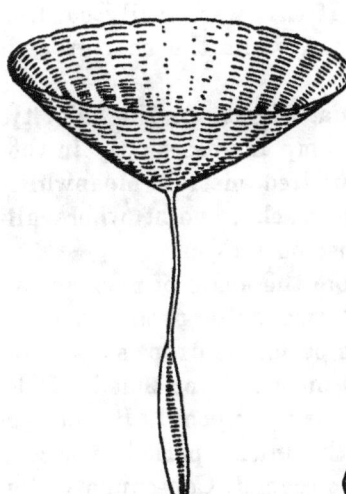

obstacle, when the seed discards its parachute and drops to the ground. That is why we see the thistle so often near walls and fences. But there are other cases, when the seed is attached permanently to its parachute."

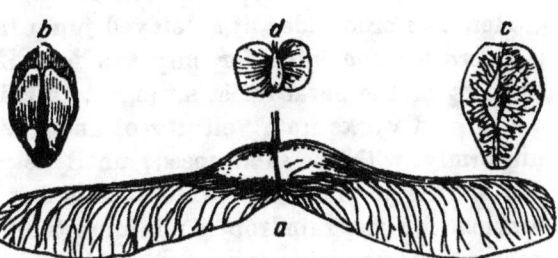

Fig. 33. Fruit of "goat's beard"

Fig. 34. Winged seeds of a) maple, b) pine-tree, c) elm, and d) birch

Figs. 33 and 34 show some seeds and fruits that have a gliding mechanism. As a matter of fact these plant "gliders" beat man-made ones on many points. They can lift a load which may be much greater than their own weight and automatically stabilise it. Thus if the seed of the Indian jasmine should chance to turn over, it will automatically regain its initial position with its convex side bottom-most, but when it meets an obstacle it doesn't capsize and drop like a plummet, but coasts down instead.

DELAYED PARACHUTE JUMPING

This, naturally, brings to mind the brave jumps parachutists sometimes make. They bail out at altitudes of some ten kilometres and pull the ripcord only after plummeting like a stone without opening their parachutes for quite a distance. Many think that in this delayed jump the parachutist falls as if in empty space. If this were really so, the delayed jump would be a much shorter affair, while the near-ground velocity would be tremendous.

However, atmospheric resistance prevents acceleration. The velocity of the falling parachutist during a delayed jump increases only in the first ten seconds, only for the first few hundred metres. Meanwhile atmospheric resistance increases, to finally reach a point where all further acceleration stops and the falling becomes even.

Here is a crude idea of a delayed jump from the angle of mechanics. *Acceleration* continues for only the first 12 seconds or even less, depending on the parachutist's weight. In this period he drops some 400-450 m and works up a velocity of about 50 m/sec. After that he falls uniformly, with the same speed, until he pulls the ripcord. Raindrops fall similarly. The only difference is that the initial period of acceleration for the raindrop is no more than a second. Consequently its near-ground velocity is not so great as in a delayed parachute jump, being between 2 and 7 metres a second, depending on its size. (Read my *Mechanics for Entertainment* for more about raindrop velocity and my *Do You Know Your Physics?* for more about delayed parachute jumping.)

THE BOOMERANG

For long this ingenious weapon, the most perfect technical device primitive man ever invented, had scientists wonderstruck. Indeed, the queer tangled trajectory the boomerang traces (*Fig. 35*) can tease any mind. Nowadays we have an elaborate theory to explain the boomerang; it is no longer a wonder. This theory is too intricate to explain at length. Let me merely note that boomeranging is the combined result of three factors: firstly, the initial throw; secondly, the boomerang's own rotation, and thirdly, atmospheric resistance. The Australian aborigine instinctively knows how to combine all three, deftly changing the boomerang's tilt and direction, and he throws it with a greater or smaller force to obtain the desired result.

You, too, can acquire some knack in boomerang-throwing. To make one for indoors, cut it out of cardboard, in the form shown in *Fig. 36*. Each arm is about 5 cm long and a little less than a centimetre

Fig. 35. Australian aborigine throwing a boomerang. The dotted line shows the trajectory of the boomerang, should it miss its target

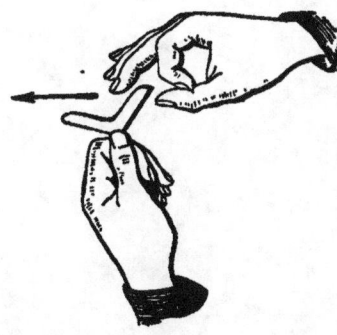

Fig. 36. A cardboard boomer-ang and how to "throw" it

Fig. 37. Another cardboard boomerang (real size)

wide. Press it under the nail of your thumb and flick it forwards and a bit upwards. It will fly some five metres, loop, and return to your feet, provided it doesn't hit anything on the way. You can make a still better boomerang by copying the one given in *Fig. 37*, and also by twisting it to look somewhat like a propeller (as shown at the bottom of *Fig. 37*). After some experience you should be able to make it describe intricate curves and loops before it returns to your feet.

In conclusion let me note that the boomerang is not at all exclusively an Australian missile as is usually thought. It was employed in India and according to extant murals it was once commonly used by Assyrian warriors (see *Fig. 38*). It was also familiar in ancient Egypt and Nubia. The Australian boomerang's only distinguishing feature is the propeller-like twist that we mentioned, sending it into such a maze of whirls and loops, returning it to the thrower, *should he miss*.

Fig. 38. Ancient Egyptian warrior throwing a boomerang

CHAPTER FOUR

ROTATION. "PERPETUAL MOTION" MACHINES

HOW TO TELL A BOILED AND RAW EGG APART?

How can we find out whether an egg is boiled or not, without breaking the shell?

Mechanics gives us the answer. The whole trick is that a boiled egg spins differently than a raw one. Take the egg, place it on a flat plate and twirl it (*Fig. 39*). A cooked egg, especially a hard-boiled one, will revolve *much faster and longer* than a raw one; as a matter of fact, it is hard even to make the raw egg turn. A hard-boiled egg spins so quickly that it takes on the hazy form of a flat white ellipsoid. If flicked sharply enough, it may even rise up to stand on its narrow end.

The explanation lies in the fact that while a hard-boiled egg revolves as one whole, a raw egg doesn't; the latter's liquid contents do not

Fig. 39. Spinning an egg

Fig. 40. Telling a boiled egg from a raw one.

have the motion of rotation imparted at once and so act as a brake, retarding by force of inertia the spinning of the solid shell. Then boiled and raw eggs stop spinning differently. When you touch a twirling boiled egg with a finger, it stops *at once*. But a raw egg will resume spinning for a while after you take your finger away. Again the force of inertia is responsible. The liquid contents of the raw egg still continue moving after the solid shell is brought to a state of rest. Meanwhile the contents of the boiled egg stop spinning together with the outer shell.

Here is another test, similar in character. Snap rubber bands around a raw egg and a boiled one, along their "meridian", as it were, and hang them up by two identical pieces of string (*Fig. 40*). Twist the strings. giving the same number of turns, and then let them go. You will spot the difference between the two eggs at once. Inertia causes the boiled egg to overshoot its starting position and give the string some more twists in the opposite direction; then the string unwinds again with the egg again giving several turns; this continues for some time, the number of twists gradually diminishing until the egg comes to rest. The raw egg, on the other hand, scarcely overshoots its initial position at all; it will give but one or two turns and stop long before the boiled egg does. As we already know, this is due to its liquid contents which impede its movement.

WHIRLIGIG

Open an umbrella, stand it up with its top on the floor and twist the handle. You can easily make it revolve rather quickly. Now throw a little ball or a crumpled piece of paper into the umbrella. It won't stay there; it will be shot out by what has wrongly come to be called the "centrifugal force" but which is actually nothing but a manifestation of the force of inertia. The ball or piece of paper will be thrown off, not along the continuation of the radius but at a tangent to the circular motion.

At some public parks one may find an amusement (*Fig. 41*) based on this principle of rotation, where you may try out the law of inertia on yourself. This is a sort of whirligig with a round floor on which people either stand, sit, or lie. A concealed motor starts the floor revolving,

Fig. 41. A whirligig. Centrifugal forces are hurling the boys off

increasing its speed till inertia makes everybody on it slither or slide towards its edge. At first this is hardly noticeable, but the further away one gets from the centre, the more noticeable do both speed and, consequently, inertia grow. You try hard to hold on, but it is to no avail and finally you are hurled off.

The Earth itself is, in point of fact, a huge whirligig. Though it doesn't hurl us off, it does reduce our weight. At the equator, where rotation is fastest, one can "shed" a 300th of one's weight in this manner. This, plus another factor, the Earth's compression, reduces weight at the equator by about 0.5% or 1/200th. An adult person will consequently weigh 300 grammes less at the equator than at any of the poles.

INKY WHIRLWINDS

Make a teetotum, as shown in life size in *Fig. 42*, out of white cardboard and a match sharpened at one end. No particular knack is needed to twirl it—it's something any child can do. But though a child's toy, it can be very instructive. Do the following. Spill a few drops of ink on it and set it spinning before the ink dries. When it stops, look

Fig. 42. Ink drop traces on a twirling teetotum

to see what has happened to the ink drops. They will have drawn whorls—a miniature whirlwind.

Incidentally, this resemblance is not accidental. The whorls on the teetotum trace the movement of the ink drops, which undergo exactly what you experienced on the revolving floor. As the drop shoots away from the centre due to centrifugal forces, it reaches a place on the teetotum having a greater speed of rotation than the speed of the drop itself. Here the disc spins faster than the drop which seems to glide away, lagging behind the radial "spokes", as it were. That is why the drops curve, and we see the trace of curvilinear motion.

The same is true for air currents diverging from a centre of high atmospheric pressure (in "anticyclones"), or converging in a centre of low atmospheric pressure (in "cyclones"). The ink whorls depict these stupendous whirlwinds in miniature.

THE DELUDED PLANT

The centrifugal force produced by fast rotation may even outvie gravity, a point that was demonstrated by the British botanist Knight more than a hundred years ago. It is common knowledge that a young plant always directs its stem contrary to gravity, or, in plain language,

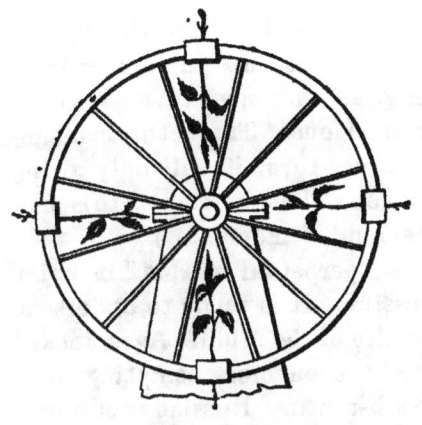

Fig. 43. Seeds germinating on the rim of a spinning wheel stem towards the axle and send their roots outwards

grows upwards. Knight, however, caused seeds to sprout inwards, from the outer rim of a quickly-spun wheel. The roots, on the other hand, were directed outwards (*Fig. 43*). He was able to fool the plant, as it were, substituting centrifugal force for gravity. The artificial gravity proved to be more powerful than the earth's natural pull—by the by, the modern theory of gravity does not present any objections, in principle, to this explanation.

"PERPETUAL MOTION" MACHINES

"Perpetual motion" is a topic that comes in for frequent mention, but I don't think all realise what it actually means. The "perpetual motion" machine is an imagined mechanism which continues its motion without end and meanwhile can also do some useful work, as lifting a load, for instance. It has never been constructed, though attempts have been made since ancient times. The futility of this task gave rise to the firm conviction that a "perpetual motion" machine is impossible, and to the law of the conservation of energy—fundamental for modern science. "Perpetual motion" as such is endless motion without any work done.

Fig. 44 depicts one of the oldest projects of a "perpetual motion" machine which certain cranks try to revive even now. Attached to the rim of the wheel are rods with weights at their ends. In any position of the wheel the weights on the right-hand side are farther from the centre than those on the left-hand side. Consequently, the right-hand weights should always outweigh the left side, thus compelling the wheel to turn. Hence the wheel should spin for ever, or at least until its axis wears through. That at any rate was what its inventor thought. Don't try to make such a machine. It will never turn. Why?

Though the right-hand weights are always farther from the centre, you are sure to have a position when they will be less in number than those on the left-hand side. Look at *Fig. 44* once again. You see only four right-hand weights and eight left-hand ones. The entire arrangement is thus balanced. The wheel will never turn; it will only swing a bit and then come to rest in this position. (The motion of this machine is explained by the so-called theorem of momenta.)

It has been proved beyond doubt that a "perpetual motion" machine as a source of energy is absolutely impossible. It is futile to undertake this task, which alchemists of yore, especially of the Middle Ages, racked their brains in vain to solve, thinking it even more tempting than the "philosopher's stone". The famous 19th-century Russian poet Pushkin describes such a dreamer, one Berthold, in his *Chivalrous Episodes*.

"'What is *perpetuum mobile*?' Martin inquired.

"'*Perpetuum mobile*,' Berthold returned, 'is perpetual motion. If I find perpetual motion I see no bounds to man's creative endeavour. For, my good Martin, while the making of gold is entrancing, a discovery perhaps, both curious and profitable, the finding of *perpetuum mobile*.... Ah, how grand that would be!'"

Hundreds of "perpetual motion" machines were invented, but none ever moved. Every inventor invariably omitted something that "upset the apple-cart".

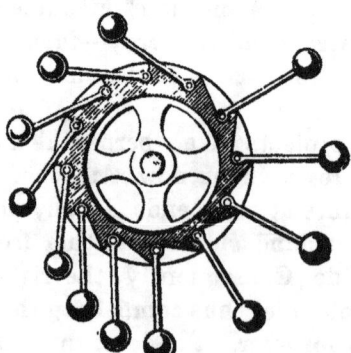

Fig. 44. An "everlastingly" moving wheel of the Middle Ages

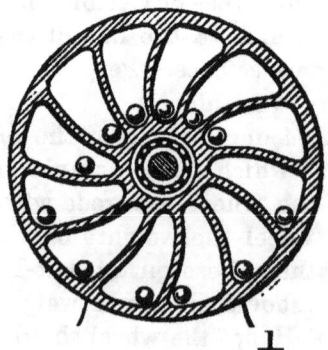

Fig. 45. A "perpetual motion" machine with balls rolling in compartments

Fig. 46. Fake *perpetuum mobile* as an advertisement
for a Los Angeles café

Fig. 45 depicts another supposed "perpetual motion" machine—a
wheel with heavy balls rolling in compartments between the outer
rim and hub. The idea was that the balls closer to the outer rim on
one side of the wheel would compel the wheel to turn by their weight.

But this will never happen—for the same reason as the wheel in *Fig.
44* doesn't turn. Still, in Los Angeles a tremendous wheel of this nature
(*Fig. 46*) was built to advertise a café. Actually it was a fake, being

turned by an artfully concealed mechanism—though people thought it was spun by the heavy balls rolling in the compartments. Other such fake "perpetual motion" machines, all set in motion by electricity, were placed in the windows of watchmaker's shops to attract the eye of the public.

Incidentally, one ad of this nature impressed my students so greatly that they wouldn't believe me when I told them that perpetual motion was impossible. Seeing is believing, they say, and when my students saw the balls rolling and turning the wheel, it seemed far more convincing than anything I could say. I told them that the fake "wonder" machine was driven by electricity from the city mains but that didn't help either. Then I recalled that on Sundays the electricity was cut off. So I advised my pupils to call on the shop on a Sunday.

"Did you see the 'perpetual motion' machine in action?" I asked afterwards.

"No," they replied, their heads ahanging, "it was covered up with a newspaper."

The law of the conservation of energy regained their confidence and they never lost faith in it again.

"THE SNAG"

Many ingenious home-taught Russian inventors tackled the fascinating problem of a "perpetual motion" machine. One, the Siberian peasant Alexander Shcheglov, is described under the name of Bürgher Prezentov by the well-known 19th-century Russian satirist Saltykov-Shchedrin in his *Modern Idyll*. Below the writer describes a visit to the inventor's workshop:

"Bürgher Prezentov was a man of some 35 summers, gaunt and pale of face. He had large pensive eyes and long hair which fell in strands onto his neck. Half of his rather roomy cottage was taken up by a big flywheel and we barely managed to squeeze in. It was a spoked wheel and had a rather large outer rim of boards nailed together like a box. Inside it was empty, and held the mechanism, the inventor's secret. There was nothing particularly cunning about it—merely bags of sand which were to balance one another. A stick in the spokes kept the wheel stationary.

"'We've heard that you've applied the law of perpetual motion in practice. Is that true?' I began.

"'I really don't know how to put it,' he returned in confusion. 'I think I've done it.'

"'Can we take a look?'

"'Pray, do! I'll be delighted.'

"He led us up to the wheel and then took us around to the other side. It was a wheel all right, from either side.

"'Does it turn?'

"'Well, it should. But it's a bit capricious.'

"'Can you take the stick out?'

"Prezentov removed it, but the wheel stood still.

"'It's up to its tricks again!' he repeated. 'It needs an impetus.'

"He gripped the rim with both hands, swung it back and forth several times, then pushed it with all his might. The wheel began to turn. It made several turns rather quickly and smoothly. One could hear the bags of sand inside the rim banging against the boards and sliding away. Then the wheel began to turn more and more slowly. We heard a rasping and a creaking and, finally, the wheel stopped altogether.

"'Must be a snag somewhere,' the inventor explained in confusion as he strained and swung the wheel again. But the result was the same.

"'Perhaps you forgot friction?'

"'I didn't.... Friction you say? It's not because of that. Friction's nothing. Sometimes it makes you happy and then, bang, it's up to its tricks, gets ornery, and that's that. If the wheel were made of real stuff, not scraps!'"

It was of course not the "snag" or the "real stuff" that was at fault, but the wrong principle at the root. The wheel turned for a time owing to the impetus that the inventor gave it, but was bound to stop when friction exhausted the imparted outside energy.

"IT'S THEM BALLS THAT DO IT"

The writer Karonin (the pen-name of N. Y. Petropavlovsky) describes another Russian "perpetual motion" machine inventor in his story "*Perpetuum Mobile*". This was Lavrenti Goldyrev, a peasant from

Perm Gubernia who died in 1884. Karonin, who changed the name in the story to Pykhtin, describes the machine in great detail.

"Before us was a large queer machine resembling at first glance the sort of thing a blacksmith uses to shoe horses on. We could see some badly planed wooden pillars and beams and a whole system of flywheels and gear wheels. It was all a very clumsy-looking affair, rough and ugly. Several iron balls lay on the floor underneath the machine and there was a whole pile of them a bit to the side.

"'Is that it?' the major-domo asked.

"'That's it.'

"'Well, does it turn?'

"'How else?'

"'Have you got a horse to turn it?'

"'A horse? What for? It turns by itself,' Pykhtin returned and began to demonstrate the monster's workings.

"The main role was played by iron balls heaped up nearby.

"'It's them balls that do it. Look. First it goes whack into this scoop. Then it flies like lightning along that groove, is scooped up by that scoop, flies like mad back to that wheel and again gives it a good push so hard that it even begins to whine. Meanwhile another ball is on its way. Again it flies along and goes whack here. From here it dashes along the groove and strikes that scoop, skips to the wheel, and again whack! That's how it goes. Wait, I'll start it off.'

"Pykhtin darted to and fro, hastily collecting the scattered balls. Finally, after heaping them up into a pile by his feet, he picked one up and threw it with all his might at the nearest scoop on the wheel. Then he quickly picked up a second, then a third. The noise was something unimaginable. The balls clanked against the iron scoops, the wheel creaked, the pillars groaned. An infernal whine and racket filled this gloomy place."

Karonin claims that Goldyrev's machine moved. But this was patently a misunderstanding. The wheel could have turned only while the balls were dropping down—at the expense of the potential energy accumulated when lifted, much in the manner of the weights of a pendulum clock. However, it couldn't have turned long because when all the lifted balls had "whacked" against the scoops and had slipped down, it would stop—provided it hadn't stopped before by the counter-effect of all the balls it was supposed to lift.

Later on, Goldyrev became disappointed in his invention when at an exhibition in Yekaterinburg, where he showed it, he saw real industrial machines. When asked about his "perpetual motion" contraption, he dejectedly replied: "The devil take it! Tell 'em to chop it up for firewood."

UFIMTSEV'S ACCUMULATOR

Ufimtsev's so-called accumulator of kinetic energy well illustrated the pitfalls that may trap a cursory observer of a "perpetual motion" machine. Ufimtsev, an inventor from Kursk, devised a new kind of windmill power station with a cheap flywheel type of "inertia accumulator". In 1920 he built a model of it, shaped as a disc that spun round a vertical axis set on ball bearings inside an air-free jacket. When revved to 20,000 r.p.m., the disc was able to turn for 15 days on end. The unthinking observer could well believe that he had before him a real "perpetual motion" machine.

"A MIRACLE, YET NOT A MIRACLE"

The futile search for a "perpetual motion" machine clouded many lives. I once knew a factory worker who sank into absolute destitution, spending all his earnings and savings in the delusion that he could make a "perpetual motion" machine. Poorly clad and always hungry, he would beg everyone he met to give him some money to make the "finished model", which would "certainly move". It was a great pity to see this man suffering so much only because of his ignorance of the rudiments of physics.

It is curious to note that whereas the search for a "perpetual motion" machine was always abortive, the profound realisation of its impossibility, on the contrary, often led to discoveries of great value.

A wonderful illustration in point is the method which the remarkable Dutch scientist Stevin, who lived at the turn of the 16th century, evolved to establish the law of the equilibrium of forces on an inclined plane. He deserves far greater fame than befell him for his many major discoveries that we now constantly address ourselves to. These are decimal fractions, the introduction of denominators in algebra, and the establishment of the hydrostatic law that Pascal rediscovered later.

Stevin evolved the law of the equilibrium of forces on an inclined plane without invoking the rule of the parallelogram of forces. He proved it with the aid of a drawing, which is reproduced in *Fig. 47*. A chain of fourteen identical spheroids is slipped round a three-sided prism. What happens to it? The bottom, which droops garland-like, is in a state of balance, as you see. But do the other two [parts balance each other? In other words, do the two spheroids on the right offset the four on the left? The answer is yes. Otherwise the chain would keep on rolling of its own accord from right to left for ever. Otherwise other spheroids

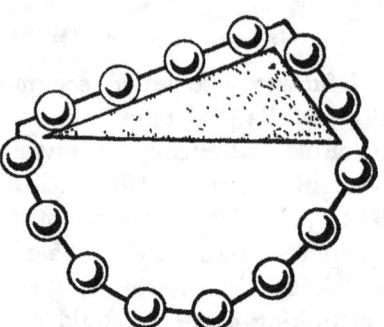

Fig. 47. "A miracle, yet not a miracle"

take the place of those that slide ˥off and ˥equilibrium would never be restored. But we know that a chain disposed in this fashion does not move of its own accord at all It is quite obvious that the two spheroids on the right really offset the four on the left.

It seems a minor miracle, doesn't it? Two spheroids pull with the same force as four! This enabled Stevin to deduce an important law of mechanics. This is how he reasoned. The two parts—the long one and the short one—possess a different weight, one being as many times heavier than the other as the longer side of the prism is longer than the short side. Consequently, any two linked loads in general balance on tilted planes, provided their weight is directly proportional to the length of these planes.

When the short plane is vertical we get a well-known law of mechanics, which is: to hold a body in place on a tilted plane we must act in the direction of this plane with a force as many times less the weight of the body as the length of the plane is greater than its height. So did the idea that a "perpetual motion" machine is impossible led to an important discovery in the realm of mechanics.

MORE "PERPETUAL MOTION" MACHINES

Fig. 48 shows a heavy chain fitted around wheels in such a way that the right-hand part is always longer than the left-hand part, whatever its position. The inventor thought that since the right-hand part would always weigh more than the left-hand part, it would always outweigh the left-hand part and thus cause the entire arrangement to keep going. But does this really happen? Of course not. You already know that the heavier part of a chain may be offset by the lighter part, provided they are pulled by forces acting at different angles. In this particular system, the left-hand part of the chain droops vertically down, while the right-hand part is inclined. So, though it is heavier, still it cannot pull over the left-hand part and we do not achieve the "perpetual motion" expected.

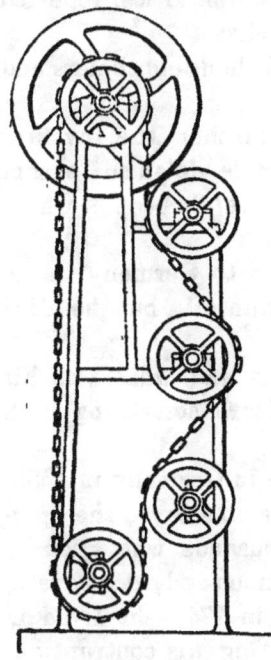

Fig. 48. Is this a "perpetual motion" machine?

I think the cleverest "perpetual motion" machine ever invented was one displayed at the Paris Exposition in the 1860's. It consisted of a large wheel with balls rolling about in its compartments. The inventor claimed that nobody would ever be able to stop the wheel. Many visitors tried to stop it but it went on turning as soon as they took their hands off it. Not a single person realised that the wheel turned precisely because of the effort he made to stop it. The backward push he gave to stop it wound up the spring of an artfully concealed mechanism.

THE "PERPETUAL MOTION" MACHINE PETER
THE GREAT WANTED TO BUY

Preserved in archives is a bulky correspondence which Peter the Great of Russia carried on between 1715 and 1722, when he wanted to buy a "perpetual motion" machine that had been devised in Germany by one Councillor Orffyreus. This man whose "self-moving wheel" won him nation-wide fame consented to sell it to the tsar only for a princely sum. Peter the Great's librarian Schumacher, whom the tsar had sent to Western Europe to collect rare oddities, reported the following, when asked to negotiate the purchase:

"The inventor's last words were: One hundred thousand thalers and you get the machine."

As for the machine itself, according to Schumacher, the inventor claimed that it was no fake and that it could not be defamed "except out of malice, and the whole world is full of spiteful people whom one cannot believe".

In January 1725 Peter the Great decided to go to Germany to see this notorious "perpetual motion" machine himself, but he died before he could accomplish his purpose.

Who was this mysterious Councillor Orffyreus and what was his "famous machine" really like? I was able to learn something both about the Councillor himself and his machine.

Orffyreus's real name was Bessler. He was born in Germany in 1680. He studied theology, medicine and painting before he essayed the "perpetual motion" machine. Among the many thousands who tried to invent such a machine he is probably the most famous and, at any rate, the luckiest. Till the end of his days—he died in 1745—he lived in comfort on the income he netted by demonstrating his contraption.

Fig. 49 is a reproduction of a drawing from an old book depicting Orffyreus's machine as seen in 1714. It shows a large wheel which apparently not only turned by itself, but even lifted a heavy load to quite a height.

The fame of this "miracle" machine, which the learned councillor first exhibited at various market fairs, quickly spread throughout Germany. Soon Orffyreus acquired powerful patrons. The Polish

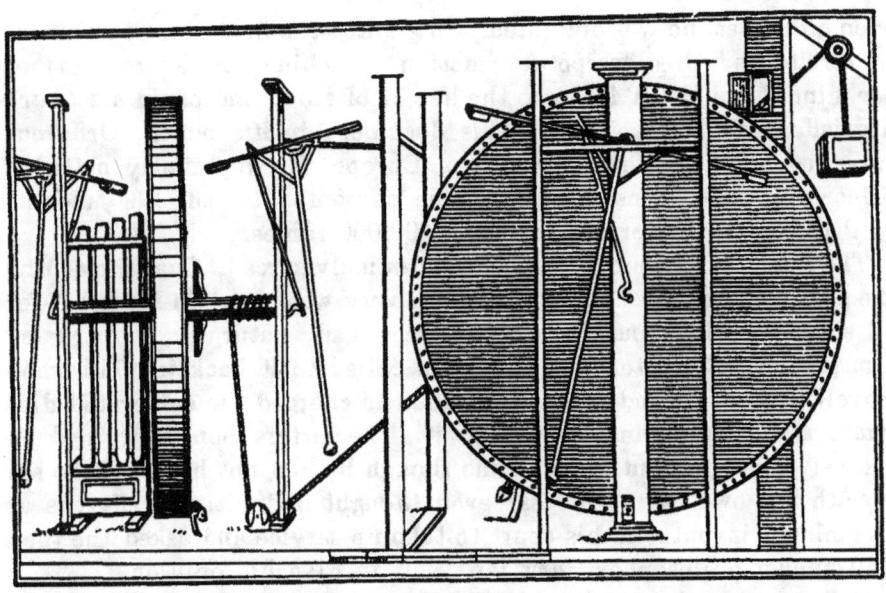

Fig. 49. Orffyreus's self-moving wheel which Peter the Great wanted to buy. (From an old drawing.)

king displayed interest and then the Landgrave of Hesse-Cassel patronised the inventor, placing his castle at the latter's disposal and subjecting the machine to every kind of trial.

On November 12, 1717, the machine was placed in a room all apart and set into motion. The room was then locked and sealed, and two grenadiers were posted outside. For a whole fortnight, until the seal was broken on November 26, no one dared to come near. Then the room was unlocked and the Landgrave and his retinue entered. The wheel was still spinning "with undiminishing speed". It was stopped, inspected carefully, and again set going. Now the room was locked and sealed for 40 days on end with grenadiers again stationed at the door. The seal was broken on January 4, 1718. A commission of experts entered and found that the wheel was still going. But this did not satisfy the Landgrave and he staged a third trial, locking up the machine for two whole months at a stretch. When he found the wheel still going even after that, he was delighted. He granted the inventor a parchment to certify that his "perpetual motion" machine did 50 revolutions per minute, could lift 16 kg to the height of 1.5 m and could also work a grinder and bellows. With this document in his pouch, Orffyreus travelled the length and breadth of Europe. He apparently netted a princely income, considering that he consented to sell his machine to Peter the Great for not less than 100,000 rubles.

The fame of the councillor's marvel quickly spread, finally reaching the ears of Peter the Great, who had a very weak spot in his heart for all sorts of curious and cunning artifices, and, naturally, it intrigued him greatly. His attention had been called to it back in 1715 when travelling abroad, and it was then that he charged the celebrated diplomat A. I. Ostermann to inspect it. The latter soon forwarded an extensive report about the machine though he had not been able to see it with his own eyes. The tsar even thought of inviting Orffyreus as an eminent inventor to his court to take up service and asked the then well-known philosopher Christian Wolf to give his opinion.

Orffyreus was showered with offers, one better than the other. Kings and princes bestowed munificent awards. Poets composed odes in honour of his wonder-wheel. But there were some who thought him a charlatan. The more daring openly accused him, even offering 1,000 marks to anyone who would come forth and expose the councillor. One lampoon against him gave a drawing which is reproduced in *Fig. 50* and which provides a rather simple explanation for the mystery—a cunningly hidden person who pulled at a rope wound round that part of the axle which was concealed in the pillars supporting the wheel.

The trick was bared by chance only because the councillor had had a tiff with his wife and maid who had both been initiated into the secret. Otherwise we would probably still be guessing. It seemed that the notorious machine was indeed turned by a hidden person—Orffyreus's brother, or maid—pulling at a slender cord. But the councillor did not lose face, persistently assuring all and sundry even on his deathbed that his wife and maid had maligned him out of spite. However, trust in him was shattered. No wonder he tried to drum into the head of the tsar's envoy, Schumacher, the point that human beings were full of malice.

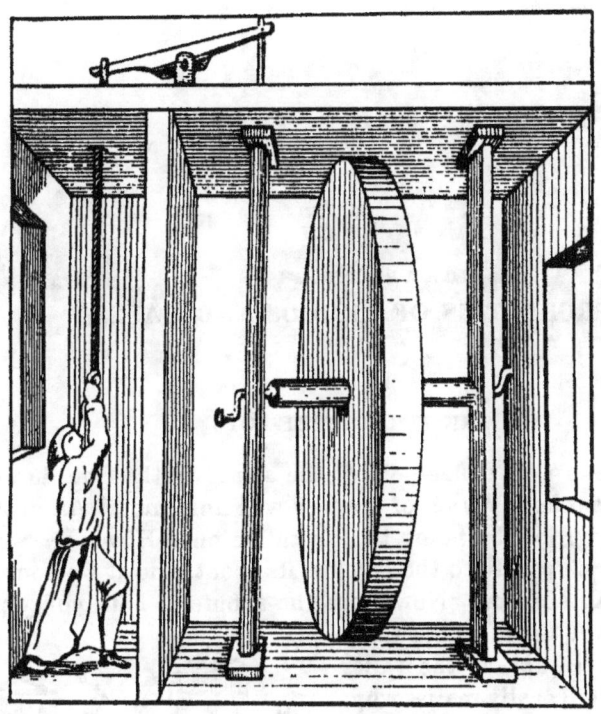

Fig. 50. The secret of Orffyreus's machine.
(From an old drawing.)

Around the same time there also lived in Germany another renowned "perpetual motion" machine inventor, one Hertner. Schumacher wrote of his contraption the following: "Herr Hertner's *perpetuum mobile*, which I saw in Dresden, consists of tarpaulin filled with sand and a grinder-like machine which turns forwards and backwards by itself. However the inventor says it cannot be made larger." Undoubtedly this machine, too, gave no "perpetual motion", being at best an artfully contrived device with a just as artfully concealed—living—but by no means "perpetual motion" machine. Schumacher was right when he wrote to Peter the Great that French and English scholars "mock these *perpetuum mobiles* as objectionable to principles of mathematics".

CHAPTER FIVE
PROPERTIES OF LIQUIDS AND GASES

THE TWO COFFEE-POTS

Fig. 51 shows two coffee-pots of the same width. One, however, is taller than the other. Which of the two will hold more? An unthinking person would probably point to the taller one. However, we would be able to fill it up only to the level of its spout, and if we poured more in, it would all spill out Now since the spouts of both coffee-pots are on the same level, the lower one takes just as much liquid as the taller one does. You will easily realise why. The coffee-pot and its spout are two communicating vessels and hence inside both the liquid should be at an identical level, even though the liquid in the spout weighs much less than that in the coffee-pot proper. Unless the spout is high

Fig. 51. Which coffee-pot takes more?

enough, you will never be able to fill the coffee-pot up to the top; the water will simply keep on spilling out. Usually the spout is even a bit higher than the top of the coffee-pot to enable one to incline it without spilling out its contents.

IGNORANCE OF ANCIENTS

Romans today still use what is left of the aqueducts that their forefathers built. Though the Roman slaves of old did a very good job, we can't say that of the Roman engineers in charge. Their knowledge of elementary physics was plainly inadequate. *Fig. 52* reproduces a picture preserved at the German Museum in Munich. As you see, the Romans did not sink their water systems in the ground but placed

Fig. 52. The aqueducts of ancient Rome

them on high supports of masonry. Why? Aren't underground pipes of the type we use today simpler? Roman engineers of old had a very hazy notion, however, of the laws of communicating vessels. They feared that in two reservoirs connected by a very long pipe, the water would not rise to the same level. Furthermore, if the pipes were laid in the ground and followed the natural relief, in some places the water would have to flow upwards, and this was something the Romans were afraid it would not do. That is why their aqueducts usually slope all along the way. They often had either to take the pipes on a roundabout route or erect tall arches. One Roman aqueduct. known as the Aqua Marcia, is 100 km long, though it is half the distance between its two points as the crow flies. As you see, the ancient Romans' ignorance of an elementary law of physics caused 50 km of extra masonry to be built.

Even people who have never studied physics know that liquids press down on the bottom of the vessels holding them and sideways at the walls. Many, however, have never suspected that liquids also press upwards. An ordinary lamp-glass will easily reveal this. Cut out of a piece of thick cardboard a disc large enough to cover the top of the lamp-glass. Cover the top of the glass with it and then dip the glass into a jar of water as shown in *Fig. 53*. To prevent the disc from slipping off when the lamp is immersed, tie a piece of thread to it and hold it as shown, or simply press it down with your finger. After you have dipped the glass far enough, you can let the thread, or your finger, go. The disc will remain where it is, being kept in place by the water pressing up on it.

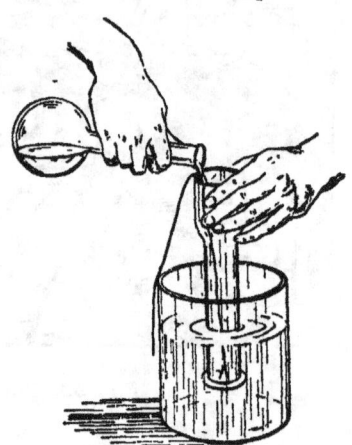

Fig.53. A simple way to demonstrate that liquids press upwards

If you want to, you can even gauge the value of this upward pressure. Carefully pour some water into the glass. As soon as the level of the water in the glass reaches that of the water in the jar, the disc slips off, because the pressure exerted by the water on the disc from below is offset by the pressure exerted on it from above by the column of water in the glass, the height of which is equal to the depth to which the glass has been dipped. Such is the law concerning the pressure that a liquid exerts on any immersed body. This incidentally results in that "loss" of weight in liquids of which Archimedes's famous principle speaks.

With the help of several lamp-glasses of different shapes but with tops of one and the same size you may test another law dealing with liquids: that the pressure a liquid exerts on the bottom of the containing vessel depends only on the size of the bottom and the height of the "column" of liquid; it does not depend at all on the vessel's shape. This is how you test this law. Take different glasses and dip them to one and the same depth. To see that no mistakes occur, first glue strips of paper to the glasses at equal heights from the bottom. The cardboard disc you used in the first experiment will slip off every time you pour in water to the same level (*Fig. 54*). Consequently the pressure exerted by columns of water of different shapes is the same as long as the bottom and height are the same. Note that it is the *height*, and not the *length*, that is important, because a long but *inclined* column exerts exactly the same

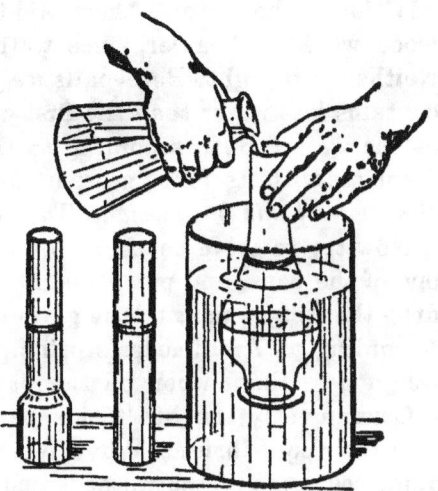

Fig. 54. The pressure liquid exerts on the bottom of the vessel depends only on the area of the base and the liquid's height. The drawing shows you how to check this

pressure on the bottom as is exerted by a shorter but perpendicular column as *high* as the inclined one—provided, of course, the bottom of each is the same.

WHICH IS HEAVIER?

Fig. 55. Both pails are full to the rim. One has a piece of wood in it. Which is heavier?

Place a pail of water, full up to the rim, on one pan of a pair of scales. Then put on the other pan another pail of water, *also full up to the rim*, but with a piece of wood floating in it (*Fig. 55*). Which of the two is heavier? I asked this of different people and got contradictory answers. Some said the pail with the piece of wood in it would be heavier because it held a piece of wood in addition to the water. Others said the pail of water without the piece of wood would be heavier, since water generally weighs more than wood. Neither were right. Both pails *weigh the same*. The second pail, true, contains less water than the first one, because the wood displaces some of the water. But, according to the related law, every *floating* body displaces with its immersed part exactly *as much liquid* (in weight) *as the whole of this body weighs*. That is why the scales balance.

Now try to solve another problem. Take a glass of water, put it on one of the pans, and put a weight next to it. Balance the scales. Then drop the weight next to the glass into it. What happens to the scales? According to Archimedes's principle, in the water the weight should weigh less than when on the pan.

Consequently, oughtn't this pan rise? However, the pans maintain their equilibrium. Why? When dropped into the glass the weight displaced some of the water which then rose to a level higher than before. This added to the pressure exerted on the bottom of the vessel, which thus sustained an additional force equivalent to the weight lost by the weight.

A LIQUID'S NATURAL SHAPE

We are used to thinking that liquids have no shape of *their own*. That is not true.

The natural shape of any liquid is that of a sphere. As a rule, gravity prevents liquids from assuming this shape. A liquid either spreads in a thin layer if spilled out of a vessel, or takes the vessel's shape. But when inclosed in another liquid of the same specific weight, it, according to Archimedes's principle, "loses" its weight, seeming to weigh nothing; now gravity has no effect on it and it assumes its natural spherical shape.

Since olive oil floats in water but sinks in alcohol we can mix the two in such proportions that the oil will neither sink nor float in this mixture. An odd thing happens when we drip in a little oil with the help of an eyedropper. The oil collects into a large round drop which neither floats nor sinks, but hangs suspended (*Fig. 56*). To get a true image of the sphere, you should do the experiment in a flat-walled vessel—or in one of any shape but placed inside a flat-walled vessel full of water.

You must do this experiment patiently and carefully, because otherwise you will get several smaller drops instead of a large one. Don't feel disheartened if it doesn't work out; even then it's sufficiently illuminating.

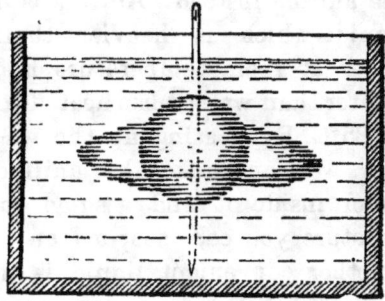

Fig. 56. Oil inside diluted alcohol collects into a drop which neither sinks nor floats. (Plateau's experiment.)

Fig. 57. A ring is given off when the oil drop in the alcohol is spun by means of a rod

Let's carry this experiment further. Take a long stick or a piece of wire and transfix the oil drop. Start turning. The drop also participates in this revolution. You get still better results by attaching to the stick or wire a small cardboard disc soaked in oil and inserting it fully in the drop you are twirling. The spin compels the drop to compress and then give off a ring a few seconds later (*Fig. 57*). As it breaks up the ring creates new drops which continue to revolve round the central one.

The Belgian physicist Plateau was the first to conduct this instructive experiment, of which I have given you the classical description. It would be much easier—and just as instructive—to do this experiment in another way. Take a small tumbler, rinse it with water, and fill it with olive oil. Place it on the bottom of a larger glass. Then carefully pour into the glass enough alcohol to cover the tumbler. Gradually add a little water with the help of a spoon. Do this very carefully, so that the water drips down the walls of the glass. The top of the oil in the tumbler starts to bulge, and when enough water has been poured in, the oil rises up from the tumbler in a rather large drop to hang suspended in this mixture of alcohol and water (*Fig. 58*).

For want of alcohol you can use aniline instead. Aniline is a liquid which is heavier than water at room temperature but lighter than water when heated to 75-85°C. By heating up the water, we can make the aniline

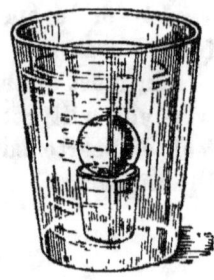

Fig. 58. Plateau's experiment simplified

swim inside it and assume the form of a large drop. At room temperature you can suspend an aniline drop in a solution of table salt. Another convenient liquid is the dark-crimson orthotoluidine, which at 24°C has the same density as salt water, into which it is poured.

WHY IS SHOT ROUND?

I noted earlier that any liquid will assume its natural spherical shape when gravity ceases to act on it. You need only remember what I said before about a falling body having no weight and discount the negligible atmospheric resistance when a body starts to fall (raindrops accelerate only when they start to fall; by the second half of the first second the fall already becomes *uniform* and the drop's weight is offset by atmospheric resistance which grows together with the velocity of the falling drop) to realise that falling portions of liquid should also take on a spherical form.

That is really so. Falling raindrops are indeed round in shape. Shot is nothing but solidified drops of molten lead which in the process of making are dropped from a great height into a cold water bath where they solidify in the shape of absolutely right spheres. Shot is also called "tower" shot because in its making it is dropped from the top of a tall "shot tower" (*Fig. 59*). These towers are metal structures 45 m high. At the top they have a shot-pouring shop with boilers for melting the lead, and at the bottom—a water bath. The ready shot is then graded and processed. The drop of molten lead solidifies into shot while falling. The water bath is needed merely to soften the impact and to prevent the shot from losing its spherical shape (shot with a diameter of more than 6 mm, so-called canister shot, is made differently, by chopping off pieces of wire, which are then rolled into balls).

THE "BOTTOMLESS" WINEGLASS

Fill a wineglass with water right up to the very rim. Take some pins. Do you think place could be found in the wineglass for a couple of them? Try and see.

Throw the pins in and count them as you do. But be careful about it. Take the pin by its head and dip its point into the water. Then carefully let go, without pushing it or exerting any pressure, so that water is not spilled out. As you drop the pins in, they fall to the bottom, but the level of the water is the same. You drop in ten, then another ten, and then another ten. The water does not spill out. You can go on till there are a hundred at the bottom of the glass. But still no water has spilled out (*Fig. 60*). Nor, for that matter, has it risen to any noticeable degree above the rim.

Add some more pins. Now you can even count them in hundreds. You may have as many as 400 pins in the glass, but still no water spills out. However, now you see that the surface is bulging above the rim. Therein lies the answer to this so far incomprehensible phenomenon. Water scarcely wettens glass as long as it is a little greased, and the rim

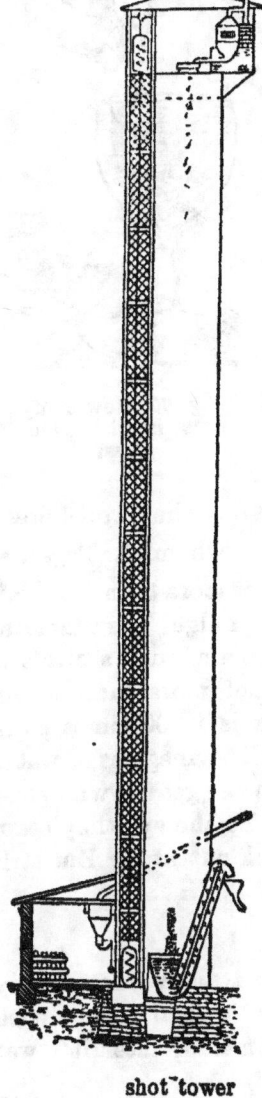

shot tower

of the wineglass—like all the chinaware and glassware we use for that matter—is sure to have some traces of grease which are left when we touch it with our fingers. And as it doesn't wetten the rim the water displaced by the pins bulges. You can't see it, but if you went to the pains of reckoning the volume of one pin and of comparing it with the volume of the bulge above the rim of the wineglass you would realise that the former volume is hundreds of times smaller than the latter, which explains why a "full" wineglass will still have enough room for another few hundred pins.

The wider the mouth of the wineglass is, the more pins it can take, because there is a larger bulge. A rough reckoning will make the point clear. A pin is about 25 mm long and half a millimetre thick. You can easily

Fig. 60. How many pins in the wineglass?

reckon the volume of this cylinder by invoking the well-known geometrical formula $\left(\frac{\pi d^2 h}{4}\right)$; it will be equal to 5 mm³. Together with the head, the pin will have a total volume of not more than 5.5 mm³. Let us now reckon the volume of the water in the bulge. The diameter of the wineglass mouth is 9 cm, or 90 mm. The area of such a circle is about 6,400 mm². Assuming that the bulge is not more than 1 mm high, we thus get a volume of 6,400 mm³, which is 1,200 times more than the volume of the pin. In other words, a "full" wineglass of water can take more than a thousand pins. And indeed we can get the wineglass to take a thousand pins if we are careful enough. To the eye they seem to occupy the whole of the wineglass and even stick out of it. But still no water spills out.

UNPLEASANT PROPERTY

Anyone who has ever had to handle a kerosene lamp most likely knows what annoying surprises it can spring on one. You fill a tank with it and then wipe the tank dry on the outside. An hour later it's wet again. You have only yourself to blame. You probably didn't screw on the burner tight enough, and the kerosene, trying to spread along the glass, seeped out. To avert such "surprises", screw the burner on as tight as you can. But when you do that, don't forget to see that the tank is not full up to the brim. When it warms up, kerosene expands rather considerably—increasing in volume by a tenth every time the temperature rises by another 100°. So if you don't want the tank to explode, you must leave some room for the kerosene to expand.

The property of kerosene to seep through causes unpleasant things aboard ships whose engines burn kerosene or oil. When due precautions are not taken, it is absolutely impossible to use such ships to carry any other cargoes except kerosene or oil, because when they seep out through unnoticeable crevices in the tanks these liquids spread not only to the metal surfaces of the tanks but literally everywhere, even to the clothing of the passengers to which they impart a smell that nothing will kill.

Attempts to fight this evil are often to no avail. Jerom K. Jerome, the British humorist, wasn't guilty of much of an exaggeration when in his *Three Men in a Boat* he wrote of paraffin oil, which is remarkably alike kerosene.

"I never saw such a thing as paraffin oil is to ooze. We kept it in the nose of the boat, and, from there, it oozed down to the rudder, impregnating the whole boat and everything in it on its way, and it oozed over the river, and saturated the scenery and spoilt the atmosphere. Sometimes a westerly oil wind blew, and at other times an easterly oil wind, and sometimes it blew a northerly oil wind, and maybe a southerly oil wind; but whether it came from the Arctic snows, or was raised in the waste of the desert sands, it came alike to us laden with the fragrance of paraffin oil.

"And that oil oozed up and ruined the sunset; and as for the moonbeams, they positively reeked of paraffin....

"We left the boat by the bridge, and took a walk through the town to escape it, but it followed us. The whole town was full of oil." (Actually it was only the clothing of the travellers that reeked of paraffin oil.)

The property kerosene has of wettening the outer surface of tanks led people to wrongly think that kerosene could ooze through metal and glass.

THE UNSINKABLE COIN

It's to be found not only in fairy tales. A few easy experiments will show you that such things really exist. Start with a small object—a needle, for instance. It seems impossible to make a steel needle float, doesn't it? But it isn't really so hard to do. Place a strip of cigarette paper on top of the water in a glass and an absolutely dry needle on top of the paper. Carefully remove the cigarette paper in the following

way. Take another needle or a pin and, gradually working to the middle, gently push the strip of paper into the water. When the strip is soaked through, it will sink, but the needle will continue to float (*Fig. 61*). By moving a magnet at water level from outside the glass you can even make the floating needle spin round.

With a little experience, you can dispense with the cigarette paper entirely. All you need do is to take the needle by the middle and, holding it parallel to the water, drop it from a small height. You can make a pin, which like the needle must not be thicker than 2 mm, a light button, or some small metal object float in the same way. When you have got the knack of it, try a coin.

All these metal objects float because water hardly wettens metal covered with a very thin film of grease from our hands. You can even see the depression a floating needle makes on the surface of the water. Trying to regain its original position, the surface film buoys up the needle which is

Fig. 61. A floating needle. Top: a cross-section of the needle (2 mm thick) and the depression it makes (a twofold magnification). Bottom: how to make the needle float by using a strip of paper

also buoyed up by a force equal to the weight of the water displaced by the needle. The easiest way of making a needle float is, of course, to cover it with grease. Then it will never sink.

CARRYING WATER IN A SIEVE

Neither is this something that can be done only in a fairy tale. Physics can help us to undertake this seemingly impossible task. Take a wire sieve of 15 cm across with holes not smaller than 1 mm in diameter and dip it into melted paraffin, to cover it with a thin, barely discernible film.

Your sieve remains a sieve; it still has holes in it through which a pin can go quite freely, but now you can carry water in it—even quite a lot of it. Only be careful when pouring the water in and see to it that you don't jolt the sieve while doing that.

Why doesn't the water drip through? Failing to wetten the paraffin, the water forms a thin film which bulges through the holes of the sieve; it is this film that keeps the water from dripping through (*Fig. 62*). This waxed sieve will even float, which means that you can not only carry water in a sieve, but also use it as a boat.

This seemingly paradoxical experiment explains several ordinary things to which we are too accustomed to ever think of why they are done. The tarring of barrels and boats, the greasing of corks and stoppers with fat, the painting of roofs with oil paint and, generally, the coating with oily substances of everything we want to make impervious to water, as well as the rubberising of cloth, is the same as making the sieve we just described, with the exception that the sieve, of course, seems exceedingly unusual.

Fig. 62. Why the sieve carries water

The experiment of the floating steel needle or copper coin bears some resemblance to a process employed in mining to "enrich" ores, i.e., to increase the content of the minerals in them. Engineers know many methods for dressing ores, but the one we have in mind and which is called "flotation" is best; it is successful when all other methods fail.

Flotation consists in the following. Finely ground ore is loaded into a bath containing water and oily substances that inclose the mineral particles in a very thin film which water cannot wet. Air is then blown in to form a foam composed of a multitude of tiny bubbles. The greased particles of the mineral attach themselves to the air bubbles and rise up with them much in the same way as an air balloon lifts a gondola (*Fig. 63*). The particles of ore gangue that have no grease envelope cannot attach themselves to the air bubbles and sink. Note that the air bubbles in the foam are much bigger than the particles they carry and are well able to lift the solid speck up. As a result, nearly all the particles of

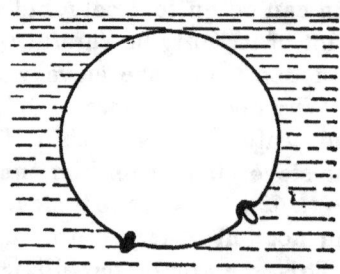

Fig. 63. The essence of flotation

the mineral are floated on top in the foam which is skimmed off for further processing, during which the so-called concentrated ore—which is dozens of times richer in content than the original ore—is separated. Flotation techniques are so well elaborated that a judicious choice of reagents will separate the mineral from the ore gangue in every particular case.

Incidentally, we have a chance accident, and no theory, to thank for the flotation method. One day, at the end of the past century, Carrie Everson, an American schoolmistress, was washing greasy sacks that had been used to stack copper pyrites. She happened to notice that the pyrite particles left in the sacks floated together with the lather. It was this that suggested the flotation method.

FAKE "PERPETUAL MOTION" MACHINE

You will sometimes find the following contraption (*Fig. 64*) described as a genuine "perpetual motion" machine. Oil (or water) poured into a vessel is soaked up by wicks at first into one vessel and then by more wicks into another vessel still higher up. The top vessel has a grooved outlet through which the oil pours onto a paddle wheel, causing it to turn. From the bottom tank the oil is again soaked up by wicks to the top. Thus, the oil supposedly never stops pouring onto the paddle wheel, making the wheel turn for ever and ever.

Fig. 64. Non-existent "perpetual motion" machine

If the people who described this thing were to take the pains to make it, they would realise that not a single drop of oil would ever reach the upper vessel, let alone make the wheel go. Incidentally, we don't necessarily have to make this contraption to realise that this is so. Indeed, why should the inventor think the oil should necessarily flow off the upper bent portion of the wick? It is quite true that capillary forces, having overcome gravity, lift the oil up the wick. But it is these same forces that prevent the oil in the pores of the soaked wick from oozing off. Even supposing for a moment that the oil will reach the upper vessel of our fake "perpetual motion" machine due to capillary forces, we shall have to admit that the same wicks which supposedly lift the oil up would themselves lower it to the bottom tank.

The contraption we have just mentioned resembles another water-driven one, invented by the Italian mechanic Strada the Elder way back in 1575. *Fig. 65* shows you this amusing device. As it turns, an Archimedes's screw lifts water to the upper tank, from which it pours out through a groove to strike at the paddles of the tank-filling wheel

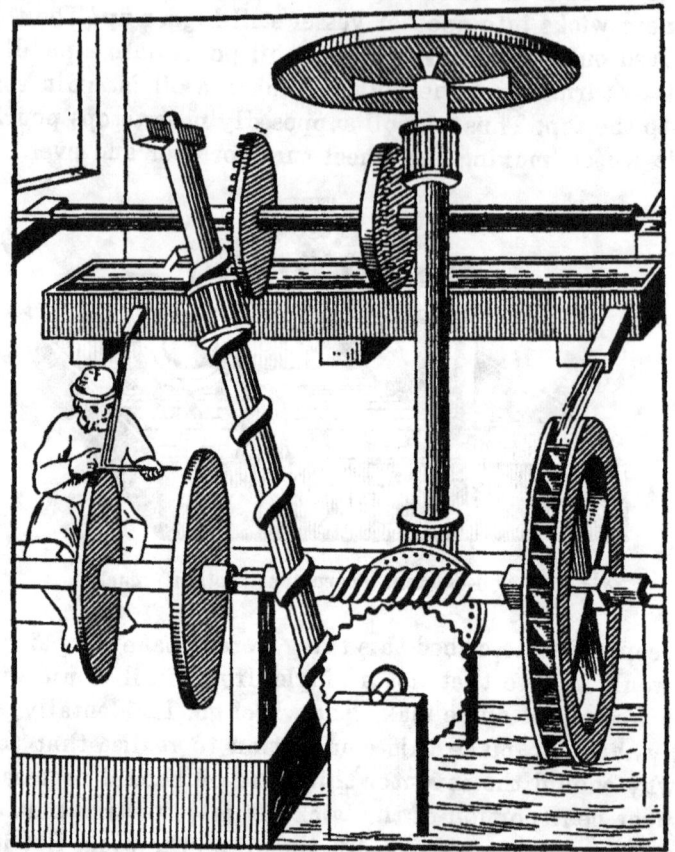

Fig. 65. An ancient design of a water-driven "perpetual motion" machine to turn a grinding stone

shown in the bottom right-hand corner. This wheel turns a grinder and simultaneously operates by means of several gears the same Archimedes's screw which lifts the water to the upper tank. To make a long story short, the screw turns the wheel and the wheel turns the screw! If such contraptions were possible, the simplest thing would be to throw a rope over a pulley and tie identical weights to each end. As one weight fell it would lift the other one, which, dropping in turn, would lift the first one. Wouldn't that be a fine "perpetual motion" machine?

BLOWING SOAP BUBBLES

Do you know how to blow soap bubbles? It is not so simple as it seems. I, too, thought there was nothing particular in it until I saw for myself that the ability to blow big beautiful bubbles is in its way an art that needs some experience. But is it really worth while doing such a seemingly silly thing as blowing soap bubbles? After all, they have won a rather bad reputation among the laymen. Physicists have other views, however. "Blow a soap bubble," said the great British physicist Kelvin, "and observe it. You may study it all your life, and draw one lesson after another in physics from it."

Indeed, that magic iridescence on the slimmest of soap films enables the physicist to gauge the length of light waves, while a study of the tension of these gossamer films helps him to formulate the laws governing the interaction of forces between particles—those self-same forces of cohesion without which the world would be but a cloud of the finest dust.

The few experiments described below do not have such serious aims; they are given simply to provide instructive entertainment and to teach you how to blow soap bubbles. In his book *Soap Bubbles and the Forces Which Mould Them*, the British physicist Charles Boys describes at length many different experiments that one can stage with these bubbles. So if you are interested in them, let me refer you to this wonderful book.

Below you will find only a few of the simplest experiments. Ordinary laundry soap will do—toilet soaps are less suitable for the purpose. But you can also use pure olive-oil or almond-oil soap, which is best for obtaining large and beautiful bubbles. Carefully dissolve a cake of soap in pure cold water till you get a rather thick lather. Pure rain water or melted snow is best but you may use cooled boiled water instead. To prolong the life of the bubbles Plateau suggests adding glycerin to the lather in a mixture of one part to every three. Skim the

92

froth and the small bubbles off with a spoon and then dip in the lather a slender clay pipe, with its end preliminarily soaped both on the inside and outside. Good results can be achieved also by using straws of about 10 cm long, that are split at the bottom in the form of a cross.

This is how you blow the bubble. Dip the pipe into the lather, holding it vertically so that it becomes covered with film. Then gently blow at the other end. As the bubble is filled with warm air from our lungs—which is lighter than the air in the room—it will float up at once as long as you can blow a bubble of some 10 cm across; otherwise you must add more soap until you can blow bubbles of this diameter. This alone is not enough; there is another test that you must make. After you blow the bubble, dip your finger in the lather and try to pierce the bubble with it. If it doesn't burst you can start experimenting. If it does—add a little more soap. Do the experiments slowly, with care, and without undue haste. The room must be well lit; otherwise the proper iridescence will be lacking. Now for a few entertaining experiments.

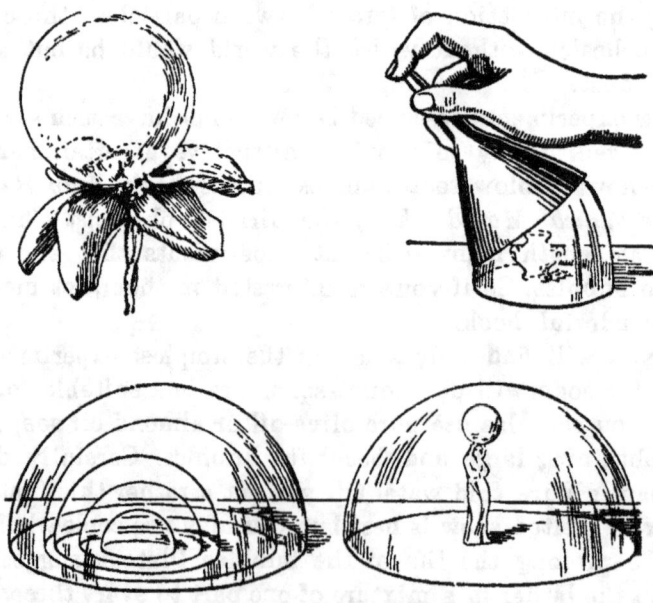

Fig. 66. Soap bubbles

1) *A flower in a bubble*. Pour the lather three millimetres deep into a plate or tray. Then place a flower or a little vase in the middle and cover it with a glass funnel. Slowly lift the funnel, blowing meanwhile in its narrow end to get a soap bubble. When the bubble is large enough, tilt the funnel as shown in *Fig. 66* and release the bubble. Your flower or vase will be under a transparent, semicircular, iridescent soap bubble. You can take a *statuette* instead of a flower and crown it with a small soap bubble as shown in *Fig. 66*. To get the smaller bubble, you must spill a little lather on top of the statuette before you blow the big bubble. Then pierce the big bubble with a pipe and blow out the small bubble inside.

2) *A nest of bubbles* (*Fig. 66*). Take the funnel you used for the previous experiment and blow a large bubble as you did before. Then take a straw and dip it into the lather, leaving only the very end, which you blow through, dry. Gently pierce the wall of the first bubble till you get to the middle. Then slowly draw the straw back without bringing it out, and blow out a second bubble inside the first. Repeat to get a third bubble inside the second, a fourth inside the third, and so on.

3) *A cylindrical bubble* (*Fig. 67*). For this purpose you must have two wire rings. Blow an ordinary round bubble onto one of them, the lower one. Then take the second ring, wet it and attach it to the top of the bubble. Lift it until the bubble assumes a cylindrical shape. Note that if you lift the upper ring to a height more than the ring's circumference, half of the cylinder will contract and the other half will bulge until the bubble divides into two.

Fig. 67. How to make a cylindrical soap bubble

The film of the soap bubble, which is continually in a state of tension, presses on the enclosed air; by directing the narrow end of the funnel at the flame of a candle you will see that the strength of this very thin film is not so negligible as you might think—the flame wavers quite noticeably (*Fig. 68*).

It is interesting to observe a bubble floating out of a warm room into a cold one. It shrinks noticeably. On the other hand, it expands when brought from a cold room into a warm one. This, naturally, depends on the contraction and expansion of the air inside. If you were to blow a bubble of 1,000 cm³ in a sub-zero frost of 15 °C and then bring it into a room where the temperature is 15 °C above zero, it would increase in volume by roughly 110 cm³ ($1,000 \times 30 \times \frac{1}{273}$).

Fig. 68. The air forced out by the walls of the soap bubble causes the candle-flame to waver

I must note that a soap bubble is not always as short-lived as is usually thought. When handled with care it can be preserved for some ten days, if not more. The British physicist Dewar, who won fame for his studies of the liquefaction of air, preserved soap bubbles in special bottles, well shielded from dust, dryness, and shock, and was able to keep some bubbles for a month and more. The American Lawrence kept soap bubbles under a bell-glass for years on end.

THINNEST OF ALL

Few probably know that the film of a soap bubble is one of the thinnest things you can see with the unaided eye. The customary comparisons we draw upon to express thinness are very thick compared with the film of a soap bubble. A thing "as thin as a hair" or "as thin as cigarette paper" is very thick compared with the walls of a soap bubble, which are 5,000 times thinner than a hair or cigarette paper. A human hair magnified 200 times is about a centimetre thick. If we magnified the cross-section of the film of a soap bubble the same number of times, we still wouldn't be able to see it. We would have to magnify it another 200 times to see it as a slender line. Then a hair—magnified 40,000 times—would be more than two metres thick. *Fig. 69* well illustrates this.

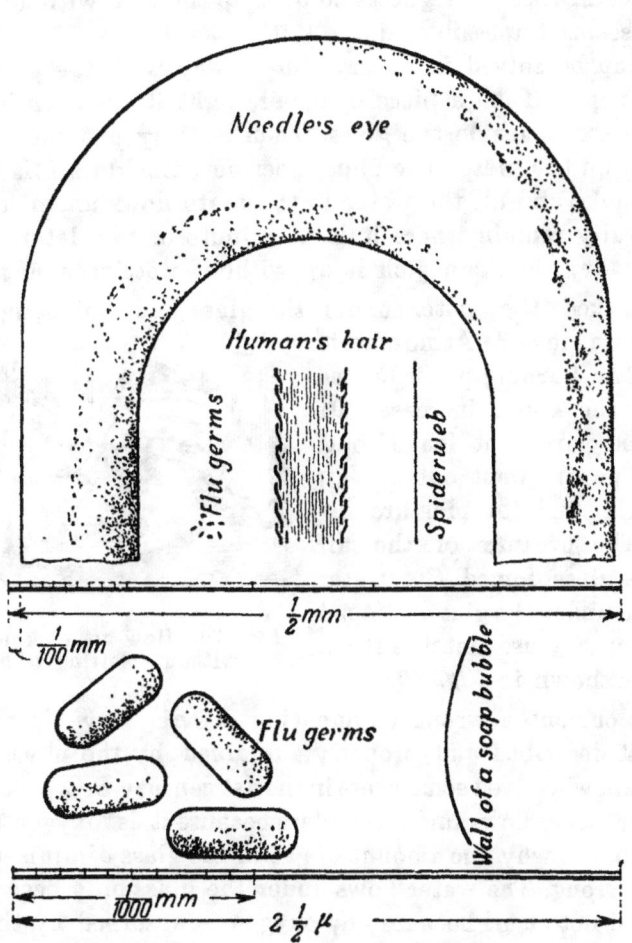

Fig. 69. Top: the eye of a needle, a human hair, germs, and a spiderweb magnified two hundred times. Bottom: germs and the wall of a soap bubble magnified 40,000 times

WITHOUT WETTING A FINGER

Take a large plate and put a coin on it. Then add enough water to cover the coin. Ask your guests to pick up the coin without wetting a finger. It seems impossible, doesn't it?

But it can be solved in a very simple way with the aid of a glass and some paper. Take a piece of paper, light it and, while it is still burning, place it inside the glass. Then quickly put the glass down, bottom up, on the plate. The paper goes out, the glass fills with white wisps of smoke and all the water in the plate flows under it. The coin will naturally remain where it is. A minute or two later, as soon as the coin is dry, you can pick it up without wetting a finger.

What sucked the water under the glass and maintained it there at a certain level? Atmospheric pressure. The burning paper heated the air in the glass, increased its pressure and part of it leaked out. When the paper went out the air cooled again, and its pressure decreased. The pressure of the air outside the glass forced the water in the plate under the glass. Instead of paper you may use matches stuck in a cork as shown in *Fig. 70.*

Fig. 70. How to pick up the coin without wetting a finger

There is current a wrong explanation of this very old experiment (it was first described and properly explained by the physicist Philo of Byzantium who lived somewhere in the 1st century B.C.). Some people say that the water flows under the glass because it is "oxygen that burns out", and that is why the amount of gas in the glass diminishes. This is absolutely wrong. The water flows under the glass only because the air is *heated* and not at all because any oxygen is absorbed by the burning paper. You can check this statement in the following way. Heat up the glass by pouring boiling water into it, thus dispensing with the burning piece of paper. Then, if you take instead of paper a piece of cotton wool soaked in alcohol, which burns longer and heats up the air better, the water will rise up to almost the middle of the glass;

note that oxygen comprises only a fifth of the air in volume. Note, finally, that instead of the allegedly "consumed" oxygen, you have carbon dioxide and water vapour. While the first dissolves in water, vapour remains, replacing part of the oxygen.

HOW WE DRINK

Can this pose a problem? It can. When drinking we bring a glass or a spoonful of liquid up to our lips and suck in the contents. It is this simple thing we are so used to, that we have to explain. Indeed why does the liquid rush into our mouth? What makes it do that? When we drink, our chest expands, thus rarefying the air in our mouth. *The pressure of the outer air forces* the liquid to rush into the place where pressure is less; so does it find itself in the mouth. Liquids in communicating vessels would behave in exactly the same way were we to rarefy the air above one of them. Atmospheric pressure would compel the liquid in this particular vessel to rise. If you enclose the mouth of a bottle with your lips you will fail to suck in the water as the pressure of the air in your mouth and above the water will be the same. So, strictly speaking, we drink not only with our mouths, but also with out lungs, since it is chest expansion that makes the liquid rush into our mouths.

A BETTER FUNNEL

Those who have ever poured liquids into a bottle through a funnel know that from time to time you have to lift the funnel a little because otherwise the liquid will stay in it. This is because the air in the bottle fails to find an outlet and so blocks up the liquid in the funnel. A little of the liquid will drip in so that the air in the bottle is slightly compressed by the liquid's pressure. However, the cramped air will become resilient enough to offset the weight of the liquid in the funnel by its own pressure. By lifting the funnel, we give the compressed air a chance to escape. Then the liquid begins to flow in again. So, to make a better funnel, the narrower part should have ridges outside to prevent the funnel from fitting tightly in the mouth of the bottle.

98

A TON OF WOOD AND A TON OF IRON

What is heavier—a ton of wood or a ton of iron? Some heedlessly answer that the ton of iron is heavier, thus raising a laugh at their expense. The questioner would probably laugh still louder were he told that the ton of wood is heavier. This seems absolutely incredible, but it is true, strictly speaking.

The point is that Archimedes's principle can be applied not only to liquids but also to gases In the air, every object "loses" in weight as much as the volume of displaced air weighs. Wood and iron also lose a part of their weight, and to get their true weight, you must add the loss. Consequently, the true weight of the wood in our case is one ton plus the weight of the air it displaces.

The true weight of the iron is also one ton plus the weight of the air that the iron displaces. However, a ton of wood occupies a much larger space—about 15 times more—than a ton of iron. Hence, the true weight of a ton of wood is more than that of a ton of iron. Rather should we say that the true weight of the amount of wood which weighs a ton in the air is more than the true weight of iron which also weighs a ton in the air.

Since a ton of iron occupies a volume of $1/8$ m^3 and a ton of wood a volume of about 2 m^3, the difference in the weight of the displaced air should be about 2.5 kg. It is by this amount that a ton of wood is really heavier than a ton of iron.

THE MAN WHO WEIGHED NOTHING

To be as light as a feather—incidentally, in spite of the popular notion, a feather is really hundreds of times heavier than air, and only hovers because due to its rather great "wing-spread" the atmospheric resistance it encounters is much greater than its weight—and even lighter than air, to rid oneself of the fetters of gravity and freely soar into the skies, has been the dream of many a child and even grown-up. But they forget that they can walk around with ease only because they are *heavier than air.*

"We live at the bottom of an ocean of air," Torricelli once said. If we were suddenly to grow a thousand times lighter, lighter than air, we would inevitably float up to the top of this ocean of air. We would rise miles up until we reached regions where the density of the rarefied air would be the same as that of our body. Our dream of hovering in free flight above the hills and vales would be shattered; we would have freed ourselves of gravity but would have been captured by other forces—those of the air currents.

H. G. Wells tells a story in which a very fat man wanted to rid himself of his fatness. The person who tells the story was the possessor of the recipe of a miraculous brew which could rid people of excessive weight. The fat man made the brew according to the recipe and drank it. And this is what happened.

"For a long time the door didn't open.

"I heard the key turn. Then Pyecraft's voice said, 'Come in.'

"I turned the handle and opened the door. Naturally I expected to see Pyecraft.

"Well, you know, he wasn't there!

"I never had such a shock in my life. There was his sitting-room in a state of untidy disorder, plates and dishes among the books and writing things, and several chairs overturned, but Pyecraft—

"'It's all right, o'man; shut the door,' he said, and then I discovered him.

"There he was right up close to the cornice in the corner by the door, as though someone had glued him to the ceiling. His face was anxious and angry. He panted and gesticulated. 'Shut the door,' he said. 'If that woman gets hold of it—'

Fig. 71. "There he was right up close to the cornice"

"I shut the door, and went and stood away from him and stared.

"'If anything gives way and you tumble down,' I said, 'you'll break your neck, Pyecraft.'

"'I wish I could,' he wheezed.

"'A man of your age and weight getting up to kiddish gymnastics—'

"'Don't,' he said, and looked agonised.

"'I'll tell you,' he said, and gesticulated.

"'How the deuce,' said I, 'are you holding on up there?'

"And then abruptly I realised that he was not holding on at all, that he was floating up there—just as a gas-filled bladder might have floated in the same position. He began a struggle to thrust himself away from the ceiling and to clamber down the wall to me. 'It's that prescription,' he panted, as he did so. 'Your great-gran—'

"He took hold of a framed engraving rather carelessly as he spoke and it gave way, and he flew back to the ceiling again, while the picture smashed on the sofa. Bump he went against the ceiling, and I knew then why he was all over white on the more salient curves and angles of his person. He tried again more carefully, coming down by way of the mantel.

"It was really a most extraordinary spectacle, that great, fat, apoplectic-looking man upside down and trying to get from the ceiling to the floor. 'That prescription,' he said. 'Too successful.'

"'How?'

"'Loss of weight—almost complete.'

"And then, of course, I understood.

"'By Jove, Pyecraft,' said I, 'what you wanted was a cure for *fatness*! But you always called it *weight*. You would call it weight.'

"Somehow I was extremely delighted. I quite liked Pyecraft for the time. 'Let me help you!' I said, and took his hand and pulled him down. He kicked about, trying to get foothold somewhere. It was very like holding a flag on a windy day.

"'That table,' he said pointing, 'is solid mahogany and very heavy. If you can put me under that—'

"I did, and there he wallowed about like a captive balloon, while I stood on his hearthrug and talked to him.

" ...'There's one thing pretty evident,' I said, 'that you mustn't do. If you go out of doors you'll go up and up....'

" ...I suggested he should adapt himself to his new conditions. So we came to the really sensible part of the business. I suggested that it would not be difficult for him to learn to walk about on the ceiling with his hands—

"'I can't sleep,' he said.

"But that was no great difficulty. It was quite possible, I pointed out, to make a shake-up under a wire mattress, fasten the under things on with tapes, and have a blanket, sheet, and coverlet to button at the side. He would have to confide in his housekeeper, I said; and after some squabbling he agreed to that. (Afterwards it was quite delightful to see the beautifully matter-of-fact way with which the good lady took all these amazing inversions.) He could have a library ladder in his room, and all his meals could be laid on the top of his bookcase. We also hit on an ingenious device by which he could get to the floor whenever he wanted, which was simply to put the *British Encyclopaedia* (tenth edition) on the top of his open shelves. He just pulled out a couple of volumes and held on, and down he came. And we agreed there must be iron staples along the skirting, so that he could cling to those whenever he wanted to get about the room on the lower level.... (Then, you know, my fatal ingenuity got the better of me.) I was sitting by his fire drinking his whisky, and he was up in his favourite corner by the cornice, tacking a Turkey carpet to the ceiling, when the idea struck me. 'By Jove, Pyecraft!' I said, 'all this is totally unnecessary.'

"And before I could calculate the complete consequences of my notion I blurted it out. 'Lead underclothing,' said I, and the mischief was done.

"Pyecraft received the thing almost in tears. 'To be right ways up again—' he said.

"I gave him the whole secret before I saw where it would take me. 'Buy sheet lead,' I said, 'stamp it into discs. Sew 'em all over your underclothes until you have enough. Have lead-soled boots, carry a bag of solid lead, and the thing is done! Instead of being a prisoner here you may go abroad again, Pyecraft; you may travel—'

"A still happier idea came to me. 'You need never fear a shipwreck. All you need do is just slip off some or all of your clothes, take the necessary amount of luggage in your hand, and float up in the air—'"

At first glance this all seems quite in conformity with the laws of physics. But objections can be made. Firstly, even if Pyecraft had lost his weight, he wouldn't have risen up to the ceiling at all. Recall Archimedes's principle. Pyecraft should have "floated" up to the ceiling only if his clothes and everything in his pockets would have weighed less than the air displaced by his fat body. We can easily reckon the weight of this volume of the air. We weigh almost the same as a similar volume of water—some 60 kg. Air of the usual density is 770 times lighter than water, so the amount we would displace would weigh only 80 gr. However fat Mr. Pyecraft was, he could have scarcely weighed much more than 100 kg; consequently, he must have displaced not more than 130 gr of air. There is no question that Pyecraft's suit, shoes, watch, wallet and all his other belongings weighed more. In that case the fat man should have remained on the floor. He would have felt rather shaky, true, but he certainly would not have "ballooned" up to the ceiling. That would have happened only if he had been stark naked. Dressed, he must have been like a man tied to a bouncing balloon. A small effort, a little jump and he would be up in the air, to smoothly descend again, provided, of course, there was no wind. (See Chapter 4 of my *Mechanics for Entertainment* for more about bouncing balloons.)

"PERPETUAL" CLOCK

You already know a few things about "perpetual motion" machines and of the futility of trying to invent them. Let me now tell you about what I shall call a "gift-power" machine, as it can work indefinitely without human interference, drawing its motive power from the inexhaustible sources of energy in nature. Everybody has most likely seen a barometer, a mercury or aneroid one. In the first one the mercury rises or falls depending on the changes in atmospheric pressure. And it is atmospheric pressure again that causes the arrow to swing in the aneroid barometer.

One 18th-century inventor availed himself of this arrangement to produce a self-winding clock that would never stop. The well-known British mechanic and astronomer James Ferguson saw it in 1774 and this is how he describes it. "I saw this clock," he says, "which is made to go without stopping by the endless rising and falling of the mercury in a curiously arranged barometer. We have no reason to think that the clock would ever stop as the accumulated motive power is enough to make it go for a whole year, even if the barometer were removed. To be frank, I must say that this clock which I examined in detail is the cleverest mechanism I have ever seen, both in design and execution."

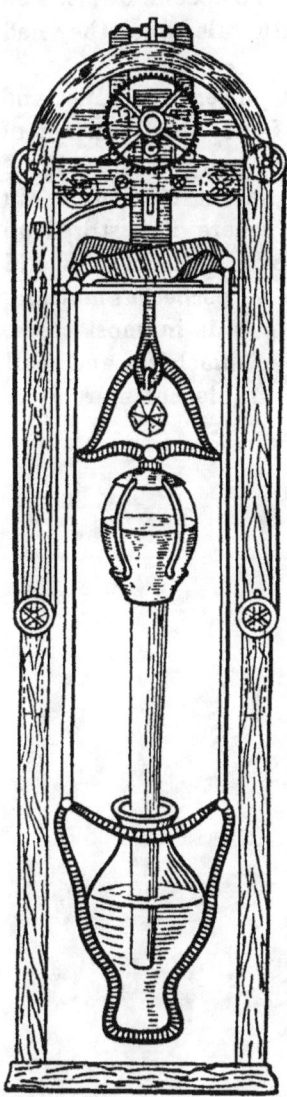

Fig. 72. An 18th-century "gift-power" machine

Unfortunately the clock was stolen and nobody knows what has become of it. Luckily enough, Ferguson made some drawings of it, so it can be reproduced.

Its mechanism consists of a large mercurial barometer, which has about 150 kg of mercury in two glass vessels, one with its mouth in the other, and both suspended in a frame. Both vessels move separately; when atmospheric pressure rises an ingenious system of levers lowers the top vessel and lifts the bottom one. When atmospheric pressure falls, the reverse takes place. This compels a small gear-wheel to turn always in one and the same direction. It doesn't turn only when the atmospheric pressure is steady. However, in these intervals the clockwork is operated by the accumulated potential energy. And though it isn't easy to make the weights rise simultaneously and wind the spring when they drop, the watchmakers of old were ingenious enough.

It even happened that the energy produced by the changes in atmospheric pressure was far more than was needed, causing the weights to rise before they had managed to drop to the bottom. So a special device had to be made to switch off the weights at regular intervals, when they had gone up all the way.

The fundamental difference between such "gift-power" machines and "perpetual motion" machines is obvious. Energy is not produced out of nothing—which was what the inventors of the "perpetual motion" machines sought to achieve. It is supplied from an outside source—in our particular case, the surrounding atmosphere where it is stored up by sunlight. To all practical intents a "gift-power" machine would give the same advantage as could be derived from a "perpetual motion" one—if ever invented—were it not so costly, as it is in most cases.

Later I shall deal with other kinds of "gift-power" machines and shall illustrate why such things are absolutely unprofitable commercially.

CHAPTER SIX

HEAT

WHEN IS THE OKTYABRSKAYA RAILWAY LONGER?

When asked how long the Oktyabrskaya Railway is one person gave this answer: "It's 640 km on the average. But in summer it's about 300 m longer than in winter."

Now this is not so absurd as it may seem. If we meant by the length of a railway the length of its rails, it should indeed be longer in summer than in winter. Don't forget that heat causes steel rails to expand—by more than 100,000th of their length to every one degree Centigrade. On a blazing summer day the temperature of rails might reach 30-40 °C and more. Sometimes rails are so hot that they burn the hand. In winter rails may cool down to 25 °C below zero and even lower. Supposing that the summer-winter difference in temperature is 55°; by multiplying the railway's total length (640 km) by 0.00001 and again by 55, we get about a third of a kilometre. So in summer the Moscow-Leningrad railway is indeed the third of a kilometre, i. e., roughly 300 m, longer than in winter.

It is, of course, not the length of the railway that changes but merely the sum-total of the lengths of all the rails. This is not one and the same thing, because the rails of a railway track do not directly abut one another. Small spaces are left between their joints for the rails to freely expand when they heat up. (This gap—in the case of 8-metre rails—should be 6 mm at zero. To fully bridge it by expansion the temperature of the rails should rise by 65 °C. For certain technical reasons we cannot leave gaps in tramway rails. Usually the rails don't curve, because they are sunk in the ground, temperature fluctuation is not so great and the method used to spike the rails prevents them from curving. However, on a very hot day tram rails do curve, as *Fig. 73*, the reproduction of an actual photograph, well illustrates. Sometimes the same thing hap-

106

Fig. 73. Tram rails bend on very hot days

pens to the rails of a railway track. On downgrades the train pulls at the rails—sometimes even together with the sleepers. As a result, the gaps often disappear on such sections and the rails directly abut one another.) The calculation we have made shows that the total length of all the rails increases at the expense of the total length of these gaps; on a hot summer day the total length in our particular case is 300 metres more than in a winter frost. So to sum up: the rails of the Oktyabrskaya Railway are indeed 300 m longer in summer than in winter.

UNPUNISHED THEFT

On the Moscow-Leningrad line several hundred metres of costly telephone and telegraph wire vanish without trace every winter. Nobody is ever worried; all know who the culprit is. I suppose you, too, have guessed by now. The thief, of course, is the frost. What is true for rails is true for wire too. The only difference is that copper telephone wires expand 1.5 times more than steel, when heated. And since we have no gaps here we can really say, without any reservations whatsoever, that *in winter the Moscow-Leningrad telephone line is indeed 500 m shorter than in summer*. Every winter the frost steals nearly half a kilometre of wire and gets away with it! But it doesn't disrupt telephone or telegraph communications. All that is stolen is dutifully refunded when warmer days set in.

But when bridges, not wires, contract due to frosts the consequences are pretty bad. Newspapers had this to report in December 1927: "The unusual frosts France has been having lately have seriously damaged the bridge across the Seine in the heart of Paris. Due to frosts the bridge's steel framework contracted, causing the road blocks to fly out. The bridge has been temporarily closed to traffic."

HOW HIGH IS THE EIFFEL TOWER?

If I were to ask you now how high the Eiffel Tower is, before saying "300 metres", you would probably want to know in what weather—cold or warm? After all, the height of such an enormous steel structure could not be the same at all temperatures. We know that a steel rod 300 m long expands by 3 mm when heated by 1° C. The height of the Eiffel Tower should increase by roughly the same amount when the temperature rises by 1°. In warm sunny weather the steel framework of the tower might warm up in Paris to 40°C above zero, whereas on a cold rainy day its temperature might fall to 10°C and in winter down to zero and even to as much as 10° below (heavy frosts are rare in Paris). The temperature fluctuation is as much as 40° and more. This means that the height of the Eiffel Tower may be 3 × 40=120 mm=12 cm more or less.

Direct measurement has disclosed that the Eiffel Tower is still more sensitive to temperature fluctuations than the air itself. It warms up and cools quicker and reacts sooner to the sun's sudden appearance on a cloudy day. The changes in the height of the Eiffel Tower were detected by using a wire made of a special nickel steel on whose length temperature fluctuations have practically no effect. This wonderful alloy is called invar from the word invariable.

So, on a hot day the Eiffel Tower is taller than on a cold day by a bit equal to 12 cm and made of iron, which, incidentally, doesn't cost a sou.

FROM TEA GLASS TO WATER GAUGE

Before pouring tea into a glass, the experienced housewife puts in a tea spoon, especially a silver one, to prevent the glass from cracking. Practice has suggested the proper solution.

But what is its basic principle? Why does hot water crack a tea glass?

Because of the uneven expansion of the glass. When you pour hot water into a glass, not all its walls warm up at once. At first the inner layer warms up, the outer one remaining cold. The heated inner layer expands at once. Meanwhile, since the outer one does not expand, it feels a strong pressure from inside. It snaps and the glass breaks.

Don't think you can safeguard yourself against this by using thick-walled glasses. They, on the contrary, are liable to crack sooner than thin-walled ones. This is because a thin wall heats up faster and its temperature and expansion even out sooner. A thick-walled glass, on the other hand, warms up slowly.

One thing you mustn't forget when buying thin-walled glassware— make sure that the bottom of the glass is thin too, because it is the bottom that chiefly heats up. A thick-bottomed glass will crack, however thin its walls. So do glasses and china cups with thick-rimmed bottoms.

The thinner-walled a glass vessel is, the safer it is for heating. Chemists use very thin-walled vessels in which they boil water right over the burner.

The ideal vessel is one that wouldn't expand at all when heated. Quartz almost has this property: it expands 15-20 times less than glass. A thick-walled vessel of transparent quartz will never crack when heated, even if immersed red-hot in a bath of ice (vessels of quartz are good for laboratory work because it melts only at 1,700°C). This is also partially because quartz conducts heat much better than glass.

Tea glasses crack not only when warmed up quickly but also when cooled quickly. Now it is uneven contraction that is to blame. As it cools, the outer layer contracts and exerts a strong pressure on the inner layer, which has not cooled and contracted yet. A prudent housewife should not put a jar of hot jam out in the cold or into cold water.

But back to the tea spoon. How does it protect the glass from cracking? The difference in the expansion of the inner and outer layers is great only when very hot water is poured into the glass at once. Warm water, however, doesn't make glasses crack. What happens when you put a tea spoon in? As it pours in, the hot water loses part of its heat to the metal spoon, which is, contrary to glass, a good conductor of heat. Its temperature drops and it becomes almost harmless, because now it is only warm. Meanwhile the glass has warmed up and more hot water won't crack it.

In a nutshell, a metal tea spoon, especially a heavy one, offsets the uneven heating of the glass and prevents it from cracking.

But why is a silver spoon still better? Because silver is a very good conductor of heat. It can take away the heat from the water sooner than a copper spoon. A silver spoon in a glass of hot tea burns the fingers. Since a copper spoon doesn't do that, you can easily tell the material the spoon is made of.

The uneven expansion of glass walls is a menace not only to tea glasses but also to very important elements of boilers—the water gauges which give the height of the water in the boiler. As the hot steam and water heat them up, their inner layers—they are tubes of glass—expand more than their outer layers. Add to this the great pressure exerted in the tubes by the steam and water, and you will realise why they may so easily burst. To prevent this, they are sometimes made of two layers of different kinds of glass, the inner one having a smaller expansion factor than the outer one.

THE BOOT IN THE BATHHOUSE

"Why in winter is the day short and the night long, and in summer the other way round? The winter day is short because like all other visible and invisible things it contracts due to cold; meanwhile the night expands—it is warmed up when lights and lamps are lit." How comically silly this "explanation", afforded by Chekhov's retired Don

110

Cossack sergeant, is. However, people who ridicule such "learned" reasoning sometimes father theories which are just as stupid. Have you ever heard the story of the boot which won't go on in the bathhouse because "the heated foot has grown larger"? A classical instance, but with a totally wrong explanation.

In the first place one's temperature hardly rises at all when one is in a bathhouse—never by more than one degree Centigrade. Only a Turkish bath will make it go up two degrees. Our body successfully resists the surrounding heat, maintaining its temperature at a definite level. Furthermore, this "rise" in our body temperature increases the volume of our body by such a negligible fraction that one doesn't notice it when drawing on a boot. The expansion factor of our bones and flesh is never more than a few ten-thousandths. Consequently, the sole and the instep could bulge only by a hundredth of a centimetre—no more. Boots and shoes are never sewn with such accuracy. After all, a hundredth of a centimetre is but the thickness of a hair!

Still it remains a fact that it is hard to draw a boot on after a hot bath. However, this is not because our foot expands due to heat but because the blood rushes to the foot, the skin swells, is damp, and grows tender—in a word, because of things that have nothing at all in common with expansion due to heat.

HOW TO WORK MIRACLES

Hero of Alexandria, the ancient Greek mathematician who invented the fountain that bears his name, has left the description of two artful methods which enabled Egyptian priests to take in worshippers by their "miracles".

Fig. 74 shows one such device consisting of a hollow metal altar which stood in front of the temple doors, and of the mechanism, hidden beneath the flagstones, that caused the temple doors to open. When incense was burned, the heated air inside the hollow altar exerted a greater pressure on the water in the vessel hidden below the floor, thus causing it to flow through a pipe into a pail which lowered and set in motion the door-opening mechanism (*Fig. 75*). The worshippers saw, of course, what they thought to be a "miracle"—the temple doors swung

111

Fig. 74. Egyptian temple "miracle" explained. The doors open when incense is burned on the altar

Fig. 75. Diagram showing how the temple doors swing open. (Compare with *Fig. 74.*)

Fig. 76. Another fake miracle of the ancient priests. How incense "everlastingly" drips into the sacrificial flame

open of their own accord as soon as incense and prayers were offered by the priests. They, naturally, knew nothing of the hidden mechanism.

Another fake "miracle" which the priests staged is shown in *Fig. 76*. As soon as incense is burned the expanding air forces more of it to flow out of the cistern below the floor into pipes concealed inside the figures of the priests. The worshippers beheld the "miracle" of an undying flame. However, when the priest in charge considered the offerings too scanty, he unnoticeably removed the stopper in the lid of the cistern. This stopped the flow of incense, because now the superfluous air could find a free outlet.

SELF-WINDING CLOCK

At the close of the previous chapter I described a self-winding clock; its working principle was based on the changes in atmospheric pressure. Now I shall tell you about similar self-winding clocks, the principle of which is based on heat expansion. *Fig. 77* depicts the mechanism of one of them. The central element consists of rods Z_1 and Z_2 which are made of a special alloy with a considerable coefficient of expansion. Upon *expansion* rod Z_1 engages the teeth of wheel X, turning it. Upon *contraction*, on the other hand, rod Z_2 engages the teeth of the wheel Y, turning it in the same direction. Both wheels are set on shaft W_1 which also revolves a large wheel with scoops on it. These scoops lift the mercury from the lower inclined tank R_1 to another contrarily-inclined tray R_2

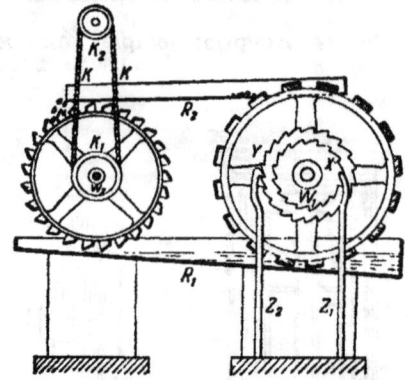

Fig. 77. Diagram of a self-winding clock

down which it flows towards the left-hand wheel also with scoops. As these scoops fill, the wheel turns, setting in motion chain KK, looped around wheel K_1, which is set on the same shaft W_2 as the big wheel, and around wheel K_2, which winds up the clock. Meanwhile the scoops of the left-hand wheel spill out the mercury into the inclined tank R_1, down which it flows to reach the right-hand wheel, and the cycle begins all over again.

This clock, apparently, would go on ticking, while rods Z_1 and Z_2 expand and contract. All we need to wind the clock is an alternate rise and fall in air temperature, which is something that takes place without our interference. Could we call this clock a "perpetual motion" machine

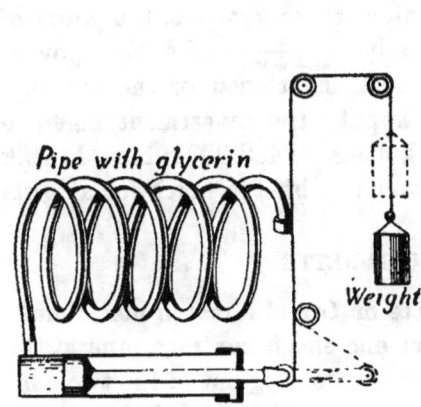

Fig. 78. Diagram of another self-winding clock

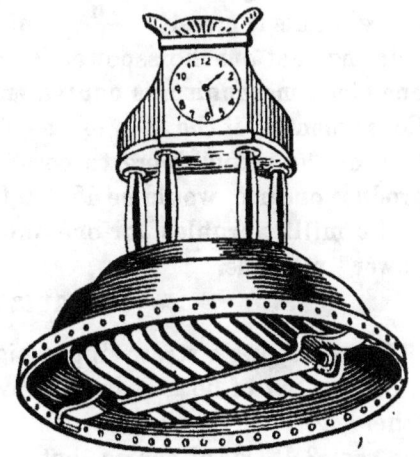

Fig. 79. Self-winding clock. The pipe with the glycerin is hidden in the base of the clock

then? Of course not. The clock will tick indefinitely until its mechanism wears out, but what makes it go is the heat of the surrounding air. The clock stores up the work of heat expansion and expends it portion after portion to turn its hands. This is really a "gift-power" machine since it does not require care or outlay. But it doesn't create energy out of nothing; its primary source is the heat of the sun, which warms up the earth.

Another specimen of a self-winding clock with a similar arrangement is given in *Figs. 78* and *79*. Its basic element is glycerin, which expands when the temperature of the air rises and causes a small weight to rise. The lowering of this weight makes the clock go. Since glycerin solidifies only at 30°C below zero and boils at 290°C above, this mechanism is quite suitable for town clocks. A 2° temperature fluctuation is already enough to keep it going. One such clock was tested for a whole year, and proved to be quite satisfactory.

Can any advantage be derived by designing other bigger machines of this kind? At first glance, such a "gift-power" machine might seem very economical. Let us see, though, whether this is really so. To wind up an ordinary clock to run for 24 hours one requires only 1/7 kgm of energy. This is merely $\frac{1}{600,000}$ of a kilogramme-metre per second. Considering that one horsepower is equivalent to 75 kgm/sec, the power of one clock mechanism is equivalent to only $\frac{1}{45,000,000}$ of a horsepower. Consequently, if the rods in the first clock mentioned or the contraption of the second were to cost one kopek, the investment made to produce one h.p. would be 45,000,000 kopeks, or 450,000 rubles. I think half a million rubles for one horsepower is a bit too much for a "gift-power" machine.

INSTRUCTIVE CIGARETTE

Fig. 80 shows a straw-tipped cigarette on top of a match box. Smoke is curling out of both ends. However, at one end it *curls up*, and at the other *down*. Why? After all, isn't the smoke coming out of the two ends the same? It is, of course, but above the smouldering end there is an ascending current of warm air which carries the particles of smoke up. Meanwhile the air carrying the smoke through the straw tip cools off and no longer rises upward; since the particles of smoke are heavier than air, they float down.

ICE THAT DOES N'T MELT IN BOILING WATER

Take a test tube, fill it with water, and put a lump of ice in. To keep the ice down at the bottom—since it is lighter than water, it floats—press it down by some small weight, seeing to it that the water can get at the lump of ice. Now heat the test tube on a spirit lamp so that the flame licks

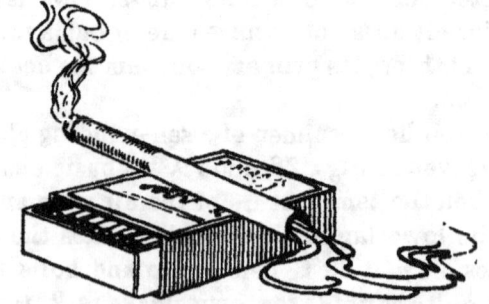

Fig. 80. Why does the smoke curl up from one end, and down from the other?

Fig. 81. The water at the top boils, but the ice at the bottom doesn't melt

only at the tube's upper part as shown in *Fig. 81.* The water will soon boil and send out steam. Oddly enough, the ice at the bottom of the tube doesn't melt. A minor miracle, one would think—ice that doesn't melt in boiling water!

The trick is that at the bottom of the tube the water doesn't boil at all; it remains *cold.* Actually we have not "ice in boiling water" but "ice beneath boiling water". As it expands due to heat, the water becomes lighter; it does not descend to the bottom and stays in the upper part of the tube. There is warm water and a mixture of warm and cold layers of water only in the tube's upper part. Heat can be transferred down only by a conductor, but water is a very poor conductor of heat.

ON TOP OR BENEATH?

When we want to heat water, we put the vessel that contains it right above the flame and not to the side of it. This is the right thing to do since the heated air which grows lighter is forced out from beneath the vessel *upwards* and thus envelops the vessel. So by placing the object we want to heat up right above the flame we use the source of heat in the most advantageous way.

But what should we do *to cool* something with ice? Many put the thing they want to cool—a jug of milk, for example—on top of the ice. This is the wrong thing to do; as the air above the ice cools it *descends,* its place being taken by the warmer surrounding air. So if you want to cool a drink or a dish, *don't put it on top* of the ice but rather the *ice on top of it.*

Let me make the point clearer. When we put a jar of water on top of ice, it is only the bottom layer that cools. The rest of the water is surrounded by uncooled air. But if we put the ice *on* the lid, the water will cool much faster. The cooled upper layers will descend, their place being taken by the warm layers rising from the bottom; the process goes on until all the water has cooled (note that pure water will cool not to zero but only to 4°C above—the temperature at which it possesses the greatest density. After all we never really cool drinks down to zero). Meanwhile the cooled air around the ice will also descend and envelop the vessel.

DRAUGHT FROM CLOSED WINDOW

We often feel a draught coming from a window that is closed tight and hasn't a single crack in it. Though it seems odd there is nothing at all surprising in it.

The air inside a room is practically never in a state of rest. An invisible current circulates as the air warms or cools. As the air warms it rarefies and grows lighter. As it cools it becomes denser and heavier.

The cold heavy air near the windows and outer wall descends to the floor, forcing the warm light air to rise to the ceiling. A toy balloon reveals this circulation at once. Tie a small weight to it, light enough to keep it suspended in mid-air. Release the balloon near the stove or radiator. You will see it travel around the room, being carried by the invisible current from the fireplace or radiator up to the ceiling and towards the window, and from there down to the floor and back to the fireplace. Here it again sets out on the same journey. That is why we feel the draught, especially around the feet, coming from the window though it is closed tight in winter.

MYSTERIOUS TWIRL

Take some thin cigarette paper and cut out a piece in the form of a rectangle. Fold it down the middle and then straighten it again. The fold will tell you where the centre of gravity is. Now stick a needle upright into the table and place the piece on the other end so that it is set on its centre of gravity and, hence, balanced. So far there is nothing

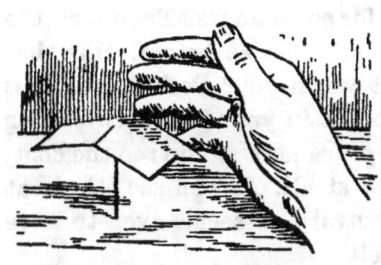

Fig. 82. Why does this piece of paper spin?

mysterious about it. Bring up your hand as is shown on *Fig. 82.* Do this gently though, otherwise the piece of paper will be blown off by the rush of air. The paper will start to spin. At first it gyrates slowly but then it picks up speed. Take your hand away and gyration stops. Bring your hand up again and gyration resumes.

This mysterious gyration once—in the 1870's—caused many to believe that we, or rather our bodies, were endowed with some supernatural properties. Mystics thought this confirmed their wild theories about the strange fluids the human body was supposed to possess. Actually, there is nothing unnatural in it; as a matter of fact, everything is as simple as pie. When you bring your hand up, the air near it, which is warmed by its proximity, rises and, pressing against the piece of paper, causes it to spin. It revolves because it is slightly folded, thus acting the same role as a curled piece of paper suspended above a lamp.

A closer look will show you that the piece of paper always gyrates in one and the same direction—from the wrist towards the finger-tips. This is because the finger-tips are always colder than the palm of the hand; consequently, the palm gives rise to a stronger ascending air current than the finger-tips. Incidentally, when one is feverish, or happens to be running a high temperature, the paper gyrates much faster. You might be interested to learn that this twirling, which once mystified so many, was the subject of a communication made to the Moscow Medical Society in 1876 (*The Gyration of Light Bodies Caused by the Heat of the Hand*, by N. P. Nechayev).

DOES A WINTER COAT WARM YOU?

If I told you that your fur coat *does not warm* you at all, you would probably think I was pulling your leg. But suppose I prove it? Stage the following experiment. Take the reading of an ordinary thermometer.

Then wrap it in your fur coat and let it be for some hours. Then read the thermometer again. It will be exactly the same as before. Has that convinced you that your fur coat doesn't warm you? Perhaps, it *cools* you then? Take two bags of ice and wrap one in your fur coat, leaving the other in a dish. When this second bag of ice melts, unwrap the coat. The ice in the first bag has hardly melted at all. As you see, the coat has not warmed it in the least; on the contrary, it seems even to have cooled it, since the ice took longer to melt!

So, does a winter coat warm you? No, if by warming we mean the *communication of heat.* A lamp does. So does a stove. And so does our body. They are all sources of heat. Your fur coat is not a source of heat; *it doesn't have any warmth of its own to give. It merely prevents our body from shedding its own warmth.* That is why a warm-blooded animal —whose body is actually a source of heat—feels much warmer in a coat of fur than without one. However, since the thermometer we took for our experiment is not a source of heat its reading naturally could not change simply because we wrapped it in the fur coat. The ice in the coat also took longer to melt because the coat is a rather poor conductor of heat and blocks any intake of surrounding warmth.

The snow on the ground is also like a fur coat; it is a poor conductor of heat—like all powdery bodies—and thus prevents the ground beneath from shedding its heat. The temperature of the ground beneath a protective layer of snow is often some 10°C higher than at a bare spot.

So the answer to the question "Does a winter coat warm you?" is: it merely helps us to warm ourselves; rather we ourselves warm the coat instead.

THE SEASON UNDERFOOT

It is summer on the ground and above it. What season of the year is it three metres down? You think it's summer? You're wrong! It's not at all the same season as one might think. The point is that the ground is a very poor conductor of heat. In Leningrad water mains don't burst even in the grimmest of frosts, because they are two metres deep. Above-surface temperature fluctuations reach the different subsoil strata with great delay. Direct measurements conducted in the town of Slutsk, Leningrad Region, showed that at three metres down the warmest time of the year comes 76 days late, while the coldest period is 108 days late. If the hottest day above the ground is July 25th, at three metres down the hottest day will come only on October 9th. On the other hand, if the coldest day is January 15th, at the depth given the coldest day will come only in May. At greater depths the delay is still greater.

The further down we go, the weaker the temperature fluctuations become, to fade to an everlasting constant at a certain depth; here you have one and the same temperature round the year for centuries on end. This temperature is the mean annual temperature of the place in question. In the cellars of the Paris observatory, 28 metres below the ground, there is a thermometer which Lavoisier stored away there more than 150 years ago. The mercury has not budged a hair since, giving all the time one and the same temperature of 11.7°C above zero.

To sum up: underfoot we never have the season of the year we have above the ground. When it is winter for us it is still autumn three metres down—of course not the autumn we had, as the fall in temperature is not so pronounced. On the other hand, when it is summer for us, deep down we still have faint repercussions of winter frosts. One must always bear this important point in mind whenever one is dealing with the conditions of life underground—for plant tubers and roots, and for cockchafer grub, for instance. It should not be surprising, for instance, that in tree roots the cells multiply in winter and that the tissue called the cambium ceases to function for practically the whole of summer, in contrast to the tissue of the above-ground tree-trunk.

PAPER POT

Look at *Fig. 83*. An egg is boiling in water in a paper cup. Won't the paper burn through and the water spill out and extinguish the flame? Try to do it yourself. boiling the egg in some stiff parchment paper attached fast to a piece of wire (or better make the paper box shown in *Fig. 84*). Nothing happens to the paper! The reason is that one can warm water only up to boiling point—100°C. The water—it has a great capacity for absorbing heat—absorbs the paper's extra heat and prevents it from warming to much more than 100°C, that is, to

120

a point where it could burst into flame. The paper won't burn even if licked by the flame.

It is the same property of water that prevents a kettle from

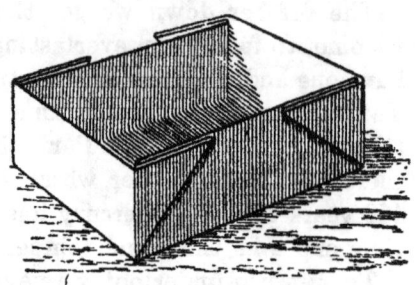

Fig. 83. An egg boiling in a paper pot   Fig. 84. A paper box for boiling water

going to pieces—which is what would happen were we absent-minded enough to put the kettle on to boil without any water in it. For the same reason you must not put soldered pots on the fire unless they have water in them. The water used to cool the old Maxim machine guns saved the barrel from melting.

By using a little box made from a playing card, you can melt a lead pellet. To do this, put the lead in the box right above the flame. Since lead is a good conductor of heat it rapidly absorbs the heat of the box, preventing the box from heating up to way above its melting point—335 °C—which is too little yet for the box to break into flame.

Fig. 85 gives another simple experiment. Take a thick nail or an iron—or better copper—rod and *tightly* curl screw-wise a narrow strip

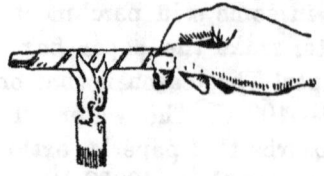

Fig. 85. Paper that doesn't burn

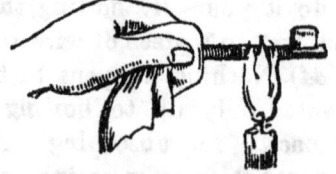

Fig. 86. The thread that doesn't burn

of paper around it. Then apply a flame. The flame will lick at the paper and even smoke it; but it'll start burning only when the rod grows red-hot. Again the metal's good heat conductivity is the reason. A glass stick, for instance, wouldn't do at all for this experiment. *Fig. 86* shows a similar experiment in which we have a "non-inflammable" piece of thread wound *tightly* round a key.

WHY IS ICE SLIPPERY?

One slips on a smoothly polished floor much more easily than on one that isn't polished. Now, shouldn't *smooth* ice be much more slippery than bumpy ice? However, contrary to expectation, a sled goes much more easily over bumpy ice than over smooth ice—which you may have noticed yourself if you have ever happened to pull a sled. How come that bumpy ice is more slippery than glossy ice? Ice is slippery not because it's smooth but because its melting point drops when pressure is increased.

Let's see what happens when we sled or skate. On skates we bring the whole weight of our body to bear down on a very small area, of but a few square millimetres. Recall Chapter 2 of this book. You will realise that a person on skates exerts a considerable pressure on the ice. Under strong pressure ice melts at a lower temperature. For instance, if the temperature of the ice is 5°C below zero and the skater's pressure has lowered the melting point of the ice beneath his skates by 6 or 7°, this ice will melt. This gives rise to a thin layer of water between the blades and the ice. No wonder the skater slides, or rather slips, along. And as soon as he moves further, the same thing repeats itself. The skater continually slides over a thin layer of water. It is only ice that has this property. One Soviet physicist even called it "nature's sole slippery body". All other bodies are smooth but not slippery.

Back now to our first point. Why is bumpy ice more slippery than smooth ice? We already know that one and the same weight exerts a stronger pressure when it rests on a smaller area. When does a man exert more pressure? On smooth ice? Or on bumpy ice? It is quite obvious that he exerts more pressure on bumpy ice because in this case he is supported only by a few bumps in the ice. The greater the pressure exerted, the more readily does ice melt and, consequently, the more slippery does ice become—provided the sled runners are wide enough (this will not apply to the thin skate blades as the energy of motion is expended to slice off the bumps).

This pressure-induced lowering of the melting point of ice explains many other things that we see around us. This is why separate lumps of ice freeze into one when strongly pressed together. Boys throwing snowballs unconsciously avail themselves of this property; the separate snowflakes stick together because the pressure exerted to form the snow-ball lowers their melting point. To make a snowman we again apply this principle. (I suppose I needn't explain, though, why in strong frosts we are unable to mould good snowballs and snowmen.) Under the pressure of the many feet walking along the pavement snow gradually turns into one solid icy mass.

It has been theoretically calculated that to lower the melting point of ice by one degree Centigrade we must exert the rather considerable pressure of 130 kg/cm². Here one must bear in mind that in the process of melting both ice and water are subjected to one and the same pressure. In the instances described it was only the ice that was subjected to strong pressure; the water the ice melted into is subjected to atmospheric pressure; consequently, in this case the effect pressure has on the melting point of ice is much greater.

THE ICICLES PROBLEM

Have you ever stopped to wonder how the icicles we see drooping from eaves form? And when do they form? During a thaw or during a frost? And if during a thaw, then how does water freeze at an above-zero temperature? On the other hand, if during a frost, then where, in general, does the water that freezes come from?

As you see, the problem is not so simple as you may have thought. To produce icicles you need *two temperatures* simultaneously—one above zero for melting and the other below zero for freezing. That is really what happens. The snow on slanting rooftops melts because it is warmed by the sun to an *above-zero* temperature. Meanwhile the drops of water dripping off the eaves freeze, because here we have a *sub-zero*

Fig. 87. The sun heats the slanted roof more than the ground

temperature. (We don't mean the icicles that form because of the warmth exuded by the heated room under the roof.)

Try to imagine the following picture. It's a clear and sunny day. The temperature is just one or two degrees Centigrade below zero. Everything is bathed in sunlight. The sun's slanting rays are not strong enough to melt the snow on the ground. But since they strike the *inclined* rooftop facing the sun *at an angle closer to a right angle*, they warm up the roof and. melt the snow on it. Sunshine gives more light and warmth the wider the angle between the line of the rays and the plane on which they are incident. It acts in direct ratio to the *sine* of this angle. As for the case in *Fig. 87*, the snow on the rooftop gets 2.5 times more warmth than the snow on the ground, because the sine of 60° is 2.5 times more than the sine of 20°. The melting snow drips off the eaves. But since the temperature beneath the eaves is a *sub-zero* one the drops of water—cooled furthermore by evaporation—freeze. Another drop drips onto the frozen one and also freezes. Then comes a third, a fourth and so on, gradually producing a tiny pendant of ice. A couple of days later, or maybe a week later, we have the same kind of weather again. The pendant grows, producing a larger and larger icicle—in much the same way as lime stalactites form in underground caverns. That is how icicles form on the eaves of sheds and other unheated premises.

The changing angle of incidence of the sun's rays produces far grander phenomena. The different climatic zones and seasons are largely due to that—but not wholly; another major factor is the varying day-length, or the time during which the sun warms the earth, which, like the seasons, is due to one and the same astronomical cause, the inclination of the earth's axis of rotation to the plane of the ecliptic. In winter the sun is practically as far away from us as in summer; it is just as far away from the poles as it is from the equator—the difference is so insignificant that it can be totally ignored. However, at the equator the angle of incidence of the sun's rays is wider than at the poles; in summer again, the angle of incidence is wider than in winter. This phenomenon gives rise to a pronounced variation in temperatures, and consequently in nature in general.

CHAPTER SEVEN

LIGHT

TRAPPED SHADOWS

Our forefathers did find some use for their shadows even though they weren't able to catch them. This was the making of silhouettes, or shadow images. Today we go to the photographer's if we want our pictures or the pictures of friends and relatives taken. But in the 18th century there were no photographers. Portrait-painters asked a stiff price for their work and only the rich could afford it. That is why *silhouettes* were so widespread; in some measure they did for our present snapshots.

Silhouettes are actually trapped shadows. They were obtained mechanically and in this we can draw a certain parallel between them and their opposites—photographs; while photographers draw on *light* ("photos" is Greek for light) to make pictures, our ancestors used *shadows* for the same purpose.

Fig. 88 shows you how silhouettes were made. The sitter turned his head to cast a characteristic profile and this profile was traced with a pencil. Then the inside of the outline was blacked, cut out, and glued onto a white ground. This was the silhouette. Whenever necessary, the silhouette was reduced by means of a special device called the pantograph (*Fig. 89*).

Don't think that this simple black outline could not give a notion of the characteristic features and profile of its prototype. A good silhouette is sometimes amazingly like the original.

This property intrigued some artists, who began to paint in this manner, thus starting a whole school. The very origin of the word is of interest. It derives from Etienne de Silhouette, an 18th-century

Fig. 88. An old way of making shadow portraits

Fig. 89. How to reduce
a silhouette

Fig. 90. A silhouette
of Schiller (1790)

French Minister of Finance, who urged his extravagant compatriots to show thrift and reproached the French aristocracy for wasting money on pictures and portraits. The cheapness of shadow likeness thus suggested the name—portraits "à la Silhouette".

THE CHICK IN THE EGG

The properties shadows possess will enable you to stage an amusing parlour trick. Take a piece of greased paper and make a screen by sticking it on top of a square hole cut in a piece of cardboard. Put two unshaded table lamps behind this screen and seat your friends in front of it. Switch on the left lamp. Place an oval piece of cardboard mounted on a piece of wire between the lit lamp and the screen. Your friends will naturally see the outline of an egg. The second lamp is still not on. Now tell your friends that you have an X-ray machine that will detect the chick inside the egg. Hey, presto! and your friends see the egg's shadow pale and the rather distinct outline of a chick appear in the middle (*Fig. 91*).

It is really all very simple. Just switch on the right lamp which has a cardboard chick between it and the screen. Part of the oval shadow upon which the chick's shadow is superimposed is illumined by the right lamp. That is why its fringes are lighter. Since your friends don't see your manipulations, those ignorant of physics and anatomy may really think that you have X-rayed the egg.

Fig. 91. A fake X-ray

PHOTOGRAPHIC CARICATURES

Many of you might not know that you can make a camera in which an ordinary small round hole will take the place of the lens. True, you get a fainter image in this case. An interesting modification of this "lensless" camera is the "slit" camera which has two criss-crossing slits instead of the round aperture. This camera has in its front part two small slats, one having a vertical slit and the other a horizontal slit. When the two slats are superimposed the image obtained is the same as produced by the aperture camera. In other words, the likeness is not distorted. But when the slats are moved apart—they are specially arranged so that this can be done—the image produced becomes distorted (*Figs. 92* and *93*), resembling a caricature rather than a photograph.

Fig. 92. A caricature obtained by means of a "slit" camera. The image is distended horizontally

Fig. 93. A similar caricature distended vertically

Why does this happen? Let us take the case when the slat with the horizontal slit is placed in front of that with the vertical slit (*Fig. 94*). The rays coming from the vertical line of figure D (a cross) pass through the first slit C as through any ordinary aperture; meanwhile slit B does not alter their course at all. Consequently, on the ground-glass screen A you get an image of the vertical line on a scale corresponding to the distance between A and C. However, this disposition of the slats produces an entirely different image of D's horizontal line. The rays pass through the horizontal slit without hindrance and don't cross until they reach the vertical slit B, which they pass as any round aperture to produce on screen *A* an image on a scale corresponding to the distance between *A* and *B*.

In short, the vertical lines are taken care of by slit *C* only, and the horizontal lines, on the contrary, by slit *B* only. Since slit *C* is further away from the screen all vertical dimensions are reproduced on glass *A* on a scale larger than that of the horizontal dimensions. In other

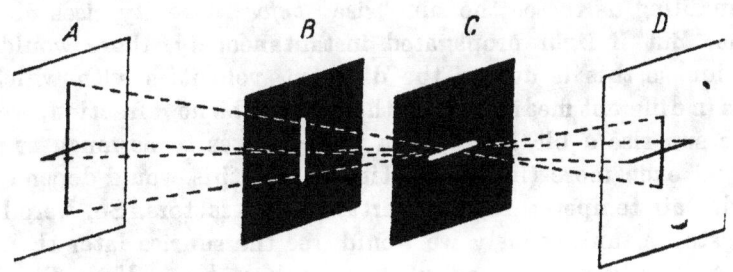

Fig. 94. Why the "slit" camera produces distorted images

words the image is distended vertically. A redisposition of the slats will produce a horizontally distended likeness (compare *Figs. 92 and 93*). A *slantwise* disposition will distort the likeness in still another way.

This camera can be employed not only to get caricatures. It can also serve a more serious purpose, as, for instance, to vary architectural embellishments, carpet and wallpaper patterns, and in general any ornamental motif that may be distended or condensed at will in a definite direction.

THE SUNRISE PROBLEM

Suppose you get up exactly at 5 o'clock early in the morning to watch the sunrise. Since light does not propagate instantaneously some time must pass before the light reaches your eye from its source. So my question is: At what time would you see the sunrise were light able to propagate instantaneously?

Since it takes eight minutes for the light to travel from the sun to us here on Earth, one might think that if light propagated *instantaneously* one would see the sun rise eight minutes earlier—at 4:52 a.m. You're in for a surprise if you think so; that answer is absolutely wrong. The sun "rises" when the Earth turns to face the *space that is already lit.* Therefore even if light propagated instantaneously we would still see the sunrise only at 5 a.m.

If we take what is called "atmospheric refraction" into consideration we get a still more startling result. Refraction curves the path of light, thus enabling us to see the sun "rise" *before* it really rises above the horizon. But if light propagated instantaneously, there would be no refraction as this is due to the different velocities with which light travels in different media. And as there would be no refraction, we would see the sun rise a bit *later*—from two minutes to as much as several days and even more (in polar latitudes), as this would depend on the latitude, air temperature, and certain other factors. So, were light to propagatej instantaneously we would see the sunrise later than we do now. A most curious paradox! (See *Do You Know Your Physics?* for further detail.)

It would be quite different, of course, if you were observing the appearance of a solar protuberance in a telescope. Then—that is, if light propagated instantaneously—you would see it eight minutes earlier.

CHAPTER EIGHT

REFLECTION AND REFRACTION

SEEING THROUGH WALLS

In the 1890's one could buy a curious contraption pompously called an "X-ray apparatus". I remember how puzzled I was when I, a schoolboy at the time, saw this ingenious device for the first time. It enabled me to see light through opaque objects—not only thick paper but even a knife blade, which is impenetrable to real X-rays. *Fig. 95*, which shows the prototype of the contraption I just mentioned, "lets the cat out of the bag". It has four small mirrors, each slanted at the angle of 45°, to reflect and rereflect the rays coming from the object and thus lead them around the opaque obstacle.

Fig. 95. A sham X-ray apparatus

The military extensively employ a similar device—the periscope (*Fig. 96*)—enabling them to follow the enemy's movements without exposing themselves to the hazard of enemy fire. The further away the

Fig. 96. The periscope

Fig. 97. Diagram of a submarine periscope

object is from the periscope, the smaller the observer's field of vision is. A special arrangement of optical lenses is used to enlarge the field of vision. But since the lenses absorb part of the light that enters the periscope, the image obtained is blurred. This limits the height of a

133

periscope, with some twenty metres being already close to the "ceiling". Taller periscopes give a very small field of vision and a blurred image, especially in cloudy weather.

Submarine commanders also use periscopes to watch the ships they attack. Though a far more complicated affair than the army periscope, this periscope, which juts out of the water when the submarine submerges, is the same in principle, having a similar arrangement of mirrors (or prisms). (*Fig. 97.*)

THE SPEAKING HEAD

This frequent side-show "marvel" dumbfounds the uninitiated. It does, indeed, astound one to see on a plate a live, seemingly severed human head, which rolls its eyes, speaks, and eats. And though you can't walk right up to the table on which it lies, you "quite perfectly" see that there is nothing underneath. If you ever see this side show, make a paper ball and throw it under the table. Strangely enough, it bounces back. The mystery is no longer a mystery—it has bounced off a mirror. Even if it doesn't reach the table it will show you that there is a mirror there because you will see its reflection (*Fig. 98*).

It is quite enough to have a mirror stretching from one table-leg to the other to give one the illusion that there is nothing beneath the table—provided, of course, that the mirror doesn't reflect the furnishings of the room or the audience. That is why it is absolutely necessary for the room to be bare and its walls all alike. The floor too should be in one tone, devoid of all ornamental design, and the audience must be kept at a respectful distance. As you see, the "secret" is as simple as pie, but until you're in the know, you just gape.

Sometimes the trick is still fancier. First the conjuror shows you a bare table, with nothing on top or beneath it. Then a closed box that is supposed to have the "live head" inside, but which is really empty, is brought onto the stage. The conjuror puts the box on the table and opens up

Fig. 98. The secret of the lopped-off head

the front flap. And lo! a speaking head appears. You've most likely guessed that the upper board of the table has sort of a trap-door in it through which the man squatting under the table behind the mirror pokes his head when the bottomless empty box is placed on the table. There are other ways of doing this trick. You'll probably be able to work it out for yourself.

IN FRONT OR BEHIND

There are many household things which are not used properly. You already know that some don't use ice properly to cool a drink; they place it on top of the ice instead of *beneath* the ice. Nor does everyone know how to use a mirror properly. Quite often one may put a lamp behind oneself to "light up" one's reflection in the mirror instead of throwing the light on one's own person. Since there are many women who do that, I hope the women among my readers will put the lamp in front of themselves when they want to use a mirror.

IS A MIRROR VISIBLE?

There, again, is proof that what we know about the ordinary mirror is not enough, because most answer this question wrongly, even though all use mirrors every day. Those who think that they can see a mirror are mistaken. A good, clean mirror is invisible. You can see its frame, its rim and everything reflected in it, but you'll never see the mirror itself unless it's dirty. In contrast to a *dispersing* surface—one that scatters light in all directions—every *reflecting* surface is invisible. In ordinary practices a reflecting surface is a polished one, and a dispersing surface, a dull one. All tricks and optical illusions using mirrors— the "speaking head", for instance—are based precisely on their invisibility. All that you do see is the reflection in the mirror of different objects.

IN THE LOOKING-GLASS

When we look in the looking-glass we see ourselves, many will say, adding that what we see is the exact copy of our own person down to the minutest detail.

Let's test that statement. Suppose you have a mole on your right cheek. The person you see in the mirror has a mole on his left cheek. You may be brushing your hair to the *right*; your double in the mirror will be doing it to the *left*. Your right brow may be a bit higher and thicker than your left one; with your copy in the mirror it's the other way round. You keep your watch in your right waistcoat pocket and your wallet in the left pocket; your double has quite opposite habits. Note the

Fig. 99. Use a mirror

dial of his watch. Your watch isn't at all like that. The figures and their arrangement are most unusual. You see an eight marked as it has never been marked before—as IIX—and standing where the twelve ought to be. Meanwhile there is no twelve at all. After a six comes a five, a four and so on. The hands of the watch's double in the mirror move the other way.

To cap it all, he has a physical handicap which you most likely don't have. He's left-handed. He writes, sews and eats with his left hand. And he'll stretch out his left hand to shake your right one. Then, does he know his letters? At any rate his knowledge is of a most peculiar brand. I greatly doubt whether you will ever be able to read a single line in the book he holds or make out a single word in his left-handed scribble. Such is the person who claims to be your exact copy, the person you claim is exactly like you!

But joking apart, if you really think that by looking in the mirror you are observing yourself, you are mistaken. The face, body and clothing of most people are not strictly symmetrical, but usually we don't notice that. The right side is not quite the same as the left side. In the looking-glass your left side assumes all the peculiar features of your right side and vice versa, so that you actually have a reflection that often produces quite a different impression than you do yourself.

MIRROR DRAWING

The fact that you and your reflection are not totally alike stands out still more when you do the following. Sit down at a table facing an upright mirror. Then take a piece of paper and try to draw, say, a rectangle with intersecting diagonals, by looking at the reflection of your hand. This seemingly simple task becomes incredibly difficult.

Fig. 100. Drawing in front of a looking-glass

As we grow up our visual impressions and motive sensations reach a definite degree of accord. The mirror violates this harmony as it gives us a distorted visual image of our hands in motion. Force of habit cries out against every move you make: you want to draw a line towards the right, but your hand pulls the pencil towards the left. You get still stranger results when you try in this manner to draw still more intricate figures or write something. You are bound to make a most comical mess of things.

The inky imprints on blotting paper are also a mirror-like symmetrical reflection of your handwriting. But try to read them. You won't be able to make out a single word, even when the letters seem quite distinct. The writing will be slanted leftwise and all the strokes are topsy-turvy. However, as soon as you try to read this muddle in a mirror, everything straightens itself out and you recognise your own customary handwriting. Actually, the mirror gives you a symmetrical reflection of what in itself is a symmetrical reflection of your own handwriting.

SHORTEST AND FASTEST

In a homogeneous medium light propagates rectilinearly, that is in the fastest way possible. Light again picks the fastest route when reflecting from a mirror. Let us trace its passage. In *Fig. 101 A* is the source

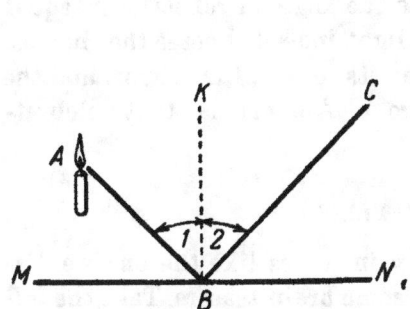

Fig. 101. The angle of reflection 2 is equal to the angle of incidence 1

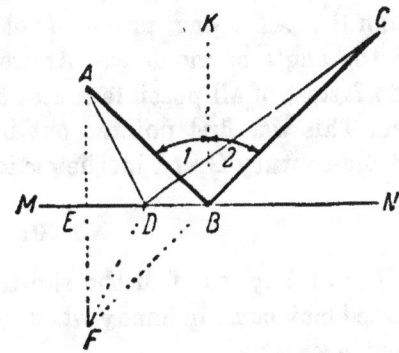

Fig. 102. Reflecting light chooses the shortest path

of light, a candle, *MN*—a mirror, and *ABC*—the ray's passage from *A* to the eye *C*. The straight line *KB* is perpendicular to *MN*.

According to the laws of optics, the angle of reflection *2* is equal to the angle of incidence *1*. Once we know this, we can easily prove that of all possible routes from *A* to *C*, that bounce off the mirror *MN*, *ABC* is the shortest. To prove that this is so, let us compare *ABC* with some other route—for example, *ADC (Fig. 102)*. Drop the perpendicular *AE* from point *A* onto *MN* and continue it further until it intersects with the continuation of the ray *BC* at point *F*. Then join points *F* and *D* by a straight line. Now let us see first whether the two triangles ABE and EBF are equal. They are both right triangles and both have the side EB adjacent to the right angle. Besides that, the angles EFB and EAB are equal as they are respectively equal to the angles 2 and 1. Consequently, AE is equal to EF. Hence, the right triangles AED and EDF are equal because their respective sides adjacent to the right angles are equal. Consequently, AD is equal to DF.

We can thus replace the route ABC by the equal CBF route—since AB is equal to FB—and the ADC route by the CDF route. Comparing CBF and CDF, we see that the straight line CBF is shorter than the broken line CDF. Consequently, the ABC route is shorter than the ADC one. Q.E.D.!

Wherever point D may be, the ABC route will always be shorter than the ADC one, provided of course the angle of reflection is equal to the angle of incidence. As we see, light indeed chooses the shortest and fastest of all possible routes between its source, the mirror, and the eye. This was first pointed out by Hero of Alexandria, that celebrated 3rd-century Greek mathematician.

AS THE CROW FLIES

The ability to find the shortest way in cases like the one we discussed may come in handy when solving some brain-teasers. Take the following case.

Fig. 103. The problem of the crow. Find the shortest line of flight to the ground and to the fence

Fig. 104. The solution of the problem of the crow

A crow is perched on a branch, and there are some grains of millet scattered on the ground below. The crow swoops down, pecks at the millet and then flies up to perch on the fence. The question is: Where should the crow peck in order to take the shortest possible route? (*Fig. 103.*) This is an absolutely similar problem to the one just discussed. So we can easily supply the right answer: the crow should follow the path of the ray of light. In other words, it should fly so that angle *1* is equal to angle *2* (*Fig. 104*), which, as we already know, is the shortest way possible.

THE KALEIDOSCOPE

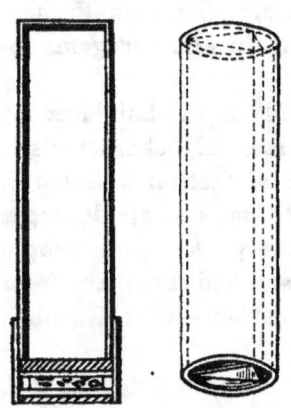

Fig. 105. A kaleidoscope

I suppose you all know what the kaleidoscope is. This amusing toy has a handful of various coloured bits of glass which are placed between two or three flat mirrors. They form extremely beautiful figures which change symmetrically with the slightest twist of the kaleidoscope. Though a very common toy, few suspect the tremendous assortment of different patterns 'one can get. Imagine that you have a kaleidoscope with 20 bits of glass inside and turn it to get ten new patterns every minute. How much time would you need to see all the patterns these 20 bits of glass could form? Even the wildest of imaginations would never provide the right answer. The oceans would dry and the mountains crumble before you saw all; you would need at least 500,000 million years to see every figure produced!

The infinitely different and eternally changing patterns that this toy provides have long intrigued art designers, whose combined imaginations will never match the inexhaustible ingenuity with which it suggests lovely ornamental motifs for wallpaper, carpets and other fabrics. But among the general public it no longer excites the interest it did a hundred years ago when it was a fascinating novelty and when poets composed odes in its honour.

140

The kaleidoscope was invented in England in 1816. Some twelve to eighteen months later it was already arousing universal admiration. In the July 1818 issue of the Russian magazine *Blagonamerenni (Loyal)*, the fabulist A. Izmailov wrote about it: "Neither poetry nor prose can describe all that the kaleidoscope shows you. The figures change with every twist, with no new one alike. What beautiful patterns! How wonderful for embroidering! But where would one find such bright silks? Certainly a most pleasant relief from idle boredom—much better than to play patience at cards.

"They say that the kaleidoscope was known way back in the 17th century. At any rate, some time ago it was revived and perfected in England to cross the Channel a couple of months ago. One rich Frenchman ordered a kaleidoscope for 20,000 francs, with pearls and gems instead of coloured bits of glass and beads."

Izmailov then provides an amusing anecdote about the kaleidoscope and finally concludes on a melancholic note, extremely characteristic of that backward time of serfdom: "The imperial mechanic Rospini, who is famed for his excellent optical instruments, makes kaleidoscopes which he sells for 20 rubles a piece. Doubtlessly, far more people will want them than to attend the lectures on physics and chemistry from which—to our regret and surprise—that loyal gentleman, Mr. Rospini, has derived no profit."

For long the kaleidoscope was nothing more than an amusing toy. Today it is used in pattern designing. A device has been invented to photograph the kaleidoscope figures and thus mechanically provide sundry ornamental patterns.

PALACES OF ILLUSIONS AND MIRAGES

I wonder what sort of a sensation we would experience if we became midgets the size of the bits of glass and slipped into the kaleidoscope? Those who visited the Paris World Fair in 1900 had this wonderful opportunity. The so-called "Palace of Illusions" was a major attraction there—a place very much like the insides of a huge rigid kaleidoscope. Imagine a hexagonal hall, in which each of the six walls was a large, beautifully polished mirror. In each corner it had architectural embellishments—columns and cornices—which merged with the sculptural adornments of the ceiling. The visitor thought he was one of a teeming crowd of people, looking all alike, and filling an endless enfilade of columned halls that stretched on every side as far as the eye could see. The halls shaded horizontally in *Fig. 106* are the result of a single reflection, the next twelve, shaded perpendicularly, the result of a double reflection, and the next eighteen, shaded slantwise, the result of a triple reflection. The halls multiply in number with each new mul-

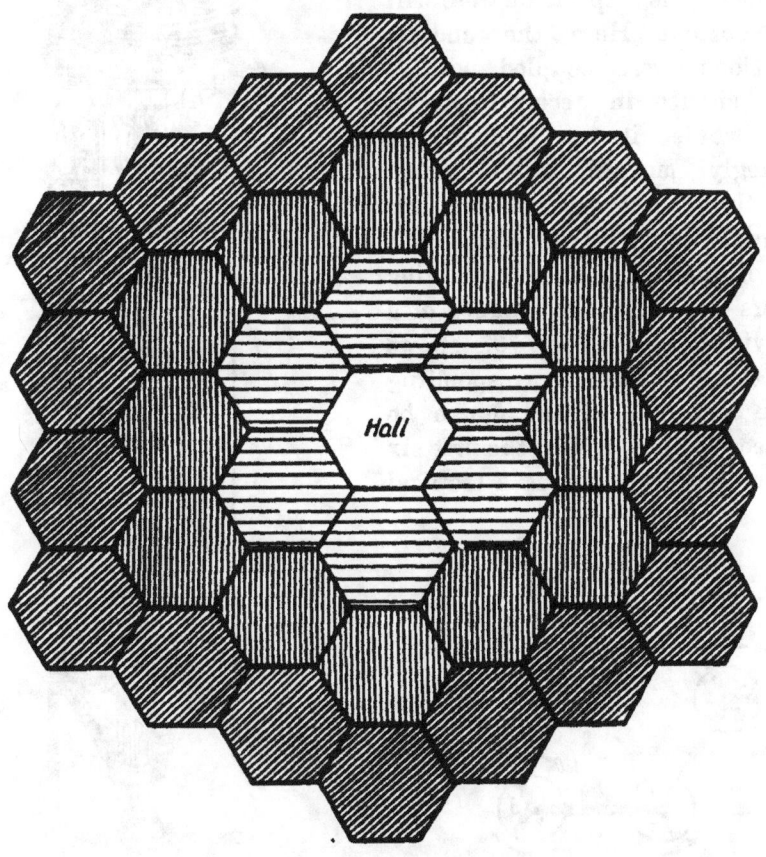

Fig. 106. A three-fold reflection from the walls of the central hall produces 36 halls

tiple reflection, depending, naturally, on how perfect the mirrors are and whether they are disposed at exact parallels. Actually, one could see only 468 halls—the result of the 12th reflection.

Everybody familiar with the laws that govern the reflection of light will realise how the illusion is produced. Since we have here three pairs of parallel mirrors and ten pairs of mirrors set at angles to each other, no wonder they give so many reflections.

The optical illusions produced by the so-called Palace of Mirages at the same Paris Exposition were still more curious. Here the endless reflections were coupled with a quick change in decorations. In other words, it was a huge but seemingly movable kaleidoscope, with the spectators inside. This was achieved by introducing in the hall of mirrors hinged revolving corners—much in the manner of a revolving stage. *Fig. 107* shows that three changes, corresponding to the corners *1*, *2* and *3*, can be effected. Supposing that the first six corners are decorated as a tropical

Fig. 108. The secret of the "Palace of Mirages"

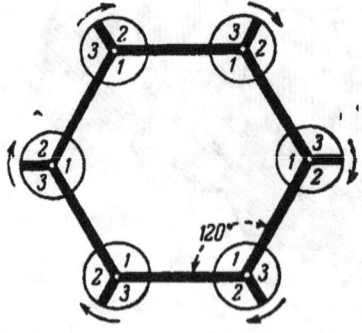

Fig. 107

forest, the next six corners as the interior of a sheikh's palace, and the last six as an Indian temple. One turn of the concealed mechanism would be enough to change a tropical forest into a temple or palace. The entire trick is based on such a simple physical phenomenon as light reflection.

WHY LIGHT REFRACTS AND HOW

Many think the fact that light refracts when passing from medium to medium is one of Nature's whims. They simply can't understand why

Fig. 109. Refraction of light explained

light does not keep on in the same direction as before but has to strike out obliquely. Do you think so too? Then you'll probably be delighted to learn that light behaves just as a marching column of soldiers does when they step from a paved road to one full of ruts.

Here is a very simple and instructive illustration to show how light refracts. Fold your tablecloth and lay it on the table as shown in *Fig. 109.* Incline the table-top slightly. Then set a couple of wheels on one axle—from a broken toy steam engine or some other toy—rolling down it. When its path is set at right angles to the tablecloth fold there is no refraction, illustrating the optical law, according to which light falling perpendicularly on the boundary between two different media does not bend. But when its path is set obliquely to the tablecloth fold the direction changes at this point—the boundary between two different media, in which we have a change in velocity.

144

When passing from that part of the table where velocity is greater (the uncovered part) to that part where velocity is less (the covered part), the direction ("the ray") is nearer to the "normal incidence". When rolling the other way the direction is farther away from the normal.

This, incidentally, explains the substance of refraction as due to the change in light velocity in the new medium. The greater this change is, the wider the angle of refraction is, since the "refractive index", which shows how greatly the direction changes, is nothing but the ratio of the two velocities. If the refractive index in passing from air to water is 4/3, it means that light travels through the air roughly 1.3 times faster than through water. This leads us to another instructive aspect of light propagation. Whereas, when reflecting, light follows the *shortest* route, when refracting, it chooses the *fastest* way; no other route will bring it to its "destination" sooner than this crooked road.

LONGER WAY FASTER

Can a crooked route really bring us sooner to our destination than the straight one? Yes—when we move with different speeds along different sections of our route. Villagers living between two railway stations *A* and *B*, but closer to *A*, prefer to walk or cycle to station *A* and board the train there for station *B*, if they want to get to station *B faster*, than to take the *shorter* way which is straight to station *B*.

Another instance. A cavalry messenger is sent with despatches from point *A* to the command post at point *C* (*Fig. 110*). Between him and the command post lie a strip of turf and a strip of soft sand, divided by the straight line *EF*. We know that it takes twice the time to cross sand than it does to cross turf. Which route would the messenger choose to deliver the despatches sooner?

At first glance one might think it to be the straight line between *A* and *C*. But I don't think a single

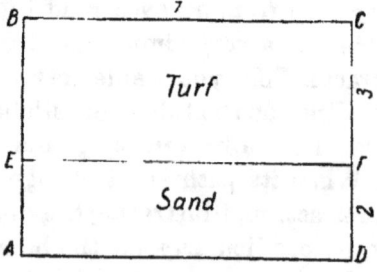

Fig. 110. The problem of the cavalry messenger. Find the fastest way from *A* to *C*

horseman would pick that route. After all, since it takes a longer time
to cross sand, a cavalryman would rightly think it better to cut the time
spent by crossing the sand less obliquely. This would naturally length-
en his way across the turf. But since the horse would take him
across it twice as fast, this longer distance would actually mean less
time spent. In other words, the horseman should follow a road that
would *refract* on the boundary between sand and turf, moreover, with
the path across the turf forming a wider angle with the perpendicular
to this boundary than the path across the sand.

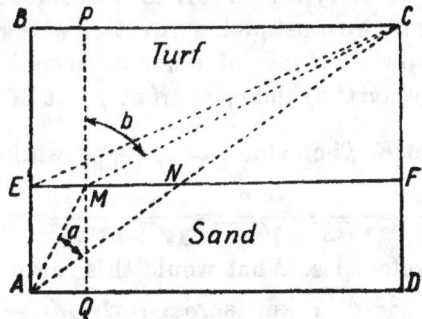

Fig. 111. The problem of the cavalry
messenger and its solution. The fastest way is AMC

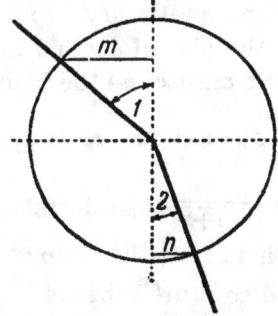

Fig. 112. What is the sine? The relation of *m* to the radius is the sine
of angle *1*, while the relation of *n*
to the radius is the sine of angle *2*

Anyone will realise that the straight path AC is actually not the quickest way and that considering the different width of the two strips and
the distances as given in *Fig. 110*, the messenger will reach his destination sooner if he takes the crooked road AEC (*Fig. 111*). *Fig. 110* gives
us a strip of sand two kilometres wide, and a strip of turf three kilometres wide. The distance BC is seven kilometres. According to Pythagoras,
the entire route from A to C (*Fig. 111*) is equal to $\sqrt{5^2 + 7^2} = \sqrt{74} =$
$= 8.6$ km. Section AN — across the sand — is two-fifths of this, or 3.44 km.
Since it takes twice as long to cross sand than it does to cross turf, the
3.44 km of sand mean from the time angle 6.88 km of turf. Hence the
8.6 km straight-line route AC is equivalent to 12.04 km across turf. Let
us now reduce to "turf" the crooked AEC route. Section AE is two kilo-

metres, which corresponds to four kilometres in time across turf. Section EC is equal to $\sqrt{3^2 + 7^2} = \sqrt{58} = 7.6$ km, which, added to four kilometres, results in a total of 11.6 km for the crooked AEC route.

As you see, the "short" straight road is 12 km across turf, while the "long" crooked road only 11.6 km across turf, which thus saves 12.00— —11.60=0.40 km, or nearly half a kilometre. But this is still not the *quickest* way. This, according to theory, is that—we shall have to invoke trigonometry—in which the ratio of the sine of angle b to the sine of angle a is the same as the ratio of the velocity across turf to that across sand, i. e., a ratio of 2:1. In other words, we must pick a direction along which the sine of angle b would be twice the sine of angle a. Accordingly, we must cross the boundary between the sand and turf at point M, which is one kilometre away from point E. Then sine $b = \dfrac{6}{\sqrt{3^2 + 6^2}}$, while sine $a = \dfrac{1}{\sqrt{1 + 2^2}}$, and the ratio of $\dfrac{\sin b}{\sin a} = \dfrac{6}{\sqrt{45}} : \dfrac{1}{\sqrt{5}} = \dfrac{6}{3\sqrt{5}} : \dfrac{1}{\sqrt{5}} = 2$, which is exactly the ratio of the two velocities. What would this route, reduced to "turf", be? $AM = \sqrt{2^2 + 1^2} = 4.47$ km across turf. $MC = \sqrt{3^2 + 6^2} = 6.49$ km. This adds up to 10.96 km, which is 1.08 km *shorter* than the straight road of 12.04 km across turf.

This instance illustrates the advantage to be derived in such circumstances by choosing a crooked road. Light naturally takes this fastest route because the law of light refraction strictly conforms to the proper mathematical solution. The ratio of the sine of the angle of refraction to the sine of the angle of incidence is the same as the ratio of the velocity of light propagation in the new medium to that in the old medium; this ratio is the refractive index for the specified media. Wedding the specific features of reflection and refraction we arrive at the "Fermat principle"—or the "principle of least time" as physicists sometimes call it—according to which light *always takes the fastest route.*

When the medium is heterogeneous and its refractive properties change gradually—as in our atmosphere, for instance—again "the principle of least time" holds. This explains the slight curvature in light as it comes from the celestial objects through our atmosphere. Astronomers call this "atmospheric refraction". In our atmosphere, which becomes denser and denser the closer we get to the ground, light bends in such a way that the inside of the bend faces the earth. It spends more time in higher atmospheric layers, where there is less to retard its progress, and less time in the "slower" lower layers, thus reaching its destination more quickly than were it to keep to a strictly rectilinear course.

The Fermat principle applies not only to light. *Sound* and all *waves* in general, whatever their nature, travel in accord with this principle. Since you probably want to know why, let me quote from a paper which the eminent physicist Schrödinger read in 1933 in Stockholm when receiving the Nobel Prize. Speaking of how light travels through a medium with a gradually changing density, he said:

"Let the soldiers each firmly grasp one long stick to keep strict breast-line formation. Then the command rings out: Double! Quick! If the ground gradually changes, first the right end, and then the left end will move faster, and the breast-line will swing round. Note that the route covered is not straight but crooked. That it strictly conforms to the shortest, as far as the time of arrival at the destination over this particular ground is concerned, is quite clear, as each soldier tried to run as fast as he could."

THE NEW CRUSOES

If you have read Jules Verne's *Mysterious Island*, you might remember how its heroes, when stranded on a desert isle, lit a fire though they had no matches and no flint, steel and tinder. It was lightning that helped Defoe's Robinson Crusoe; by pure accident it struck a tree and set fire to it. But in Jules Verne's novel it was the resourcefulness of an educated engineer and his knowledge of physics that stood the heroes in good stead. Do you remember how amazed that naïve sailor Pencroft was when, coming back from a hunting trip, he found the engineer and the reporter seated before a blazing bonfire?

"'But who lighted it?' asked Pencroft.

"'The sun!'

"Gideon Spilett was quite right in his reply. It was the sun that had furnished the heat which so astonished Pencroft. The sailor could scarce ly believe his eyes, and he was so amazed that he did not think of questioning the engineer.

"'Had you a burning-glass, sir?' asked Herbert of Harding.

"'No, my boy,' replied he, 'but I made one.'

"And he showed the apparatus which served for a burning-glass. It was simply two glasses which he had taken off his own and the reporter's watch. Having filled them with water and rendered their edges adhesive by means of a little clay, he thus fabricated a regular burning-glass, which, concentrating the solar rays on some very dry moss, soon caused it to blaze."

I dare say you would like to know why the space between the two watch glasses had to be filled with water. After all, wouldn't an air-filling focus the sun's rays well enough? Not at all. A watch glass is bounded by two—outer and inner—parallel (concentric) surfaces. Physics tells us that when light passes through a medium bounded by such surfaces it hardly changes its direction at all. Nor does it bend when passing through the second watch glass. Consequently, the rays of light cannot be focussed on one point. To do this we must fill up the empty space between the glasses with a transparent substance that would refract rays better than air does. And that is what Jules Verne's engineer did.

Any ordinary ball-shaped water-filled carafe will act as a burning-glass. The ancients knew that and also noticed that the water didn't warm up in the process. There have been cases when a carafe of water inadvertently left to stand in the sunlight on the sill of an open window set curtains and tablecloths on fire and charred tables. The big spheres of coloured water, which were traditionally used to adorn the show-windows of chemist's shops, now and again caused fires by igniting the inflammable substances stored nearby.

A small round retort—12 cm in diameter is quite enough—full of water will do to boil water in a watch glass. With a focal distance of 15 cm (the focus is very close to the retort), you can produce a temperature of 120° C. You can light a cigarette with it just as easily as with a glass. One must note, however, that a glass lens is much more effective than a water-filled one, firstly, because the refractive index of water is much less, and, secondly, because water intensively absorbs the infra-red rays which are so very essential for heating bodies.

It is curious to note that the ancient Greeks were aware of the igni-tion effect of glass lenses a thousand odd years before eyeglasses and spyglasses were invented. Aristophanes speaks of it in his famous com-edy *The Cloud*. Socrates propounds the following] problem to Strep-tiadis:

"Were one to write a promissory note on you for five talents, how would you destroy it?

"*Streptiadis*: I have found a way which you yourself will admit to be very artful. I suppose you have seen the wondrous, transparent stone that burns and is sold at the chemist's?

"*Socrates*: The burning-glass, you mean?

"*Streptiadis*: That is right.

"*Socrates*: Well, and what?

"*Streptiadis*: While the notary is writing I shall stand behind him and focus the sun on the promissory note and melt all he writes."

I might explain that in Aristophanes's days the Greeks used to write on waxed tablets which easily melted.

ICE HELPS TO LIGHT FIRE

Even ice, provided it is transparent enough, can serve as a convex lens and consequently as a burning-glass. Let] me note, furthermore, that in this process the ice does not warm up and melt. Its refractive index is a wee bit less than that of water, and since a spherical water-filled vessel can be used as a burning-glass, so can a similarly shaped lump of ice. An ice "burning-glass" enabled Dr. Clawbonny in Jules Verne's *The Adventures of Captain Hatteras* to light a fire when the travellers found themselves stranded without a fire or anything to light it in terri-bly cold weather, with the mercury at 48° C below zero.

"'This is terrible ill-luck,' the captain said.

"'Yes,' replied the doctor.

"'We haven't even a spyglass to make a fire with!'

"'That's a great pity,' the doctor remarked, 'because the sun is strong enough to light tinder.'

"'We'll have to eat the bear raw, then,' said the captain.

"'As a last resort, yes,' the doctor pensively replied. 'But why not....'

"'What?' Hatteras inquired.

"'I've got an idea.'

"'Then we're saved,' exclaimed the bosun.

"'But...' the doctor was hesitant.

"'What is it?' asked the captain.

"'We haven't got a burning-glass, but we can make one.'

"'How?' asked the bosun.

"'From a piece of ice!'

"'And you think....'

"'Why not? We must focus the sun's rays on the tinder and a piece of ice can do that. Fresh-water ice is better though—it's more transparent and less liable to break.'

Fig. 113. "The doctor focussed the sun's bright rays on the tinder"

"'The ice boulder over there,' the bosun pointed to a boulder some hundred steps away, 'seems to be what we need.'

"'Yes. Take your axe and let's go.'

"The three walked over to the boulder and found that it was indeed of fresh-water ice.

"The doctor told the bosun to chop off a chunk of about a foot in diameter, and then he ground it down with his axe, his knife, and finally polished it with his hand and produced a very good, transparent burning-glass. The doctor focussed the sun's bright rays on the tinder which began to blaze a few seconds later."

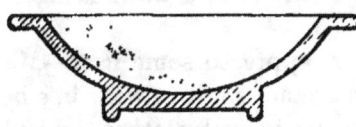

Fig. 114. A bowl for making an ice burning-glass

Jules Verne's story is not an impossibility. The first time this was ever done with success was in England in 1763. Since then ice has been used more than once for the purpose. It is, of course, hard to believe that one could make an ice burning-glass with such crude tools as an axe and knife and "one's hand" in a frost of 48° C below zero. There is, however, a much simpler way: pour some water into a bowl of the proper shape, freeze it, and then take out the ice by slightly heating the bottom of the bowl. Such a "burning-glass" will work only in the open air on a clear and frosty day. Inside a room behind closed windows it is out of the question, because the glass panes absorb much of the solar energy and what is left of it is not strong enough.

HELPING SUNLIGHT

Here is one more experiment which you can easily do in wintertime. Take two pieces of cloth of the same size, one black and the other white, and put them on the snow out in the sun. An hour or two later you will find the black piece half-sunk, while the white piece is still where it was. The snow melts sooner under the black piece because cloth of this colour absorbs most of the solar rays falling on it, while white cloth disperses most of the solar rays and consequently warms up much less.

This very instructive experiment was first performed by Benjamin

Franklin, the American scientist of War for Independence fame, who won immortality for his invention of the lightning conductor.

"I took a number of little square pieces of broad cloth from a tailor's pattern card, of various colours. There were black, deep blue, lighter blue, green, purple, red. yellow, white, and other colours, or shades of colours. I laid them all out upon the snow in a bright sunshiny morning. In a few hours (I cannot now be exact as to the time), the black, being warmed most by the sun, was sunk so low as to be below the stroke of the sun's rays; the dark blue almost as low, the lighter blue not quite so much as the dark, the other colours less as they were lighter; and the quite white remained on the surface of the snow, not having entered it at all.

"What signifies philosophy that does not apply to some use?—May we not learn from hence, that black clothes are not so fit to wear in a hot sunny climate or season, as white ones; because in such clothes the body is more heated by the sun when we walk abroad, and we are at the same time heated by the exercise, which double heat is apt to bring on putrid dangerous fevers?... That summer hats for men or women' should be white, as repelling that heat which gives headaches to many, and to some the fatal stroke that the French call the *coup de soleil*?... That fruit walls being blacked may receive so much heat from the sun in the daytime, as to continue warm in some degree through the night, and thereby preserve the fruit from frosts, or forward its growth?—with sundry other particulars of less or greater importance, that will occur from time to time to attentive minds?"

The benefit that can be drawn from this knowledge was well illustrated during the expedition to the South Pole that the Germans made aboard the good ship *Hauss* in 1903. The ship was jammed by icepacks and all methods usually applied in such circumstances—explosives and ice-saws—proved abortive. Solar rays were then invoked. A two-kilometre long strip, a dozen metres in width, of dark ash and coal was strewn from the ship's bow to the nearest rift. Since this happened during the Antarctic summer, with its long and clear days, the sun was able to accomplish what dynamite and saws had failed to do. The ice melted and cracked all along the strip, releasing the ship from its clutches.

MIRAGES

I suppose you all know what causes a mirage. The blazing sun heats up the desert sands and lends to them the property of a mirror because the density of the hot surface layer of air is less than the strata higher up. Oblique rays of light from a remote object meet this layer of air and curve upwards from the ground as if reflected by a mirror after striking it at a very obtuse angle. The desert-traveller thus thinks he is seeing a sheet of water which reflects the objects standing on its banks (*Fig. 115*).

Fig. 115. Desert mirages explained. This drawing, usually given in textbooks, shows too steeply the ray's course towards the ground

Rather should we say that the hot surface layer of air reflects not like a mirror but like the surface of water when viewed from a submarine. This is not an ordinary reflection but what physicists call total reflection, which occurs when light enters the layer of air at an extremely obtuse angle, far greater than the one in the figure. Otherwise the "critical angle" of incidence will not be exceeded.

154

Please note—to avoid misunderstanding—that a denser strata must be above the rarer layers. However, we know that denser air is heavier and always seeks to descend to take the place of lighter lower layers and force them upwards. Why, in the case of a mirage, is the denser air above the rarer air? Because air is in constant motion. The heated surface air keeps on being forced up by a new replacing lot of heated air. This is responsible for some rarefied air always remaining just above the hot sand. It need not be the same rarefied air all the time—but that is something that makes no difference to the rays.

This phenomenon has been known from times immemorial. (A somewhat different mirage appearing in the air at a higher level than the observer is caused by reflection in upper rarefied layers.) Most people think this classical type of mirage can be observed only in the blazing southern deserts and never in more northerly latitudes. They are wrong. This is frequently to be observed in summer on asphalted roads which, because they are dark, are greatly heated by the sun. The dull road's surface seems to look like a pool of water able to reflect distant objects. Fig. 116 shows the path light takes in this case. A sufficiently observant person will see these mirages oftener than one might think.

There is one more type of mirage—a side one—which people usually do not have the faintest suspicion about. This mirage, which has been

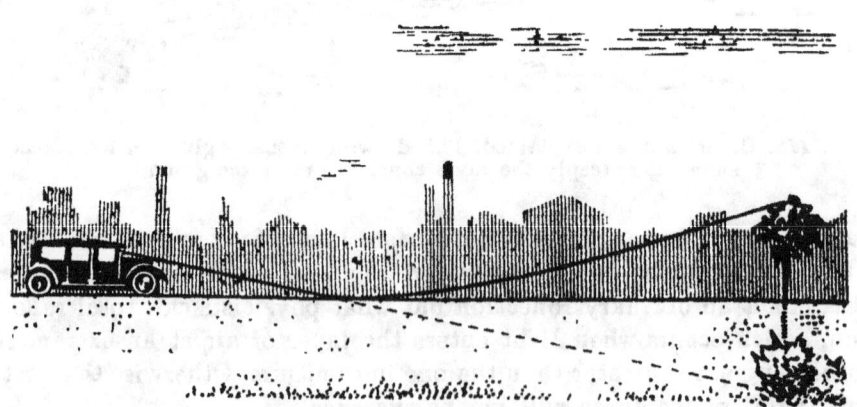

Fig. 116. Mirage on paved highway

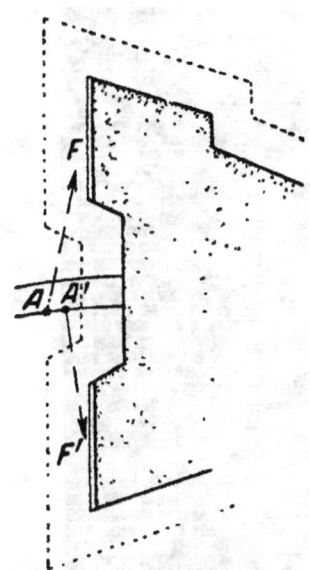

Fig. 117. Ground plan of the fortress where the mirage was seen. Wall F seemed polished from point A, and wall F' from point A'

described by a Frenchman, was produced by reflection from a heated sheer wall. As he drew near to the wall of a fortress he noticed it suddenly glisten like a polished mirror and reflect the surrounding landscape. Taking a few steps he saw a similar change in another wall. He concluded that this was due to the walls having heated up considerably under the blazing sun. *Fig. 117* gives the position of the walls (*F* and *F''*) and the spots (*A* and *A'*) where the observer stood.

The Frenchman found that the mirage recurred every time the wall was hot enough and even managed to photograph the phenomenon.

Fig 118 depicts, on the left, the fortress wall *F*, which suddenly turned into the glistening mirror on the right, as photographed from point *A'*. The ordinary grey concrete wall on the left naturally cannot reflect the two soldiers near it. But the same wall, miraculously transformed into a mirror on the right, does *symmetrically* reflect the closer of the two soldiers. Of course it isn't the wall itself that reflects him, but its surface layer of hot air. If on a hot summer day you pay notice to walls of big buildings, you might spot a mirage of this kind.

"THE GREEN RAY"

"Have you ever seen the sun dip into the horizon at sea? No doubt, you have. Have you ever watched it until the upper rim touches the horizon and then disappears? Probably you have. But have you ever noticed what happens on the instant when our brilliant luminary sheds its last ray—provided the sky is a cloudless, pellucid blue? Probably not. Don't miss this opportunity. You will see, instead of a red ray, one of an exquisite green that no artist could ever reproduce and that nature

Fig. 118. Rough, grey wall (left) suddenly seems to act like
a polished mirror (right)

herself never displays either in the variously tinted plants or in the
most transparent of seas."

This note published in an English newspaper sent the young heroine
of Jules Verne's *The Green Ray* in raptures and made her roam the world
solely to see this phenomenon with her own eyes. Though, according to
Jules Verne, this Scottish girl failed to see the lovely work of nature,
still it exists It is no myth, though many legends are associated with
it. Any lover of nature can admire it, provided he takes the pains to
hunt for it.

Where does the green ray or flash come from? Recall what you saw
when you looked at something through a prism. Try the following. Hold
the prism at eye level with its broad horizontal plane turned downwards
and look through it at a piece of paper tacked to the wall You will see
the sheet firstly loom and secondly display a violet-blue rim at the
top and a yellow-red edge at the bottom. The elevation is due to refraction, while the coloured rims owe their origin to the property of glass

to refract differently light of different colours. It bends violets and blues more than any other colour. That is why we see a violet-blue rim on top. Meanwhile, since it bends reds least, the bottom edge is precisely of this colour.

So that you comprehend my further explanations more easily, I must say something about the origin of these coloured rims. A prism breaks up the white light emitted by the paper into all the colours of the spectrum, giving many coloured images of the paper, disposed in the order of their refraction and often superimposed, one on the other. The combined effect of these superimposed coloured images produces white light (the composition of the spectral colours) but with coloured fringes at top and bottom. The famous poet Goethe who performed this experiment but failed to grasp its real meaning thought that he had debunked Newton's colour theory. Later he wrote his own *Theory of Colours* which is based almost entirely on misconceptions. But I suppose you won't repeat his blunder and expect the prism to colour everything anew.

We see the earth's atmosphere as a vast prism of air, with its base facing us. Looking at the sun on the horizon we see it through a prism of gas. The solar disc has a blue-green fringe on top and a yellow-red one at the bottom. While the sun is above the horizon, its disc's brilliant colour outshines all other less bright bands of colour and we don't see them at all. But during the sunrises and sunsets, when practically the entire disc of the sun is below the horizon, we may spot the blue double-tinted fringe on the upper rim, with an azure blue right on top and a paler blue—produced by the mixing of green and blue—below it. When the air near the horizon is clear and translucent, we see a blue fringe, or the "blue ray". But often the atmosphere disperses the blues and we see only the remaining green fringe—the "green ray". However, most often a turbid atmosphere disperses both blues and greens and then we see no fringe at all, the setting sun assuming a crimson red.

The Pulkovo astronomer G.A. Tikhov, who devoted a special monograph to the "green ray", gives us some tokens by which we may see it. "When the setting sun is crimson-hued and it doesn't hurt to look at it with the naked eye you may be sure that there will be no green flash." This is clear enough: the fact of a red sun means that the atmosphere

intensively disperses blues and greens, or, in other words, the whole
of the upper rim of the solar disc. "On the other hand," he continues,
"when the setting sun scarcely changes its customary whitish yellow
and is very bright [in other words, when atmospheric absorption of light
is insignificant—*Y.P.*]—you may quite likely expect the green flash.
However, it is important for the horizon to be a distinct straight line
with no uneven relief, forests or buildings. We have all these conditions at sea, which explains why seamen are familiar with the green
flash."

To sum up: to see the "green ray", you must observe the sun when
setting or rising and when the sky is extremely clear. Since the sky
at the horizon in southern climes is much more translucent than in
northern latitudes, one is liable to see the "green ray" there much oftener. But neither in our latitudes is it so rare as many think—most
likely, I suppose, because of Jules Verne. You will detect the "green ray"
sooner or later as long as you look hard enough. This phenomenon has
been seen even in a spyglass.

Here is how two Alsatian astronomers describe it:

"During the very last minute before the sun sets, when, consequently,
a goodly part of its disc is still to be seen, a green fringe hems the waving
but clearly etched outline of the sun's ball. But until the sun sets altogether, it cannot be seen with the naked eye. It will be seen only when
the sun disappears completely below the horizon. However, should one
use a spyglass with a powerful enough magnification—of roughly 100—
one will see the entire phenomenon very well. The green fringe is seen
some ten minutes before the sun sets at the latest. It incloses the disc's
upper half, while a red fringe hems the lower half. At first the fringe
is extremely narrow, encompassing at the outset but a few seconds of an
arc. As the sun sets, it grows wider, sometimes reaching as much as half
a minute of an arc. Above the green fringe one may often spot similarly
green prominences, which, as the sun gradually sinks, seem to slide along
its rim up to its apex and sometimes break away entirely to shine independently a few seconds before fading" (*Fig. 119*).

Usually this phenomenon lasts a couple of seconds. In extremely
favourable conditions, however, it may last much longer. A case of more
than 5 minutes has been registered; this was when the sun was setting

Fig. 119. Protracted observation of the "green ray"; it was seen beyond the mountain range for 5 minutes. Top right-hand corner: the "green ray" as seen in a spyglass. The Sun's disc has a ragged shape. 1. The Sun's blinding glare prevents us from seeing the green fringe with the unaided eye. 2. The "green ray" can be seen with the unaided eye when the Sun has almost completely set

behind a distant mountain and the quickly walking observer saw the green fringe as seemingly sliding down the hill (*Fig. 119*).

The instances recorded when the "green ray" has been observed during a sunrise—that is, when the upper rim of our celestial luminary peeps out above the horizon—are extremely instructive, as they debunk the frequent suggestion that the phenomenon is presumably nothing more than an optical illusion to which the eye succumbs owing to the fatigue caused by looking at the brilliant setting sun. Incidentally, the sun is not the only celestial object that sheds the "green ray". Venus has also produced it when setting. (You will find more about mirages and the green flash in M. Minaert's superb book *Light and Colour in Nature.*)

CHAPTER NINE

VISION

BEFORE PHOTOGRAPHY WAS INVENTED

Photography is so ordinary nowadays that we find it hard to imagine how our forefathers, even in the past century, got along without it. In his *Posthumous Papers of the Pickwick Club* Charles Dickens tells us the amusing story of how British prison officers took a person's likeness some hundred or so years ago. The action takes place in the debtors' prison where Pickwick has been brought. Pickwick is told that he'll have to sit for his portrait.

"'Sitting for my portrait!' said Mr. Pickwick.

"'Having your likeness taken, sir,' replied the stout turnkey. 'We're capital hands at likeness here. Take 'em in no time, and always exact. Walk in, sir, and make yourself at home.'

"Mr. Pickwick complied with the invitation, and sat himself down: when Mr. Weller, who stationed himself at the back of the chair, whispered that the sitting was merely another term for undergoing an inspection by the different turnkeys, in order that they might know prisoners from visitors.

"'Well, Sam,' said Mr. Pickwick. 'Then I wish the artists would come. This is rather a public place.'

"'They won't be long, sir, I des-say,' replied Sam. 'There's a Dutch clock, sir.'

"'So I see,' observed Mr. Pickwick.

"'And a bird-cage, sir,' says Sam. 'Veels within veels, a prison in a prison. Ain't it, sir?'

"As Mr. Weller made this philosophical remark, Mr. Pickwick was aware that his sitting had commenced. The stout turnkey having been relieved from the lock, sat down, and looked at him carelessly, from time to time, while a long thin man who had relieved him, thrust his hands beneath his coat-tails, and planting himself opposite, took a good long view of him. A third, rather surly-looking gentleman: who had apparently been disturbed at his tea, for he was disposing of the last remnant of a crust and butter when he came in: stationed himself close to Mr. Pickwick; and, resting his hands on his hips, inspected him narrowly; while two others mixed with the group, and studied his features with most intent and thoughtful faces. Mr. Pickwick winced a good deal under the operation, and appeared to sit very uneasily in his chair; but he made no remark to anybody while it was being performed, not even to Sam, who reclined upon the back of the chair, reflecting, partly on the situation of his master, and partly on the great satisfaction it would have afforded him to make a fierce assault upon all the turnkeys there assembled, one after the other, if it were lawful and peaceable so to do.

"At length the likeness was completed, and Mr. Pickwick was informed, that he might now proceed into the prison."

Still earlier it was a list of "features" that did for such memorised "portraits". In his *Boris Godunov*, Pushkin tells us how Grigory Otrepyev was described in the tsar's edict: "Of short stature, and broad chest; one arm is shorter than the other; the eyes are blue and hair ginger; a wart on one cheek and another on the forehead." Today we needn't do that; we simply provide a photograph instead.

WHAT MANY DON'T KNOW HOW TO DO

Photography was introduced in Russia in the 1840's, first as daguerreotypes—prints on metal plates that were called so after their inventor, Daguerre. It was a very inconvenient method; one had to pose for quite a long stretch—for as long as fourteen minutes or more. "My grandfather," Prof. B.P. Weinberg, the Leningrad physicist, told me, "had to sit for 40 minutes before the camera to get just one daguerreotype, from which, moreover, no prints could be made."

Still the chance to have one's portrait made without the artist's intervention seemed such a wonderful novelty that it took the general public quite a time to get used to the idea. One old Russian magazine for 1845 contains quite an amusing anecdote on the score:

"Many still cannot believe that the daguerreotype acts by itself. One gentleman came to have his portrait done. The owner [the photographer —Y.P.] begged him to be seated, adjusted the lenses, inserted a plate, glanced at his watch, and retired. While the owner was present, the gentleman sat as if rooted to the spot. But he had barely gone out when the gentleman thought it no longer necessary to sit still; he rose, took a pinch of snuff, examined the camera from every side, put his eye to the lens, shook his head, mumbled, 'How ingenious,' and began to meander up and down the room.

"The owner returned, stopped short in surprise at the doorway, and exclaimed: 'What are you doing? I told you to sit still!'

"'Well, I did. I got up only when you went out.'

"'But that was exactly when you should have sat still.'

"'Why should I sit still for nothing?' the gentleman retorted."

We're certainly not so naïve today.

Still, there are some things about photography that many do not know. Few, incidentally, know how one should *look* at a photograph. Indeed, it's not so simple as one might think, though photography has been in existence for more than a century now and is as common as could be. Nevertheless, even professionals don't look at photographs *in the proper way*.

HOW TO LOOK AT PHOTOGRAPHS

The camera is based on the same optical principle as our eye. Everything projected onto its ground-glass screen depends on the distance between the lens and the object. The camera gives a perspective, which we would get with *one eye*—note that!—were our eye to replace the lens. So, if you want to obtain from a photograph the same visual impression that the photographed object produced, we must, firstly, look at the photograph with *one eye* only, and, secondly, hold it *at the proper distance away*.

After all, when you look at a photograph with *both eyes* the picture you get is flat and not three-dimensional. This is the fault of our own vision. When we look at something solid the image it causes on the retina of either eye is not the same (*Fig. 120*). This is mainly why we see objects in relief. Our brain blends the two different images into one that springs into *relief*—this is the basic principle of the stereoscope. On the other hand, if we are looking at something that is flat—a wall, for instance—both eyes get an identical sensory picture telling our brain that the object we are looking at is really flat.

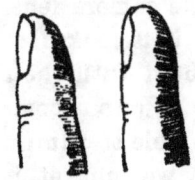

Fig. 120. A finger as !seen separately by the left and right eye when held close to the face ˙

Now you should realise the mistake we make when we look at a photograph with both eyes. In this manner we compel ourselves to believe that the picture we have before us is flat. When we look with *both eyes* at a photograph which is really intended only for *one eye*, we prevent ourselves from seeing the picture that the photograph really shows, and thus destroy the illusion which the camera produces with such perfection.

HOW FAR TO HOLD A PHOTOGRAPH

The second rule I mentioned—that of holding the photograph *at the proper distance away* from the eye—is just as important, for otherwise we get the wrong perspective. How far away should we hold a photograph? To recreate the proper picture we must look at the photograph from the same angle of vision from which the camera lens reproduced the image on the ground-glass screen, or in the same way as it "saw" the object being photographed (*Fig. 121*). Consequently, we must hold the photograph at such a distance away from the eye that would be as many times less the distance between the object and the lens as the size of the image on the photograph is less its actual size. In other words,

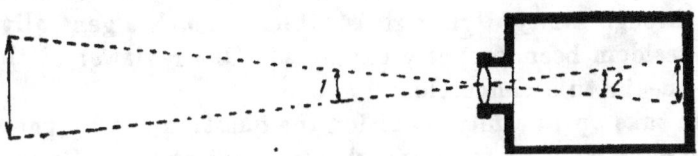

Fig. 121. In a camera angle *1* is equal to angle *2*

we must hold the photograph at a distance which is roughly the same as the focal length of the camera lens.

Since most cameras have a focal length of 12-15 cm (the author has in mind the cameras that were in use when he wrote his *Physics for Entertainment—Ed.*), we shall never be able to get the proper distance for the photographs they give, as the focal length of a normal eye at best (25 cm) is nearly twice the indicated focal length of the camera lens. A photograph tacked on a wall also seems flat because it is looked at from a still greater distance away. Only the short-sighted with their short focal length of vision, as well as children, who are able to accommodate their vision to see objects very close up, will be able to admire the effect that an ordinary photograph produces when we look at it properly with one eye, because when they hold a photograph 12-15 cm away, they get not a flat image but one in relief—the kind of image a stereoscope produces.

I suppose you will now agree with me in noting that it is only due to ignorance that we do not derive the pleasure a photograph can give, and that we often unjustly blame them for being lifeless.

QUEER EFFECT OF MAGNIFYING GLASS

The short-sighted easily see ordinary photographs in relief. What should people with normal eyesight do? Here a magnifying glass will help. By looking at photographs through a magnifying glass with a two-fold power, people with normal eyesight will derive the indicated advantage of the short-sighted, and see them in relief without straining their eyesight.

There is a tremendous difference between the effect thus produced and the impression we get when we look at a photograph with both eyes from quite a distance. It almost amounts to the stereoscopic effect. Now we know why photographs often spring into relief when looked at with one eye through a magnifying glass, which, though a generally known fact, has seldom been properly explained. One reviewer of this book wrote to me in this connection:

"Please take up in a future edition the question of why photographs appear in relief when viewed through a magnifying glass. Because I contend that the involved explanation provided of the stereoscope holds no water at all. Try to look in the stereoscope with one eye. The picture appears in relief despite all that theory has to say."

I am sure you will agree that this does not pick any holes in the theory of stereoscopic vision.

The same principle lies at the root of the curious effect produced by the so-called panoramas, that are sold at toy shops. This is a small box, in which an ordinary photograph—a landscape or a group of people—is placed and viewed through a magnifying glass with one eye, which in itself already gives a stereoscopic effect. The illusion is usually enhanced by some of the objects in the foreground being cut out and placed separately in front of the photograph proper. Our eye is very sensitive to the solidity of objects close by; as far as distant objects are concerned, the impression is much less perceptible.

ENLARGED PHOTOGRAPHS

Can we make photographs so that people with *normal* eyesight are able to see them properly, without using a magnifying glass? We can, merely by using cameras having lenses with a long focal length. You already know that a photograph obtained with the aid of a lens having a focal distance of 25-30 cm will appear in relief when viewed with one eye from the usual distance away.

One can even obtain photographs that won't seem flat even when looked at with *both eyes* from quite a distance. You also know that our brain blends two identical retinal images into one flat picture. However, the greater the distance away from the object, the less our brain is able to do that. Photographs taken with the aid of a lens having a focal distance of 70 cm can be looked at with both eyes without losing the sense of depth.

Since it is incommoding to resort to such lenses, let me suggest another method, which is to *enlarge* the picture you take with any ordinary camera. This increases the distance at which you should look at photographs to get the proper effect. A four- or fivefold enlargement of a photograph taken with a 15 cm lens is already quite enough to obtain the desired effect—you can look at it with both eyes from 60 to 75 centimetres away. True, the picture will be a bit blurred but this is barely discernible at such a distance. Meanwhile, as far as the stereoscopic effect and depth are concerned, you only stand to gain.

BEST SEAT IN MOVIE-HOUSE

Cinema-goers have most likely noticed that some films seem to spring into unusually clear relief—to such an extent at times that one seems to see real scenery and real actors. This depends not on the film, as is often thought, but on where you take your seat. Though motion pictures are taken with cameras having lenses with a very short focal length, their projection on the screen is a hundred times larger—and you can see them with both eyes from quite a distance (10 cm × 100 = 10 m). The effect of relief is best when you look at the picture from the same angle of vision as the movie camera "looked" when it was shooting the film.

How should one find the distance corresponding to such an optimal angle of vision? Firstly, one must choose a seat *right opposite the middle of the screen*. Secondly, one's seat must be away from the screen at a distance which is as many times the screen's width as the focal length of the movie-camera lens is greater than the width of the film itself. Movie-camera lens usually have a focal length of 35 mm, 50 mm, 75 mm, or 100 mm, depending on the subject being shot. The standard width of film is 24 mm. For a focal length of 75 mm, for instance, we get the proportion:

$$\frac{\text{the distance}}{\text{screen width}} = \frac{\text{focal length}}{\text{film width}} = \frac{75}{24} \approx 3.$$

So, to find how far away you should seat yourself from the screen, you should multiply the width of the screen, or rather the projection onto the screen, by three. If the width is six of your steps, then the best seat would be 18 steps away from the screen. Keep this in mind when trying various devices offering a stereoscopic effect, because one may easily ascribe to the invention what is really due to the tioned.

FOR READERS OF PICTORIAL MAGAZINES

Reproductions in books and magazines naturally have the same properties as the original photographs from which they were made; they also spring into relief when looked at with one eye from the proper distance. But since different photographs are taken by cameras having lenses with different focal lengths, one can find the proper distance only by trial and error. Cup one eye with your hand and hold the illustration at arm's length. Its plane must be perpendicular to the line of vision and your open eye must be right opposite the middle of the picture. Gradually bring the picture closer, steadily looking at it meanwhile; you easily catch the moment when it appears in clearest relief.

Many illustrations that seem blurred and flat when you look at them in your habitual way acquire depth and clearness when viewed as I suggest. One will even catch the sparkle of water and other such purely stereoscopic effects.

It's amazing that few people know these simple things though they were all explained in popular-science books more than half a century ago. In his *Principles of Mental Physiology, with Their Application to the Training and Discipline of the Mind, and the Study of Its Morbid Conditions*, William Carpenter has the following to say about how one should look at photographs.

"It is remarkable that the effect of this mode of viewing photographic pictures is not limited to bringing out the solid forms of objects; for other features are thus seen in a manner more true to the reality, and therefore more suggestive of it. This may be noticed especially with regard to the representation of *still water*, which is generally one of the most unsatisfactory parts of a photograph; for although, when looked at with *both* eyes, its surface appears opaque, like white wax, a wonderful depth and transparence are often given to it by viewing it with only *one*. And the same holds good also in regard to the characters of *surfaces* from which light is reflected—as bronze or ivory; the material of the object from which the photograph was taken being recognised much more certainly when the picture is looked at with *one* eye, than when *both* are used (unless in stereoscopic combination)."

There is one more thing we must note. Photographic enlargements, as we have seen, are more lifelike; photographs of a reduced size are not. True, the smaller-size photograph gives a better contrast; but it is flat and fails to give the effect of depth and relief. You should now be able to say why: it also reduces the corresponding perspective—which is usually too little as it is.

HOW TO LOOK AT PAINTINGS

All I have said of photographs applies in some measure to paintings as well. They appear best also at the proper distance away, for only then do they spring into relief. It is better, too, to view them with but one eye, especially if they are small.

"It has long been known," Carpenter wrote in the same book, "that if we gaze steadily at a picture, whose perspective projection, lights and shadows, and general arrangement of details, are such as accurately correspond with the reality which it represents, the impression it produces will be much more vivid when we look with *one* eye only, than when we use both; and that the effect will be further heightened, when we carefully shut out the surroundings of the picture, by looking through a tube of appropriate size and shape. This fact has been commonly accounted for in a very erroneous manner. 'We see more exquisitely,' says Lord Bacon, 'with one eye than with both, because the vital spirits thus unite themselves the more and become the stronger'; and other writers, though in different language, agree with Bacon in attributing the result to the *concentration* of the visual power, when only one eye is used. But the fact is, that when we look with *both* eyes at a picture within a moderate distance, we are *forced* to recognise it as a flat surface; whilst, when we look with only *one*, our minds are at liberty to be acted on by the suggestions furnished by the perspective, chiaroscuro, etc.; so that, after we have gazed for a little time, the picture may begin to start into relief, and may even come to possess the solidity of a model."

Reduced photographic reproductions of big paintings often give a greater illusion of relief than the original. This is because the reduced size lessens the ordinarily long distance from which the painting should be looked at, and so the photograph acquires relief, even close up.

THREE DIMENSIONS IN TWO

All I have said about looking at photographs, paintings and drawings, while being true, should not be taken in the sense that there is no other way of looking at flat pictures to get the effect of depth and relief. Every artist, whatever his field—painting, the graphic arts, or photography—strives to produce an impression on the spectator regardless of his "point of view". After all he can't count on everybody viewing his creations with hands cupped over one eye and sizing up the distance for every piece.

Every artist, including the photographer, has an extensive arsenal of means to draw upon to give in two dimensions objects possessing three. The different retinal images produced by distant objects are not the only token of depth. The "aerial perspective" painters employ grading tones and contrasts to make the background blurred and seemingly veiled by diaphanous mist of air, plus their use of linear perspective produces the illusion of depth. A good specialist in art photography will follow the same principles, cleverly choosing lighting, lenses, and also the appropriate brand of photographic paper to produce perspective.

Proper focussing is also very important in photography. If the foreground is sharply contrasted and the remoter objects are "out of focus", this alone is already enough, in many cases, to create the impression of depth. On the contrary, when you reduce the aperture and give both foreground and background in the same contrast, you achieve a flat picture with no depth to it. Generally speaking, the effect a picture produces on the spectator—thanks to which he sees three dimensions in two, irrespective of physiological conditions for visual perception and sometimes in violation of geometrical perspective—depends largely, of course, on the artist's talent.

STEREOSCOPE

Why is it that we see solid objects as things having three dimensions and not two? After all the retinal image is a flat one. So why do we get a sensory picture of geometrical solidity? For several reasons. Firstly, the different lighting of the different parts of objects enables us to perceive their shape. Secondly, the strain we feel when accommodating our eye to get a clear perception of the different distance of the object's different parts also plays a role; this is not a flat picture in which every part of the object depicted is set at the same distance away. And thirdly—the most important cause—is that the two retinal images are different, which is easy enough to demonstrate by looking at some close object, shutting alternately the right and left eye (*Figs. 120* and *122*).

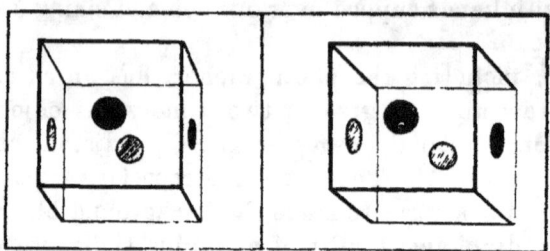

Fig. 122. A spotted glass cube as seen with the
left and right eye

Imagine now two drawings of one and the same object, one as seen by the left eye, and the other, as seen by the right eye. If we look at them so that each eye sees only its "own" drawing, we get instead of two separate flat pictures one in relief. The impression of relief is greater even than the impression produced when we look at a solid object with one eye only.

There is a special device, called the stereoscope, to view these pairs. Older types of stereoscopes used mirrors and the later models convex glass prisms to superimpose the two images. In the prisms—which slightly enlarge the two images, because they are convex—the light coming from the pair is refracted in such a way that its imagined continuation causes this superimposition.

As you see, the stereoscope's basic principle is extremely simple; all the more amazing, therefore, is the effect produced. I suppose most of you have seen various stereoscopic pictures. Some may have used the stereoscope to learn stereometry more easily. However, I shall proceed to tell you about applications of the stereoscope which I presume many of you do not know.

Actually we can—provided we accustom our eyes to it—dispense with the stereoscope to view such pairs, and achieve the same effect, with the sole difference that the image will not be bigger than it usually is in a stereoscope. Wheatstone, the inventor of the stereoscope, made use of this arrangement of nature. Provided here are several stereoscopic drawings—, graded in difficulty—that I would advise you to try viewing without a stereoscope. Remember that you will achieve results only if you exercise. (Note that not all can see stereoscopically, even in a stereoscope: some—the squint-eyed or people used to working with one eye—are utterly incapable of adjustment to binocular vision; others achieve results only after prolonged exercise. Young people, however, quickly adapt themselves, after a quarter of an hour.)

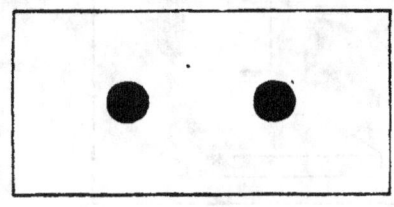

Fig. 123. Stare at the space between the two dots for several seconds. The dots seem to merge

Start with *Fig. 123* which depicts two black dots. Stare several seconds at the space between them, meanwhile trying to look at an imagined object behind. Soon you will be seeing double, seeing four dots instead of two. Then the two

Fig. 124. Do the same, after which turn to the next exercise

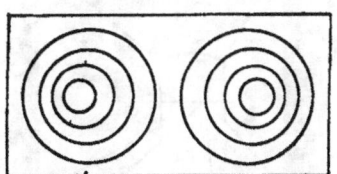

Fig. 125. When these images merge you will see something like the inside of a pipe receding into the distance

extreme dots will swing far apart, while the two innermost dots will close up and become one. Repeat with *Figs. 124* and *125* to see something like the inside of a long pipe receding into the distance.

Then turn to *Fig. 126* to see geometrical bodies seemingly suspended in mid-air. *Fig. 127* will appear as a long corridor or tunnel. *Fig. 128* will produce the illusion of transparent glass in an aquarium. Finally, *Fig. 129* gives you a complete picture, a seascape.

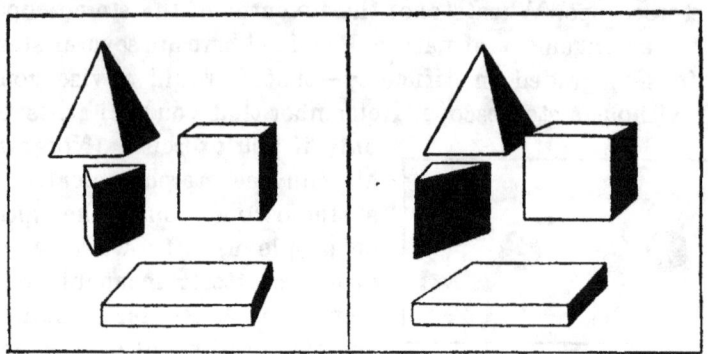

Fig. 126. When these four geometrical bodies merge, they seem to hover in mid-air

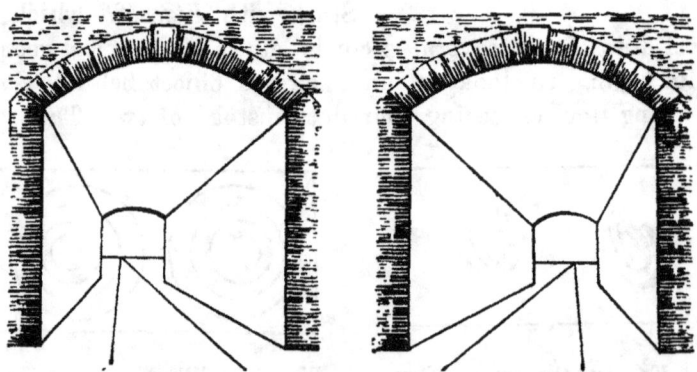

Fig. 127. This pair gives a long corridor receding into the distance

It is easy to achieve results. Most of my friends learned the trick very quickly, after a few tries. The short-sighted and far-sighted needn't take off their glasses; they view the pairs just as they look at any picture. Catch the proper distance at which they should be held by trial and error. See that the lighting is good—this is important.

Now you can try to view stereoscopic pairs in general without a stereoscope. You might try the pairs in *Figs. 130* and *133* first. Don't

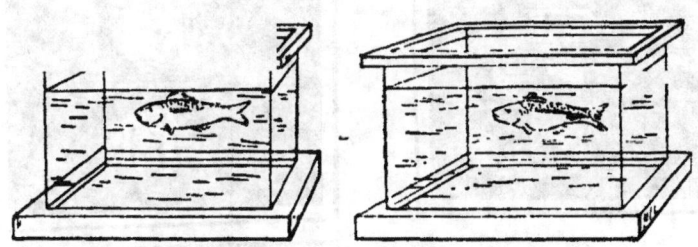

Fig. 128. A fish in an aquarium

Fig. 129. A stereoscopic seascape

overdo this so as not to strain your eyesight. If you fail to acquire the knack, you may use lenses for the far-sighted to make a simple but quite serviceable stereoscope. Mount them side by side in a piece of cardboard so that only their inner rims are available for viewing. Partition off the pairs with a diaphragm.

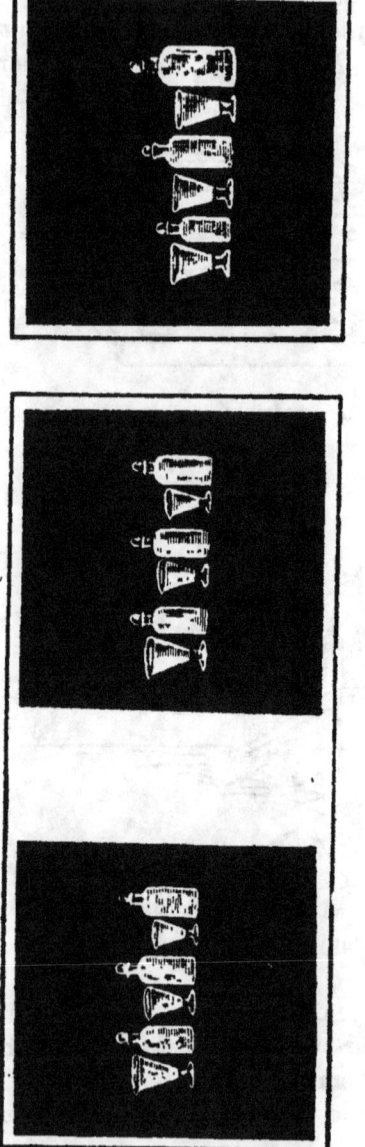

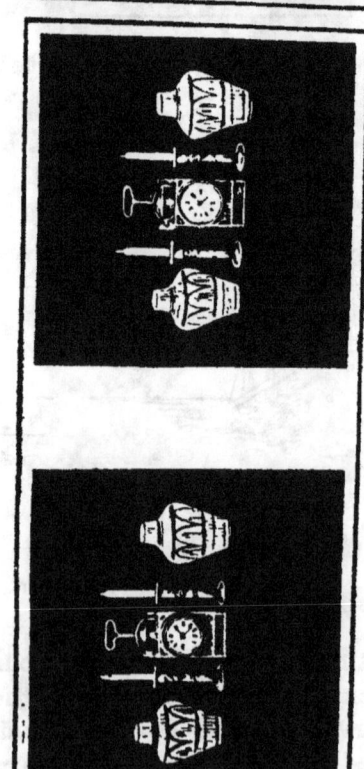

Fig. 130

Fig. 130 (the upper left-hand corner) gives two photographs of three bottles of presumably one and the same size. However hard you look you cannot detect any difference in size. But there is a difference, and, moreover, a significant one. They seem alike only because they are not set at one and the same distance away from the eye or camera. The bigger bottle is further away than the smaller ones. But which of the three is the bigger bottle? Stare as much as you may, you will never get the answer. But the problem is easily solved by using a stereoscope or exercising binocular vision. Then you clearly see that the left-hand bottle is furthest away, and the right-hand bottle closest. The photo in the upper right-hand corner shows the real size of the bottles.

The stereoscopic pair at the bottom of *Fig. 130* provides a still bigger teaser. Though the vases and candlesticks seem identical there is a great difference in size between them. The left-hand vase is nearly twice as tall as the right-hand one, while the left-hand candlestick, on the contrary, is much smaller than the clock and the right-hand candlestick. Binocular vision immediately reveals the cause. The objects are not in one row; they are placed at different distances, with the bigger objects being further away than the smaller articles. A fine illustration of the great advantage of binocular "two-eyed" vision over "one-eyed" vision!

DETECTING FORGERY

Suppose you have two absolutely identical drawings, of two equal black squares, for instance. In the stereoscope they appear as one square which is exactly alike either of the twin squares. If there is a white dot in the middle of each square, it is bound to show up on the square in the stereoscope. But if you shift the dot on one of the squares slightly off centre, the stereoscope will show one dot—however, it will appear either *in front* of, or *beyond*, the square, not on it. The slightest of differences already produces the impression of depth in the stereoscope. This provides a simple method for revealing forgeries. You need only put the suspected bank-bill next to a genuine one in a stereoscope, to detect the forged one, however cunningly made. The slightest discrepancy, even in one teeny-weeny line, will strike the eye at once—appearing either in front of, or behind, the banknote. (The idea, which was first suggested by Dove in the mid-19th century, is not applicable—for reasons of printing technique—to all currency notes issued today. Still his method will do to distinguish between two proofs of a book-page, when one is printed from newly-composed type.)

AS GIANTS SEE IT

When an object is very far away, more than 450 metres distant, the stereoscopic impression is no longer perceptible. After all the 6 centimetres at which our eyes are set apart are nothing compared with such a distance as 450 metres. No wonder buildings, mountains, and landscapes that are far away seem flat. So do the celestial objects all appear to be at the same distance, though, actually, the moon is much closer than the planets, while the planets, in turn, are very much closer than the fixed stars. Naturally, a stereoscopic pair thus photographed will not produce the illusion of relief in the stereoscope.

There is an easy way out, however. Just photograph distant objects from two points, taking care that they be further apart than our two eyes. The stereoscopic illusion thus produced is one that we would get were our eyes set much further apart than they really are. This is actually how stereoscopic pictures of landscapes are made. They are usually viewed through magnifying (convex) prisms and the effect is most amazing.

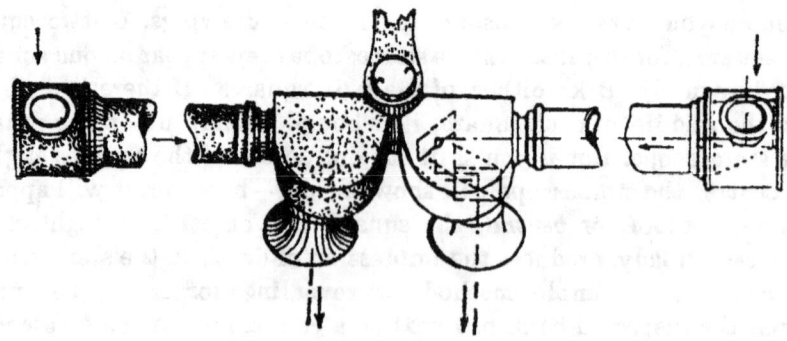

Fig. 131. Telestereoscope

You have probably guessed that we could arrange two spyglasses to present the surrounding scenery in its real relief. This instrument, called a telestereoscope, consists of two telescopes mounted further apart than eyes normally are. The two images are superimposed by means of reflecting prisms (*Fig. 131*).

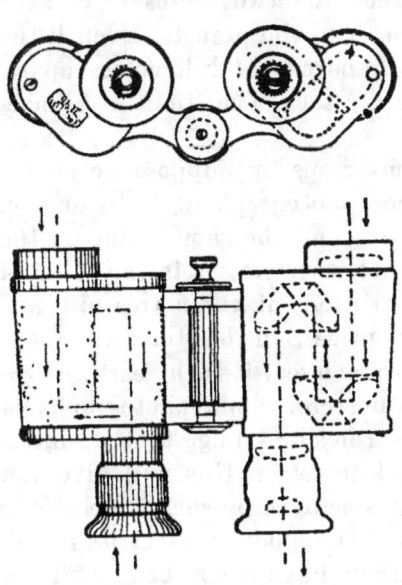

Fig. 132. Prism binoculars

Words fail to convey the sensation one experiences when looking through a telestereoscope, it is so unusual. Nature is transformed; distant mountains spring into relief; trees, rocks, buildings and ships at sea appear in all three dimensions. No longer is everything flat and fixed; the ship, that seems a stationary spot on the horizon in an ordinary spyglass, is moving. That is most likely how the legendary giants saw surrounding nature. When this device has a tenfold power and the distance between its lenses is six times the interocular distance (6.5×6=39 cm), the impression of relief is enhanced 60-fold (6×10), compared with the impression obtained by the naked eye. Even objects 25 kilometres away still appear in discernible relief. For land surveyors, seamen, gunners and travellers this instrument is a godsend, especially if equipped with a range-finder. The Zeiss prism binoculars produces the same effect, as the distance between its lenses is greater than the normal interocular distance (*Fig. 132*). The opera glass, on the contrary, has its lenses set not so far apart, to reduce the illusion of relief—so that the décor and settings present the intended impression.

UNIVERSE IN STEREOSCOPE

If we direct our telestereoscope at the moon or any other celestial object we shall fail to obtain any illusion of relief at all. This is only natural, as celestial distances are too big even for such instruments. After all, the 30-50 cm distance between the two lenses is nothing compared with the distance from the earth to the planets. Even if the two telescopes were mounted tens and hundreds of kilometres apart, we would get no results, as the planets are tens of millions of kilometres away.

This is where stereoscopic photography steps in. Suppose we photograph a planet today and take another photograph of it tomorrow. Both photographs will be taken from one and the same point on the globe, but from different points in the solar system, as in the space of 24 hours the earth will have travelled millions of kilometres in orbit. Hence the two photographs won't be identical. In the stereoscope, the pair will produce the illusion of relief. As you see, it is the earth's orbital motion that enables us to obtain stereoscopic photographs of celestial objects. Imagine a giant with a head so huge that its interocular distance ranges into millions of kilometres; this will give you a notion of the unusual effect astronomers achieve by such stereoscopic photography. Stereoscopic photographs of the moon present its mountains in relief so distinct that scientists have even been able to *measure* their height. It seems as if the magic chisel of some supercolossal sculptor has breathed life into the moon's flat and lifeless scenery.

The stereoscope is used today to discover the asteroids which swarm between the orbits of Mars and Jupiter. Not so long ago the astronomer considered it a stroke of good fortune if he was able to spot one of these asteroids. Now it can be done by viewing stereoscopic photographs of this part of space. The stereoscope immediately reveals the asteroid; it "sticks" out.

In the stereoscope we can detect the difference not only in the *position* of celestial objects but also in their *brightness*. This provides the astronomer with a convenient method for tracking down the so-called *variable* stars whose light periodically fluctuates. As soon as a star exhibits a dissimilar brightness the stereoscope detects at once the star possessing that varying light.

Astronomers have also been able to take stereoscopic photographs of the nebulae (Andromeda and Orion). Since the solar system is too small for taking such photographs astronomers availed themselves of our system's displacement amidst the stars. Thanks to this motion in the universe we always see the starry heavens from new points. After the lapse of an interval long enough, this difference may even be detected by the camera. Then we can make a stereoscopic pair, and view it in the stereoscope.

THREE-EYED VISION

Don't think this a slip of the tongue on my part; I really mean three eyes. But how can one see with three eyes? And can one really acquire a third eye?

Science cannot give you or me a third eye, but it can give us the magic power to see an object as it would appear to a three-eyed creature. Let me note first that a one-eyed man can get from stereoscopic photographs that impression of relief which he can't and doesn't get in ordinary life. For this purpose we must project onto a screen in rapid sequence the photographs intended for right and left eyes that a normal person sees with both eyes simultaneously. The net result is the same because a rapid sequence of visual images fuses into one image just as two images seen simultaneously do. (It is quite likely that the surprising "depth" of movie films at times, in addition to the causes mentioned, is due also to this. When the movie camera sways with an even motion—as often happens because of the film-winder— the stills will not be identical and, as they rapidly flit onto the screen will appear to us as one 3-dimensional image.)

In that case couldn't a two-eyed person simultaneously watch a rapid sequence of two photographs with one eye and a third photograph, taken from yet another angle, with the other eye? Or, in other words, a stereoscopic "trio"? We could. One eye would get a single image, but in relief, from a rapidly alternating stereoscopic pair, while the other eye would look at the third photograph. This "three-eyed" vision enhances the relief to the extreme.

STEREOSCOPIC SPARKLE

The stereoscopic pair in *Fig. 133* depicts polyhedrons, one in white against a black background and the other in black against a white background. How would they appear in a stereoscope? This is what Helmholtz says:

"When you have a certain plane in white on one of a stereoscopic pair and in black on the other, the combined image seems to sparkle,

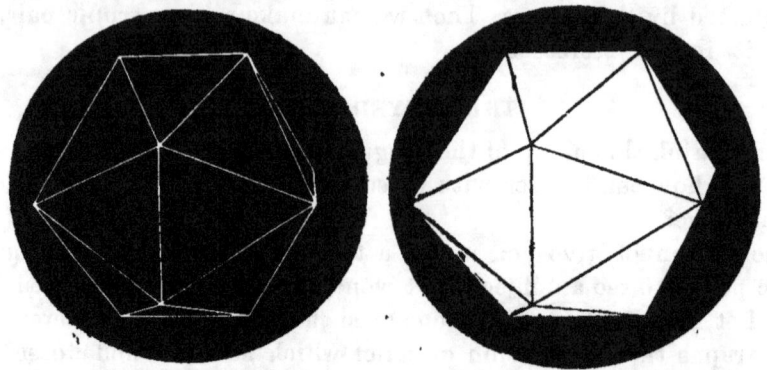

Fig. 133. Stereoscopic sparkle. In the stereoscope this pair produces a sparkling crystal against a black background

even though the paper used for the pictures is dull. Such stereoscopic drawings of models of crystals produce the impression of glittering graphite. The sparkle of water, the glisten of leaves and other such things are still more noticeable in stereoscopic photographs when this is done."

In an old but far from obsolete book, *The Physiology of the Senses. Vision*, which the Russian physiologist Sechenov published in 1867, we find a wonderful explanation of this phenomenon.

"Experiments artificially producing stereoscopic fusion of differently lighted or differently painted surfaces repeat the actual conditions in which we see sparkling objects. Indeed, how does a dull surface differ from a glittering polished one? The first one reflects and diffuses light and so seems identically lighted from every point of observation, while the polished surface reflects light in but one definite direction.

Therefore you can have instances when with one eye you get many reflected rays, and with the other practically none—these are precisely the conditions that correspond to the stereoscopic fusion of a white surface with a black one. Evidently there are bound to be instances in looking at glistening polished surfaces when reflected light is unevenly distributed between the eyes of the observer. Consequently, the stereoscopic sparkle proves that experience is paramount in the act during which images fuse bodily. The conflict between the fields of vision immediately yields to a firm conception, as soon as the experience-trained apparatus of vision has the chance to attribute the difference to some familiar instance of actual vision."

So the reason we see things *sparkle*—or at least one of the reasons—is that the two retinal images are not the same. Without the stereoscope we would have scarcely guessed it.

TRAIN WINDOW OBSERVATION

I noted a little earlier that different images of one and the same object produce the illusion of relief when in rapid alternation they perceptibly fuse. Does this happen only when we see moving images and stand still ourselves? Or will it also take place when the images are standing still but we are moving? Yes, we get the same illusion, as was only to be expected. Most likely many have noticed that movies shot from an express train spring into unusually clear relief—just as good as in the stereoscope. If we pay heed to our visual perceptions when riding in a fast train or car we shall see this ourselves. Landscapes thus observed spring into clear relief with the foreground distinctly separate from the background. The "stereoscopic radius" of our eyes increases appreciably to far beyond the 450-metre limit of binocular vision for stationary eyes.

Doesn't this explain the pleasant impression we derive from a landscape when observing it from the window of an express train? Remote objects recede and we distinctly see the vastness of the scenic panorama unfolding before us. When we ride through a forest we stereoscopically perceive every tree, branch, and leaf; they do not blend into one flat picture as they would to a stationary observer. On a mountain road fast driving again produces the same effect. We seem to sense tangibly the dimensions of the hills and valleys.

One-eyed people will also see this—and I'm sure it will afford a startlingly novel sensation, as this is tantamount to the rapid sequence of pictures producing the illusion of relief, a point mentioned before. (This, incidentally, accounts for the noticeable stereoscopic effect produced by movie films shot from a train taking a bend, when the objects being photographed lie in the radius of this bend. This track "effect" is well-known to cameramen.)

It is as easy as pie to check my statements. Just be mindful of your visual perceptions when riding in a car or a train. You might also notice another amazing circumstance which Dove remarked upon some hundred years ago—what is well forgotten is indeed novel!—that the closer objects flashing by seem smaller in size. The cause has little to do with binocular vision. It's simply because our estimate of distance is wrong. Our subconscious mind suggests that a closer object should really be smaller than usually, to seem as big as always. This is Helmholtz's explanation.

THROUGH TINTED EYEGLASSES

Looking through red-tinted eyeglasses at a *red* inscription on *white* paper you see nothing but a plain red background. The letters disappear entirely from view, merging with the red background. But look through the same red-tinted glasses at *blue* letters on *white* paper and the inscription distinctly appears in *black*—again on a red background. Why black? The explanation is simple. Red glass does not pass blue rays; it is red because it can pass red rays only. Consequently, instead of the blue letters you see the absence of light, or black letters.

The effect produced by what are called colour *anaglyphs*—the same as produced by stereoscopic photographs—is based precisely on this property of tinted glass. The anaglyph is a picture in which the two stereoscopic images for the right and left eye respectively are *superimposed*; the two images are coloured differently—one in blue and the other in red.

The anaglyphs appear as one black but three-dimensional image when viewed through differently-tinted glasses. Through the red glass the right eye sees only the *blue* image—the one intended for the right eye—and sees it, moreover, in black. Meanwhile the left eye sees through the blue glass only the *red* image which is intended for the left eye—again in black. Each eye sees only one image, the one intended for it. This repeats the stereoscope and, consequently, the result is the same—the illusion of depth.

"SHADOW MARVELS"

The "shadow marvels" that were once shown at the cinemas are also based on the above-mentioned principle. Shadows cast by moving figures on the screen appear to the viewer, who is equipped with differently-tinted glasses, as objects in three dimensions. The illusion is achieved by bicoloured stereoscopy. The shadow-casting object is placed between the screen and two adjacent sources of light, red and green. This produces two partially superimposed coloured shadows which are viewed through viewers matching in colour.

The stereoscopic illusion thus produced is most amusing. Things seem to fly right your way; a giant spider creeps towards you; and you involuntarily shudder or cry out. The apparatus required is extremely simple. *Fig. 134* gives the idea. In this diagram G and R stand for

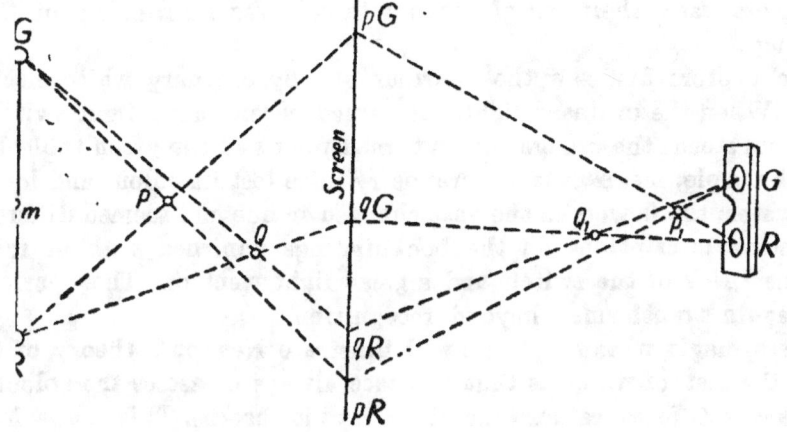

Fig. 134. The "shadow marvel" explained

the green and red lamps (left); P and Q represent the objects placed between the lamps and screen; pG, qG, pR and qR are the tinted shadows that these objects cast on the screen; P_1 and Q_1 show where the viewer looking through the differently-tinted glasses—G is the green glass, and R, the red one—sees these objects. When the "spider" behind the screen is shifted from Q to P the viewer thinks it to be creeping from Q_1 to P_1.

Generally speaking, every time the object behind the screen is moved towards the source of light, thus causing the shadow cast on the screen to grow larger, the viewer thinks the object to be moving from the screen towards him.

Everything the viewer thinks is moving towards him from the screen is actually moving—on the other side of the screen—in the opposite direction—from the screen to the source of light.

MAGIC METAMORPHOSES

I think it would be appropriate at this stage to describe a series of illuminating experiments conducted at the Science for Entertainment Pavilion of a Leningrad recreation park. A corner of the pavilion was furnished as a parlour. Its furniture was covered with dark-orange antimacassars, the table was laid with green baize, on which there stood a decanter full of cranberry juice and a vase with flowers in it, and there was a shelf full of books with coloured inscriptions on their bindings.

The visitors first saw the "parlour" lit by ordinary white electric light. When the ordinary light was turned off and a *red* light switched on in its stead, the orange covers turned pink and the green tablecloth a dark purple; meanwhile the cranberry juice lost its colour and looked like water; the flowers in the vase changed in hue and seemed different; and some inscriptions on the bookbindings vanished without trace. Another flick of the switch and a *green* light went on. The "parlour" was again transformed beyond recognition.

These magic metamorphoses will illustrate Newton's theory of colour, the gist of which is that a surface always possesses the colour of the rays it diffuses, rather than of the rays it absorbs. This is how Newton's compatriot, the celebrated British physicist John Tyndall, formulates the point.

"Permitting a concentrated beam of white light to fall upon fresh leaves in a dark room, the sudden change from green to red, and from red back to green, when the violet glass is alternately introduced and withdrawn, is very surprising ... question of absorption."

Consequently the green tablecloth shows up as green in white light because it diffuses primarily the rays of the green and adjacent spectral bands and absorbs most of all the other rays. If we direct a mixed red and violet light at this green tablecloth, it will diffuse only the violet and absorb most of the red, thus turning purple. This is the main explanation for all the other colour metamorphoses in the "parlour".

But why does the cranberry juice lose all colour when a red light is directed at it? Because the decanter stands on a white runner laid across the green baize. Once we remove the runner the cranberry juice turns red. It loses its colour (in red lighting) only against the background of the runner, which, though it turns red, we ourselves *continue to regard as white*, both by force of habit and due to the contrast it presents to the purple tablecloth. Since the juice has the same colour as the runner, which we imagine to be white, we involuntarily think the juice to be white too. That is why it appears no longer as red juice but as colourless water. You may derive the same impressions by viewing the surroundings through tinted glasses. (See my *Do You Know Your Physics?* for more about this effect.)

HOW TALL IS THIS BOOK?

Ask a friend to show you how high the book he is holding would be from the floor, if he stood it up on one edge. Then check his statement. He is sure to guess wrongly: the book will actually be half as tall. Furthermore, better ask him not to bend down to show how high the book would come up to, but provide the answer in so many words, with you assisting. You can try this with any other familiar object—a table lamp, say, or a hat. However, it should be one you have grown accustomed to seeing at the level of your eyes. The reason why people err is because every object diminishes in size when looked at edgeways.

We constantly make the same mistake when we try to estimate the size of objects that are way above our heads, especially tower clocks. Even though we know that these clocks are very large, our estimates of their size are much less than the actual size. *Fig. 135* shows how large the dial of the famous Westminster Tower clock in London looks when brought down to the road below. Ordinary human beings look like midgets next to it. Still it fits the orifice in the clock tower shown in the distance—believe it or not!

BLACK AND WHITE

Look from *afar* at *Fig. 136* and say how many black spots would fit in between the bottom spot and any of the top spots. Four or five? I daresay your answer will be: "Well, there's not enough room for five but there's certainly enough for four."

Believe it or not—you can check it!—there's just enough

Fig. 135. The size of the Westminster Tower clock

room for three, no more! This illusion, owing to which dark patches seem smaller than white patches of the same size, is known as "irradiation". This comes from an imperfection of our eye, which, as an optical instrument, does not quite measure up to strict optical requirements. Its refrangible media do not cast on the retina that sharply-etched outline which one gets on the ground-glass screen of a well-focussed camera. Owing to what is called *spherical aberration*, every light patch has a light fringe which enlarges the retinal image. That is why light areas always seem bigger than dark areas of equal size.

In his *Theory of Colours* the great poet Goethe—who, though an observant student of nature, was not always a prudent enough physicist—has the following to say about this phenomenon:

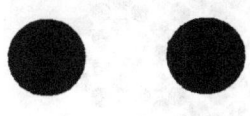

"A dark object seems smaller than a light object of the same size. If we look simultaneously at a white spot on a black background and at a black spot of the same diameter but against a white background, the latter will seem about a fifth smaller than the former. If we render the black spot correspondingly larger, the two spots will seem identical. The crescent moon seems part of a circle the diameter of which would be larger than that of the moon's darker portion—which we sometimes see [the ashen light witnessed when the "old moon is in the new moon's arms"—*Y.P.*]. In dark dress we seem slimmer than in clothes of light tones. Light coming over the rim of something seems to make a depression in it. A ruler from behind which we see a candle flame seems to have a notch in it at this point. The rising and setting sun seem to make a depression in the horizon."

Fig. 136. The gap between the bottom spot and each of the top two seems more than the distance between the outer edges of the two top spots. Actually, they are identical

Goethe was right on every point, with the sole exception that a white spot does not always seem larger than a black spot of equal size by one and the same fraction. This depends solely on from how far away you look at the spots. Why? Just move *Fig. 136* still further away. The illusion is still more striking, because the additional fringe we mentioned is always of the same width. Close up, the fringe enlarges the white area by 10%; further away, it takes up from 30 to even 50% of the white area, because the actual image of the spot is already smaller itself. This also explains why we see the circular white spots in *Fig. 137* as hexagons when viewed from two or three steps away. From six to eight steps away this figure will already seem a typical honeycomb.

To say that irradiation is responsible for this illusion is an explanation that has not quite satisfied me, ever since I noticed that *black* dots

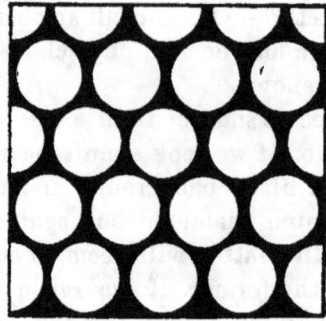

Fig. 137. From a distance the circular white spots seem hexagonal

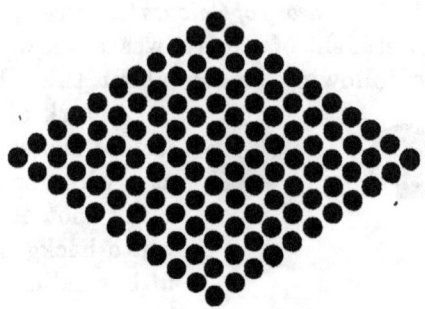

Fig. 138. From a distance the black dots appear as hexagons

on a white background (*Fig. 138*) also seem hexagonal from far away, though irradiation does not enlarge, but, on the contrary, reduces the dots in size. One must note that the explanations afforded for optical illusions in general are not completely satisfactory. As a matter of fact, most illusions still have to be explained. (For more on the topic see my *Optical Illusions* album.)

WHICH IS BLACKER?

Fig. 139 introduces us to another imperfection of the eye—"astigmatism" this time. Look at it with one eye. Not all four letters will seem identical in blackness. Note which is the blackest and turn the drawing sideways. The letter you thought the blackest will suddenly go grey, and now another letter will seem the blackest. Actually, all

Fig. 139. Look at this word with one eye. One letter will seem blacker than the rest

four letters are identical in blackness; they are merely shaded in different directions. If our eyes were just as perfect and faultless as expensive glass lenses, this would have no effect on the blackness of the letters; but since our eyes do not refract light identically in different directions, we cannot see vertical, horizontal, and slanting lines just as distinctly.

Very seldom is the eye absolutely free of this shortcoming. With some people *astigmatism* is so great that it noticeably lessens the acuteness of vision and they have to wear special glasses to correct this. Our eyes also have other imperfections, which opticians know how to avoid. This is what Helmholtz had to say about them:

"If an optician were to dare sell me an instrument with such imperfections, I would most roundly chide him and demonstratively return the instrument."

Besides these illusions which our eyes succumb to due to certain imperfections in them, there are many other illusions to which they fall victim for totally different reasons.

STARING PORTRAIT

You have most likely seen at one time or another portraits that not only look you square in the eye, but even follow you with their eyes wherever you go. This was noticed long ago and has always baffled many, giving some the jitters. The great Russian writer Nikolai Gogol provides a wonderful description of this in his "Portrait":

"The eyes dug right into him and seemed wanting to watch only him and nothing else. The portrait stared right past everything else, straight at him and into him."

Quite a number of superstitions and legends are associated with this mysterious stare. Actually it is nothing more than an optical illusion. The trick is that on these portraits the pupil is placed square in the middle of the eye—just as we would see it in the eye of anybody looking at us point-blank. When a person looks past us, the pupil and the entire iris are no longer in the centre of the eye; they shift sideways.

On the portrait, however, the pupil stays right in the centre of the eye whichever way we step. And since we continue to see the face in the same position in relation to us, we, naturally, think that the man in the portrait has turned his head our way and is watching us. This explains the odd sensation we derive from other such pictures—the horse seems to be charging straight at us however hard we try to dodge it; the man's finger keeps pointing straight at us, and so on and so forth. *Fig. 140* is one such picture. They are often used to advertise or for propaganda purposes.

Fig. 140. The mysterious portrait

MORE OPTICAL ILLUSIONS

There doesn't seem to be anything out of the ordinary in the set of pins in *Fig. 141*, does there? However, lift the book to eye level and, cupping one eye, look at the pins so that your line of vision slides along them, as it were. Your eye must be at the point where the imagined continuations of these pins cross. Then the pins will seem to be stuck in the paper upright. When you shift your head sideways, the pins seem to sway in the same direction.

This illusion is governed by laws of perspective. The drawing is of upright pins projected on paper as they appear to the observer when viewed from the given point.

Our ability to succumb to optical illusions should not at all be regarded as just an imperfection of our eyesight. This ability presents a definite advantage, often overlooked, which is that without it we would have no painting; nor, in general, would we derive any pleasure from the fine arts. Artists draw extensively on these imperfections of our vision.

"The whole art of painting is based on this illusion," the brilliant 18th-century scholar Euler wrote in his famous *Letters on Various*

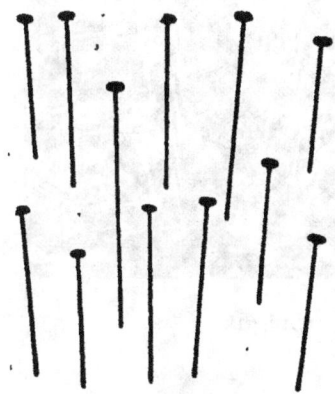

Fig. 141. Fix one eye (have the other shut) at the point where the imagined continuations of the pins would converge. The pins will seem to be stuck in the paper upright. By gently shifting the book from side to side, you get the impression that the pins are swaying

Physical Subjects. "If we passed judgement on things by what they really were, this art (painting) could not exist and we would be blind. The painter would strive in vain to mix his colours, for we would say here is red and there is blue, here is black and there are dashes of white. Everything would be contained in one plane; no difference in distance would be observed and no object could be depicted. Whatever the painter would want to show would all seem to us as writing on paper. And given this perfection, would we not be deserving of pity on being robbed of the delight such pleasant and useful artistry affords us daily?"

There are very many optical illusions, enough to fill albums (the *Optical Illusions* album mentioned earlier contains more than sixty). Many are common, others are less known. I shall give you some of the more curious instances that are less known. The illusions provided by *Figs. 142* and *143*, with lines on a checkered background, are particularly effective. One simply can't believe that the letters in *Fig. 142* are straight and it is still harder to believe that the circles in *Fig. 143* are not one spiral. The only way to check it is to apply a pencil and trace the circles. Only a pair of compasses will tell us that the straight line *AC* in *Fig. 144* is just as long as *AB* and not shorter, as it appears to be. The other illusions in *Figs. 145, 146, 147,* and *148* are explained in the captions. The following curious incident shows how effective the illusion *Fig. 147* provides is. When the publisher of a previous edition of this book was examining the cliché, he thought it badly done and was about to return it to the printshop to have the grey splotches at the intersection of the white lines scraped off, when I chanced to intervene and explained the matter.

Fig. 142. The letters are upright

Fig. 143. This seems a spiral; actually the curves are circles, which you can see for yourself by following the lines with a pointed pencil

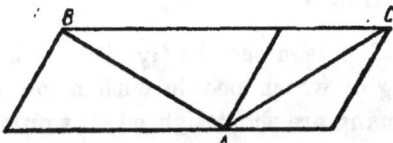

Fig. 144. *AB* is equal to *AC*, though *AB* seems longer

Fig. 147. Tiny faint grey squares seem to appear and disappear where the white strips cross, though the strips are really white throughout, as can be demonstrated by closing up the black squares with a piece of paper. The illusion is due to contrasts

Fig. 145. The slanting line seems broken

Fig. 146. The white and black squares are identical, as, too, are the round white and black spots

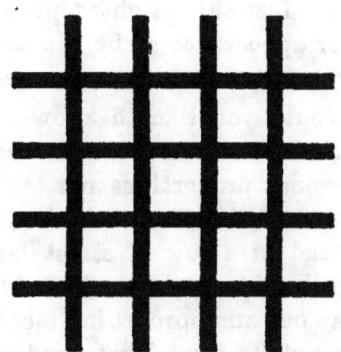

Fig. 148. Faint grey squares seem to appear and disappear where the black strips cross

SHORT-SIGHTED VISION

With his spectacles off, a short-sighted person sees badly. But what he sees and how he sees it is something of which people with normal eyesight have a very hazy notion. Since many are short-sighted, it would not be without interest to learn how they see.

Firstly, to the short-sighted person everything seems blurred. What to a person with normal eyesight are leaves and twigs—all clearly etched against the sky—are to the short-sighted merely an amorphous mass of green. He misses the minor details. Human faces seem younger and more attractive; crow's feet and other minor blemishes are not seen; the coarse ruddiness that may be the product of nature or make-up appears as a delicate flush. He may miscalculate age, being as much as 20 years out. He has odd taste—to those with good eyesight—of the beautiful. He may be considered tactless when he looks a person straight in the eye but seems reluctant to recognise him. He is not to blame. It is his near-sightedness that is the culprit.

"At the Lycée," the 19th-century Russian poet Delvig wrote, "I was forbidden to wear spectacles and my female acquaintances seemed exquisite creatures. How shocked I was after graduation!" When your short-sighted friend (minus his spectacles) chats with you he doesn't see your face, or, at any rate, what you think he sees. His image of you is blurred. No wonder he fails to recognise you an hour later. Most short-sighted people recognise others not so much by their outer appearance as by the sound of their voices. Inadequate vision is compensated for by acuter hearing.

Would you like to know what the near-sighted sees at night? All bright objects—street lanterns, lamps, lighted windows, etc.—assume enormous proportions and transform the world around into a chaotic jumble of shapeless bright splotches and dark and misty silhouettes. Instead of a row of street lamps the short-sighted sees two or three huge bright patches, which blot out the rest of the street. He cannot make out an approaching motor car; instead he sees just the two bright halos of its front lights and a dark mass behind. Even the sky seems different. The short-sighted sees stars of only the first three or four stellar magnitudes. Consequently, instead of several thousand stars he sees only a few hundred, which seem to be as large as lamps. The moon seems tremendous and very close, while a crescent moon takes on a phantastic form.

The fault lies in the structure of the eye; the eye-ball is too deep, so much so that its changed refractive power causes images from distant objects to be focussed before they reach the retina. The blurred retinal image is produced by diverging beams of light.

CHAPTER TEN

SOUND AND HEARING

HUNTING THE ECHO

Mark Twain tells a very funny story of the misadventures of a man whose hobby was to collect—you'll never guess—echoes! This eccentric spared no effort to buy up every tract of land that would have a multiple echo or some other extraordinary natural echo.

"His first purchase was an echo in Georgia that repeated four times; his next was a six-repeater in Maryland; his next was a thirteen-repeater in Maine; his next was a nine-repeater in Kansas; his next was a twelve-repeater in Tennessee, which he got cheap, so to speak, because it was out of repair, a portion of the crag which reflected it having tumbled down. He believed he could repair it at a cost of a few thousand dollars, and, by increasing the elevation with masonry, treble the repeating capacity; but the architect who undertook the job had never built an echo before, and so he utterly spoiled this one. Before he meddled with it, it used to talk back like a mother-in-law, but now it was only fit for the deaf and dumb asylum."

Joking apart, there are some wonderful discrete multiple echoes in various—primarily mountainous—spots, some of which are of long-standing universal fame. The following are some of the better-known echoes. The Woodstock castle echo in England repeats seventeen syllables quite distinctly. The ruins of the Derenburg castle near Halberstadt echoed 27 syllables before one of its walls was blown up. There is one definite place in the rocky cirque near Adersbach in Czechoslovakia which echoes seven syllables thrice; however, a few steps aside even a gunshot will fail to produce any response. There was a castle near Milan that had a very fine repeater-echo before it was demolished.

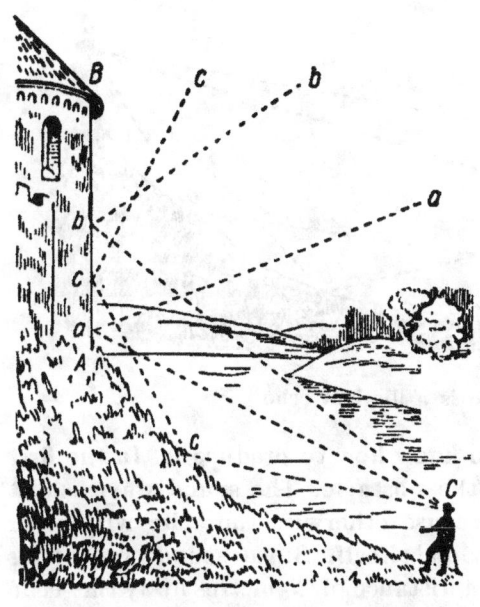

Fig. 149. There is no echo

A shot fired from a window in one of its wings echoed back 40 to 50 times, and a word said in a loud voice, some 30 times.

It is not so easy to find even a discrete single echo. The U.S.S.R. is a bit better off in this respect as it has many open plains ringed by woods and many forest clearings, where a shout will already produce a response in the form of a more or less distinct echo coming back from the wall of forest. In mountain land echoes are more varied than in plains, but occur much more seldom and are harder to catch. Why is this so? Because an echo is nothing but a train of sound waves reflected back by some obstacle. Sound abides by the same laws as light: its angle of incidence is equal to its angle of reflection.

Imagine yourself at the foot of a hill (*Fig. 149*), with the sound-reflecting barrier *AB* higher than you are. Naturally the sound waves propagating along the lines *Ca*, *Cb* and *Cc* will not reflect back to your ear but into the air along the directions *aa*, *bb* and *cc*. On the other hand, when the sound-reflecting barrier is at the same level with you or even a bit lower, as in *Fig. 150*, you will hear an echo. The sound travels down along *Ca* and *Cb* and returns along the broken lines *CaaC* or *CbbC*, bouncing off the ground once or twice. The pocket between the points acts like a concave mirror and makes the echo still more distinct. Were the ground between the two points *C* and *B* a bulge the echo would be very faint and might not reach you at all, because it would diffuse sound just as a convex mirror diffuses light.

You must develop a certain knack to detect an echo on uneven ter-

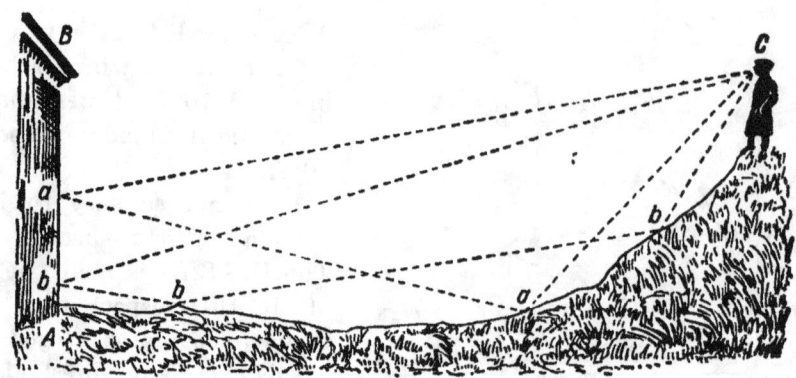

Fig. 150. There is a distinct echo

rain, and even then you must also know how to produce it. In the first place, don't stand too close to the obstacle. The sound waves must travel a long enough distance because otherwise the echo will occur too early and merge with the sound itself. Since sound propagates with the speed of 340 m/sec, at a distance of 85 metres away the echo should be heard exactly half a second later. Though every sound has its echo, not every echo is as distinct, depending on whether it is a beast roaring in a forest, a bugle blowing, thunder reverberating, or a girl singing. The more abrupt and louder the sound, the more distinct is the echo. A hand-clap is best. The human voice is less suitable, especially a man's voice. The high-pitched voices of women and children furnish a more distinct echo.

SOUND AS RULER

Sometimes one can use one's knowledge of the velocity with which sound travels in air to measure the distance to an inaccessible object. Jules Verne provides a case in point in his *Journey to the Centre of Earth*, where in the course of their subterranean exploration the two travellers, the professor and his nephew, lost each other. They hallooed, and when they finally heard each other the following conversation took place between them.

"'Uncle,' I [the nephew] spoke.

"'My boy,' was his ready answer.

14*

"'It is of the utmost consequence that we should know how far we are asunder.'

"'That is not difficult.'

"'You have your chronometer at hand?' I asked.

"'Certainly.'

"'Well, take it into your hand. Pronounce my name, noting exactly the second at which you speak. I will reply as soon as I hear your words—and you will then note exactly the moment at which my reply reaches you.'

"'Very good; and the mean time between my question and your answer will be the time occupied by my voice in reaching you....'

"'Are you ready?'

"'Yes.'

"'Well, I am about to pronounce your name,' said the professor.

"I applied my ear close to the sides of the cavernous gallery, and as soon as the word 'Harry' reached my ear, I turned round and, placing my lips to the wall, repeated the sound.

"'Forty seconds,' said my uncle. 'There has elapsed forty seconds between the two words. The sound, therefore, takes twenty seconds to travel. Now, allowing a thousand and twenty feet for every second, we have twenty thousand four hundred feet—a league and a half and one-eighth.'"

Now you should be able to answer this question: How far away is the train engine if I hear its toot one and a half seconds after I see the wisp of smoke rise from the whistle?

SOUND MIRRORS

A forest wall, high fence, building, mountain, or any echo-producing obstacle in general is nothing but a sound *mirror*, as it reflects sound in the same way as an ordinary flat mirror reflects light.

You can also have a concave sound mirror that would focus the wave-trains of sound. With two soup dishes and a watch you can stage the following illuminating experiment. Put one dish on the table and hold the watch a few centimetres above its bottom. Hold the other dish near your ear as shown in *Fig. 151*. If you gauge the position of all three objects right—do this by trial and error—the ticking of the watch will seem to come from the dish near your ear. But shutting your eyes you enhance the illusion and your ear alone will not tell you in which hand you are holding the watch.

Mediaeval castle-builders often played tricks with sound, by placing a bust either at the focus of a concave sound mirror or at the tail end of a speaking pipe cunningly concealed in the wall. *Fig. 152*, which has been taken from a 16th-century book, shows these arrangements. The vaulted ceiling reflects to the bust's lips all sounds coming in through the speaking pipe; the huge bricked-in speaking

Fig. 151. Concave sound mirrors

pipes carry sounds from the courtyard to the marble busts placed near the walls in one of the galleries, etc. The illusion of whispering or singing busts is thus produced.

Fig. 152. Whispering busts (from a book by Athanasius Kircher. 1560)

SOUND IN THEATRE

The theatre- and concert-goer knows very well that there are halls with good acoustics and bad acoustics. In some speech and music carry distinctly to quite a distance; in others they are muted even quite near.

Not so long ago the good acoustics of one or another theatre was considered simply a stroke of good luck. Now builders have found ways and means of successfully suppressing objectionable reverberation. Though I shall not expand on this point as it can interest only the architect, let me note that the main way of avoiding acoustical defects is to create surfaces to absorb superfluous sounds.

An open window absorbs sound best—just as any aperture is best for absorbing light. Incidentally, a square metre of open window has been accepted as the standard unit to estimate sound absorption. The audience itself is a good sound-absorber, with every person being equivalent to roughly half a square metre of open window. "The audience literally absorbs what the speaker says," one physicist said; it is just as true that the absence of an absorbing audience is literally a great annoyance for a speaker.

When too much sound is absorbed, this is also bad, as, firstly, it mutes speech and music, and, secondly, suppresses reverberation so much that the sounds seem ragged and brittle. As we see, some measure—neither too long, nor too short—of reverberation is desirable. This measure cannot be the same for all halls and must be quantitatively estimated by the designing architect.

There is another place in the theatre of interest from the angle of physics. This is the prompt box. It always has the same shape—have you ever noticed that? Physics is responsible. Its ceiling—a concave sound mirror—serves a dual purpose: firstly, to prevent what the prompter is saying from reaching the audience, and, secondly, to reflect his voice towards the actors on the stage.

SEA-BOTTOM ECHO

Echoes were useless until a method was devised to sound sea and ocean depths with their help. We stumbled upon this invention by accident. When in 1912 the huge ocean liner *Titanic* ran afoul of

202

an iceberg and went down with nearly all its passengers, navigators thought of employing the echo during fogs or at night to detect obstacles in a ship's way. Though this failed to achieve its original purpose, it suggested a fine method, whereby sea depths could be sounded by the echo from the sea-bottom.

Fig. 153 shows you how this is done. By igniting a detonator against the ship's skin near the keel a sharp signal is sent. The sound pierces the water, reaches the sea-bottom and echoes back. This echo, the reflected signal, is recorded by a sensitive device placed against the ship's skin. An accurate timepiece gauges the time interval between the sending of the signal and the reception of the echo. Knowing how fast sound travels in water, we can easily reckon the distance to the reflecting barrier, or, in other words, ascertain the depth.

Echo depth-sounding completely revolutionised sounding practices. To use the old methods one had to stop the ship; and, in general, they were a very tedious affair. The line was payed out very slowly at the rate of 150 metres a minute and it took the same amount of time to rewind it. For instance, it took about 45 minutes to sound a depth of three kilometres. Echo sounding produces

Fig. 153. Echo depth-sounding

the same result in but a few seconds. Furthermore, we don't have to stop the ship to do it and the result is incomparably more accurate, being never more than a quarter of a metre out—provided the time is gauged with an accuracy down to the three-thousandths of a second.

Whereas the exact sounding of great depths is important for oceanography, the quick, reliable and accurate ascertainment of

203

shallow depths is essential for safe steering, especially in offshore waters.

To take soundings today, people employ not ordinary sounds but extremely intensive "ultra-sounds", which we will never hear as their frequency ranges into several million vibrations a second. These sounds are produced by the vibrations of a quartz plate (a piezo-quartz) which is placed in a quickly alternating electric field.

WHY DO BEES BUZZ?

Indeed, why? After all most insects have no special organ for the purpose. The buzzing, which is heard only while the insect is flying, is produced by the flapping of the insect's wings, which vibrate with a rapidity of several hundred times a second. The wings act the role of a vibrating plate, and any plate vibrating with sufficiently great rapidity—more than 16 times a second—produces a tone of a definite pitch.

It is this that reveals to scientists how many times a second an insect moves its wings in flight. To determine the number of times, it is enough to ascertain the pitch of the insect's buzzing, because each tone has its own vibration frequency. With the aid of the slow-motion camera (mentioned in Chapter One) scientists proved that each insect vibrates its wings with practically the same rapidity on every occasion; to regulate its flight it modifies only the "amplitude" of its wing movement and the angle at which the wing is inclined; it increases the number of wing movements per second only in cold weather. That is why the tone of the buzz remains on one level. The ordinary house fly, for instance—its buzz gives the tone F—vibrates its wings 352 times a second. The bumblebee moves its wings 220 times. A honey bee vibrates its wings 440 times a second (tone A) when not burdened with honey, and only 330 times (tone B) when carrying it. Beetles, whose buzzing is lower-pitched, move their wings much less nimbly. Mosquitoes, on the other hand, vibrate their wings between 500 and 600 times a second. Let me note for the sake of comparison that an airplane propeller averages only some 25 revolutions a second.

AUDITORY ILLUSIONS

Once we for some reason imagine the source of a slight noise to be far away, the noise will seem *much louder*. We frequently succumb to these illusions but rarely pay any heed to them. The following curious instance was described by the American scientist William James in his *Psychology.*

"Sitting reading, late one night, I suddenly heard a most formidable noise proceeding from the upper part of the house, which it seemed to fill. It ceased, and in a moment renewed itself. I went into the hall to listen, but it came no more. Resuming my seat in the room, however, there it was again, low, mighty, alarming, like a rising flood or the *avant-courier* of an awful gale. It came from all space. Quite startled, I again went into the hall, but it had already ceased once more. On returning a second time to the room, I discovered that it was nothing but the breathing of a little Scotch terrier which lay asleep on the floor. The noteworthy thing is that as soon as I recognised what it was, I was compelled to think it a different sound, and could not then hear it as I had heard it a moment before." Has anything of the sort ever happened to you? Most likely it has; I, for one, have observed such things more than once.

WHERE'S THE GRASSHOPPER?

We very often err in determining not how far away the sound is, but the direction from which it comes. We can distinguish pretty well by ear whether the shot was fired to the right or left of us (*Fig. 154*), but we are often unable to determine whether it was fired in front of us or behind us (*Fig. 155*). We often hear a shot fired in front of us as one coming from behind. All we can say in such cases is whether it is near or far—depending on how loud the shot is.

Here is a very instructive experiment. Blindfold your friend and seat him in the middle of a room. Ask him to sit still and not turn his head. Then take two coins and click them against each other, standing meanwhile in the imagined vertical plane that passes between your friend's eyes, and ask him to guess where the sound was made. Surprisingly enough, he will point anywhere except at you. But as soon

Fig. 154. Where was the shot fired? On the right or on the left?

as you leave that plane of symmetry which I mentioned, his guessing will be much better, because his ear closest to you will hear the sound a bit earlier and a bit louder.

Fig. 155. Where was the shot fired? In front? Or behind?

This experiment, incidentally, explains why it is so difficult to spot a chirring grasshopper. You hear this shrill singing some two steps away on your right. You turn your head but see nothing, and now hear the grasshopper on your left. Again you turn your head, only to hear the singing come from some other spot. The quicker you turn your head, the nimbler our invisible musician seems. Actually, the grasshopper hasn't moved; you've only imagined it to be hopping about. You have fallen victim to an auditory illusion.

Your mistake is that you turn your head so that the grasshopper occupies its symmetrical plane. As you already know, this readily causes you to blunder in determining the direction. So, if you want to find the grasshopper, the cuckoo, or any other similar distant source of sound, turn your head not in the direction from which it comes but away from it, which, incidentally, is exactly what one does when one "pricks up one's ears".

THE TRICKS OUR EARS PLAY

When we nibble at a rusk we hear a noise that is simply deafening. But for some reason our neighbour makes hardly any noise though he is doing the same. How come? The noise we make is one that only we ourselves can hear and it doesn't annoy our neighbours. The point is that like all solid elastic bodies, the bones of our head are very good conductors of sound. The denser the medium through which sound travels, the louder it is. The sound our neighbour makes when nibbling a rusk is a very light one as it travels through air, but this same sound turns into thunder when it reaches the auditory nerve via the solid bones of your head.

Do the following. Grip the strap-ring of your pocket watch between your teeth and stop up your ears. The bones of your head will amplify the ticking so greatly that you seem to hear the pounding of heavy hammers.

The deaf Beethoven, the story goes, could hear a piano being played by placing one end of his walking stick on it and gripping the other end between his teeth. In the same way deaf people can dance to music, provided there is nothing wrong with their internal ear. The music reaches the auditory nerve via the floor and the bones of the head.

Ventriloquism and the "marvels" it works are all based on the peculiar properties of hearing that have just been described.

The illusion that ventriloquism produces depends wholly on our inability to determine both where the voice is coming from and how far away it is. Ordinarily we can do this only approximately. As soon as we find ourselves in unusual circumstances we already make the crudest of blunders in trying to say where the sound comes from. I, too, couldn't rid myself of the illusion when I was listening to a ventriloquist, even though I very well knew what the matter was.

99 QUESTIONS

1. How much slower is a snail than you are?
2. How fast do modern aircraft fly?
3. Can you overtake the Sun?
4. How do we get slow-motion films?
5. When do we move round the Sun faster?
6. Why are the upper spokes of a rolling wheel blurred and the lower spokes seen distinctly?
7. Which points in a train going forward move backwards?
8. What is aberration of light?
9. Why do we lean forwards or shove our feet under a chair when we get up?
10. Why does a sailor waddle?
11. What is the difference between running and walking?
12. How should one jump off a moving car? Explain.
13. Baron Munchausen, that famous teller of "tall stories", claimed he had caught flying cannon balls with his hands. Could he have done that?
14. Would you like to have presents tossed at you when you're driving a car?
15. Does a body weigh more or less when falling than when at rest?
16. Must everything fall back to Earth?
17. Is Jules Verne's description of life inside the projectile, that set off for the Moon, right?
18. Can you weigh things right on faulty scales with correct weights, or on a properly calibrated balance, but with wrong weights?
19. Are the bones of our arm advantageous levers?
20. Why doesn't a skier sink into soft snow?
21. Why is it pleasant to loaf in a hammock?
22. How was Paris shelled in the First World War?
23. Why does a kite fly?
24. Does a stone continue accelerating all the time it drops?
25. What is the greatest speed a parachutist making a delayed jump can achieve?
26. Why does a boomerang boomerang?

208

27. Can we find out whether an egg is boiled without cracking it open?

28. Where is a thing heavier? Closer to the equator or to the poles?

29. When a seed germinates on the rim of a spinning wheel, in which direction does it stem?

30. What is *perpetuum mobile?*

31. Has a "perpetual motion" machine ever been made?

32. Where does a body immersed in a liquid experience the greatest pressure? From the top, the sides, or the bottom?

33. What happens when a small weight suspended on a piece of thread is dipped into a jar of water balanced on a pair of scales?

34. What shape does liquid take when it weighs nothing? Can you prove this experimentally?

35. Why are raindrops round?

36. Is it true that kerosene oozes through glass or metal? Why do people think it does?

37. Can you make a steel needle float?

38. What is flotation?

39. Why does soap wash dirt off?

40. Why does a soap bubble rise? And where does it rise faster—in a cold or warm room?

41. What is thinner? The human hair or the film of a soap bubble? How many times is one thinner than the other?

42. Water gathers under a glass when it's placed, with a burning piece of paper in it, bottom up on a tray of water. Why does this happen?

43. Why does a liquid rise when sipped through a straw?

44. A stick is balanced by weights on a pair of scales. Will the equilibrium be disturbed if the scales are placed under an evacuated bell?

45. What happens to these scales if placed in liquified air?

46. If you lost your weight, but your clothes didn't, would you fly up into the air?

47. What difference is there between a "perpetual motion" machine and a "gift-power" machine? Have any "gift-power" machines been made?

48. What happens to tram rails on a very hot day or on a very cold one? And why is the weather not so dangerous for railway tracks?

49. When do telegraph and telephone wires sag most?

50. What sort of tumblers crack more often because of hot or cold water?

51. Why do lemonade glasses have a thick bottom and why are they no good as tea glasses?

52. What sort of transparent material that wouldn't crack because of heat or cold is best for tableware?

53. Why is it hard to draw a boot on after a hot bath?

54. Can we make a self-winding clock?

55. Can the self-winding principle be used for bigger machines?

56. Why does smoke curl up?

57. What would you do if you wanted to ice a bottle of lemonade?

58. Will ice melt sooner if wrapped in fur?

59. Is it true that the snow warms the Earth?

60. Why doesn't water freeze in underground pipes in winter?

61. Where is it winter in the Northern Hemisphere in July?

62. Why can you boil water in a welded vessel, without fearing that it might come to pieces?

63. Why does a sled cross snow with difficulty in a heavy frost?

64. When can we roll good snowballs?

65. How do icicles form?

66. Why is it warmer at the equator than at the poles?

67. When would we see the Sun rise if light propagated instantaneously?

68. What would happen to telescopes and microscopes if light propagated instantaneously in any medium?

69. Can we make light circumvent obstacles?

70. How is a periscope made?

71. Where should you place a lamp to see yourself better in a mirror?

72. Are you and your reflection in a mirror completely identical?

73. Is the kaleidoscope of any benefit?

74. How can we use ice to light a fire?

75. Can you see mirages in the temperate zones?

76. What is the "green ray"?

77. How should one look at photographs?

78. Why do photographs acquire relief and depth when looked at through a magnifying glass or in a concave mirror?

79. Why is it best to seat oneself in the middle of a movie-house?

80. Why is it better to look at a painting with one eye?

81. How does a stereoscope work?

82. How can we see things like the giants in fairy tales?

83. What is a telestereoscope?

84. Why do things sparkle?

85. Why does the landscape acquire deeper relief when viewed from a passing train?

86. How are stereoscopic photographs of celestial objects taken?

87. What is the effect of the so-called "shadow marvel" based on?

88. What colour does a red flag assume in blue light?

89. What is irradiation and astigmatism?

90. What kind of pictures follow you with their eyes? And why?

91. Is it a person with normal eyesight or a short-sighted person who thinks the bright stars bigger?

92. When you hear the echo 1.5 seconds after you clap your hands, how far away is the sound barrier?

93. Are there such things as sound mirrors?

94. Where does sound propagate faster, in air or in water?

95. To what technical uses can echoes be put?

96. Why does a bee buzz?

97. Why is it so hard to spot a chirring grasshopper?

98. What transmits sound better—air or some denser medium?

99. What is ventriloquism based upon?

* * *

There is a second part to this book. However, both can be read independently of each other.

THE INVENTIONS

RESEARCHES AND WRITINGS

OF

NIKOLA TESLA

TO HIS COUNTRYMEN

IN EASTERN EUROPE THIS RECORD OF
THE WORK ALREADY ACCOMPLISHED BY

NIKOLA TESLA

IS RESPECTFULLY DEDICATED

Nikola Tesla

THE INVENTIONS

RESEARCHES AND WRITINGS

OF

NIKOLA TESLA

WITH SPECIAL REFERENCE TO HIS WORK IN POLYPHASE
CURRENTS AND HIGH POTENTIAL LIGHTING

BY

THOMAS COMMERFORD MARTIN

Editor THE ELECTRICAL ENGINEER; Past-President American Institute Electrical Engineers

1894

THE ELECTRICAL ENGINEER

NEW YORK

D. VAN NOSTRAND COMPANY,

NEW YORK.

Press of McIlroy & Emmet, 36 Cortlandt St., N. Y.

PREFACE.

THE electrical problems of the present day lie largely in the economical transmission of power and in the radical improvement of the means and methods of illumination. To many workers and thinkers in the domain of electrical invention, the apparatus and devices that are familiar, appear cumbrous and wasteful, and subject to severe limitations. They believe that the principles of current generation must be changed, the area of current supply be enlarged, and the appliances used by the consumer be at once cheapened and simplified. The brilliant successes of the past justify them in every expectancy of still more generous fruition.

The present volume is a simple record of the pioneer work done in such departments up to date, by Mr. Nikola Tesla, in whom the world has already recognized one of the foremost of modern electrical investigators and inventors. No attempt whatever has been made here to emphasize the importance of his researches and discoveries. Great ideas and real inventions win their own way, determining their own place by intrinsic merit. But with the conviction that Mr. Tesla is blazing a path that electrical development must follow for many years to come, the compiler has endeavored to bring together all that bears the impress of Mr. Tesla's genius, and is worthy of preservation. Aside from its value as showing the scope of his inventions, this volume may be of service as indicating the range of his thought. There is intellectual profit in studying the push and play of a vigorous and original mind.

Although the lively interest of the public in Mr. Tesla's work is perhaps of recent growth, this volume covers the results of full ten years. It includes his lectures, miscellaneous articles and discussions, and makes note of all his inventions thus far known, particularly those bearing on polyphase motors and the effects obtained with currents of high potential and high frequency. It will be seen that Mr. Tesla has ever pressed forward, barely pausing for an instant to work out in detail the utilizations that have at once been obvious to him of the new principles he has elucidated. Wherever possible his own language has been employed.

It may be added that this volume is issued with Mr. Tesla's sanction and approval, and that permission has been obtained for the re-publication in it of such papers as have been read before various technical societies of this country and Europe. Mr. Tesla has kindly favored the author by looking over the proof sheets of the sections embodying his latest researches. The work has also enjoyed the careful revision of the author's friend and editorial associate, Mr. Joseph Wetzler, through whose hands all the proofs have passed.

DECEMBER, 1893. T. C. M.

CONTENTS.

PART I.

POLYPHASE CURRENTS.

CHAPTER VII.

CHAPTER VIII.

CHAPTER IX.

CHAPTER X.

CHAPTER XI.

CHAPTER XII.

CHAPTER XIII.

CHAPTER XIV.

CHAPTER XV.

CHAPTER XVI.

CHAPTER XVII.

PART II.

THE TESLA EFFECTS WITH HIGH FREQUENCY AND
HIGH POTENTIAL CURRENTS.

CHAPTER XXVII.

CHAPTER XXVIII.

CHAPTER XXIX.

CHAPTER XXX.

CHAPTER XXXI.

CHAPTER XXXII.

PART III.

MISCELLANEOUS INVENTIONS AND WRITINGS.

CHAPTER XXXIII.

CHAPTER XXXIV.

CHAPTER XXXV.

PART IV.

APPENDIX: EARLY PHASE MOTORS AND THE TESLA OSCILLATORS.

PART I.

—

POLYPHASE CURRENTS.

CHAPTER I.

Biographical and Introductory.

As an introduction to the record contained in this volume of Mr. Tesla's investigations and discoveries, a few words of a biographical nature will, it is deemed, not be out of place, nor other than welcome.

Nikola Tesla was born in 1857 at Smiljan, Lika, a borderland region of Austro-Hungary, of the Serbian race, which has maintained against Turkey and all comers so unceasing a struggle for freedom. His family is an old and representative one among these Switzers of Eastern Europe, and his father was an eloquent clergyman in the Greek Church. An uncle is to-day Metropolitan in Bosnia. His mother was a woman of inherited ingenuity, and delighted not only in skilful work of the ordinary household character, but in the construction of such mechanical appliances as looms and churns and other machinery required in a rural community. Nikola was educated at Gospich in the public school for four years, and then spent three years in the Real Schule. He was then sent to Carstatt, Croatia, where he continued his studies for three years in the Higher Real Schule. There for the first time he saw a steam locomotive. He graduated in 1873, and, surviving an attack of cholera, devoted himself to experimentation, especially in electricity and magnetism. His father would have had him maintain the family tradition by entering the Church, but native genius was too strong, and he was allowed to enter the Polytechnic School at Gratz, to finish his studies, and with the object of becoming a professor of mathematics and physics. One of the machines there experimented with was a Gramme dynamo, used as a motor. Despite his instructor's perfect demonstration of the fact that it was impossible to operate a dynamo without commutator or brushes, Mr. Tesla could not be convinced that such accessories were necessary or desirable. He had already seen with quick intuition that a way could be found to dispense with them; and from that time he may be said to have begun work on the ideas that fructified ultimately in his rotating field motors.

In the second year of his Gratz course, Mr. Tesla gave up the notion of becoming a teacher, and took up the engineering curriculum. His studies ended, he returned home in time to see his father die, and then went to Prague and Buda-Pesth to study languages, with the object of qualifying himself broadly for the practice of the engineering profession. For a short time he served as an assistant in the Government Telegraph Engineering Department, and then became associated with M. Puskas, a personal and family friend, and other exploiters of the telephone in Hungary. He made a number of telephonic inventions, but found his opportunities of benefiting by them limited in various ways. To gain a wider field of action, he pushed on to Paris and there secured employment as an electrical engineer with one of the large companies in the new industry of electric lighting.

It was during this period, and as early as 1882, that he began serious and continued efforts to embody the rotating field principle in operative apparatus. He was enthusiastic about it; believed it to mark a new departure in the electrical arts, and could think of nothing else. In fact, but for the solicitations of a few friends in commercial circles who urged him to form a company to exploit the invention, Mr. Tesla, then a youth of little worldly experience, would have sought an immediate opportunity to publish his ideas, believing them to be worthy of note as a novel and radical advance in electrical theory as well as destined to have a profound influence on all dynamo electric machinery.

At last he determined that it would be best to try his fortunes in America. In France he had met many Americans, and in contact with them learned the desirability of turning every new idea in electricity to practical use. He learned also of the ready encouragement given in the United States to any inventor who could attain some new and valuable result. The resolution was formed with characteristic quickness, and abandoning all his prospects in Europe, he at once set his face westward.

Arrived in the United States, Mr. Tesla took off his coat the day he arrived, in the Edison Works. That place had been a goal of his ambition, and one can readily imagine the benefit and stimulus derived from association with Mr. Edison, for whom Mr. Tesla has always had the strongest admiration. It was impossible, however, that, with his own ideas to carry out, and his own inventions to develop, Mr. Tesla could long remain in even the most delightful employ; and, his work now attracting attention, he left the Edison ranks to join a company intended to make and sell an arc lighting system based on some of his inventions in that branch of the art. With unceasing diligence he brought the system to perfection, and saw it placed on the market. But the thing which most occupied his time and thoughts, however, all through this period, was his old discovery of the rotating field principle for alternating current work, and the application of it in motors that have now become known the world over.

Strong as his convictions on the subject then were, it is a fact that he stood very much alone, for the alternating current had no well recognized place. Few electrical engineers had ever used it, and the majority were entirely unfamiliar with its value, or even its essential features. Even Mr. Tesla himself did not, until after protracted effort and experimentation, learn how to construct alternating current apparatus of fair efficiency. But that he had accomplished his purpose was shown by the tests of Prof. Anthony, made in the of winter 1887–8, when Tesla motors in the hands of that distinguished expert gave an efficiency equal to that of direct current motors. Nothing now stood in the way of the commercial development and introduction of such motors, except that they had to be constructed with a view to operating on the circuits then existing, which in this country were all of high frequency.

The first full publication of his work in this direction—outside his patents—was a paper read before the American Institute of Electrical Engineers in New York, in May, 1888 (read at the suggestion of Prof. Anthony and the present writer), when he exhibited motors that had been in operation long previous, and with which his belief that brushes and commutators could be dispensed with, was triumphantly proved to be correct. The section of this volume devoted to Mr. Tesla's inventions in the utilization of polyphase currents will show how thoroughly from the outset he had mastered the fundamental idea and applied it in the greatest variety of ways.

Having noted for years the many advantages obtainable with alternating currents, Mr. Tesla was naturally led on to experiment with them at higher potentials and higher frequencies than were common or approved of. Ever pressing forward to determine in even the slightest degree the outlines of the unknown, he

was rewarded very quickly in this field with results of the most
surprising nature. A slight acquaintance with some of these
experiments led the compiler of this volume to urge Mr. Tesla
to repeat them before the American Institute of Electrical Engineers. This was done in May, 1891, in a lecture that marked,
beyond question, a distinct departure in electrical theory and
practice, and all the results of which have not yet made themselves fully apparent. The New York lecture, and its suc-
cessors, two in number, are also included in this volume, with a
few supplementary notes.

Mr. Tesla's work ranges far beyond the vast departments of
polyphase currents and high potential lighting. The "Miscellaneous" section of this volume includes a great many other inventions in arc lighting, transformers, pyro-magnetic generators,
thermo-magnetic motors, third-brush regulation, improvements
in dynamos, new forms of incandescent lamps, electrical meters,
condensers, unipolar dynamos, the conversion of alternating into
direct currents, etc. It is needless to say that at this moment
Mr. Tesla is engaged on a number of interesting ideas and inventions, to be made public in due course. The present volume
deals simply with his work accomplished to date.

CHAPTER II.

A NEW SYSTEM OF ALTERNATING CURRENT MOTORS AND TRANSFORMERS.

THE present section of this volume deals with polyphase currents, and the inventions by Mr. Tesla, made known thus far, in which he has embodied one feature or another of the broad principle of rotating field poles or *resultant attraction* exerted on the armature. It is needless to remind electricians of the great interest aroused by the first enunciation of the rotating field principle, or to dwell upon the importance of the advance from a single alternating current, to methods and apparatus which deal with more than one. Simply prefacing the consideration here attempted of the subject, with the remark that in nowise is the object of this volume of a polemic or controversial nature, it may be pointed out that Mr. Tesla's work has not at all been fully understood or realized up to date. To many readers, it is believed, the analysis of what he has done in this department will be a revelation, while it will at the same time illustrate the beautiful flexibility and range of the principles involved. It will be seen that, as just suggested, Mr. Tesla did not stop short at a mere rotating field, but dealt broadly with the shifting of the resultant attraction of the magnets. It will be seen that he went on to evolve the "multiphase" system with many ramifications and turns; that he showed the broad idea of motors employing currents of differing phase in the armature with direct currents in the field; that he first described and worked out the idea of an armature with a body of iron and coils closed upon themselves; that he worked out both synchronizing and torque motors; that he explained and illustrated how machines of ordinary construction might be adapted to his system; that he employed condensers in field and armature circuits, and went to the bottom of the fundamental principles, testing, approving or rejecting, it would appear, every detail that inventive ingenuity could hit upon.

Now that opinion is turning so emphatically in favor of lower frequencies, it deserves special note that Mr. Tesla early recognized the importance of the low frequency feature in motor work. In fact his first motors exhibited publicly—and which, as Prof. Anthony showed in his tests in the winter of 1887–8, were the equal of direct current motors in efficiency, output and starting torque—were of the low frequency type. The necessity arising, however, to utilize these motors in connection with the existing high frequency circuits, our survey reveals in an interesting manner Mr. Tesla's fertility of resource in this direction. But that, after exhausting all the possibilities of this field, Mr. Tesla returns to low frequencies, and insists on the superiority of his polyphase system in alternating current distribution, need not at all surprise us, in view of the strength of his convictions, so often expressed, on this subject. This is, indeed, significant, and may be regarded as indicative of the probable development next to be witnessed.

Incidental reference has been made to the efficiency of rotating field motors, a matter of much importance, though it is not the intention to dwell upon it here. Prof. Anthony in his remarks before the American Institute of Electrical Engineers, in May, 1888, on the two small Tesla motors then shown, which he had tested, stated that one gave an efficiency of about 50 per cent. and the other a little over sixty per cent. In 1889, some tests were reported from Pittsburgh, made by Mr. Tesla and Mr. Albert Schmid, on motors up to 10 H. P. and weighing about 850 pounds. These machines showed an efficiency of nearly 90 per cent. With some larger motors it was then found practicable to obtain an efficiency, with the three wire system, up to as high as 94 and 95 per cent. These interesting figures, which, of course, might be supplemented by others more elaborate and of later date, are cited to show that the efficiency of the system has not had to wait until the present late day for any demonstration of its commercial usefulness. An invention is none the less beautiful because it may lack utility, but it must be a pleasure to any inventor to know that the ideas he is advancing are fraught with substantial benefits to the public.

CHAPTER III.

The Tesla Rotating Magnetic Field.—Motors with Closed Conductors.—Synchronizing Motors.—Rotating Field Transformers.

The best description that can be given of what he attempted, and succeeded in doing, with the rotating magnetic field, is to be found in Mr. Tesla's brief paper explanatory of his rotary current, polyphase system, read before the American Institute of Electrical Engineers, in New York, in May, 1888, under the title " A New System of Alternate Current Motors and Transformers." As a matter of fact, which a perusal of the paper will establish, Mr. Tesla made no attempt in that paper to describe all his work. It dealt in reality with the few topics enumerated in the caption of this chapter. Mr. Tesla's reticence was no doubt due largely to the fact that his action was governed by the wishes of others with whom he was associated, but it may be worth mention that the compiler of this volume—who had seen the motors running, and who was then chairman of the Institute Committee on Papers and Meetings—had great difficulty in inducing Mr. Tesla to give the Institute any paper at all. Mr. Tesla was overworked and ill, and manifested the greatest reluctance to an exhibition of his motors, but his objections were at last overcome. The paper was written the night previous to the meeting, in pencil, very hastily, and under the pressure just mentioned.

In this paper casual reference was made to two special forms of motors not within the group to be considered. These two forms were : 1. A motor with one of its circuits in series with a transformer, and the other in the secondary of the transformer. 2. A motor having its armature circuit connected to the generator, and the field coils closed upon themselves. The paper in its essence is as follows, dealing with a few leading features of the Tesla system, namely, the rotating magnetic field, motors with closed conductors, synchronizing motors, and rotating field transformers :—

The subject which I now have the pleasure of bringing to your notice is a novel system of electric distribution and transmission of power by means of alternate currents, affording peculiar advantages, particularly in the way of motors, which I am confident will at once establish the superior adaptability of these currents to the transmission of power and will show that many results heretofore unattainable can be reached by their use ; results which are very much desired in the practical operation of such systems, and which cannot be accomplished by means of continuous currents.

Before going into a detailed description of this system, I think it necessary to make a few remarks with reference to certain conditions existing in continuous current generators and motors, which, although generally known, are frequently disregarded.

In our dynamo machines, it is well known, we generate alternate currents which we direct by means of a commutator, a complicated device and, it may be justly said, the source of most of the troubles experienced in the operation of the machines. Now, the currents so directed cannot be utilized in the motor, but they must—again by means of a similar unreliable device— be reconverted into their original state of alternate currents. The function of the commutator is entirely external, and in no way does it affect the internal working of the machines. In reality, therefore, all machines are alternate current machines, the currents appearing as continuous only in the external circuit during their transit from generator to motor. In view simply of this fact, alternate currents would commend themselves as a more direct application of electrical energy, and the employment of continuous currents would only be justified if we had dynamos which would primarily generate, and motors which would be directly actuated by, such currents.

But the operation of the commutator on a motor is twofold : first, it reverses the currents through the motor, and secondly, it effects automatically, a progressive shifting of the poles of one of its magnetic constituents. Assuming, therefore, that both of the useless operations in the systems, that is to say, the directing of the alternate currents on the generator and reversing the direct currents on the motor, be eliminated, it would still be necessary, in order to cause a rotation of the motor, to produce a progressive shifting of the poles of one of its elements, and the question presented itself—How to perform this operation by the direct action of alternate currents? I will now proceed to show how this result was accomplished.

In the first experiment a drum-armature was provided with

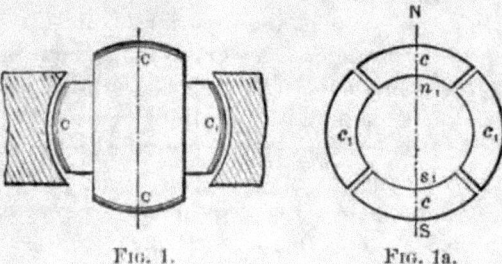

Fig. 1.   Fig. 1a.

two coils at right angles to each other, and the ends of these coils were connected to two pairs of insulated contact-rings as usual. A ring was then made of thin insulated plates of sheet-iron and wound with four coils, each two opposite coils being connected together so as to produce free poles on diametrically opposite sides of the ring. The remaining free ends of the coils were then connected to the contact-rings of the generator armature so as to form two independent circuits, as indicated in Fig. 9. It may now be seen what results were secured in this combination, and with this view I would refer to the diagrams, Figs. 1 to 8a. The field of the generator being independently excited, the rotation of the armature sets up currents in the coils c c_1, varying in

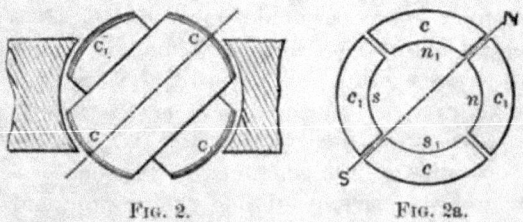

Fig. 2.   Fig. 2a.

strength and direction in the well-known manner. In the position shown in Fig. 1, the current in coil c is nil, while coil c_1 is traversed by its maximum current, and the connections may be such that the ring is magnetized by the coils c_1 c_1, as indicated by the letters N s in Fig. 1a, the magnetizing effect of the coils

c c being nil, since these coils are included in the circuit of coil c.

In Fig. 2, the armature coils are shown in a more advanced position, one-eighth of one revolution being completed. Fig. 2a illustrates the corresponding magnetic condition of the ring. At this moment the coil c_1 generates a current of the same di-

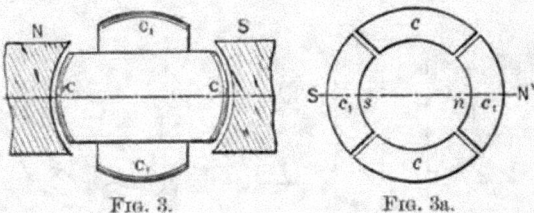

FIG. 3.   FIG. 3a.

rection as previously, but weaker, producing the poles n_1 s_1 upon the ring; the coil c also generates a current of the same direction, and the connections may be such that the coils c c produce the poles n s, as shown in Fig. 2a. The resulting polarity is indicated by the letters N s, and it will be observed that the poles of the ring have been shifted one-eighth of the periphery of the same.

In Fig. 3 the armature has completed one quarter of one revolution. In this phase the current in coil c is a maximum, and of such direction as to produce the poles N s in Fig. 3a, whereas the current in coil c_1 is nil, this coil being at its neutral position.

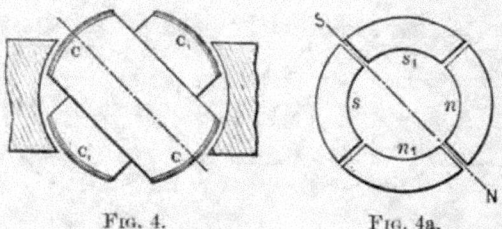

FIG. 4.   FIG. 4a.

The poles N s in Fig. 3a are thus shifted one quarter of the circumference of the ring.

Fig. 4 shows the coils c c in a still more advanced position, the armature having completed three-eighths of one revolution. At that moment the coil c still generates a current of the same direction as before, but of less strength, producing the comparatively weaker poles n s in Fig. 4a. The current in the coil c_1 is of the same strength, but opposite direction. Its effect is, therefore, to produce upon the ring the poles n_1 s_1 as indicated, and a polarity, N s, results, the poles now being shifted three-eighths of the periphery of the ring.

In Fig. 5 one half of one revolution of the armature is com-

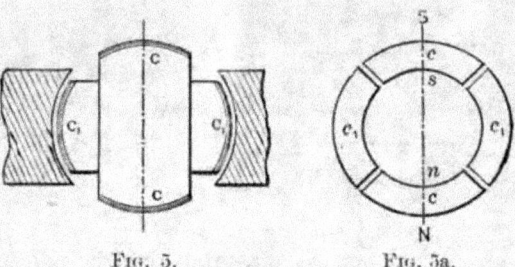

FIG. 5.   FIG. 5a.

pleted, and the resulting magnetic condition of the ring is indicated in Fig. 5a. Now the current in coil c is nil, while the coil c_1 yields its maximum current, which is of the same direction as previously; the magnetizing effect is, therefore, due to the coils, c_1 c_1 alone, and, referring to Fig. 5a, it will be observed that the poles N s are shifted one half of the circumference of the ring. During the next half revolution the operations are repeated, as represented in the Figs. 6 to 8a.

A reference to the diagrams will make it clear that during one

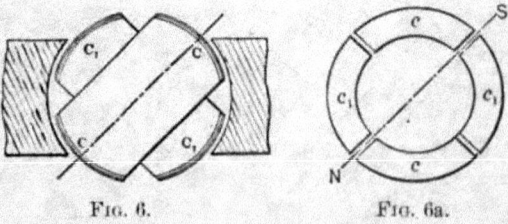

FIG. 6.   FIG. 6a.

revolution of the armature the poles of the ring are shifted once around its periphery, and, each revolution producing like effects, a rapid whirling of the poles in harmony with the rotation of the armature is the result. If the connections of either one of the circuits in the ring are reversed, the shifting of the poles is made to progress in the opposite direction, but the operation is identically the same. Instead of using four wires, with like result, three wires may be used, one forming a common return for both circuits.

This rotation or whirling of the poles manifests itself in a series of curious phenomena. If a delicately pivoted disc of steel or other magnetic metal is approached to the ring it is set in rapid rotation, the direction of rotation varying with the position of

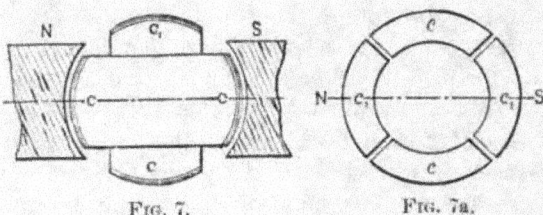

FIG. 7.   FIG. 7a.

the disc. For instance, noting the direction outside of the ring it will be found that inside the ring it turns in an opposite direction, while it is unaffected if placed in a position symmetrical to the ring. This is easily explained. Each time that a pole approaches, it induces an opposite pole in the nearest point on the disc, and an attraction is produced upon that point; owing to this, as the pole is shifted further away from the disc a tangential pull is exerted upon the same, and the action being constantly repeated, a more or less rapid rotation of the disc is the result. As the pull is exerted mainly upon that part which is nearest to the ring, the rotation outside and inside, or right and left, respectively, is in opposite directions, Fig. 9. When placed symmetrically to the ring, the pull on the opposite sides of the disc being equal, no rotation results. The action is based on the magnetic inertia of iron; for this reason a disc of hard steel is much more affected than a disc of soft iron, the latter being capable of very rapid variations of magnetism. Such a disc has proved to be a very useful instrument in all these investigations, as it has enabled me to detect any irregularity in the action. A curious effect is also produced upon iron filings. By placing some upon a paper and holding them externally quite close to the ring, they are set in a vibrating motion, remaining in the same place, although the paper may be moved back and forth; but in lifting the paper to a certain height which seems to be dependent on the intensity of the poles and the speed of rotation, they are thrown away in a direction always opposite to the supposed movement of the poles. If a paper with filings is put flat upon the ring and the current turned on suddenly, the existence of a magnetic whirl may easily be observed.

To demonstrate the complete analogy between the ring and a revolving magnet, a strongly energized electro-magnet was rotated by mechanical power, and phenomena identical in every particular to those mentioned above were observed.

Obviously, the rotation of the poles produces corresponding inductive effects and may be utilized to generate currents in a closed conductor placed within the influence of the poles. For this purpose it is convenient to wind a ring with two sets of superimposed coils forming respectively the primary and secondary circuits, as shown in Fig. 10. In order to secure the most economical results the magnetic circuit should be completely closed, and with this object in view the construction may be modified at will.

The inductive effect exerted upon the secondary coils will be mainly due to the shifting or movement of the magnetic action ; but there may also be currents set up in the circuits in consequence of the variations in the intensity of the poles. However, by properly designing the generator and determining the magnetizing effect of the primary coils, the latter element may be made to disappear. The intensity of the poles being maintained con-

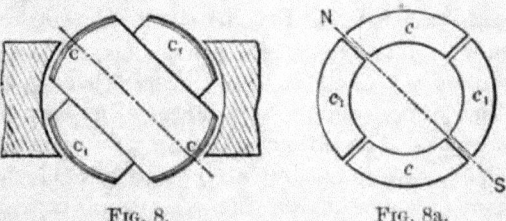

FIG. 8.   FIG. 8a.

stant, the action of the apparatus will be perfect, and the same result will be secured as though the shifting were effected by means of a commutator with an infinite number of bars. In such case the theoretical relation between the energizing effect of each set of primary coils and their resultant magnetizing effect may be expressed by the equation of a circle having its centre coinciding with that of an orthogonal system of axes, and in which the radius represents the resultant and the co-ordinates both of its components. These are then respectively the sine and cosine of the angle a between the radius and one of the axes $(O\ X)$. Referring to Fig. 11, we have $r^2 = x^2 + y^2$; where $x = r \cos a$, and $y = r \sin a$.

Assuming the magnetizing effect of each set of coils in the transformer to be proportional to the current—which may be admitted for weak degrees of magnetization—then $x = K'c$ and $y = K'c^1$, where K' is a constant and c and c^1 the current in both sets of coils respectively. Supposing, further, the field of the generator to be uniform, we have for constant speed $c^1 = K^1 \sin a$ and $c = K^1 \sin (90° + a) = K^1 \cos a$, where K^1 is a constant. See Fig. 12.

Therefore, $x = K'c = K'K^1 \cos a$;
                 $y = K'c^1 = K'K^1 \sin a$; and
                 $K'K^1 = r$.

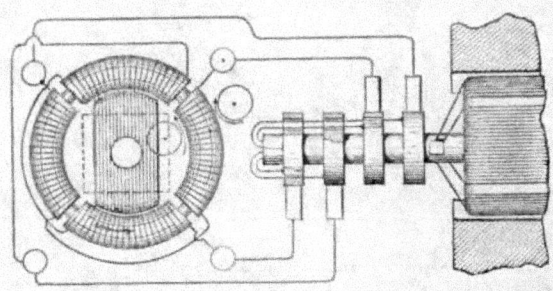

FIG. 9.

That is, for a uniform field the disposition of the two coils at right angles will secure the theoretical result, and the intensity of the shifting poles will be constant. But from $r^2 = x^2 + y^2$ it follows that for $y = 0$, $r = x$; it follows that the joint magnetizing effect of both sets of coils should be equal to the effect of one set when at its maximum action. In transformers and in a certain class of motors the fluctuation of the poles is not of great importance, but in another class of these motors it is desirable to obtain the theoretical result.

In applying this principle to the construction of motors, two typical forms of motor have been developed. First, a form having a comparatively small rotary effort at the start but maintaining a perfectly uniform speed at all loads, which motor has been termed synchronous. Second, a form possessing a great rotary effort at the start, the speed being dependent on the load.

These motors may be operated in three different ways: 1. By the alternate currents of the source only. 2. By a combined action of these and of induced currents. 3. By the joint action of alternate and continuous currents.

The simplest form of a synchronous motor is obtained by winding a laminated ring provided with pole projections with four coils, and connecting the same in the manner before indicated. An iron disc having a segment cut away on each side may be used

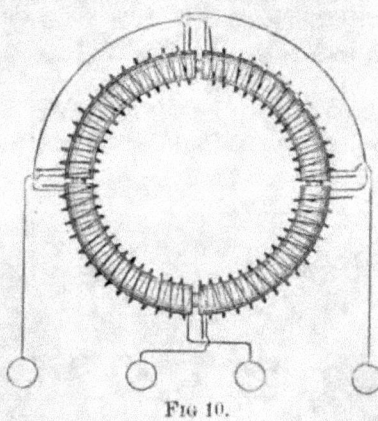

Fig 10.

as an armature. Such a motor is shown in Fig. 9. The disc being arranged to rotate freely within the ring in close proximity to the projections, it is evident that as the poles are shifted it will, owing to its tendency to place itself in such a position as to embrace the greatest number of the lines of force, closely follow the movement of the poles, and its motion will be synchronous with that of the armature of the generator; that is, in the peculiar disposition shown in Fig. 9, in which the armature produces by one revolution two current impulses in each of the circuits. It is evident that if, by one revolution of the armature, a greater number of impulses is produced, the speed of the motor will be correspondingly increased. Considering that the attraction exerted upon the disc is greatest when the same is in close proximity to the poles, it follows that such a motor will maintain exactly the same speed at all loads within the limits of its capacity.

To facilitate the starting, the disc may be provided with a coil closed upon itself. The advantage secured by such a coil is evident. On the start the currents set up in the coil strongly ener-

gize the disc and increase the attraction exerted upon the same by
the ring, and currents being generated in the coil as long as the
speed of the armature is inferior to that of the poles, considerable work may be performed by such a motor even if the speed
be below normal. The intensity of the poles being constant, no
currents will be generated in the coil when the motor is turning
at its normal speed.

Instead of closing the coil upon itself, its ends may be connected
to two insulated sliding rings, and a continuous current supplied
to these from a suitable generator. The proper way to start such
a motor is to close the coil upon itself until the normal speed is
reached, or nearly so, and then turn on the continuous current. If the disc be very strongly energized by a continuous
current the motor may not be able to start, but if it be weakly
energized, or generally so that the magnetizing effect of the ring

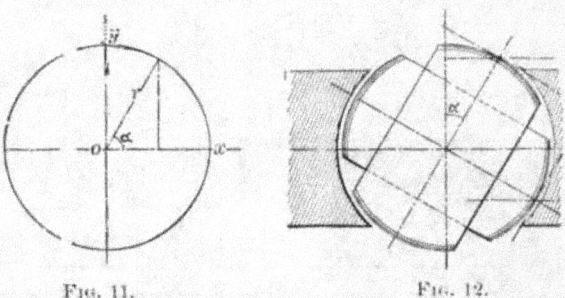

FIG. 11.   FIG. 12.

is preponderating, it will start and reach the normal speed. Such
a motor will maintain absolutely the same speed at all loads. It
has also been found that if the motive power of the generator is
not excessive, by checking the motor the speed of the generator is
diminished in synchronism with that of the motor. It is characteristic of this form of motor that it cannot be reversed by reversing the continuous current through the coil.

The synchronism of these motors may be demonstrated experimentally in a variety of ways. For this purpose it is best to
employ a motor consisting of a stationary field magnet and an
armature arranged to rotate within the same, as indicated in
Fig. 13. In this case the shifting of the poles of the armature
produces a rotation of the latter in the opposite direction. It
results therefrom that when the normal speed is reached, the
poles of the armature assume fixed positions relatively to the

field magnet, and the same is magnetized by induction, exhibiting a distinct pole on each of the pole-pieces. If a piece of soft iron is approached to the field magnet, it will at the start be attracted with a rapid vibrating motion produced by the reversals of polarity of the magnet, but as the speed of the armature increases, the vibrations become less and less frequent and finally entirely cease. Then the iron is weakly but permanently attracted, showing that synchronism is reached and the field magnet energized by induction.

The disc may also be used for the experiment. If held quite close to the armature it will turn as long as the speed of rotation of the poles exceeds that of the armature; but when the normal

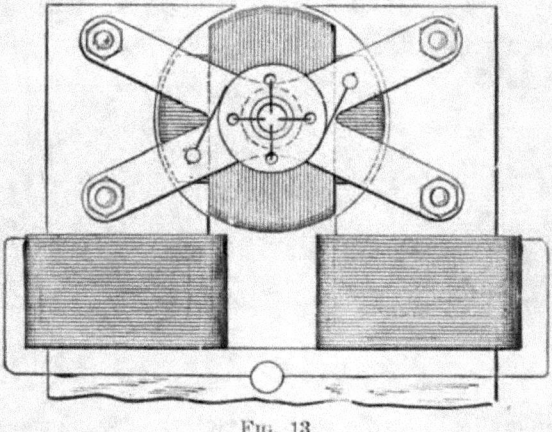

Fig. 13.

speed is reached, or very nearly so, it ceases to rotate and is permanently attracted.

A crude but illustrative experiment is made with an incandescent lamp. Placing the lamp in circuit with the continuous current generator and in series with the magnet coil, rapid fluctuations are observed in the light in consequence of the induced currents set up in the coil at the start; the speed increasing, the fluctuations occur at longer intervals, until they entirely disappear, showing that the motor has attained its normal speed. A telephone receiver affords a most sensitive instrument; when connected to any circuit in the motor the synchronism may be easily detected on the disappearance of the induced currents.

In motors of the synchronous type it is desirable to maintain the quantity of the shifting magnetism constant, especially if the magnets are not properly subdivided.

To obtain a rotary effort in these motors was the subject of long thought. In order to secure this result it was necessary to make such a disposition that while the poles of one element of the motor are shifted by the alternate currents of the source, the poles produced upon the other elements should always be maintained in the proper relation to the former, irrespective of the speed of the motor. Such a condition exists in a continuous current motor; but in a synchronous motor, such as described, this condition is fulfilled only when the speed is normal.

The object has been attained by placing within the ring a properly subdivided cylindrical iron core wound with several independent coils closed upon themselves. Two coils at right angles as

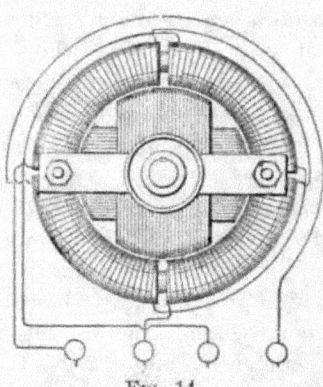

FIG. 14.

in Fig. 14, are sufficient, but a greater number may be advantageously employed. It results from this disposition that when the poles of the ring are shifted, currents are generated in the closed armature coils. These currents are the most intense at or near the points of the greatest density of the lines of force, and their effect is to produce poles upon the armature at right angles to those of the ring, at least theoretically so; and since this action is entirely independent of the speed—that is, as far as the location of the poles is concerned—a continuous pull is exerted upon the periphery of the armature. In many respects these motors are similar to the continuous current motors. If load is put on, the speed, and also the resistance of the motor, is diminished and more current is made to pass through the energizing coils, thus increasing the effort. Upon the load being taken off, the counter-electromotive force increases and less current passes through the primary or energizing coils. Without any load the speed is very nearly equal to that of the shifting poles of the field magnet.

It will be found that the rotary effort in these motors fully

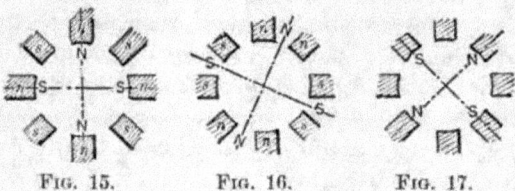

FIG. 15. FIG. 16. FIG. 17.

equals that of the continuous current motors. The effort seems to be greatest when both armature and field magnet are without any projections; but as in such dispositions the field cannot be concentrated, probably the best results will be obtained by leaving pole projections on one of the elements only. Generally, it may be stated the projections diminish the torque and produce a tendency to synchronism.

A characteristic feature of motors of this kind is their property of being very rapidly reversed. This follows from the peculiar action of the motor. Suppose the armature to be rotating and the direction of rotation of the poles to be reversed. The apparatus then represents a dynamo machine, the power to drive this machine being the momentum stored up in the armature and its speed being the sum of the speeds of the armature and the poles.

If we now consider that the power to drive such a dynamo

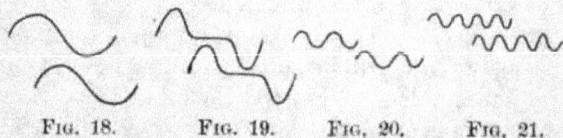

FIG. 18. FIG. 19. FIG. 20. FIG. 21.

would be very nearly proportional to the third power of the speed, for that reason alone the armature should be quickly reversed. But simultaneously with the reversal another element is brought into action, namely, as the movement of the poles with respect to the armature is reversed, the motor acts like a transformer in which the resistance of the secondary circuit would be abnormally diminished by producing in this circuit an additional electromotive force.   Owing to these causes the reversal is instantaneous.

If it is desirable to secure a constant speed, and at the same time a certain effort at the start, this result may be easily attained in a variety of ways.   For instance, two armatures, one for torque and the other for synchronism, may be fastened on the same shaft and any desired preponderance may be given to either one, or an armature may be wound for rotary effort, but a more or less pronounced tendency to synchronism may be given to it by properly constructing the iron core; and in many other ways.

As a means of obtaining the required phase of the currents in both the circuits, the disposition of the two coils at right angles is the simplest, securing the most uniform action; but the phase may be obtained in many other ways, varying with the machine employed.   Any of the dynamos at present in use may be easily adapted for this purpose by making connections to proper points of the generating coils.   In closed circuit armatures, such as used in the continuous current systems, it is best to make four derivations from equi-distant points or bars of the commutator, and to connect the same to four insulated sliding rings on the shaft.   In this case each of the motor circuits is connected to two diametrically opposite bars of the commutator.   In such a disposition the motor may also be operated at half the potential and on the three-wire plan, by connecting the motor circuits in the proper order to three of the contact rings.

In multipolar dynamo machines, such as used in the converter systems, the phase is conveniently obtained by winding upon the armature two series of coils in such a manner that while the coils of one set or series are at their maximum production of current, the coils of the other will be at their neutral position, or nearly so, whereby both sets of coils may be subjected simultaneously or successively to the inducing action of the field magnets.

Generally the circuits in the motor will be similarly disposed, and various arrangements may be made to fulfill the requirements; but the simplest and most practicable is to arrange primary circuits on stationary parts of the motor, thereby obviating, at least in certain forms, the employment of sliding contacts.   In such a case the magnet coils are connected alternately in both the circuits; that is, 1, 3, 5 .... in one, and 2, 4, 6 .... in the other, and the coils of each set of series may be connected all in the same manner, or alternately in opposition; in the latter case a motor with half the number of poles will result, and its action will be correspondingly modified. The Figs. 15, 16, and 17, show three different phases, the magnet coils in each circuit being connected alternately in opposition. In this case there will be always four poles, as in Figs. 15 and 17; four pole projections will be neutral; and in Fig. 16 two adjacent pole projections will have the same polarity. If the coils are connected in the same manner there will be eight alternating poles, as indicated by the letters n' s' in Fig. 15.

The employment of multipolar motors secures in this system an advantage much desired and unattainable in the continuous current system, and that is, that a motor may be made to run exactly at a predetermined speed irrespective of imperfections in construction, of the load, and, within certain limits, of electromotive force and current strength.

In a general distribution system of this kind the following plan should be adopted. At the central station of supply a generator should be provided having a considerable number of poles. The motors operated from this generator should be of the synchronous type, but possessing sufficient rotary effort to insure their starting. With the observance of proper rules of construction it may be admitted that the speed of each motor will be in some inverse proportion to its size, and the number of poles should be chosen accordingly. Still, exceptional demands may modify this rule. In view of this, it will be advantageous to provide each motor with a greater number of pole projections or coils, the number being preferably a multiple of two and three. By this means, by simply changing the connections of the coils, the motor may be adapted to any probable demands.

If the number of the poles in the motor is even, the action will be harmonious and the proper result will be obtained; if this is not the case, the best plan to be followed is to make a motor with a double number of poles and connect the same in the manner before indicated, so that half the number of poles result. Suppose, for instance, that the generator has twelve poles, and it would be desired to obtain a speed equal to $\frac{1}{2}$ of the speed of the generator. This would require a motor with seven pole projections or magnets, and such a motor could not be properly connected in the circuits unless fourteen armature coils would be provided, which would necessitate the employment of sliding contacts. To avoid this, the motor should be provided with fourteen magnets and seven connected in each circuit, the magnets in each circuit alternating among themselves. The armature should have fourteen closed coils. The action of the motor will not be quite as perfect as in the case of an even number of poles, but the drawback will not be of a serious nature.

However, the disadvantages resulting from this unsymmetrical form will be reduced in the same proportion as the number of the poles is augmented.

If the generator has, say, n, and the motor n_1 poles, the speed of the motor will be equal to that of the generator multiplied by $\dfrac{n.}{n_1}$

The speed of the motor will generally be dependent on the number of the poles, but there may be exceptions to this rule. The speed may be modified by the phase of the currents in the circuit or by the character of the current impulses or by intervals between each or between groups of impulses. Some of the possible cases are indicated in the diagrams, Figs. 18, 19, 20 and 21, which are self-explanatory. Fig. 18 represents the condition generally existing, and which secures the best result. In such a case, if the typical form of motor illustrated in Fig. 9 is employed, one complete wave in each circuit will produce one revolution of the motor. In Fig. 19 the same result will be effected by one wave in each circuit, the impulses being successive; in Fig. 20 by four, and in Fig. 21 by eight waves.

By such means any desired speed may be attained, that is, at least within the limits of practical demands. This system possesses this advantage, besides others, resulting from simplicity. At full loads the motors show an efficiency fully equal to that of the continuous current motors. The transformers present an additional advantage in their capability of operating motors. They are capable of similar modifications in construction, and will facilitate the introduction of motors and their adaptation to practical demands. Their efficiency should be higher than that of the present transformers, and I base my assertion on the following:

In a transformer, as constructed at present, we produce the currents in the secondary circuit by varying the strength of the primary or exciting currents. If we admit proportionality with respect to the iron core the inductive effect exerted upon the secondary coil will be proportional to the numerical sum of the variations in the strength of the exciting current per unit of time; whence it follows that for a given variation any prolongation of the primary current will result in a proportional loss. In order to obtain rapid variations in the strength of the current, essential to efficient induction, a great number of undulations are employed; from this practice various disadvantages result. These are: Increased cost and diminished efficiency of the generator; more waste of energy in heating the cores, and also diminished output of the transformer, since the core is not properly utilized, the reversals being too rapid. The inductive effect is also very small in certain phases, as will be apparent from a graphic representation, and there may be periods of inaction, if there are intervals between the succeeding current impulses or waves. In producing a shifting of the poles in a transformer, and thereby inducing currents, the induction is of the ideal character, being always maintained at its maximum action. It is also reasonable to assume that by a shifting of the poles less energy will be wasted than by reversals.

CHAPTER IV.

MODIFICATIONS AND EXPANSIONS OF THE TESLA POLYPHASE SYSTEMS.

In his earlier papers and patents relative to polyphase currents, Mr. Tesla devoted himself chiefly to an enunciation of the broad lines and ideas lying at the basis of this new work; but he supplemented this immediately by a series of other striking inventions which may be regarded as modifications and expansions of certain features of the Tesla systems. These we shall now proceed to deal with.

In the preceding chapters we have thus shown and described the Tesla electrical systems for the transmission of power and the conversion and distribution of electrical energy, in which the motors and the transformers contain two or more coils or sets of coils, which were connected up in independent circuits with corresponding coils of an alternating current generator, the operation of the system being brought about by the co-operation of the alternating currents in the independent circuits in progressively moving or shifting the poles or points of maximum magnetic effect of the motors or converters. In these systems two independent conductors are employed for each of the independent circuits connecting the generator with the devices for converting the transmitted currents into mechanical energy or into electric currents of another character. This, however, is not always necessary. The two or more circuits may have a single return path or wire in common, with a loss, if any, which is so extremely slight that it may be disregarded entirely. For the sake of illustration, if the generator have two independent coils and the motor two coils or two sets of coils in corresponding relations to its operative elements one terminal of each generator coil is connected to the corresponding terminals of the motor coils through two independent conductors, while the opposite terminals of the respective coils are both connected to one return wire. The following description deals with the modification. Fig. 22 is a diagrammatic illustration of a generator and single motor constructed and electrically connected in accordance with the invention. Fig. 23 is a diagram of the system as it is used in operating motors or converters, or both, in parallel, while Fig. 24 illustrates diagrammatically the manner of operating two or more motors or converters, or both, in series. Referring to Fig. 22, A A designate the poles of the field magnets of an alternating-current generator, the armature of which, being in this case cylindrical in form and mounted on a shaft, c, is wound

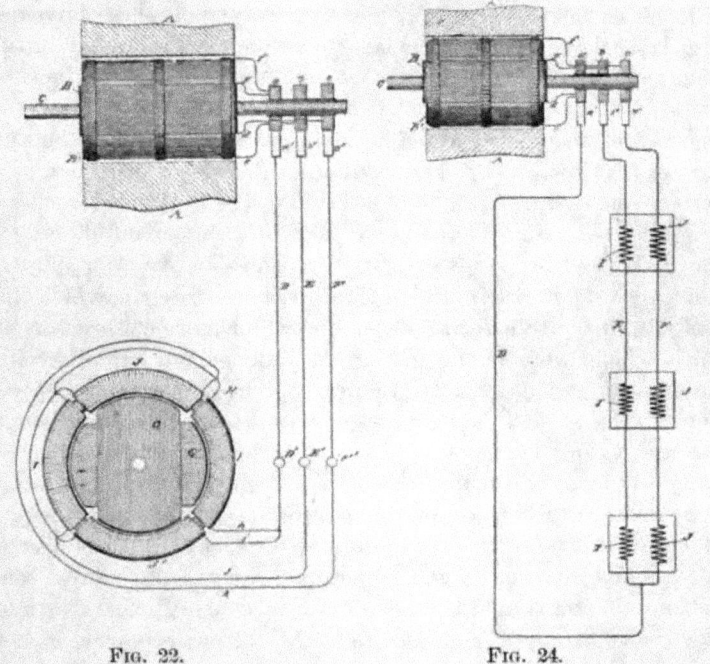

FIG. 22.   FIG. 24.

longitudinally with coils B B′. The shaft c carries three insulated contact-rings, *a b c*, to two of which, as *b c*, one terminal of each coil, as *e d*, is connected. The remaining terminals, *f g*, are both connected to the third ring, *a*.

A motor in this case is shown as composed of a ring, H, wound with four coils, I I J J, electrically connected, so as to co-operate in pairs, with a tendency to fix the poles of the ring at four points ninety degrees apart. Within the magnetic ring H is a disc or cylindrical core wound with two coils, G G′, which may be con-

nected to form two closed circuits. The terminals $j\ k$ of the two
sets or pairs of coils are connected, respectively, to the binding-
posts E′ F′, and the other terminals, $h\ i$, are connected to a single
binding-post, D′. To operate the motor, three line-wires are used
to connect the terminals of the generator with those of the motor.

So far as the apparent action or mode of operation of this arrangement is concerned, the single wire D, which is, so to speak,

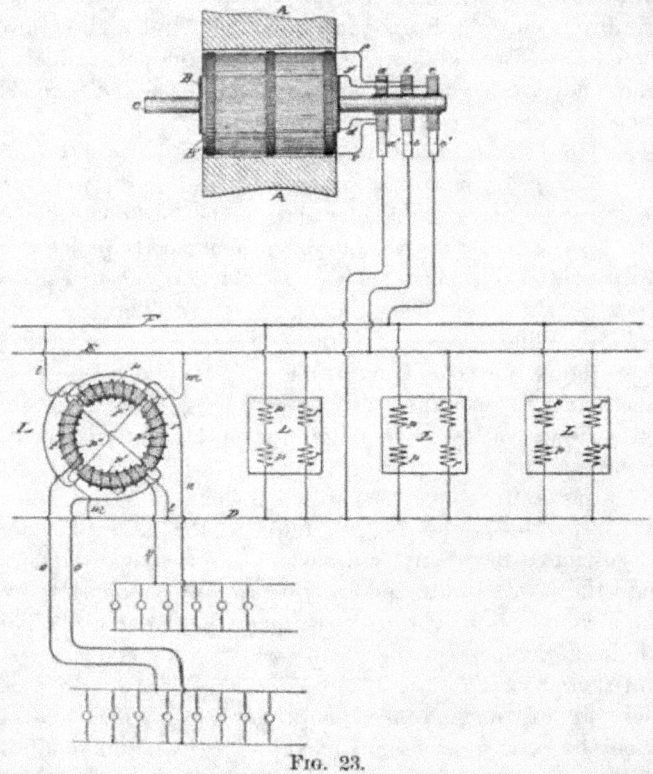

Fig. 23.

a common return-wire for both circuits, may be regarded as two
independent wires. In the illustration, with the order of connection shown, coil B′ of the generator is producing its maximum
current and coil B its minimum; hence the current which passes
through wire e, ring b, brush $b′$, line-wire E, terminal E′, wire j,
coils I I, wire or terminal D′, line-wire D, brush $a′$, ring a, and
wire f, fixes the polar line of the motor midway between the

two coils i i; but as the coil b′ moves from the position indicated it generates less current, while coil b, moving into the field, generates more. The current from coil b passes through the devices and wires designated by the letters *d, c, c′* f, f′ *k*, j j, *i,* d′, d, *a′, a,* and *g,* and the position of the poles of the motor will be due to the resultant effect of the currents in the two sets of coils— that is, it will be advanced in proportion to the advance or forward movement of the armature coils. The movement of the generator-armature through one-quarter of a revolution will obviously bring coil b′ into its neutral position and coil b into its position of maximum effect, and this shifts the poles ninety degrees, as they are fixed solely by coils b. This action is repeated for each quarter of a complete revolution.

When more than one motor or other device is employed, they may be run either in parallel or series. In Fig. 23 the former arrangement is shown. The electrical device is shown as a converter, l, of which the two sets of primary coils *p r* are connected, respectively, to the mains f e, which are electrically connected with the two coils of the generator. The cross-circuit wires *l m*, making these connections, are then connected to the common return-wire d. The secondary coils *p′ p″* are in circuits *n o*, including, for example, incandescent lamps. Only one converter is shown entire in this figure, the others being illustrated diagrammatically.

When motors or converters are to be run in series, the two wires e f are led from the generator to the coils of the first motor or converter, then continued on to the next, and so on through the whole series, and are then joined to the single wire d, which completes both circuits through the generator. This is shown in Fig. 24, in which j i represent the two coils or sets of coils of the motors.

There are, of course, other conditions under which the same idea may be carried out. For example, in case the motor and generator each has three independent circuits, one terminal of each circuit is connected to a line-wire, and the other three terminals to a common return-conductor. This arrangement will secure similar results to those attained with a generator and motor having but two independent circuits, as above described.

When applied to such machines and motors as have three or more induced circuits with a common electrical joint, the three or more terminals of the generator would be simply connected to those of the motor. Mr. Tesla states, however, that the results obtained in this manner show a lower efficiency than do the forms dwelt upon more fully above.

CHAPTER V.

Utilizing Familiar Types of Generator of the Continuous Current Type.

THE preceding descriptions have assumed the use of alternating current generators in which, in order to produce the progressive movement of the magnetic poles, or of the resultant attraction of independent field magnets, the current generating coils are independent or separate. The ordinary forms of continuous current dynamos may, however, be employed for the same work, in accordance with a method of adaptation devised by Mr. Tesla. As will be seen, the modification involves but slight changes in their construction, and presents other elements of economy.

On the shaft of a given generator, either in place of or in addition to the regular commutator, are secured as many pairs of insulated collecting-rings as there are circuits to be operated. Now, it will be understood that in the operation of any dynamo electric generator the currents in the coils in their movement through the field of force undergo different phases—that is to say, at different positions of the coils the currents have certain directions and certain strengths—and that in the Tesla motors or transformers it is necessary that the currents in the energizing coils should undergo a certain order of variations in strength and direction. Hence, the further step—viz., the connection between the induced or generating coils of the machine and the contact-rings from which the currents are to be taken off—will be determined solely by what order of variations of strength and direction in the currents is desired for producing a given result in the electrical translating device. This may be accomplished in various ways; but in the drawings we give typical instances only of the best and most practicable ways of applying the invention to three of the leading types of machines in widespread use, in order to illustrate the principle.

Fig. 25 is a diagram illustrative of the mode of applying the invention to the well-known type of "closed" or continuous cir-

cuit machines. Fig. 26 is a similar diagram embodying an armature with separate coils connected diametrically, or what is generally called an "open-circuit" machine. Fig. 27 is a diagram
showing the application of the invention to a machine the armature-coils of which have a common joint.

Referring to Fig. 25, let A represent a Tesla motor or transformer which, for convenience, we will designate as a "converter." It consists of an annular core, B, wound with four independent coils, C and D, those diametrically opposite being con-

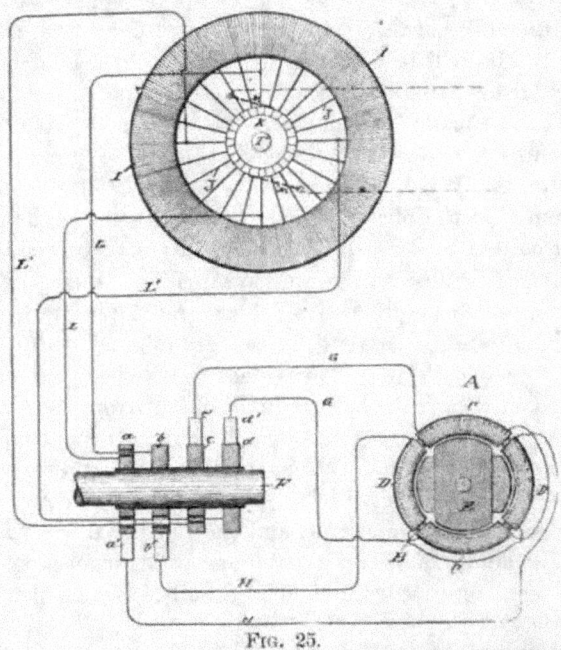

FIG. 25.

nected together so as to co-operate in pairs in establishing free
poles in the ring, the tendency of each pair being to fix the poles
at ninety degrees from the other. There may be an armature,
E, within the ring, which is wound with coils closed upon themselves. The object is to pass through coils C D currents of such
relative strength and direction as to produce a progressive shifting or movement of the points of maximum magnetic effect
around the ring, and to thereby maintain a rotary movement of
the armature. There are therefore secured to the shaft F of the
generator, four insulated contact-rings, a b c d, upon which bear

the collecting-brushes *a' b' c' d'*, connected by wires G G H H, respectively, with the terminals of coils C and D.

Assume, for sake of illustration, that the coils D D are to receive the maximum and coils C C at the same instant the minimum current, so that the polar line may be midway between the coils D D. The rings *a b* would therefore be connected to the continuous armature-coil at its neutral points with respect to the field, or the point corresponding with that of the ordinary commutator brushes, and between which exists the greatest difference of potential; while rings *c d* would be connected to two points in the coil, between which exists no difference of potential. The best results will be obtained by making these connections at points equidistant from one another, as shown. These connections are easiest made by using wires L between the rings and the loops or wires J, connecting the coil I to the segments of the commutator K. When the converters are made in this manner, it is evident that the phases of the currents in the sections of the generator coil will be reproduced in the converter coils. For example, after turning through an arc of ninety degrees the conductors L L, which before conveyed the maximum current, will receive the minimum current by reason of the change in the position of their coils, and it is evident that for the same reason the current in these coils has gradually fallen from the maximum to the minimum in passing through the arc of ninety degrees. In this special plan of connections, the rotation of the magnetic poles of the converter will be synchronous with that of the armature coils of the generator, and the result will be the same, whether the energizing circuits are derivations from a continuous armature coil or from independent coils, as in Mr. Tesla's other devices.

In Fig. 25, the brushes M M are shown in dotted lines in their proper normal position. In practice these brushes may be removed from the commutator and the field of the generator excited by an external source of current; or the brushes may be allowed to remain on the commutator and to take off a converted current to excite the field, or to be used for other purposes.

In a certain well-known class of machines known as the "open circuit," the armature contains a number of coils the terminals of which connect to commutator segments, the coils being connected across the armature in pairs. This type of machine is represented in Fig. 26. In this machine each pair of coils goes through the same phases as the coils in some of the generators already shown, and it is obviously only necessary to utilize them in pairs or sets to operate a Tesla converter by extending the segments of the commutators belonging to each pair of coils and causing a collecting brush to bear on the continuous portion of each segment. In this way two or more circuits may be taken off from the generator, each including one or more pairs or sets of coils as may be desired.

In Fig. 26 i i represent the armature coils, T T the poles of the field magnet, and F the shaft carrying the commutators, which are extended to form continuous portions a b c d. The brushes

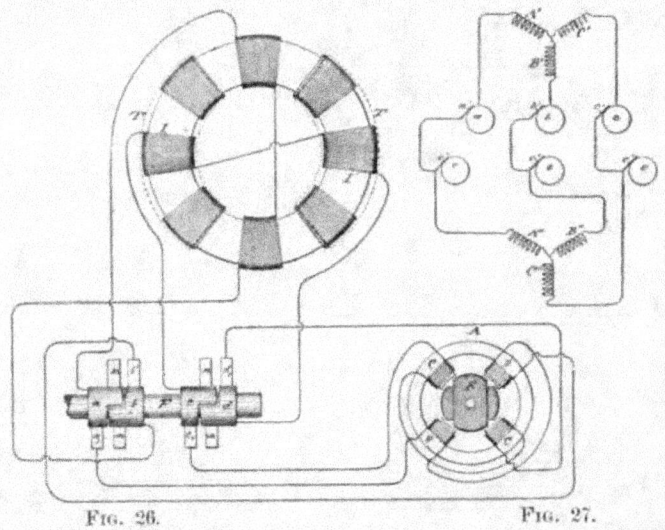

FIG. 26.   FIG. 27.

bearing on the continuous portions for taking off the alternating currents are represented by a' b' c' d'. The collecting brushes, or those which may be used to take off the direct current, are designated by M M. Two pairs of the armature coils and their commutators are shown in the figure as being utilized; but all may be utilized in a similar manner.

There is another well-known type of machine in which three or more coils, A' B' C', on the armature have a common joint, the free ends being connected to the segments of a commutator. This form of generator is illustrated in Fig. 27. In this case each terminal of the generator is connected directly or in derivation to a continuous ring, a b c, and collecting brushes, a' b' c', bearing thereon, take off the alternating currents that operate the motor. It is preferable in this case to employ a motor or transformer with three energizing coils, A″ B″ C″, placed symmetrically with those of the generator, and the circuits from the latter are connected to the terminals of such coils either directly—as when they are stationary—or by means of brushes e' and contact rings e. In this, as in the other cases, the ordinary commutator may be used on the generator, and the current taken from it utilized for exciting the generator field-magnets or for other purposes.

CHAPTER VI.

METHOD OF OBTAINING DESIRED SPEED OF MOTOR OR GENERATOR.

WITH the object of obtaining the desired speed in motors operated by means of alternating currents of differing phase, Mr. Tesla has devised various plans intended to meet the practical requirements of the case, in adapting his system to types of multipolar alternating current machines yielding a large number of current reversals for each revolution.

For example, Mr. Tesla has pointed out that to adapt a given type of alternating current generator, you may couple rigidly two complete machines, securing them together in such a way that the requisite difference in phase will be produced; or you may fasten two armatures to the same shaft within the influence of the same field and with the requisite angular displacement to yield the proper difference in phase between the two currents; or two armatures may be attached to the same shaft with their coils symmetrically disposed, but subject to the influence of two sets of field magnets duly displaced; or the two sets of coils may be wound on the same armature alternately or in such manner that they will develop currents the phases of which differ in time sufficiently to produce the rotation of the motor.

Another method included in the scope of the same idea, whereby a single generator may run a number of motors either at its own rate of speed or all at different speeds, is to construct the motors with fewer poles than the generator, in which case their speed will be greater than that of the generator, the rate of speed being higher as the number of their poles is relatively less. This may be understood from an example, taking a generator that has two independent generating coils which revolve between two pole pieces oppositely magnetized; and a motor with energizing coils that produce at any given time two magnetic poles in one element that tend to set up a rotation of the motor. A generator thus constructed yields four reversals, or impulses, in each revolution, two in each of its independent circuits; and the effect upon the motor is to shift the magnetic poles through three hundred and sixty degrees. It is obvious that if the four reversals in the same order could be produced by each half-revolution of the generator the motor would make two revolutions to the generator's one. This would be readily accomplished by adding two intermediate poles to the generator or altering it in any of the other equivalent ways above indicated. The same rule applies to generators and motors with multiple poles. For instance, if a generator be constructed with two circuits, each of which produces twelve reversals of current to a revolution, and these currents be directed through the independent energizing-coils of a motor, the coils of which are so applied as to produce twelve

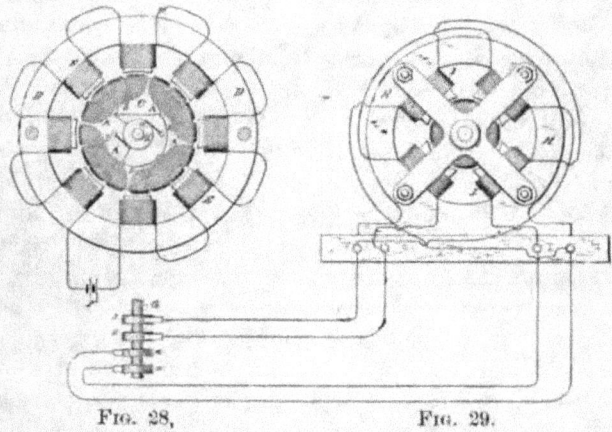

FIG. 28.   FIG. 29.

magnetic poles at all times, the rotation of the two will be synchronous; but if the motor-coils produce but six poles, the movable element will be rotated twice while the generator rotates once; or if the motor have four poles, its rotation will be three times as fast as that of the generator.

These features, so far as necessary to an understanding of the principle, are here illustrated. Fig. 28 is a diagrammatic illustration of a generator constructed in accordance with the invention. Fig. 29 is a similar view of a correspondingly constructed motor. Fig. 30 is a diagram of a generator of modified construction. Fig. 31 is a diagram of a motor of corresponding character. Fig. 32 is a diagram of a system containing a generator and several motors adapted to run at various speeds.

In Fig. 28, let C represent a cylindrical armature core wound longitudinally with insulated coils A A, which are connected up in series, the terminals of the series being connected to collecting-rings *a a* on the shaft G. By means of this shaft the armature is mounted to rotate between the poles of an annular field-magnet D, formed with polar projections wound with coils E, that magnetize the said projections. The coils E are included in the circuit of a generator F, by means of which the field-magnet is energized. If thus constucted, the machine is a well-known form of alternating-current generator. To adapt it to his system, however, Mr. Tesla winds on armature C a second set of coils B B intermediate to the first, or, in other words, in such positions that while the coils of one set are in the relative positions to the poles of the field-magnet to produce the maximum current, those of the other set will be in the position in which they produce the mininum current. The coils B are connected, also, in

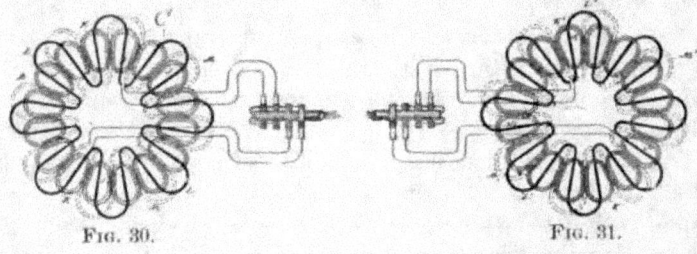

Fig. 30.   Fig. 31.

series and to two connecting-rings, secured generally to the shaft at the opposite end of the armature.

The motor shown in Fig. 29 has an annular field-magnet H, with four pole-pieces wound with coils I. The armature is constructed similarly to the generator, but with two sets of two coils in closed circuits to correspond with the reduced number of magnetic poles in the field. From the foregoing it is evident that one revolution of the armature of the generator producing eight current impulses in each circuit will produce two revolutions of the motor-armature.

The application of the principle of this invention is not, however, confined to any particular form of machine. In Figs. 30 and 31 a generator and motor of another well-known type are shown. In Fig. 30, J J are magnets disposed in a circle and wound with coils K, which are in circuit with a generator which supplies the current that maintains the field of force. In the usual construction of these machines the armature-conductor L is carried by a suitable frame, so as to be rotated in face of the magnets J J, or between these magnets and another similar set in front of them. The magnets are energized so as to be of alternately opposite polarity throughout the series, so that as the conductor C is rotated the current impulses combine or are added to one another, those produced by the conductor in any given position being all in the same direction. To adapt such a machine to his system, Mr. Tesla adds a second set of induced conductors M, in all respects similar to the first, but so placed in reference to it that the currents produced in each will differ by a quarter-phase. With such relations it is evident that as the current decreases in conductor L it increases in conductor M, and conversely, and that any of the forms of Tesla motor invented for use in this system may be operated by such a generator.

Fig. 31 is intended to show a motor corresponding to the machine in Fig. 30. The construction of the motor is identical with that of the generator, and if coupled thereto it will run synchronously therewith. J′ J′ are the field-magnets, and K′ the coils thereon. L′ is one of the armature-conductors and M′ the other.

Fig. 32 shows in diagram other forms of machine. The generator N in this case is shown as consisting of a stationary ring O, wound with twenty-four coils P P′, alternate coils being connected in series in two circuits. Within this ring is a disc or drum Q, with projections Q′ wound with energizing-coils included in circuit with a generator R. By driving this disc or cylinder alternating currents are produced in the coils P and P′, which are carried off to run the several motors.

The motors are composed of a ring or annular field-magnet S, wound with two sets of energizing-coils T T′, and armatures U, having projections U′ wound with coils V, all connected in series in a closed circuit or each closed independently on itself.

Suppose the twelve generator-coils P are wound alternately in opposite directions, so that any two adjacent coils of the same set tend to produce a free pole in the ring O between them and the twelve coils P′ to be similarly wound. A single revolution of the disc or cylinder Q, the twelve polar projections of which are of opposite polarity, will therefore produce twelve current impulses in each of the circuits W W′. Hence the motor X, which has sixteen coils or eight free poles, will make one and a half turns to the generator's one. The motor Y, with twelve coils or six poles, will rotate with twice the speed of the generator, and the motor z, with eight coils or four poles, will revolve three times as fast as the generator. These multipolar motors have a peculiarity which may be often utilized to great advantage. For ex-

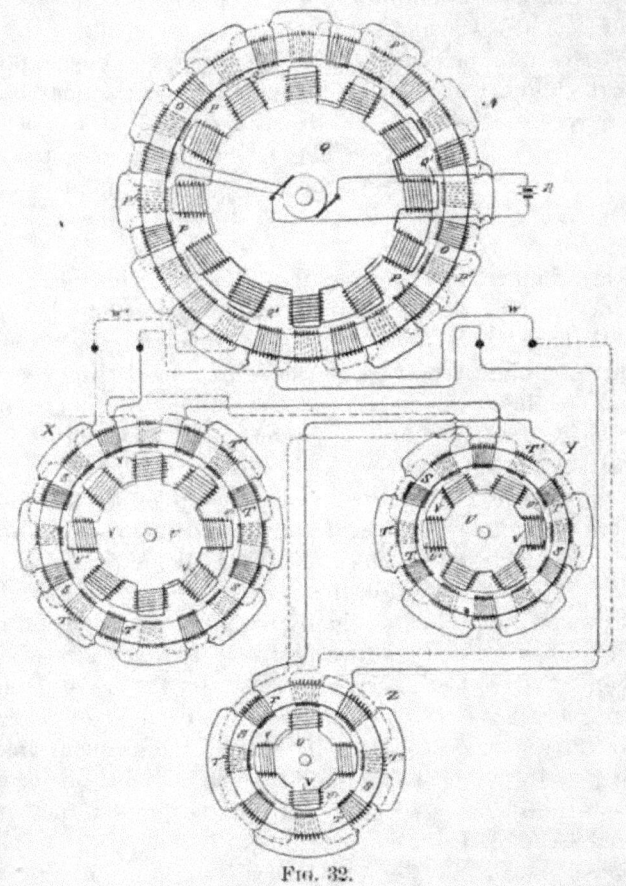

Fig. 32.

ample, in the motor x, Fig. 32, the eight poles may be either alternately opposite or there may be at any given time alternately two like and two opposite poles. This is readily attained by making the proper electrical connections. The effect of such a change, however, would be the same as reducing the number of poles one-half, and thereby doubling the speed of any given motor.

It is obvious that the Tesla electrical transformers which have independent primary currents may be used with the generators described. It may also be stated with respect to the devices we now describe that the most perfect and harmonious action of the generators and motors is obtained when the numbers of the poles of each are even and not odd. If this is not the case, there will be a certain unevenness of action which is the less appreciable as the number of poles is greater; although this may be in a measure corrected by special provisions which it is not here necessary to explain. It also follows, as a matter of course, that if the number of the poles of the motor be greater than that of the generator the motor will revolve at a slower speed than the generator.

In this chapter, we may include a method devised by Mr. Tesla for avoiding the very high speeds which would be necessary with large generators. In lieu of revolving the generator armature at a high rate of speed, he secures the desired result by a rotation of the magnetic poles of one element of the generator, while driving the other at a different speed. The effect is the same as that yielded by a very high rate of rotation.

In this instance, the generator which supplies the current for operating the motors or transformers consists of a subdivided ring or annular core wound with four diametrically-opposite coils, E E', Fig. 33. Within the ring is mounted a cylindrical armature-core wound longitudinally with two independent coils, F F', the ends of which lead, respectively, to two pairs of insulated contact or collecting rings, D D' G G', on the armature shaft. Collecting brushes d d' g g' bear upon these rings, respectively, and convey the currents through the two independent line-circuits M M'. In the main line there may be included one or more motors or transformers, or both. If motors be used, they are of the usual form of Tesla construction with independent coils or sets of coils J J', included, respectively, in the circuits M M'. These energizing-coils are wound on a ring or annular field or on pole pieces thereon, and produce by the action of the alternating currents passing through them a progressive shifting of the magnetism from pole to pole. The cylindrical armature H of the motor is wound with two coils at right angles, which form independent closed circuits.

If transformers be employed, one set of the primary coils, as N N, wound on a ring or annular core is connected to one circuit, as M′, and the other primary coils, N N′, to the circuit M. The secondary coils K K′ may then be utilized for running groups of incandescent lamps P P′.

With this generator an exciter is employed. This consists of

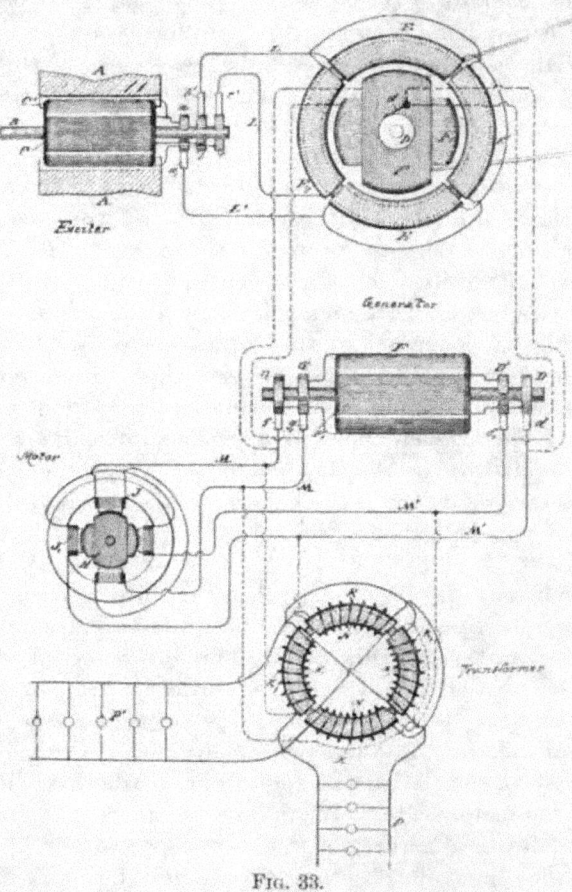

FIG. 33.

two poles, A A, of steel permanently magnetized, or of iron excited by a battery or other generator of continuous currents, and a cylindrical armature core mounted on a shaft, B, and wound with two longitudinal coils, c c′. One end of each of these coils is connected to the collecting-rings b c, respectively, while the other ends are both connected to a ring, a. Collecting-brushes b' c' bear on the rings b c, respectively, and conductors L L convey the currents therefrom through the coils E and E of the generator. L' is a common return-wire to brush a'. Two independent circuits are thus formed, one including coils c of the exciter and E E of the generator, the other coils c' of the exciter and E' E' of the generator. It results from this that the operation of the exciter produces a progressive movement of the magnetic poles of the annular field-core of the generator, the shifting or rotary movement of the poles being synchronous with the rotation of the exciter armature. Considering the operative conditions of a system thus established, it will be found that when the exciter is driven so as to energize the field of the generator, the armature of the latter, if left free to turn, would rotate at a speed practically the same as that of the exciter. If under such conditions the coils F F' of the generator armature be closed upon themselves or short-circuited, no currents, at least theoretically, will be generated in these armature coils. In practice the presence of slight currents is observed, the existence of which is attributable to more or less pronounced fluctuations in the intensity of the magnetic poles of the generator ring. So, if the armature-coils F F' be closed through the motor, the latter will not be turned as long as the movement of the generator armature is synchronous with that of the exciter or of the magnetic poles of its field. If, on the contrary, the speed of the generator armature be in any way checked, so that the shifting or rotation of the poles of the field becomes relatively more rapid, currents will be induced in the armature coils. This obviously follows from the passing of the lines of force across the armature conductors. The greater the speed of rotation of the magnetic poles relatively to that of the armature the more rapidly the currents developed in the coils of the latter will follow one another, and the more rapidly the motor will revolve in response thereto, and this continues until the armature generator is stopped entirely, as by a brake, when the motor, if properly constructed, runs at the speed with which the magnetic poles of the generator rotate.

The effective strength of the currents developed in the armature coils of the generator is dependent upon the strength of the currents energizing the generator and upon the number of rotations per unit of time of the magnetic poles of the generator; hence the speed of the motor armature will depend in all cases upon the relative speeds of the armature of the generator and of its magnetic poles. For example, if the poles are turned two thousand times per unit of time and the armature is turned eight hundred, the motor will turn twelve hundred times, or nearly so. Very slight differences of speed may be indicated by a delicately balanced motor.

Let it now be assumed that power is applied to the generator armature to turn it in a direction opposite to that in which its magnetic poles rotate. In such case the result would be similar to that produced by a generator the armature and field magnets of which are rotated in opposite directions, and by reason of these conditions the motor armature will turn at a rate of speed equal to the sum of the speeds of the armature and magnetic poles of the generator, so that a comparatively low speed of the generator armature will produce a high speed in the motor.

It will be observed in connection with this system that on diminishing the resistance of the external circuit of the generator armature by checking the speed of the motor or by adding translating devices in multiple arc in the secondary circuit or circuits of the transformer the strength of the current in the armature circuit is greatly increased. This is due to two causes: first, to the great differences in the speeds of the motor and generator, and, secondly, to the fact that the apparatus follows the analogy of a transformer, for, in proportion as the resistance of the armature or secondary circuits is reduced, the strength of the currents in the field or primary circuits of the generator is increased and the currents in the armature are augmented correspondingly. For similar reasons the currents in the armature-coils of the generator increase very rapidly when the speed of the armature is reduced when running in the same direction as the magnetic poles or conversely.

It will be understood from the above description that the generator-armature may be run in the direction of the shifting of the magnetic poles, but more rapidly, and that in such case the speed of the motor will be equal to the difference between the two rates.

CHAPTER VII.

Regulator for Rotary Current Motors.

An interesting device for regulating and reversing has been devised by Mr. Tesla for the purpose of varying the speed of polyphase motors. It consists of a form of converter or transformer with one element capable of movement with respect to the other, whereby the inductive relations may be altered, either manually or automatically, for the purpose of varying the strength of the induced current. Mr. Tesla prefers to construct this device in such manner that the induced or secondary element may be movable with respect to the other; and the invention, so far as relates merely to the construction of the device itself, consists, essentially, in the combination, with two opposite magnetic poles, of an armature wound with an insulated coil and mounted on a shaft, whereby it may be turned to the desired extent within the field produced by the poles. The normal position of the core of the secondary element is that in which it most completely closes the magnetic circuit between the poles of the primary element, and in this position its coil is in its most effective position for the inductive action upon it of the primary coils; but by turning the movable core to either side, the induced currents delivered by its coil become weaker until, by a movement of the said core and coil through 90°, there will be no current delivered.

Fig. 34 is a view in side elevation of the regulator. Fig. 35 is a broken section on line x x of Fig. 34. Fig. 36 is a diagram illustrating the most convenient manner of applying the regulator to ordinary forms of motors, and Fig. 37 is a similar diagram illustrating the application of the device to the Tesla alternating-current motors. The regulator may be constructed in many ways to secure the desired result; but that which is, perhaps, its best form is shown in Figs. 34 and 35.

A represents a frame of iron. B B are the cores of the inducing or primary coils c c. D is a shaft mounted on the side bars, D', and on which is secured a sectional iron core, E, wound with an induced or secondary coil, F, the convolutions of which are parallel with the axis of the shaft. The ends of the core are rounded off so as to fit closely in the space between the two poles and permit the core E to be turned to and held at any desired point. A handle, G, secured to the projecting end of the shaft D, is provided for this purpose.

In Fig. 36 let H represent an ordinary alternating current generator, the field-magnets of which are excited by a suitable source of current, I. Let J designate an ordinary form of electromagnetic motor provided with an armature, K, commutator L, and field-magnets M. It is well known that such a motor, if its

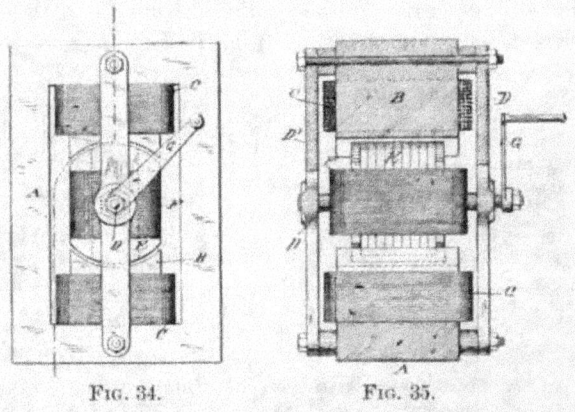

FIG. 34.   FIG. 35.

field-magnet cores be divided up into insulated sections, may be practically operated by an alternating current; but in using this regulator with such a motor, Mr. Tesla includes one element of the motor only—say the armature-coils—in the main circuit of the generator, making the connections through the brushes and the commutator in the usual way. He also includes one of the elements of the regulator—say the stationary coils—in the same circuit, and in the circuit with the secondary or movable coil of the regulator he connects up the field-coils of the motor. He also prefers to use flexible conductors to make the connections from the secondary coil of the regulator, as he thereby avoids the use of sliding contacts or rings without interfering with the requisite movement of the core E.

If the regulator be in its normal position, or that in which its magnetic circuit is most nearly closed, it delivers its maximum induced current, the phases of which so correspond with those of the primary current that the motor will run as though both field and armature were excited by the main current.

To vary the speed of the motor to any rate between the minimum and maximum rates, the core E and coils F are turned in either direction to an extent which produces the desired result, for in its normal position the convolutions of coil F embrace the maximum number of lines of force, all of which act with the same effect upon the coil; hence it will deliver its maximum current; but by turning the coil F out of its position of maximum effect the number of lines of force embraced by it is diminished. The inductive effect is therefore impaired, and the current delivered by coil F will continue to diminish in proportion to the angle at which the coil F is turned until, after passing through

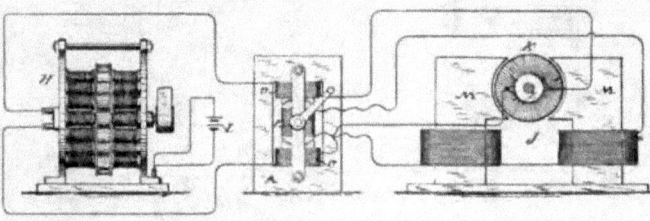

FIG. 36.

an angle of ninety degrees, the convolutions of the coil will be at right angles to those of coils c c, and the inductive effect reduced to a minimum.

Incidentally to certain constructions, other causes may influence the variation in the strength of the induced currents. For example, in the present case it will be observed that by the first movement of coil F a certain portion of its convolutions are carried beyond the line of the direct influence of the lines of force, and that the magnetic path or circuit for the lines is impaired; hence the inductive effect would be reduced. Next, that after moving through a certain angle, which is obviously determined by the relative dimensions of the bobbin or coil F, diagonally opposite portions of the coil will be simultaneously included in the field, but in such positions that the lines which produce a current-impulse in one portion of the coil in a certain direction will produce in the diagonally opposite portion a corresponding impulse in the opposite direction; hence portions of the current will neutralize one another.

As before stated, the mechanical construction of the device may be greatly varied; but the essential conditions of the principle will be fulfilled in any apparatus in which the movement of the elements with respect to one another effects the same results by varying the inductive relations of the two elements in a manner similar to that described.

It may also be stated that the core E is not indispensable to the operation of the regulator; but its presence is obviously beneficial. This regulator, however, has another valuable property in its capability of reversing the motor, for if the coil F be turned

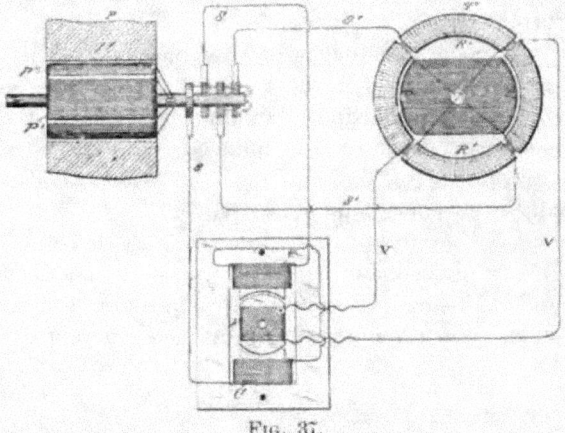

FIG. 37.

through a half-revolution, the position of its convolutions relatively to the two coils c c and to the lines of force is reversed, and consequently the phases of the current will be reversed. This will produce a rotation of the motor in an opposite direction. This form of regulator is also applied with great advantage to Mr. Tesla's system of utilizing alternating currents, in which the magnetic poles of the field of a motor are progressively shifted by means of the combined effects upon the field of magnetizing coils included in independent circuits, through which pass alternating currents in proper order and relations to each other.

In Fig. 37, let P represent a Tesla generator having two independent coils, P' and P", on the armature, and T a diagram of a motor having two independent energizing coils or sets of coils, R R'. One of the circuits from the generator, as s' s', includes one set, R' R', of the energizing coils of the motor, while the other circuit, as s s, includes the primary coils of the regulator. The secondary coil of the regulator includes the other coils, R R, of the motor.

While the secondary coil of the regulator is in its normal position, it produces its maximum current, and the maximum rotary effect is imparted to the motor; but this effect will be diminished in proportion to the angle at which the coil F of the regulator is turned. The motor will also be reversed by reversing the position of the coil with reference to the coils c c, and thereby reversing the phases of the current produced by the generator. This changes the direction of the movement of the shifting poles which the armature follows.

One of the main advantages of this plan of regulation is its economy of power. When the induced coil is generating its maximum current, the maximum amount of energy in the primary coils is absorbed; but as the induced coil is turned from its normal position the self-induction of the primary-coils reduces the expenditure of energy and saves power.

It is obvious that in practice either coils c c or coil F may be used as primary or secondary, and it is well understood that their relative proportions may be varied to produce any desired difference or similarity in the inducing and induced currents.

CHAPTER VIII.

SINGLE CIRCUIT, SELF–STARTING SYNCHRONIZING MOTORS.

In the first chapters of this section we have, bearing in mind the broad underlying principle, considered a distinct class of motors, namely, such as require for their operation a special generator capable of yielding currents of differing phase. As a matter of course, Mr. Tesla recognizing the desirability of utilizing his motors in connection with ordinary systems of distribution, addressed himself to the task of inventing various methods and ways of achieving this object. In the succeeding chapters, therefore, we witness the evolution of a number of ideas bearing upon this important branch of work. It must be obvious to a careful reader, from a number of hints encountered here and there, that even the inventions described in these chapters to follow do not represent the full scope of the work done in these lines. They might, indeed, be regarded as exemplifications.

We will present these various inventions in the order which to us appears the most helpful to an understanding of the subject by the majority of readers. It will be naturally perceived that in offering a series of ideas of this nature, wherein some of the steps or links are missing, the descriptions are not altogether sequential; but any one who follows carefully the main drift of the thoughts now brought together will find that a satisfactory comprehension of the principles can be gained.

As is well known, certain forms of alternating-current machines have the property, when connected in circuit with an alternating current generator, of running as a motor in synchronism therewith; but, while the alternating current will run the motor after it has attained a rate of speed synchronous with that of the generator, it will not start it. Hence, in all instances heretofore where these "synchronizing motors," as they are termed, have been run, some means have been adopted to bring the motors up to synchronism with the generator, or approximately so, before the alternating current of the generator is applied to drive them.

In some instances mechanical appliances have been utilized for this purpose. In others special and complicated forms of motor have been constructed. Mr. Tesla has discovered a much more simple method or plan of operating synchronizing motors, which requires practically no other apparatus than the motor itself. In other words, by a certain change in the circuit connections of the motor he converts it at will from a double circuit motor, or such as have been already described, and which will start under the action of an alternating current, into a synchronizing motor, or one which will be run by the generator only when it has reached a certain speed of rotation synchronous with that of the generator. In this manner he is enabled to extend very greatly the applications of his system and to secure all the advantages of both forms of alternating current motor.

The expression "synchronous with that of the generator," is used here in its ordinary acceptation—that is to say, a motor is said to synchronize with the generator when it preserves a certain relative speed determined by its number of poles and the number of alternations produced per revolution of the generator. Its actual speed, therefore, may be faster or slower than that of the generator; but it is said to be synchronous so long as it preserves the same relative speed.

In carrying out this invention Mr. Tesla constructs a motor which has a strong tendency to synchronism with the generator. The construction preferred is that in which the armature is provided with polar projections. The field-magnets are wound with two sets of coils, the terminals of which are connected to a switch mechanism, by means of which the line-current may be carried directly through these coils or indirectly through paths by which its phases are modified. To start such a motor, the switch is turned on to a set of contacts which includes in one motor circuit a dead resistance, in the other an inductive resistance, and, the two circuits being in derivation, it is obvious that the difference in phase of the current in such circuits will set up a rotation of the motor. When the speed of the motor has thus been brought to the desired rate the switch is shifted to throw the main current directly through the motor-circuits, and although the currents in both circuits will now be of the same phase the motor will continue to revolve, becoming a true synchronous motor. To secure greater efficiency, the armature or its polar projections are wound with coils closed on themselves.

In the accompanying diagrams, Fig. 38 illustrates the details of the plan above set forth, and Figs. 39 and 40 modifications of the same.

Referring to Fig. 38, let A designate the field-magnets of a

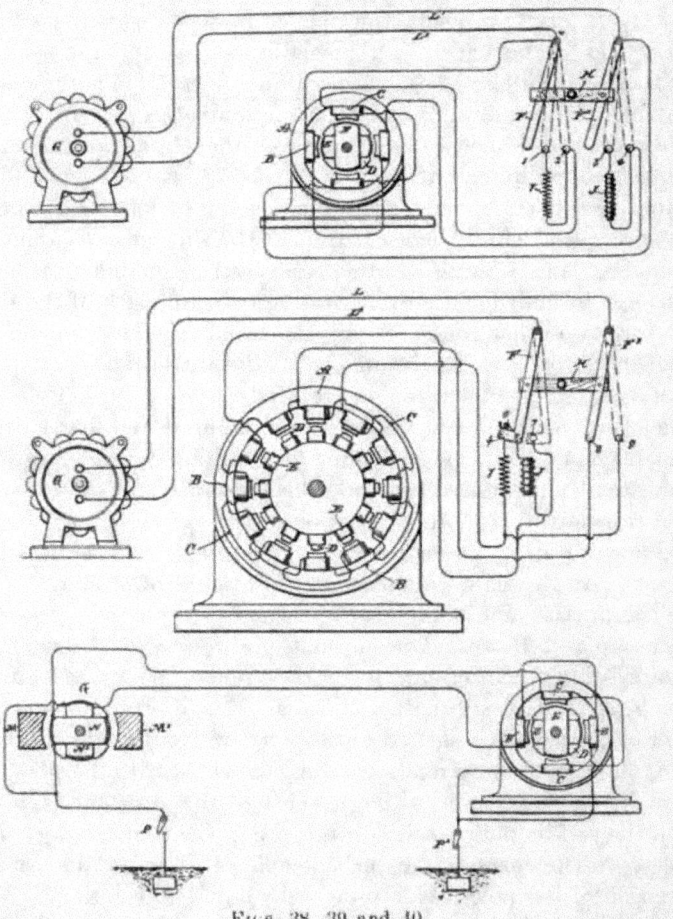

Figs. 38, 39 and 40.

motor, the polar projections of which are wound with coils B C included in independent circuits, and D the armature with polar projections wound with coils E closed upon themselves, the motor in these respects being similar in construction to those described already, but having on account of the polar projections on the armature core, or other similar and well-known features, the properties of a synchronizing-motor. L L' represents the conductors of a line from an alternating current generator G.

Near the motor is placed a switch the action of which is that of the one shown in the diagrams, which is constructed as follows: F F' are two conducting plates or arms, pivoted at their ends and connected by an insulating cross-bar, H, so as to be shifted in parallelism. In the path of the bars F F' is the contact 2, which forms one terminal of the circuit through coils C, and the contact 4, which is one terminal of the circuit through coils B. The opposite end of the wire of coils C is connected to the wire L or bar F', and the corresponding end of coils B is connected to wire L' and bar F; hence if the bars be shifted so as to bear on contacts 2 and 4 both sets of coils B C will be included in the circuit L L' in multiple arc or derivation. In the path of the levers F F' are two other contact terminals, 1 and 3. The contact 1 is connected to contact 2 through an artificial resistance, I, and contact 3 with contact 4 through a self-induction coil, J, so that when the switch levers are shifted upon the points 1 and 3 the circuits of coils B and C will be connected in multiple arc or derivation to the circuit L L', and will include the resistance and self-induction coil respectively. A third position of the switch is that in which the levers F and F' are shifted out of contact with both sets of points. In this case the motor is entirely out of circuit.

The purpose and manner of operating the motor by these devices are as follows: The normal position of the switch, the motor being out of circuit, is off the contact points. Assuming the generator to be running, and that it is desired to start the motor, the switch is shifted until its levers rest upon points 1 and 3. The two motor-circuits are thus connected with the generator circuit; but by reason of the presence of the resistance I in one and the self-induction coil J in the other the coincidence of the phases of the current is disturbed sufficiently to produce a progression of the poles, which starts the motor in rotation. When the speed of the motor has run up to synchronism with the generator, or approximately so, the switch is shifted over upon the points 2 and 4, thus cutting out the coils I and J, so that the currents in both circuits have the same phase; but the motor now runs as a synchronous motor.

It will be understood that when brought up to speed the mo

tor will run with only one of the circuits B or C connected with the main or generator circuit, or the two circuits may be connected in series. This latter plan is preferable when a current having a high number of alternations per unit of time is employed to drive the motor. In such case the starting of the motor is more difficult, and the dead and inductive resistances must take up a considerable proportion of the electromotive force of the circuits. Generally the conditions are so adjusted that the electromotive force used in each of the motor circuits is that which is required to operate the motor when its circuits are in series. The plan followed in this case is illustrated in Fig. 39. In this instance the motor has twelve poles and the armature has polar projections D wound with closed coils E. The switch used is of substantially the same construction as that shown in the previous figure. There are, however, five contacts, designated as 5, 6, 7, 8, and 9. The motor-circuits B C, which include alternate field-coils, are connected to the terminals in the following order: One end of circuit C is connected to contact 9 and to contact 5 through a dead resistance, I. One terminal of circuit B is connected to contact 7 and to contact 6 through a self-induction coil, J. The opposite terminals of both circuits are connected to contact 8.

One of the levers, as F, of the switch is made with an extension, *f*, or otherwise, so as to cover both contacts 5 and 6 when shifted into the position to start the motor. It will be observed that when in this position and with lever F' on contact 8 the current divides between the two circuits B C, which from their difference in electrical character produce a progression of the poles that starts the motor in rotation. When the motor has attained the proper speed, the switch is shifted so that the levers cover the contacts 7 and 9, thereby connecting circuits B and C in series. It is found that by this disposition the motor is maintained in rotation in synchronism with the generator. This principle of operation, which consists in converting by a change of connections or otherwise a double-circuit motor, or one operating by a progressive shifting of the poles, into an ordinary synchronizing motor may be carried out in many other ways. For instance, instead of using the switch shown in the previous figures, we may use a temporary ground circuit between the generator and motor, in order to start the motor, in substantially the manner indicated in Fig. 40. Let G in this figure represent an ordinary alternating-current generator with, say, two poles, M M′, and an armature wound with two coils, N N′, at right angles and connected in series. The motor has, for example, four poles wound with coils B C, which are connected in series, and an armature with polar projections D wound with closed coils E E. From the common joint or union between the two circuits of both the generator and the motor an earth connection is established, while the terminals or ends of these circuits are connected to the line. Assuming that the motor is a synchronizing motor or one that has the capability of running in synchronism with the generator, but not of starting, it may be started by the above-described apparatus by closing the ground connection from both generator and motor. The system thus becomes one with a two-circuit generator and motor, the ground forming a common return for the currents in the two circuits L and L′. When by this arrangement of circuits the motor is brought to speed, the ground connection is broken between the motor or generator, or both, ground-switches PP′ being employed for this purpose. The motor then runs as a synchronizing motor.

In describing the main features which constitute this invention illustrations have necessarily been omitted of the appliances used in conjunction with the electrical devices of similar systems—such, for instance, as driving-belts, fixed and loose pulleys for the motor, and the like; but these are matters well understood.

Mr. Tesla believes he is the first to operate electro-magnetic motors by alternating currents in any of the ways herein described—that is to say, by producing a progressive movement or rotation of their poles or points of greatest magnetic attraction by the alternating currents until they have reached a given speed, and then by the same currents producing a simple alternation of their poles, or, in other words, by a change in the order or character of the circuit connections to convert a motor operating on one principle to one operating on another.

CHAPTER IX.

CHANGE FROM DOUBLE CURRENT TO SINGLE CURRENT MOTOR.

A DESCRIPTION is given elsewhere of a method of operating alternating current motors by first rotating their magnetic poles until they have attained synchronous speed, and then alternating the poles. The motor is thus transformed, by a simple change of circuit connections from one operated by the action of two or more independent energizing currents to one operated either by a single current or by several currents acting as one. Another way of doing this will now be described.

At the start the magnetic poles of one element or field of the motor are progressively shifted by alternating currents differing in phase and passed through independent energizing circuits, and short circuit the coils of the other element. When the motor thus started reaches or passes the limit of speed synchronous with the generator, Mr. Tesla connects up the coils previously short-circuited with a source of direct current and by a change of the circuit connections produces a simple alternation of the poles. The motor then continues to run in synchronism with the generator. The motor here shown in Fig. 41 is one of the ordinary forms, with field-cores either laminated or solid and with a cylindrical laminated armature wound, for example, with the coils A B at right angles. The shaft of the armature carries three collecting or contact rings C D E. (Shown, for better illustration, as of different diameters.)

One end of coil A connects to one ring, as C, and one end of coil B connects with ring D. The remaining ends are connected to ring E. Collecting springs or brushes F G H bear upon the rings and lead to the contacts of a switch, to be presently described. The field-coils have their terminals in binding-posts K K, and may be either closed upon themselves or connected with a source of direct current L, by means of a switch M. The main or controlling switch has five contacts a b c d e and two levers f g, pivoted and connected by an insulating cross-bar h, so as to move in parallelism. These levers are connected to the line wires from a source of alternating currents N. Contact *a* is connected to brush G and coil B through a dead resistance R and wire P. Contact *b* is connected with brush F and coil A through a self-induction coil S and wire O. Contacts *c* and *e* are connected to brushes G F, respectively, through the wires P O, and contact *d* is directly connected with brush H. The lever *f* has a widened end, which may span the contacts *a b*. When in such position and with lever *g* on contact *d*, the alternating currents divide between the two motor-coils, and by reason of their different self-

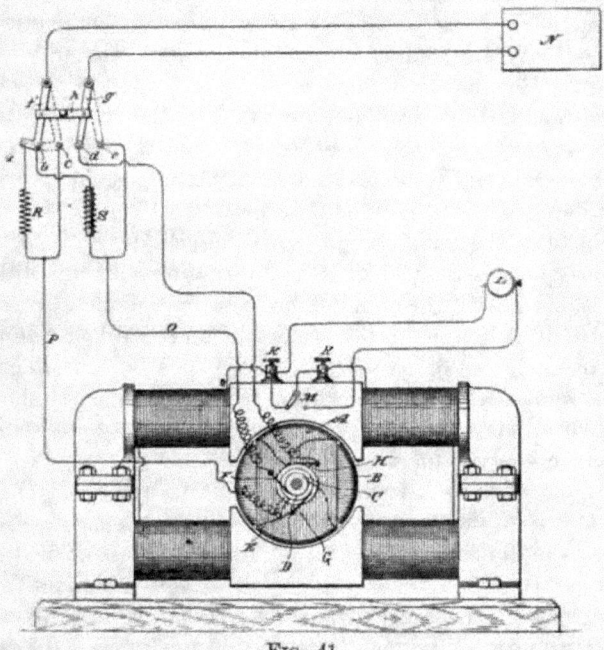

FIG. 41.

induction a difference of current-phase is obtained that starts the motor in rotation. In starting, the field-coils are short circuited.

When the motor has attained the desired speed, the switch is shifted to the position shown in dotted lines—that is to say, with the levers *f g* resting on points *e e*. This connects up the two armature coils in series, and the motor will then run as a synchronous motor. The field-coils are thrown into circuit with the direct current source when the main switch is shifted.

CHAPTER X.

MOTOR WITH "CURRENT LAG" ARTIFICIALLY SECURED.

ONE of the general ways followed by Mr. Tesla in developing his rotary phase motors is to produce practically independent currents differing primarily in phase and to pass these through the motor-circuits. Another way is to produce a single alternating current, to divide it between the motor-circuits, and to effect artificially a lag in one of these circuits or branches, as by giving to the circuits different self-inductive capacity, and in other ways. In the former case, in which the necessary difference of phase is primarily effected in the generation of currents, in some instances, the currents are passed through the energizing coils of both elements of the motor—the field and armature; but a further result or modification may be obtained by doing this under the conditions hereinafter specified in the case of motors in which the lag, as above stated, is artificially secured.

Figs. 42 to 47, inclusive, are diagrams of different ways in which the invention is carried out; and Fig. 48, a side view of a form of motor used by Mr. Tesla for this purpose.

A B in Fig. 42 indicate the two energizing circuits of a motor, and C D two circuits on the armature. Circuit or coil A is connected in series with circuit or coil C, and the two circuits B D are similarly connected. Between coils A and C is a contact-ring e, forming one terminal of the latter, and a brush a, forming one terminal of the former. A ring d and brush c similarly connect coils B and D. The opposite terminals of the field-coils connect to one binding post h of the motor, and those of the armature coils are similarly connected to the opposite binding post i through a contact-ring f and brush g. Thus each motor-circuit while in derivation to the other includes one armature and one field coil. These circuits are of different self-induction, and may be made so in various ways. For the sake of clearness, an artificial resistance R is shown in one of these circuits, and in the other a self-induction coil s. When an alternating current is passed through this motor it divides between its two energizing-circuits. The higher self-induction of one circuit produces a greater retardation or lag in the current therein than in the other. The difference of phase between the two currents effects the rotation or shifting of the points of maximum magnetic effect that secures

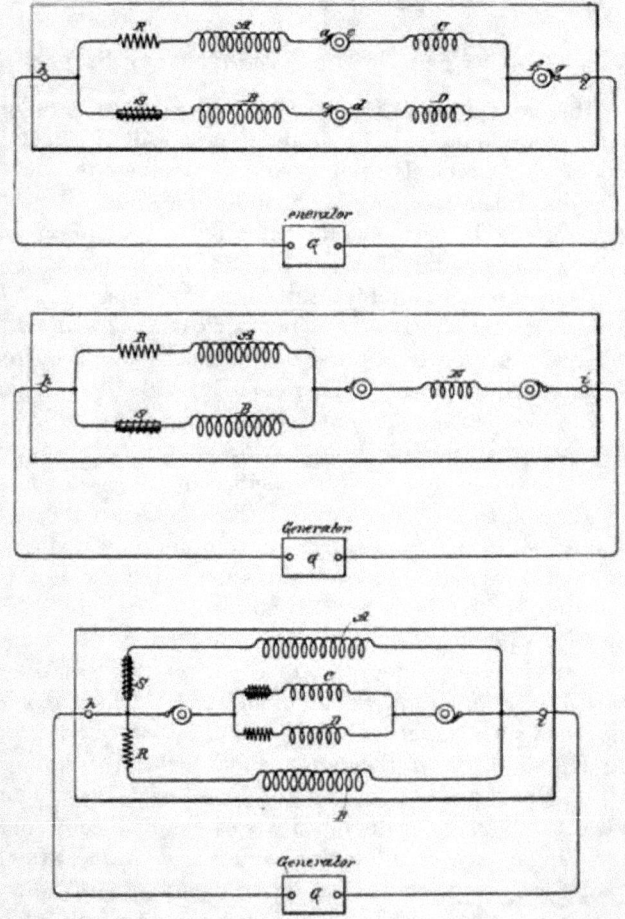

Figs. 42, 43 and 44.

the rotation of the armature. In certain respects this plan of including both armature and field coils in circuit is a marked improvement. Such a motor has a good torque at starting; yet it has also considerable tendency to synchronism, owing to the fact

that when properly constructed the maximum magnetic effects in
both armature and field coincide—a condition which in the usual
construction of these motors with closed armature coils is not
readily attained.   The motor thus constructed exhibits too, a
better regulation of current from no load to load, and there is
less difference between the apparent and real energy expended
in running it.   The true synchronous speed of this form of motor
is that of the generator when both are alike—that is to say, if
the number of the coils on the armature and on the field is x, the
motor will run normally at the same speed as a generator driving

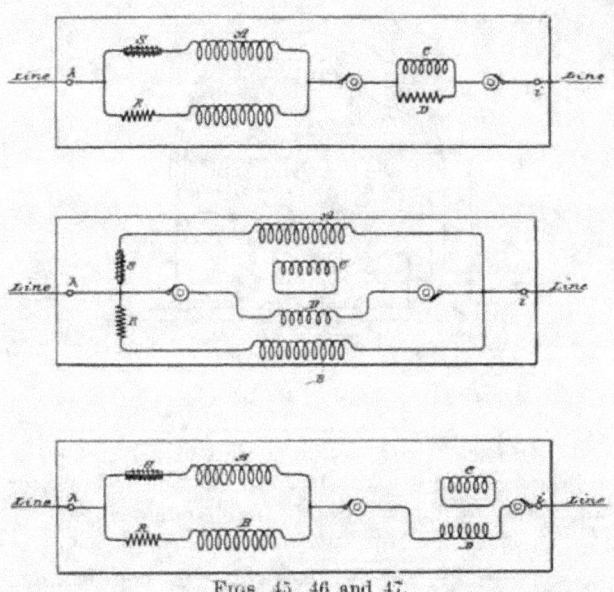

Figs. 45, 46 and 47.

it if the number of field magnets or poles of the same be also x.

Fig. 43 shows a somewhat modified arrangement of circuits.
There is in this case but one armature coil ᴇ, the winding of
which maintains effects corresponding to the resultant poles produced by the two field-circuits.

Fig. 44 represents a disposition in which both armature and
field are wound with two sets of coils, all in multiple arc to the
line or main circuit.   The armature coils are wound to correspond with the field-coils with respect to their self-induction.   A
modification of this plan is shown in Fig. 45—that is to say, the

two field coils and two armature coils are in derivation to themselves and in series with one another. The armature coils in this case, as in the previous figure, are wound for different self-induction to correspond with the field coils.

Another modification is shown in Fig. 46. In this case only one armature-coil, as D, is included in the line-circuit, while the other, as C, is short-circuited.

In such a disposition as that shown in Fig. 43, or where only one armature-coil is employed, the torque on the start is somewhat reduced, while the tendency to synchronism is somewhat

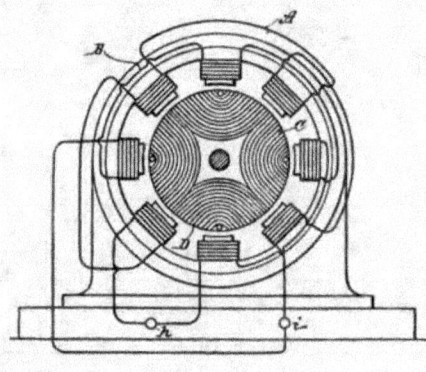

FIG. 48.

increased. In such a disposition as shown in Fig. 46, the opposite conditions would exist. In both instances, however, there is the advantage of dispensing with one contact-ring.

In Fig. 46 the two field-coils and the armature-coil D are in multiple arc. In Fig. 47 this disposition is modified, coil D being shown in series with the two field-coils.

Fig. 48 is an outline of the general form of motor in which this invention is embodied. The circuit connections between the armature and field coils are made, as indicated in the previous figures, through brushes and rings, which are not shown.

CHAPTER XI.

ANOTHER METHOD OF TRANSFORMATION FROM A TORQUE TO A SYNCHRONIZING MOTOR.

IN a preceding chapter we have described a method by which Mr. Tesla accomplishes the change in his type of rotating field motor from a torque to a synchronizing motor. As will be observed, the desired end is there reached by a change in the circuit connections at the proper moment. We will now proceed to describe another way of bringing about the same result. The principle involved in this method is as follows:—

If an alternating current be passed through the field coils only of a motor having two energizing circuits of different self-induction and the armature coils be short-circuited, the motor will have a strong torque, but little or no tendency to synchronism with the generator; but if the same current which energizes the field be passed also through the armature coils the tendency to remain in synchronism is very considerably increased. This is due to the fact that the maximum magnetic effects produced in the field and armature more nearly coincide. On this principle Mr. Tesla constructs a motor having independent field circuits of different self-induction, which are joined in derivation to a source of alternating currents. The armature is wound with one or more coils, which are connected with the field coils through contact rings and brushes, and around the armature coils a shunt is arranged with means for opening or closing the same. In starting this motor the shunt is closed around the armature coils, which will therefore be in closed circuit. When the current is directed through the motor, it divides between the two circuits, (it is not necessary to consider any case where there are more than two circuits used), which, by reason of their different self-induction, secure a difference of phase between the two currents in the two branches, that produces a shifting or rotation of the of the poles. By the alternations of current, other currents are induced in the closed—or short-circuited—armature coils and the motor has a strong torque. When the desired speed is reached, the shunt around the armature-coils is opened and the current directed through both armature and field coils. Under these conditions the motor has a strong tendency to synchronism.

In Fig. 49, A and B designate the field coils of the motor. As the circuits including these coils are of different self-induction, this is represented by a resistance coil R in circuit with A, and a

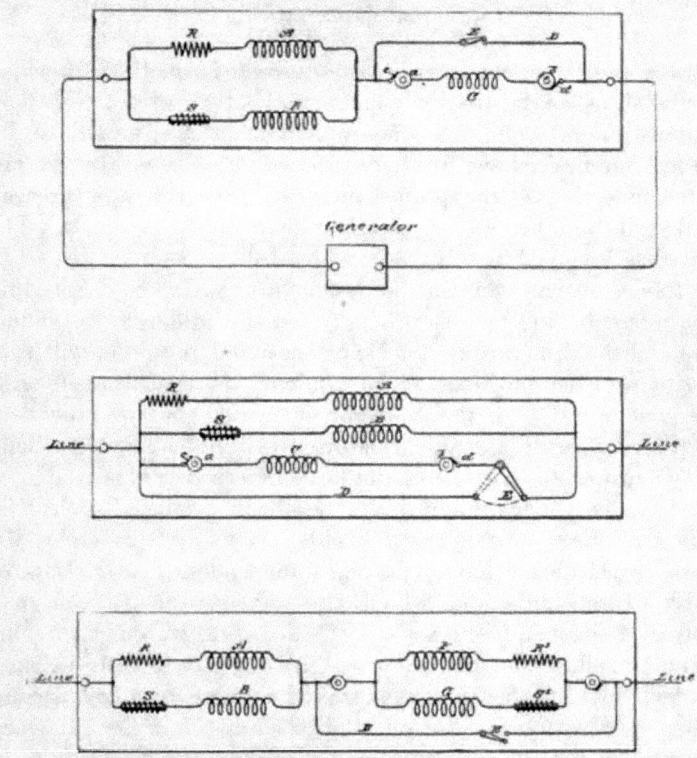

FIGS. 49, 50 and 51.

self-induction coil s in circuit with B. The same result may of course be secured by the winding of the coils. C is the armature circuit, the terminals of which are rings *a b*. Brushes *c d* bear on these rings and connect with the line and field circuits. D is the shunt or short circuit around the armature. E is the switch in the shunt.

It will be observed that in such a disposition as is illustrated in

Fig. 49, the field circuits A and B being of different self-induction, there will always be a greater lag of the current in one than the other, and that, generally, the armature phases will not correspond with either, but with the resultant of both. It is therefore important to observe the proper rule in winding the armature. For instance, if the motor have eight poles—four in each circuit —there will be four resultant poles, and hence the armature winding should be such as to produce four poles, in order to constitute a true synchronizing motor.

The diagram, Fig. 50, differs from the previous one only in respect to the order of connections. In the present case the armature-coil, instead of being in series with the field-coils, is in multiple arc therewith. The armature-winding may be similar to that of the field—that is to say, the armature may have two or more coils wound or adapted for different self-induction and

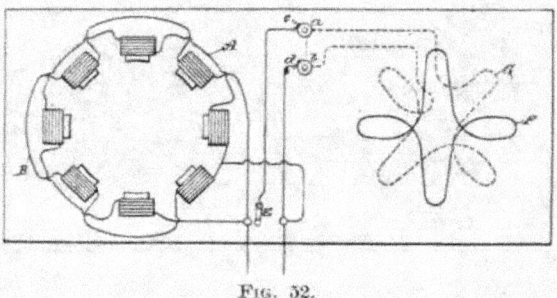

FIG. 52.

adapted, preferably, to produce the same difference of phase as the field-coils. On starting the motor the shunt is closed around both coils. This is shown in Fig. 51, in which the armature coils are F G. To indicate their different electrical.character, there are shown in circuit with them, respectively, the resistance R' and the self-induction coil S'. The two armature coils are in series with the field-coils and the same disposition of the shunt or short-circuit D is used. It is of advantage in the operation of motors of this kind to construct or wind the armature in such manner that when short-circuited on the start it will have a tendency to reach a higher speed than that which synchronizes with the generator. For example, a given motor having, say, eight poles should run, with the armature coil short-circuited, at two thousand revolutions per minute to bring it up to synchronism. It will generally happen, however, that this speed is not reached, owing to the fact that the armature and field currents do not properly correspond, so that when the current is passed through the armature (the motor not being quite up to synchronism) there is a liability that it will not "hold on," as it is termed. It is preferable, therefore, to so wind or construct the motor that on the start, when the armature coils are short-circuited, the motor will tend to reach a speed higher than the synchronous—as for instance, double the latter. In such case the difficulty above alluded to is not felt, for the motor will always hold up to synchronism if the synchronous speed—in the case supposed of two thousand revolutions—is reached or passed. This may be accomplished in various ways; but for all practical purposes the following will suffice: On the armature are wound two sets of coils. At the start only one of these is

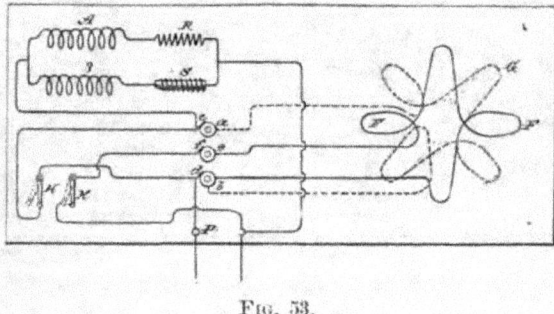

Fig. 53.

short-circuited, thereby producing a number of poles on the armature, which will tend to run the speed up above the synchronous limit. When such limit is reached or passed, the current is directed through the other coil, which, by increasing the number of armature poles, tends to maintain synchronism.

In Fig. 52, such a disposition is shown. The motor having, say, eight poles contains two field-circuits A and B, of different self-induction. The armature has two coils F and G. The former is closed upon itself, the latter connected with the field and line through contact-rings *a b*, brushes *c d*, and a switch E. On the start the coil F alone is active and the motor tends to run at a speed above the synchronous; but when the coil G is connected to the circuit the number of armature poles is increased, while the motor is made a true synchronous motor. This disposition has the advantage that the closed armature-circuit imparts to the motor torque when the speed falls off, but at the same time the conditions are such that the motor comes out of synchronism more readily. To increase the tendency to synchronism, two circuits may be used on the armature, one of which is short-circuited on the start and both connected with the external circuit after the synchronous speed is reached or passed. This disposition is shown in Fig. 53. There are three contact-rings *a b c* and three brushes *c d f*, which connect the armature circuits with the external circuit. On starting, the switch н is turned to complete the connection between one binding-post P and the field-coils. This short-circuits one of the armature-coils, as G. The other coil F is out of circuit and open. When the motor is up to speed, the switch н is turned back, so that the connection from binding-post P to the field coils is through the coil G, and switch к is closed, thereby including coil F in multiple arc with the field coils. Both armature coils are thus active.

From the above-described instances it is evident that many other dispositions for carrying out the invention are possible.

CHAPTER XII.

" Magnetic Lag " Motor.

The following description deals with another form of motor, namely, depending on " magnetic lag " or hysteresis, its peculiarity being that in it the attractive effects or phases while lagging behind the phases of current which produce them, are manifested simultaneously and not successively. The phenomenon utilized thus at an early stage by Mr. Tesla, was not generally believed in by scientific men, and Prof. Ayrton was probably first to advocate it or to elucidate the reason of its supposed existence.

Fig. 54 is a side view of the motor, in elevation. Fig. 55 is a part-sectional view at right angles to Fig. 54. Fig. 56 is an end view in elevation and part section of a modification, and Fig. 57 is a similar view of another modification.

In Figs. 54 and 55, A designates a base or stand, and B B the supporting-frame of the motor. Bolted to the supporting-frame are two magnetic cores or pole-pieces C C', of iron or soft steel. These may be subdivided or laminated, in which case hard iron or steel plates or bars should be used, or they should be wound with closed coils. D is a circular disc armature, built up of sections or plates of iron and mounted in the frame between the pole-pieces C C', curved to conform to the circular shape thereof. This disc may be wound with a number of closed coils E. F F are the main energizing coils, supported by the supporting-frame, so as to include within their magnetizing influence both the pole-pieces C C' and the armature D. The pole-pieces C C' project out beyond the coils F F on opposite sides, as indicated in the drawings. If an alternating current be passed through the coils F F, rotation of the armature will be produced, and this rotation is explained by the following apparent action, or mode of operation: An impulse of current in the coils F F establishes two polarities in the motor. The protruding end of pole-piece C, for instance, will be of one sign, and the corresponding end of pole-piece c′ will be of the opposite sign. The armature also exhibits two poles at right angles to the coils F F, like poles to those in the pole-pieces being on the same side of the coils. While the current is flowing there is no appreciable tendency to rotation developed; but after each current impulse ceases or begins to fall, the magnetism in the armature and in the ends of the pole-pieces c c′ lags or continues to manifest itself, which produces a rotation of the armature by the repellent force between the more closely approximating points of maximum magnetic effect. This effect is continued by the reversal of current, the polarities of field and armature being simply reversed. One or both of the elements—the armature or field—may be wound with

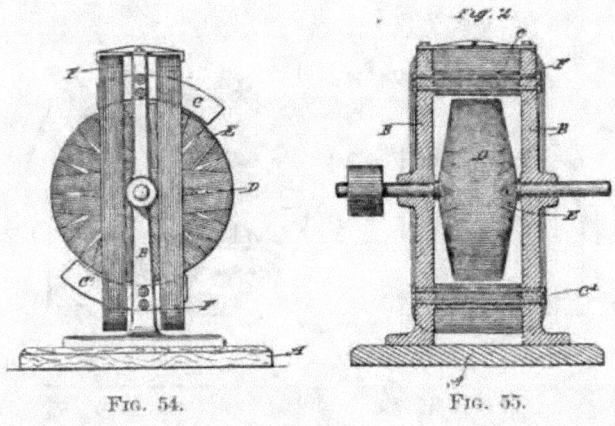

Fig. 2

Fig. 54.   Fig. 55.

closed induced coils to intensify this effect. Although in the illustrations but one of the fields is shown, each element of the motor really constitutes a field, wound with the closed coils, the currents being induced mainly in those convolutions or coils which are parallel to the coils F F.

A modified form of this motor is shown in Fig. 56. In this form G is one of two standards that support the bearings for the armature-shaft. H H are uprights or sides of a frame, preferably magnetic, the ends c c′ of which are bent in the manner indicated, to conform to the shape of the armature D and form field-magnet poles. The construction of the armature may be the same as in the previous figure, or it may be simply a magnetic disc or cylinder, as shown, and a coil or coils F F are se-

cured in position to surround both the armature and the poles
c c'. The armature is detachable from its shaft, the latter being
passed through the armature after it has been inserted in position. The operation of this form of motor is the same in principle as that previously described and needs no further explanation.

One of the most important features in alternating current
motors is, however, that they should be adapted to and capable
of running efficiently on the alternating circuits in present use,
in which almost without exception the generators yield a very
high number of alternations. Such a motor, of the type under
consideration, Mr. Tesla has designed by a development of the
principle of the motor shown in Fig. 56, making a multipolar
motor, which is illustrated in Fig. 57. In the construction of

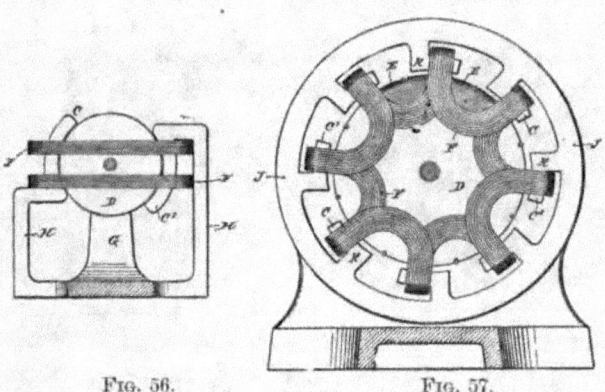

Fig. 56. Fig. 57.

this motor he employs an annular magnetic frame J, with in-
wardly-extending ribs or projections K, the ends of which all
bend or turn in one direction and are generally shaped to conform to the curved surface of the armature. Coils F F are wound
from one part K to the one next adjacent, the ends or loops of
each coil or group of wires being carried over toward the shaft,
so as to form U-shaped groups of convolutions at each end of the
armature. The pole-pieces C C', being substantially concentric
with the armature, form ledges, along which the coils are laid
and should project to some extent beyond the the coils, as shown.
The cylindrical or drum armature D is of the same construction
as in the other motors described, and is mounted to rotate within
the annular frame J and between the U-shaped ends or bends of

the coils F. The coils F are connected in multiple or in series with a source of alternating currents, and are so wound that with a current or current impulse of given direction they will make the alternate pole-pieces c of one polarity and the other pole-pieces c' of the opposite polarity. The principle of the operation of this motor is the same as the other above described, for, considering any two pole-pieces c c', a current impulse passing in the coil which bridges them or is wound over both tends to establish polarities in their ends of opposite sign and to set up in the armature core between them a polarity of the same sign as that of the nearest pole-piece c. Upon the fall or cessation of the current impulse that established these polarities the magnetism which lags behind the current phase, and which continues to manifest itself in the polar projections c c' and the armature, produces by repulsion a rotation of the armature. The effect is continued by each reversal of the current. What occurs in the case of one pair of pole-pieces occurs simultaneously in all, so that the tendency to rotation of the armature is measured by the sum of all the forces exerted by the pole-pieces, as above described. In this motor also the magnetic lag or effect is intensified by winding one or both cores with closed induced coils. The armature core is shown as thus wound. When closed coils are used, the cores should be laminated.

It is evident that a pulsatory as well as an alternating current might be used to drive or operate the motors above described.

It will be understood that the degree of subdivision, the mass of the iron in the cores, their size and the number of alternations in the current employed to run the motor, must be taken into consideration in order to properly construct this motor. In other words, in all such motors the proper relations between the number of alternations and the mass, size, or quality of the iron must be preserved in order to secure the best results.

CHAPTER XIII.

METHOD OF OBTAINING DIFFERENCE OF PHASE BY MAGNETIC SHIELDING.

In that class of motors in which two or more sets of energizing magnets are employed, and in which by artificial means a certain interval of time is made to elapse between the respective maximum or minimum periods or phases of their magnetic attraction or effect, the interval or difference in phase between the two sets of magnets is limited in extent. It is desirable, however, for the economical working of such motors that the strength or attraction of one set of magnets should be maximum, at the time when that of the other set is minimum, and conversely; but these conditions have not heretofore been realized except in cases where the two currents have been obtained from independent sources in the same or different machines. Mr. Tesla has therefore devised a motor embodying conditions that approach more nearly the theoretical requirements of perfect working, or in other words, he produces artificially a difference of magnetic phase by means of a current from a single primary source sufficient in extent to meet the requirements of practical and economical working. He employs a motor with two sets of energizing or field magnets, each wound with coils connected with a source of alternating or rapidly-varying currents, but forming two separate paths or circuits. The magnets of one set are protected to a certain extent from the energizing action of the current by means of a magnetic shield or screen interposed between the magnet and its energizing coil. This shield is properly adapted to the conditions of particular cases, so as to shield or protect the main core from magnetization until it has become itself saturated and no longer capable of containing all the lines of force produced by the current. It will be seen that by this means the energizing action begins in the protected set of magnets a certain arbitrarily-determined period of time later than in the other, and that by this means alone or in conjunction with other means or devices

heretofore employed a practical difference of magnetic phase
may readily be secured.

Fig. 58 is a view of a motor, partly in section, with a diagram illustrating the invention. Fig. 59 is a similar view of a
modification of the same.

In Fig. 58, which exhibits the simplest form of the invention,
A A is the field-magnet of a motor, having, say, eight poles or
inwardly-projecting cores B and C. The cores B form one set of
magnets and are energized by coils D. The cores C, forming
the other set are energized by coils E, and the coils are
connected, preferably, in series with one another, in two derived or branched circuits, F G, respectively, from a suitable
source of current. Each coil E is surrounded by a magnetic
shield H, which is preferably composed of an annealed, insulated,

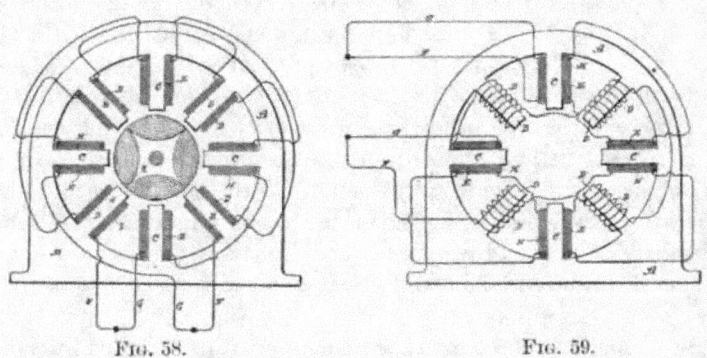

FIG. 58.   FIG. 59.

or oxidized iron wire wrapped or wound on the coils in the manner indicated so as to form a closed magnetic circuit around the
coils and between the same and the magnetic cores C. Between the pole pieces or cores B C is mounted the armature K,
which, as is usual in this type of machines, is wound with coils
L closed upon themselves. The operation resulting from this
disposition is as follows: If a current impulse be directed
through the two circuits of the motor, it will quickly energize
the cores B, but not so the cores C, for the reason that in
passing through the coils E there is encountered the influence
of the closed magnetic circuits formed by the shields H. The
first effect is to retard effectively the current impulse in circuit
G, while at the same time the proportion of current which does
pass does not magnetize the cores C, which are shielded or

screened by the shields H. As the increasing electromotive force then urges more current through the coils E, the iron wire H becomes magnetically saturated and incapable of carrying all the lines of force, and hence ceases to protect the cores c, which becomes magnetized, developing their maximum effect after an interval of time subsequent to the similar manifestation of strength in the other set of magnets, the extent of which is arbitrarily determined by the thickness of the shield H, and other well-understood conditions.

From the above it will be seen that the apparatus or device acts in two ways. First, by retarding the current, and, second, by retarding the magnetization of one set of the cores, from which its effectiveness will readily appear.

Many modifications of the principle of this invention are possible. One useful and efficient application of the invention is shown in Fig. 59. In this figure a motor is shown similar in all respects to that above described, except that the iron wire H, which is wrapped around the coils E, is in this case connected in series with the coils D. The iron-wire coils H, are connected and wound, so as to have little or no self-induction, and being added to the resistance of the circuit F, the action of the current in that circuit will be accelerated, while in the other circuit G it will be retarded. The shield H may be made in many forms, as will be understood, and used in different ways, as appears from the foregoing description.

As a modification of his type of motor with "shielded" fields, Mr. Tesla has constructed a motor with a field-magnet having two sets of poles or inwardly-projecting cores and placed side by side, so as practically to form two fields of force and alternately disposed—that is to say, with the poles of one set or field opposite the spaces between the other. He then connects the free ends of one set of poles by means of laminated iron bands or bridge-pieces of considerably smaller cross-section than the cores themselves, whereby the cores will all form parts of complete magnetic circuits. When the coils on each set of magnets are connected in multiple circuits or branches from a source of alternating currents, electromotive forces are set up in or impressed upon each circuit simultaneously; but the coils on the magnetically bridged or shunted cores will have, by reason of the closed magnetic circuits, a high self-induction, which retards the current, permitting at the beginning of each impulse but lit-

tle current to pass. On the other hand, no such opposition being
encountered in the other set of coils, the current passes freely
through them, magnetizing the poles on which they are wound.
As soon, however, as the laminated bridges become saturated
and incapable of carrying all the lines of force which the rising
electromotive force, and consequently increased current, produce, free poles are developed at the ends of the cores, which,
acting in conjunction with the others, produce rotation of the
armature.

The construction in detail by which this invention is illustrated
is shown in the accompanying drawings.

Fig. 60 is a view in side elevation of a motor embodying the
principle. Fig. 61 is a vertical cross-section of the motor. A is
the frame of the motor, which should be built up of sheets of
iron punched out to the desired shape and bolted together with

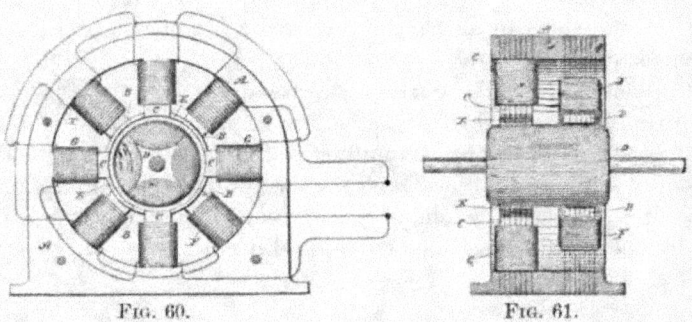

FIG. 60.   FIG. 61.

insulation between the sheets. When complete, the frame makes
a field-magnet with inwardly projecting pole-pieces B and C. To
adapt them to the requirements of this particular case these pole-
pieces are out of line with one another, those marked B surrounding one end of the armature and the others, as C, the opposite
end, and they are disposed alternately—that is to say, the pole-
pieces of one set occur in line with the spaces between those of the
other sets.

The armature D is of cylindrical form, and is also laminated in
the usual way and is wound longitudinally with coils closed upon
themselves. The pole-pieces C are connected or shunted by
bridge-pieces E. These may be made independently and attached
to the pole-pieces, or they may be parts of the forms or blanks
stamped or punched out of sheet-iron. Their size or mass is de-

termined by various conditions, such as the strength of the current to be employed, the mass or size of the cores to which they are applied, and other familiar conditions.

Coils F surround the pole-pieces B, and other coils G are wound on the pole-pieces c. These coils are connected in series in two circuits, which are branches of a circuit from a generator of alternating currents, and they may be so wound, or the respective circuits in which they are included may be so arranged, that the circuit of coils G will have, independently of the particular construction described, a higher self-induction than the other circuit or branch.

The function of the shunts or bridges E is that they shall form with the cores c a closed magnetic circuit for a current up to a predetermined strength, so that when saturated by such current and unable to carry more lines of force than such a current produces they will to no further appreciable extent interfere with the development, by a stronger current, of free magnetic poles at the ends of the cores c.

In such a motor the current is so retarded in the coils G, and the manifestation of the free magnetism in the poles c is so delayed beyond the period of maximum magnetic effect in poles B, that a strong torque is produced and the motor operates with approximately the power developed in a motor of this kind energized by independently generated currents differing by a full quarter phase.

CHAPTER XIV.

TYPE OF TESLA SINGLE-PHASE MOTOR.

UP TO this point, two principal types of Tesla motors have been described : First, those containing two or more energizing circuits through which are caused to pass alternating currents differing from one another in phase to an extent sufficient to produce a continuous progression or shifting of the poles or points of greatest magnetic effect, in obedience to which the movable element of the motor is maintained in rotation ; second, those containing poles, or parts of different magnetic susceptibility, which under the energizing influence of the same current or two currents coinciding in phase will exhibit differences in their magnetic periods or phases. In the first class of motors the torque is due to the magnetism established in different portions of the motor by currents from the same or from independent sources, and exhibiting time differences in phase. In the second class the torque results from the energizing effects of a current upon different parts of the motor which differ in magnetic susceptibility—in other words, parts which respond in the same relative degree to the action of a current, not simultaneously, but after different intervals of time.

In another Tesla motor, however, the torque, instead of being solely the result of a time difference in the magnetic periods or phases of the poles or attractive parts to whatever cause due, is produced by an angular displacement of the parts which, though movable with respect to one another, are magnetized simultaneously, or approximately so, by the same currents. This principle of operation has been embodied practically in a motor in which the necessary angular displacement between the points of greatest magnetic attraction in the two elements of the motor—the armature and field—is obtained by the direction of the lamination of the magnetic cores of the elements.

Fig. 62 is a side view of such a motor with a portion of its armature core exposed. Fig. 63 is an end or edge view of the same. Fig. 64 is a central cross-section of the same, the armature being shown mainly in elevation.

Let A A designate two plates built up of thin sections or laminæ of soft iron insulated more or less from one another and held together by bolts a and secured to a base B. The inner faces of these plates contain recesses or grooves in which a coil or coils D are secured obliquely to the direction of the laminations. Within the coils D is a disc E, preferably composed of a spirally-wound iron wire or ribbon or a series of concentric rings and mounted on a shaft F, having bearings in the plates A A. Such a device when acted upon by an alternating current is capable of rotation and constitutes a motor, the operation of which may be explained in the following manner: A current or current-impulse traversing the coils D tends to magnetize the

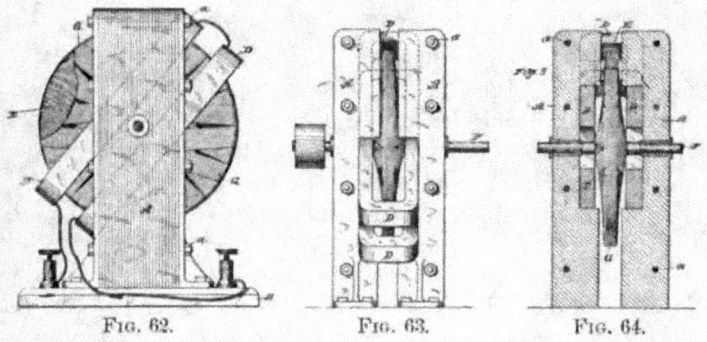

Fig. 62.   Fig. 63.   Fig. 64.

cores A A and E, all of which are within the influence of the field of the coils. The poles thus established would naturally lie in the same line at right angles to the coils D, but in the plates A they are deflected by reason of the direction of the laminations, and appear at or near the extremities of these plates. In the disc, however, where these conditions are not present, the poles or points of greatest attraction are on a line at right angles to the plane of the coils; hence there will be a torque established by this angular displacement of the poles or magnetic lines, which starts the disc in rotation, the magnetic lines of the armature and field tending toward a position of parallelism. This rotation is continued and maintained by the reversals of the current in coils D D, which change alternately the polarity of the field-cores A A. This rotary tendency or effect will be greatly increased by winding the disc with conductors G, closed upon themselves and having a radial direction, whereby the magnetic intensity of the poles of the disc will be greatly increased by the energizing effect of the currents induced in the coils G by the alternating currents in coils D.

The cores of the disc and field may or may not be of different magnetic susceptibility—that is to say, they may both be of the same kind of iron, so as to be magnetized at approximately the same instant by the coils D; or one may be of soft iron and the other of hard, in order that a certain time may elapse between the periods of their magnetization. In either case rotation will be produced; but unless the disc is provided with the closed energizing coils it is desirable that the above-described difference of magnetic susceptibility be utilized to assist in its rotation.

The cores of the field and armature may be made in various ways, as will be well understood, it being only requisite that the laminations in each be in such direction as to secure the necessary angular displacement of the points of greatest attraction. Moreover, since the disc may be considered as made up of an infinite number of radial arms, it is obvious that what is true of a disc holds for many other forms of armature.

CHAPTER XV.

MOTORS WITH CIRCUITS OF DIFFERENT RESISTANCE.

As has been pointed out elsewhere, the lag or retardation of the phases of an alternating current is directly proportional to the self-induction and inversely proportional to the resistance of the circuit through which the current flows. Hence, in order to secure the proper differences of phase between the two motor-circuits, it is desirable to make the self-induction in one much higher and the resistance much lower than the self-induction and resistance, respectively, in the other. At the same time the magnetic quantities of the two poles or sets of poles which the two circuits produce should be approximately equal. These requirements have led Mr. Tesla to the invention of a motor having the following general characteristics: The coils which are included in that energizing circuit which is to have the higher self-induction are made of coarse wire, or a conductor of relatively low resistance, and with the greatest possible length or number of turns. In the other set of coils a comparatively few turns of finer wire are used, or a wire of higher resistance. Furthermore, in order to approximate the magnetic quantities of the poles excited by these coils, Mr. Tesla employs in the self-induction circuit cores much longer than those in the other or resistance circuit.

Fig. 65 is a part sectional view of the motor at right angles to the shaft. Fig. 66 is a diagram of the field circuits.

In Fig. 66, let A represent the coils in one motor circuit, and B those in the other. The circuit A is to have the higher self-induction. There are, therefore, used a long length or a large number of turns of coarse wire in forming the coils of this circuit. For the circuit B, a smaller conductor is employed, or a conductor of a higher resistance than copper, such as German silver or iron, and the coils are wound with fewer turns. In applying these coils to a motor, Mr. Tesla builds up a field-magnet of plates C, of iron and steel, secured together in the usual manner

by bolts D.   Each plate is formed with four (more or less) long
cores E, around which is a space to receive the coil and an equal
number of short projections F to receive the coils of the resistance-
circuit.   The plates are generally annular in shape, having an
open space in the centre for receiving the armature G, which Mr.
Tesla prefers to wind with closed coils.   An alternating current
divided between the two circuits is retarded as to its phases in
the circuit A to a much greater extent than in the circuit B.   By

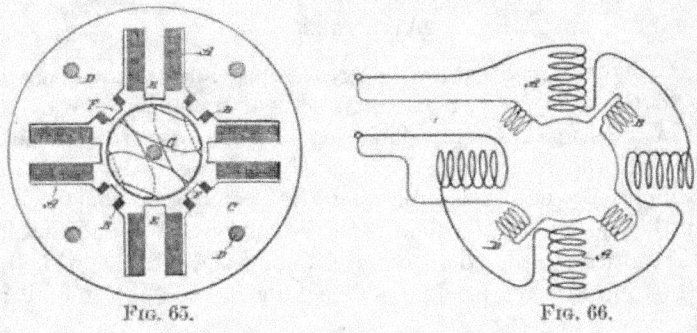

FIG. 65.   FIG. 66.

reason of the relative sizes and disposition of the cores and coils
the magnetic effect of the poles E and F upon the armature closely
approximate.

An important result secured by the construction shown here
is that these coils which are designed to have the higher self-
induction are almost completely surrounded by iron, and that the
retardation is thus very materially increased.

CHAPTER XVI.

MOTOR WITH EQUAL MAGNETIC ENERGIES IN FIELD AND ARMATURE.

LET it be assumed that the energy as represented in the magnetism in the field of a given rotating field motor is ninety and that of the armature ten. The sum of these quantities, which represents the total energy expended in driving the motor, is one hundred; but, assuming that the motor be so constructed that the energy in the field is represented by fifty, and that in the armature by fifty, the sum is still one hundred; but while in the first instance the product is nine hundred, in the second it is

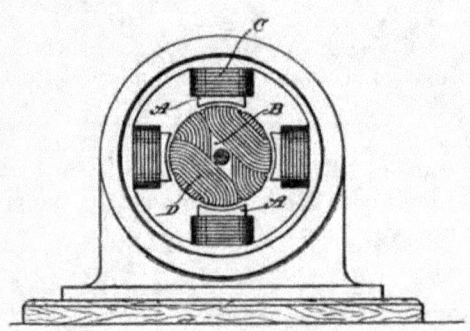

FIG. 67.

two thousand five hundred, and as the energy developed is in proportion to these products it is clear that those motors are the most efficient—other things being equal—in which the magnetic energies developed in the armature and field are equal. These results Mr. Tesla obtains by using the same amount of copper or ampere turns in both elements when the cores of both are equal, or approximately so, and the same current energizes both; or in cases where the currents in one element are induced to those of the other he uses in the induced coils an excess of copper over that in the primary element or conductor.

The conventional figure of a motor here introduced, Fig. 67, will give an idea of the solution furnished by Mr. Tesla for the specific problem. Referring to the drawing, A is the field-magnet, B the armature, C the field coils, and D the armature-coils of the motor.

Generally speaking, if the mass of the cores of armature and field be equal, the amount of copper or ampere turns of the energizing coils on both should also be equal; but these conditions will be modified in different forms of machine. It will be understood that these results are most advantageous when existing under the conditions presented where the motor is running with its normal load, a point to be well borne in mind.

CHAPTER XVII.

MOTORS WITH COINCIDING MAXIMA OF MAGNETIC EFFECT IN ARMATURE AND FIELD.

IN THIS form of motor, Mr. Tesla's object is to design and build machines wherein the maxima of the magnetic effects of the armature and field will more nearly coincide than in some of the types previously under consideration. These types are: First, motors having two or more energizing circuits of the same electrical character, and in the operation of which the currents used differ primarily in phase; second, motors with a plurality of energizing circuits of different electrical character, in or by means of which the difference of phase is produced artificially, and, third, motors with a plurality of energizing circuits, the currents in one being induced from currents in another. Considering the structural and operative conditions of any one of them—as, for example, that first named—the armature which is mounted to rotate in obedience to the co-operative influence or action of the energizing circuits has coils wound upon it which are closed upon themselves and in which currents are induced by the energizing-currents with the object and result of energizing the armature-core; but under any such conditions as must exist in these motors, it is obvious that a certain time must elapse between the manifestations of an energizing current impulse in the field coils, and the corresponding magnetic state or phase in the armature established by the current induced thereby; consequently a given magnetic influence or effect in the field which is the direct result of a primary current impulse will have become more or less weakened or lost before the corresponding effect in the armature indirectly produced has reached its maximum. This is a condition unfavorable to efficient working in certain cases—as, for instance, when the progress of the resultant poles or points of maximum attraction is very great, or when a very high number of alternations is employed—for it is apparent that a stronger tendency to rotation will be maintained if the maximum magnetic attractions or conditions in both armature and field coincide, the energy developed by a motor being measured by the product of the magnetic quantities of the armature and field.

To secure this coincidence of maximum magnetic effects, Mr. Tesla has devised various means, as explained below. Fig. 68 is a diagrammatic illustration of a Tesla motor system in which the alternating currents proceed from independent sources and differ primarily in phase.

A designates the field-magnet or magnetic frame of the motor;

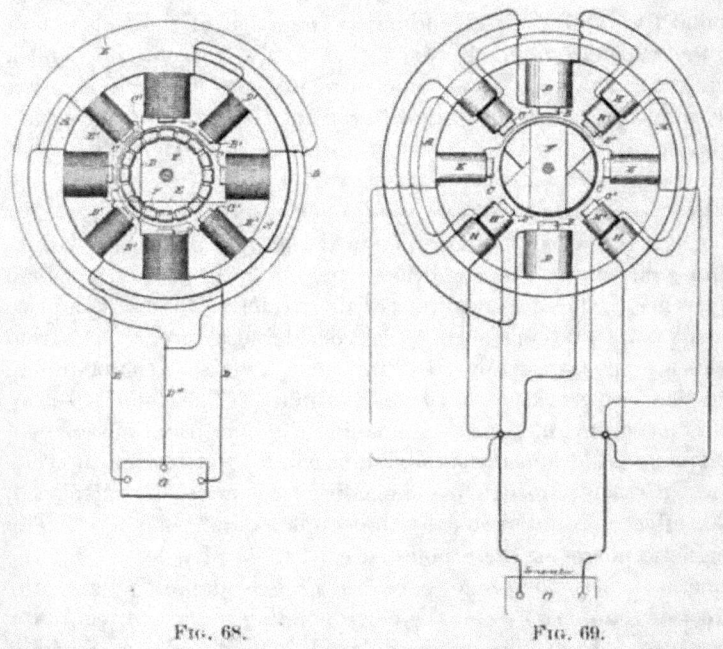

FIG. 68.   FIG. 69.

B B, oppositely located pole-pieces adapted to receive the coils of one energizing circuit; and c c, similar pole-pieces for the coils of the other energizing circuit. These circuits are designated, respectively, by D E, the conductor D″ forming a common return to the generator G. Between these poles is mounted an armature —for example, a ring or annular armature, wound with a series of coils F, forming a closed circuit or circuits. The action or operation of a motor thus constructed is now well understood. It will be observed, however, that the magnetism of poles B, for example, established by a current impulse in the coils thereon, precedes the magnetic effect set up in the armature by the induced current in coils F. Consequently the mutual attraction between the armature and field-poles is considerably reduced. The same conditions will be found to exist if, instead of assuming the poles B or C as acting independently, we regard the ideal resultant of both acting together, which is the real condition. To remedy this, the motor field is constructed with secondary poles B' C', which are situated between the others. These pole-pieces are wound with coils D' E', the former in derivation to the coils D, the latter to coils E. The main or primary coils D and E are wound for a different self-induction from that of the coils D' and E', the relations being so fixed that if the currents in D and E differ, for example, by a quarter-phase, the currents in each secondary coil, as D' E', will differ from those in its appropriate primary D or E by, say, forty-five degrees, or one-eighth of a period.

Now, assuming that an impulse or alternation in circuit or branch E is just beginning, while in the branch D it is just falling from maximum, the conditions are those of a quarter-phase difference. The ideal resultant of the attractive forces of the two sets of poles B C therefore may be considered as progressing from poles B to poles C, while the impulse in E is rising to maximum, and that in D is falling to zero or minimum. The polarity set up in the armature, however, lags behind the manifestations of field magnetism, and hence the maximum points of attraction in armature and field, instead of coinciding, are angularly displaced. This effect is counteracted by the supplemental poles B' C'. The magnetic phases of these poles succeed those of poles B C by the same, or nearly the same, period of time as elapses between the effect of the poles B C and the corresponding induced effect in the armature; hence the magnetic conditions of poles B' C' and of the armature more nearly coincide and a better result is obtained. As poles B' C' act in conjunction with the poles in the armature established by poles B C, so in turn poles C B act similarly with the poles set up by B' C', respectively. Under such conditions the retardation of the magnetic effect of the armature and that of the secondary poles will bring the maximum of the two more nearly into coincidence and a correspondingly stronger torque or magnetic attraction secured.

In such a disposition as is shown in Fig. 68 it will be observed

that as the adjacent pole-pieces of either circuit are of like polarity they will have a certain weakening effect upon one another.
Mr. Tesla therefore prefers to remove the secondary poles from
the direct influence of the others. This may be done by constructing a motor with two independent sets of fields, and with
either one or two armatures electrically connected, or by using
two armatures and one field. These modifications are illustrated
further on.

Fig. 69 is a diagrammatic illustration of a motor and system in
which the difference of phase is artificially produced. There are
two coils D D in one branch and two coils E E in another branch

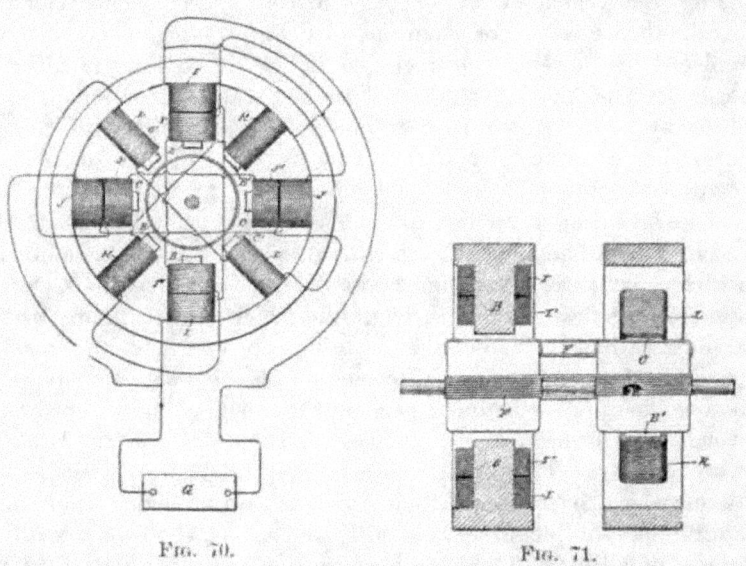

Fig. 70.   Fig. 71.

of the main circuit from the generator G. These two circuits or
branches are of different self-induction, one, as D, being higher
than the other. This is graphically indicated by making coils D
much larger than coils E. By reason of the difference in the
electrical character of the two circuits, the phases of current in
one are retarded to a greater extent than the other. Let this
difference be thirty degrees. A motor thus constructed will
rotate under the action of an alternating current; but as happens
in the case previously described the corresponding magnetic effects of the armature and field do not coincide owing to the time
that elapses between a given magnetic effect in the armature and

the condition of the field that produces it. The secondary or supplemental poles B′ c′ are therefore availed of. There being thirty degrees difference of phase between the currents in coils D E, the magnetic effect of poles B′ c′ should correspond to that produced by a current differing from the current in coils D or E by fifteen degrees. This we can attain by winding each supplemental pole B′ c′ with two coils H H′. The coils H are included in a derived circuit having the same self-induction as circuit D, and coils H′ in a circuit having the same self-induction as circuit E, so that if these circuits differ by thirty degrees the magnetism of poles B′ c′ will correspond to that produced by a current differing from that in either D or E by fifteen degrees. This is true in all other cases. For example, if in Fig. 68 the coils D′ E′ be replaced by the coils H H′ included in the derived circuits, the magnetism of the poles B′ c′ will correspond in effect or phase, if it may be so termed, to that produced by a current differing from that in either circuit D or E by forty-five degrees, or one-eighth of a period.

This invention as applied to a derived circuit motor is illustrated in Figs. 70 and 71. The former is an end view of the motor with the armature in section and a diagram of connections, and Fig. 71 a vertical section through the field. These figures are also drawn to show one of the dispositions of two fields that may be adopted in carrying out the principle. The poles B B C C are in one field, the remaining poles in the other. The former are wound with primary coils I J and secondary coils I′ J′, the latter with coils K L. The primary coils I J are in derived circuits, between which, by reason of their different self-induction, there is a difference of phase, say, of thirty degrees. The coils I′ K are in circuit with one another, as also are coils J′ L, and there should be a difference of phase between the currents in coils K and L and their corresponding primaries of, say, fifteen degrees. If the poles B C are at right angles, the armature-coils should be connected directly across, or a single armature core wound from end to end may be used; but if the poles B C be in line there should be an angular displacement of the armature coils, as will be well understood.

The operation will be understood from the foregoing. The maximum magnetic condition of a pair of poles, as B′ B′, coincides closely with the maximum effect in the armature, which lags behind the corresponding condition in poles B B.

CHAPTER XVIII.

MOTOR BASED ON THE DIFFERENCE OF PHASE IN THE MAGNETIZATION OF THE INNER AND OUTER PARTS OF AN IRON CORE.

IT IS well known that if a magnetic core, even if laminated or subdivided, be wound with an insulated coil and a current of electricity be directed through the coil, the magnetization of the entire core does not immediately ensue, the magnetizing effect not being exhibited in all parts simultaneously. This may be attributed to the fact that the action of the current is to energize first those laminæ or parts of the core nearest the surface and adjacent to the exciting-coil, and from thence the action progresses toward the interior. A certain interval of time therefore elapses between the manifestation of magnetism in the external and the internal sections or layers of the core. If the core be thin or of small mass, this effect may be inappreciable; but in the case of a thick core, or even of a comparatively thin one, if the number of alternations or rate of change of the current strength be very great, the time interval occurring between the manifestations of magnetism in the interior of the core and in those parts adjacent to the coil is more marked. In the construction of such apparatus as motors which are designed to be run by alternating or equivalent currents—such as pulsating or undulating currents generally—Mr. Tesla found it desirable and even necessary to give due consideration to this phenomenon and to make special provisions in order to obviate its consequences. With the specific object of taking advantage of this action or effect, and to render it more pronounced, he constructs a field magnet in which the parts of the core or cores that exhibit at different intervals of time the magnetic effect imparted to them by alternating or equivalent currents in an energizing coil or coils, are so placed with relation to a rotating armature as to exert thereon their attractive effect successively in the order of their magnetization. By this means he secures a result similar to that which he had previously attained in other forms or types of mo-

tor in which by means of one or more alternating currents he
has produced the rotation or progression of the magnetic poles.

This new mode of operation will now be described. Fig. 72
is a side elevation of such motor. Fig. 73 is a side elevation of
a more practicable and efficient embodiment of the invention.
Fig. 74 is a central vertical section of the same in the plane of
the axis of rotation.

Referring to Fig. 72, let x represent a large iron core, which
may be composed of a number of sheets or laminæ of soft iron
or steel. Surrounding this core is a coil y, which is connected
with a source E of rapidly varying currents. Let us consider now

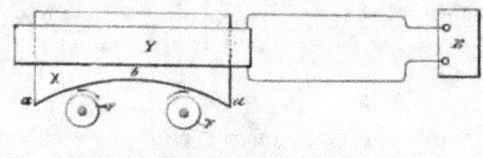

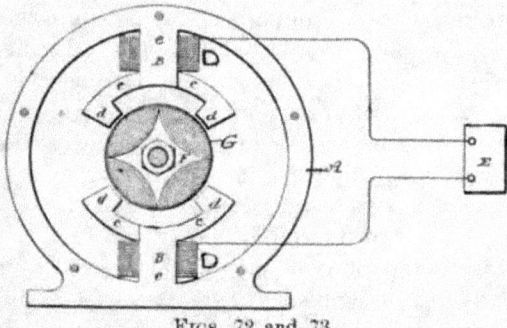

Figs. 72 and 73.

the magnetic conditions existing in this core at any point, as *b*,
at or near the centre, and any other point, as *a*, nearer the surface. When a current impulse is started in the magnetizing coil
y, the section or part at *a*, being close to the coil, is immediately
energized, while the section or part at *b*, which, to use a convenient expression, is "protected" by the intervening sections or
layers between *a* and *b*, does not at once exhibit its magnetism.
However, as the magnetization of *a* increases, *b* becomes also
affected, reaching finally its maximum strength some time later
than *a*. Upon the weakening of the current the magnetization
of *a* first diminishes, while *b* still exhibits its maximum strength;

but the continued weakening of a is attended by a subsequent weakening of b. Assuming the current to be an alternating one, a will now be reversed, while b still continues of the first imparted polarity. This action continues the magnetic condition of b, following that of a in the manner above described. If an armature —for instance, a simple disc F, mounted to rotate freely on an axis—be brought into proximity to the core, a movement of rotation will be imparted to the disc, the direction depending upon its position relatively to the core, the tendency being to turn the portion of the disc nearest to the core from a to b, as indicated in Fig. 72.

This action or principle of operation has been embodied in a practicable form of motor, which is illustrated in Fig. 73. Let A

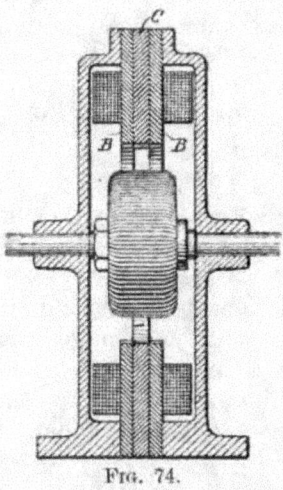

FIG. 74.

in that figure represent a circular frame of iron, from diametrically opposite points of the interior of which the cores project. Each core is composed of three main parts B, B and C, and they are similarly formed with a straight portion or body e, around which the energizing coil is wound, a curved arm or extension c, and an inwardly projecting pole or end d. Each core is made up of two parts B B, with their polar extensions reaching in one direction, and a part C between the other two, and with its polar extension reaching in the opposite direction. In order to lessen in the cores the circulation of currents induced therein, the several sections are insulated from one another in the manner usually followed in such cases. These cores are wound with coils D, which are connected in the same circuit, either in parallel or series, and supplied with an alternating or a pulsating current, preferably the former, by a generator E, represented diagrammatically. Between the cores or their polar extensions is mounted a cylindrical or similar armature F, wound with magnetizing coils G, closed upon themselves.

The operation of this motor is as follows: When a current impulse or alternation is directed through the coils D, the sections B B of the cores, being on the surface and in close proximity to the coils, are immediately energized. The sections c, on the other hand, are protected from the magnetizing influence of the coil by the interposed layers of iron B B. As the magnetism of B B increases, however, the sections c are also energized; but they do not attain their maximum strength until a certain time subsequent to the exhibition by the sections B B of their maximum. Upon the weakening of the current the magnetic strength of B B first diminishes, while the sections c have still their maximum strength; but as B B continue to weaken the interior sections are similarly weakened. B B may then begin to exhibit an opposite polarity, which is followed later by a similar change on c, and this action continues. B B and c may therefore be considered as separate field-magnets, being extended so as to act on the armature in the most efficient positions, and the effect is similar to that in the other forms of Tesla motor—viz., a rotation or progression of the maximum points of the field of force. Any armature—such, for instance, as a disc—mounted in this field would rotate from the pole first to exhibit its magnetism to that which exhibits it later.

It is evident that the principle here described may be carried out in conjunction with other means for securing a more favorable or efficient action of the motor. For example, the polar extensions of the sections c may be wound or surrounded by closed coils. The effect of these coils will be to still more effectively retard the magnetization of the polar extensions of c.

CHAPTER XIX.

ANOTHER TYPE OF TESLA INDUCTION MOTOR.

IT WILL have been gathered by all who are interested in the advance of the electrical arts, and who follow carefully, step by step, the work of pioneers, that Mr. Tesla has been foremost to utilize inductive effects in permanently closed circuits, in the operation of alternating motors. In this chapter one simple type of such a motor is described and illustrated, which will serve as an exemplification of the principle.

Let it be assumed that an ordinary alternating current generator is connected up in a circuit of practically no self-induction, such, for example, as a circuit containing incandescent lamps only. On the operation of the machine, alternating currents will be developed in the circuit, and the phases of these currents will theoretically coincide with the phases of the impressed electromotive force. Such currents may be regarded and designated as the "unretarded currents."

It will be understood, of course, that in practice there is always more or less self-induction in the circuit, which modifies to a corresponding extent these conditions; but for convenience this may be disregarded in the consideration of the principle of operation, since the same laws apply. Assume next that a path of currents be formed across any two points of the above circuit, consisting, for example, of the primary of an induction device. The phases of the currents passing through the primary, owing to the self-induction of the same, will not coincide with the phases of the impressed electromotive force, but will lag behind, such lag being directly proportional to the self-induction and inversely proportional to the resistance of the said coil. The insertion of this coil will also cause a lagging or retardation of the currents traversing and delivered by the generator behind the impressed electromotive force, such lag being the mean or resultant of the lag of the current through the primary alone and of the "unretarded current" in the entire working circuit. Next consider the conditions imposed by the association in inductive relation with the primary coil, of a secondary coil. The current generated in the secondary coil will react upon the primary current, modifying the retardation of the same, according to the amount of self-induction and resistance in the secondary circuit. If the secondary circuit has but little self-induction—as, for instance, when it contains incandescent lamps only—it will increase the actual difference of phase between its own and the primary current, first, by diminishing the lag between the primary current and the impressed electromotive force, and, second, by its own lag or retardation behind the impressed electromotive force. On the other hand, if the secondary circuit have a high self-induction, its lag behind the current in the primary is

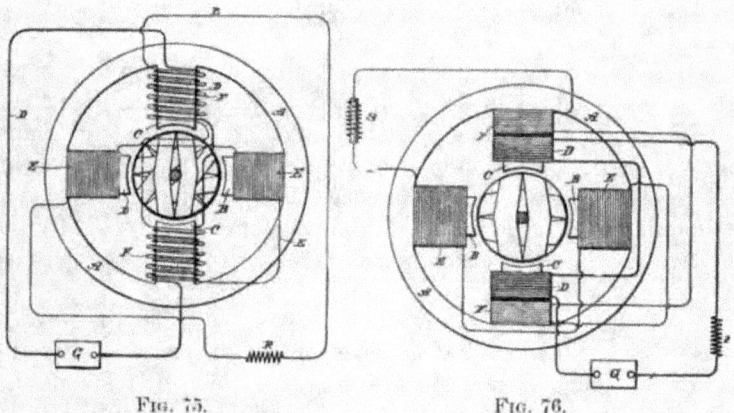

Fig. 75.   Fig. 76.

directly increased, while it will be still further increased if the primary have a very low self-induction. The better results are obtained when the primary has a low self-induction.

Fig. 75 is a diagram of a Tesla motor embodying this principle. Fig. 76 is a similar diagram of a modification of the same. In Fig. 75 let A designate the field-magnet of a motor which, as in all these motors, is built up of sections or plates. B C are polar projections upon which the coils are wound. Upon one pair of these poles, as C, are wound primary coils D, which are directly connected to the circuit of an alternating current generator G. On the same poles are also wound secondary coils F, either side by side or over or under the primary coils, and these are connected with other coils E, which surround the poles B B.

The currents in both primary and secondary coils in such a motor will be retarded or will lag behind the impressed electromotive force; but to secure a proper difference in phase between the primary and secondary currents themselves, Mr. Tesla increases the resistance of the circuit of the secondary and reduces as much as practicable its self-induction. This is done by using for the secondary circuit, particularly in the coils E, wire of comparatively small diameter and having but few turns around the cores; or by using some conductor of higher specific resistance, such as German silver; or by introducing at some point in the secondary circuit an artificial resistance R. Thus the self-induction of the secondary is kept down and its resistance increased, with the result of decreasing the lag between the impressed electro-motive force and the current in the primary coils and increasing the difference of phase between the primary and secondary currents.

In the disposition shown in Fig. 76, the lag in the secondary is increased by increasing the self-induction of that circuit, while the increasing tendency of the primary to lag is counteracted by inserting therein a dead resistance. The primary coils D in this case have a low self-induction and high resistance, while the coils E F, included in the secondary circuit, have a high self-induction and low resistance. This may be done by the proper winding of the coils; or in the circuit including the secondary coils E F, we may introduce a self-induction coil S, while in the primary circuit from the generator G and including coils D, there may be inserted a dead resistance R. By this means the difference of phase between the primary and secondary is increased. It is evident that both means of increasing the difference of phase—namely, by the special winding as well as by the supplemental or external inductive and dead resistance—may be employed conjointly.

In the operation of this motor the current impulses in the primary coils induce currents in the secondary coils, and by the conjoint action of the two the points of greatest magnetic attraction are shifted or rotated.

In practice it is found desirable to wind the armature with closed coils in which currents are induced by the action thereon of the primaries.

CHAPTER XX.

COMBINATIONS OF SYNCHRONIZING MOTOR AND TORQUE MOTOR.

IN THE preceding descriptions relative to synchronizing motors and methods of operating them, reference has been made to the plan adopted by Mr. Tesla, which consists broadly in winding or arranging the motor in such manner that by means of suitable switches it could be started as a multiple-circuit motor, or one operating by a progression of its magnetic poles, and then, when up to speed, or nearly so, converted into an ordinary synchronizing motor, or one in which the magnetic poles were simply alternated. In some cases, as when a large motor is used and when the number of alternations is very high, there is more or less difficulty in bringing the motor to speed as a double or multiple-circuit motor, for the plan of construction which renders the motor best adapted to run as a synchronizing motor impairs its efficiency as a torque or double-circuit motor under the assumed conditions on the start. This will be readily understood, for in a large synchronizing motor the length of the magnetic circuit of the polar projections, and their mass, are so great that apparently considerable time is required for magnetization and demagnetization. Hence with a current of a very high number of alternations the motor may not respond properly. To avoid this objection and to start up a synchronizing motor in which these conditions obtain, Mr. Tesla has combined two motors, one a synchronizing motor, the other a multiple-circuit or torque motor, and by the latter he brings the first-named up to speed, and then either throws the whole current into the synchronizing motor or operates jointly both of the motors.

This invention involves several novel and useful features. It will be observed, in the first place, that both motors are run, without commutators of any kind, and, secondly, that the speed of the torque motor may be higher than that of the synchronizing motor, as will be the case when it contains a fewer number of poles or sets of poles, so that the motor will be more readily and easily brought up to speed. Thirdly, the synchronizing motor may be constructed so as to have a much more pronounced tendency to synchronism without lessening the facility with which it is started.

Fig. 77 is a part sectional view of the two motors; Fig. 78 an end view of the synchronizing motor; Fig. 79 an end view and part section of the torque or double-circuit motor; Fig. 80 a diagram of the circuit connections employed; and Figs. 81, 82, 83, 84 and 85 are diagrams of modified dispositions of the two motors.

Inasmuch as neither motor is doing any work while the current is acting upon the other, the two armatures are rigidly connected, both being mounted upon the same shaft A, the field-magnets B of the synchronizing and C of the torque motor being secured to

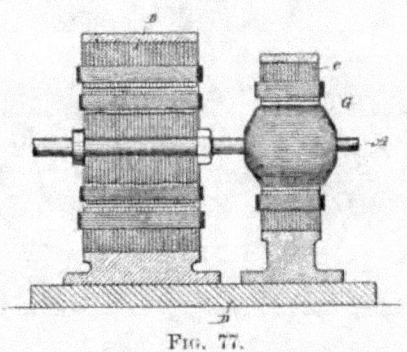

FIG. 77.

the same base D. The preferably larger synchronizing motor has polar projections on its armature, which rotate in very close proximity to the poles of the field, and in other respects it conforms to the conditions that are necessary to secure synchronous action. The pole-pieces of the armature are, however, wound with closed coils E, as this obviates the employment of sliding contacts. The smaller or torque motor, on the other hand, has, preferably, a cylindrical armature F, without polar projections and wound with closed coils G. The field-coils of the torque motor are connected up in two series H and I, and the alternating current from the generator is directed through or divided between these two circuits in any manner to produce a progression of the poles or points of maximum magnetic effect. This result is secured by connecting the two motor-circuits in derivation with the circuit from the generator, inserting in one motor circuit a dead resistance and in the other a self-induction coil, by which means a difference in phase between the two divisions of the current is secured. If both motors have the same number of field poles, the torque motor for a given number of alternations will tend to run at double the speed of the other, for, assuming the connections to be such as to give the best results, its poles are divided into two series and the number of poles is virtually reduced one-half, which being acted upon by the same number of alternations tend to rotate the armature at twice the speed. By this means the main armature is more easily brought to or above the required speed. When the speed necessary for synchronism is imparted to the main motor, the current is shifted from the torque motor into the other.

A convenient arrangement for carrying out this invention is

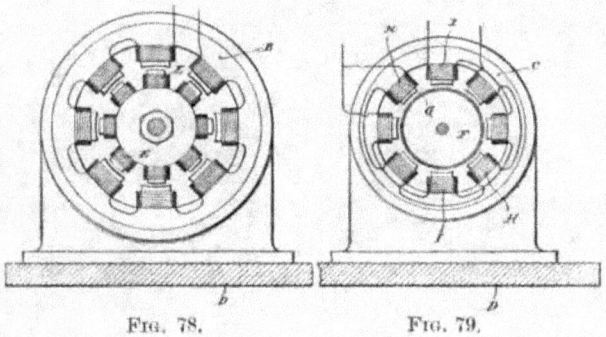

FIG. 78.   FIG. 79.

shown in Fig. 80, in which J J are the field coils of the synchronizing, and H I the field coils of the torque motor. L L' are the conductors of the main line. One end of, say, coils H is connected to wire L through a self-induction coil M. One end of the other set of coils I is connected to the same wire through a dead resistance N. The opposite ends of these two circuits are connected to the contact *m* of a switch, the handle or lever of which is in connection with the line-wire L'. One end of the field circuit of the synchronizing motor is connected to the wire L. The other terminates in the switch-contact *n*. From the diagram it will be readily seen that if the lever P be turned upon contact *m*, the torque motor will start by reason of the difference of phase between the currents in its two energizing circuits. Then when the desired speed is attained, if the lever P be shifted upon contact n the entire current will pass through the field coils of the synchronizing motor and the other will be doing no work.

The torque motor may be constructed and operated in various ways, many of which have already been touched upon. It is not necessary that one motor be cut out of circuit while the other is in, for both may be acted upon by current at the same time, and Mr. Tesla has devised various dispositions or arrangements of the two motors for accomplishing this. Some of these arrangements are illustrated in Figs. 81 to 85.

Referring to Fig. 81, let T designate the torque or multiple circuit motor and s the synchronizing motor, L L' being the line-wires from a source of alternating current. The two circuits of the torque motor of different degrees of self-induction, and designated by N M, are connected in derivation to the wire L. They are then joined and connected to the energizing circuit of the

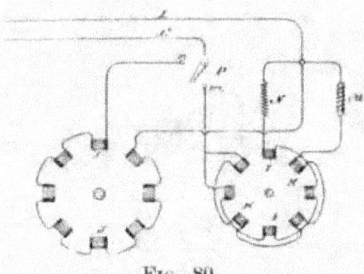

FIG. 80.

synchronizing motor, the opposite terminal of which is connected to wire L'. The two motors are thus in series. To start them Mr. Tesla short-circuits the synchronizing motor by a switch P', throwing the whole current through the torque motor. Then when the desired speed is reached the switch P' is opened, so that the current passes through both motors. In such an arrangement as this it is obviously desirable for economical and other reasons that a proper relation between the speeds of the two motors should be observed.

In Fig. 82 another disposition is illustrated. s is the synchronizing motor and T the torque motor, the circuits of both being in parallel. W is a circuit also in derivation to the motor circuits and containing a switch P''. s' is a switch in the synchronizing motor circuit. On the start the switch s' is opened, cutting out the motor s. Then P'' is opened, throwing the entire current through the motor T, giving it a very strong torque. When the desired speed is reached, switch s' is closed and the current divides

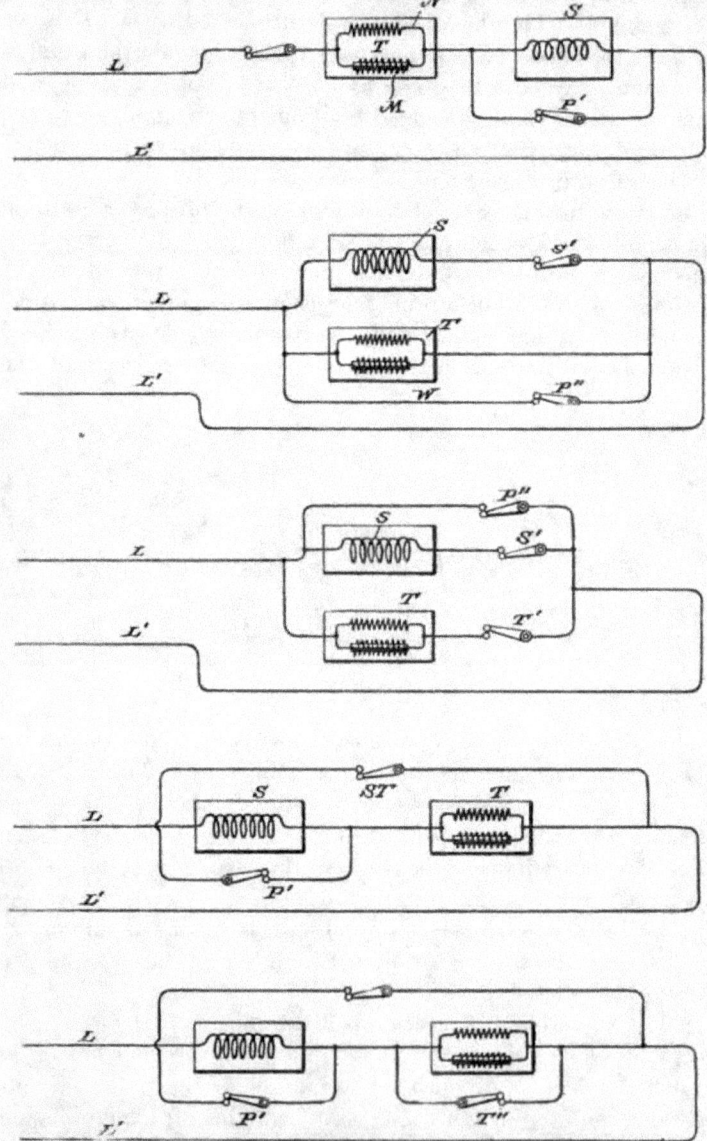

Figs. 81, 82, 83, 84 and 85.

between both motors. By means of switch P" both motors may be cut out.

In Fig. 83 the arrangement is substantially the same, except that a switch T' is placed in the circuit which includes the two circuits of the torque motor. Fig. 84 shows the two motors in series, with a shunt around both containing a switch s T. There is also a shunt around the synchronizing motor s, with a switch P'. In Fig. 85 the same disposition is shown; but each motor is provided with a shunt, in which are switches P' and T", as shown.

CHAPTER XXI.

MOTOR WITH A CONDENSER IN THE ARMATURE CIRCUIT.

WE NOW come to a new class of motors in which resort is had
to condensers for the purpose of developing the required difference of phase and neutralizing the effects of self-induction. Mr.
Tesla early began to apply the condenser to alternating apparatus, in just how many ways can only be learned from a perusal
of other portions of this volume, especially those dealing with
his high frequency work.

Certain laws govern the action or effects produced by a condenser when connected to an electric circuit through which an
alternating or in general an undulating current is made to pass.
Some of the most important of such effects are as follows: First,
if the terminals or plates of a condenser be connected with two
points of a circuit, the potentials of which are made to rise and
fall in rapid succession, the condenser allows the passage, or more
strictly speaking, the transference of a current, although its
plates or armatures may be so carefully insulated as to prevent
almost completely the passage of a current of unvarying strength
or direction and of moderate electromotive force. Second, if a
circuit, the terminals of which are connected with the plates of
the condenser, possess a certain self-induction, the condenser will
overcome or counteract to a greater or less degree, dependent
upon well-understood conditions, the effects of such self-induction. Third, if two points of a closed or complete circuit
through which a rapidly rising and falling current flows be
shunted or bridged by a condenser, a variation in the strength of
the currents in the branches and also a difference of phase of the
currents therein is produced. These effects Mr. Tesla has utilized
and applied in a variety of ways in the construction and operation
of his motors, such as by producing a difference in phase in the
two energizing circuits of an alternating current motor by connecting the two circuits in derivation and connecting up a condenser in series in one of the circuits. A further development,

however, possesses certain novel features of practical value and involves a knowledge of facts less generally understood. It comprises
the use of a condenser or condensers in connection with the induced
or armature circuit of a motor and certain details of the con-

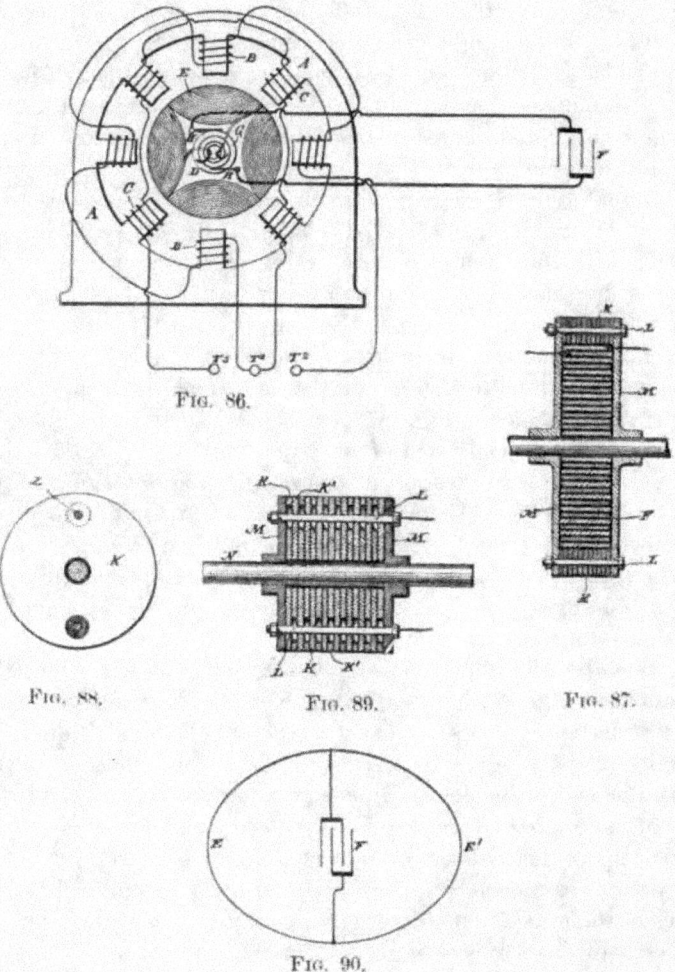

Fig. 86.

Fig. 88.   Fig. 89.   Fig. 87.

Fig. 90.

struction of such motors. In an alternating current motor of the
type particularly referred to above, or in any other which has
an armature coil or circuit closed upon itself, the latter represents not only an inductive resistance, but one which is period-

ically varying in value, both of which facts complicate and render difficult the attainment of the conditions best suited to the most efficient working conditions; in other words, they require, first, that for a given inductive effect upon the armature there should be the greatest possible current through the armature or induced coils, and, second, that there should always exist between the currents in the energizing and the induced circuits a given relation of phase. Hence whatever tends to decrease the self-induction and increase the current in the induced circuits will, other things being equal, increase the output and efficiency of the motor, and the same will be true of causes that operate to maintain the mutual attractive effect between the field magnets and armature at its maximum. Mr. Tesla secures these results by connecting with the induced circuit or circuits a condenser, in the manner described below, and he also, with this purpose in view, constructs the motor in a special manner.

Referring to the drawings, Fig. 86, is a view, mainly diagrammatic, of an alternating current motor, in which the present principle is applied. Fig. 87 is a central section, in line with the shaft, of a special form of armature core. Fig. 88 is a similar section of a modification of the same. Fig. 89 is one of the sections of the core detached. Fig. 90 is a diagram showing a modified disposition of the armature or induced circuits.

The general plan of the invention is illustrated in Fig. 86. A A in this figure represent the the frame and field magnets of an alternating current motor, the poles or projections of which are wound with coils B and C, forming independent energizing circuits connected either to the same or to independent sources of alternating currents, so that the currents flowing through the circuits, respectively, will have a difference of phase. Within the influence of this field is an armature core D, wound with coils E. In motors of this description heretofore these coils have been closed upon themselves, or connected in a closed series; but in the present case each coil or the connected series of coils terminates in the opposite plates of a condenser F. For this purpose the ends of the series of coils are brought out through the shaft to collecting rings G, which are connected to the condenser by contact brushes H and suitable conductors, the condenser being independent of the machine. The armature coils are wound or connected in such manner that adjacent coils produce opposite poles.

The action of this motor and the effect of the plan followed in its construction are as follows: The motor being started in operation and the coils of the field magnets being traversed by alternating currents, currents are induced in the armature coils by one set of field coils, as B, and the poles thus established are acted upon by the other set, as c. The armature coils, however, have necessarily a high self-induction, which opposes the flow of the currents thus set up. The condenser F not only permits the passage or transference of these currents, but also counteracts the effects of self-induction, and by a proper adjustment of the capacity of the condenser, the self-induction of the coils, and the periods of the currents, the condenser may be made to overcome entirely the effect of self-induction.

It is preferable on account of the undesirability of using sliding contacts of any kind, to associate the condenser with the armature directly, or make it a part of the armature. In some cases Mr. Tesla builds up the armature of annular plates K K, held by bolts L between heads M, which are secured to the driving shaft, and in the hollow space thus formed he places a condenser F, generally by winding the two insulated plates spirally around the shaft. In other cases he utilizes the plates of the core itself as the plates of the condenser. For example, in Figs. 88 and 89, N is the driving shaft, M M are the heads of the armature-core, and K K' the iron plates of which the core is built up. These plates are insulated from the shaft and from one another, and are held together by rods or bolts L. The bolts pass through a large hole in one plate and a small hole in the one next adjacent, and so on, connecting electrically all of plates K, as one armature of a condenser, and all of plates K' as the other.

To either of the condensers above described the armature coils may be connected, as explained by reference to Fig. 86.

In motors in which the armature coils are closed upon themselves—as, for example, in any form of alternating current motor in which one armature coil or set of coils is in the position of maximum induction with respect to the field coils or poles, while the other is in the position of minimum induction—the coils are best connected in one series, and two points of the circuit thus formed are bridged by a condenser. This is illustrated in Fig. 90, in which E represents one set of armature coils and E' the other. Their points of union are joined through a condenser F. It will be observed that in this disposition the self-induction of the two branches E and E' varies with their position relatively to the field magnet, and that each branch is alternately the predominating source of the induced current. Hence the effect of the condenser F is twofold. First, it increases the current in each of the branches alternately, and, secondly, it alters the phase of the currents in the branches, this being the well-known effect which results from such a disposition of a condenser with a circuit, as above described. This effect is favorable to the proper working of the motor, because it increases the flow of current in the armature circuits due to a given inductive effect, and also because it brings more nearly into coincidence the maximum magnetic effects of the coacting field and armature poles.

It will be understood, of course, that the causes that contribute to the efficiency of condensers when applied to such uses as the above must be given due consideration in determining the practicability and efficiency of the motors. Chief among these is, as is well known, the periodicity of the current, and hence the improvements described are more particularly adapted to systems in which a very high rate of alternation or change is maintained.

Although this invention has been illustrated in connection with a special form of motor, it will be understood that it is equally applicable to any other alternating current motor in which there is a closed armature coil wherein the currents are induced by the action of the field, and the feature of utilizing the plates or sections of a magnetic core for forming the condenser is applicable, generally, to other kinds of alternating current apparatus.

CHAPTER XXII.

Motor with Condenser in one of the Field Circuits.

If the field or energizing circuits of a rotary phase motor be both derived from the same source of alternating currents and a condenser of proper capacity be included in one of the same, approximately, the desired difference of phase may be obtained between the currents flowing directly from the source and those flowing through the condenser; but the great size and expense of condensers for this purpose that would meet the requirements of the ordinary systems of comparatively low potential are particularly prohibitory to their employment.

Another, now well-known, method or plan of securing a difference of phase between the energizing currents of motors of this kind is to induce by the currents in one circuit those in the other circuit or circuits; but as no means had been proposed that would secure in this way between the phases of the primary or inducing and the secondary or induced currents that difference—theoretically ninety degrees—that is best adapted for practical and economical working, Mr. Tesla devised a means which renders practicable both the above described plans or methods, and by which he is enabled to obtain an economical and efficient alternating current motor. His invention consists in placing a condenser in the secondary or induced circuit of the motor above described and raising the potential of the secondary currents to such a degree that the capacity of the condenser, which is in part dependent on the potential, need be quite small. The value of this condenser is determined in a well-understood manner with reference to the self-induction and other conditions of the circuit, so as to cause the currents which pass through it to differ from the primary currents by a quarter phase.

Fig. 91 illustrates the invention as embodied in a motor in which the inductive relation of the primary and secondary circuits is secured by winding them inside the motor partly upon the same cores; but the invention applies, generally, to other forms of motor in which one of the energizing currents is induced in any way from the other.

Let A B represent the poles of an alternating current motor, of which C is the armature wound with coils D, closed upon themselves, as is now the general practice in motors of this kind. The poles A, which alternate with poles B, are wound with coils of ordinary or coarse wire E in such direction as to make them of alternate north and south polarity, as indicated in the diagram by the characters N S. Over these coils, or in other inductive relation to the same, are wound long fine-wire coils F F, and in the

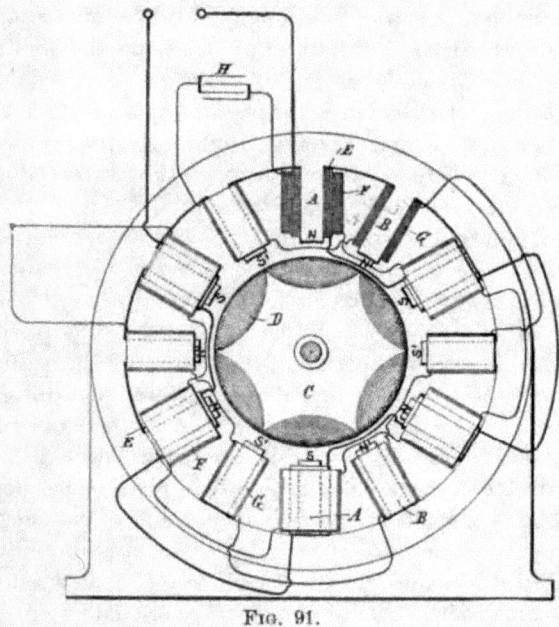

FIG. 91.

same direction throughout as the coils E. These coils are secondaries, in which currents of very high potential are induced. All the coils E in one series are connected, and all the secondaries F in another.

On the intermediate poles B are wound fine-wire energizing coils G, which are connected in series with one another, and also with the series of secondary coils F, the direction of winding being such that a current-impulse induced from the primary coils E imparts the same magnetism to the poles B as that produced

in poles A by the primary impulse. This condition is indicated
by the characters N′ S′.

In the circuit formed by the two sets of coils F and G is introduced a condenser H; otherwise this circuit is closed upon
itself, while the free ends of the circuit of coils E are connected
to a source of alternating currents. As the condenser capacity
which is needed in any particular motor of this kind is dependent upon the rate of alternation or the potential, or both, its size
or cost, as before explained, may be brought within economical
limits for use with the ordinary circuits if the potential of the
secondary circuit in the motor be sufficiently high. By giving
to the condenser proper values, any desired difference of phase
between the primary and secondary energizing circuits may be
obtained.

CHAPTER XXIII.

TESLA POLYPHASE TRANSFORMER.

APPLYING the polyphase principle to the construction of transformers as well to the motors already noticed, Mr. Tesla has invented some very interesting forms, which he considers free from the defects of earlier and, at present, more familiar forms. In these transformers he provides a series of inducing coils and corresponding induced coils, which are generally wound upon a core closed upon itself, usually a ring of laminated iron.

The two sets of coils are wound side by side or superposed or otherwise placed in well-known ways to bring them into the most effective relations to one another and to the core. The inducing or primary coils wound on the core are divided into pairs or sets by the proper electrical connections, so that while the coils of one pair or set co-operate in fixing the magnetic poles of the core at two given diametrically opposite points, the coils of the other pair or set—assuming, for sake of illustration, that there are but two—tend to fix the poles ninety degrees from such points. With this induction device is used an alternating current generator with coils or sets of coils to correspond with those of the converter, and the corresponding coils of the generator and converter are then connected up in independent circuits. It results from this that the different electrical phases in the generator are attended by corresponding magnetic changes in the converter; or, in other words, that as the generator coils revolve, the points of greatest magnetic intensity in the converter will be progressively shifted or whirled around.

Fig. 92 is a diagrammatic illustration of the converter and the electrical connections of the same. Fig. 93 is a horizontal central cross-section of Fig. 92. Fig. 94 is a diagram of the circuits of the entire system, the generator being shown in section.

Mr. Tesla uses a core, A, which is closed upon itself—that is to say, of an annular cylindrical or equivalent form—and as the efficiency of the apparatus is largely increased by the subdivision of this core, he makes it of thin strips, plates, or wires of soft iron electrically insulated as far as practicable. Upon this core are wound, say, four coils, в в в′ в′, used as primary coils, and for which long lengths of comparatively fine wire are employed. Over these coils are then wound shorter coils of coarser wire, c c c′ c′, to constitute the induced or secondary coils. The construction of this or any equivalent form of converter may be carried further, as above pointed out, by inclosing these coils with iron —as, for example, by winding over the coils layers of insulated iron wire.

The device is provided with suitable binding posts, to which

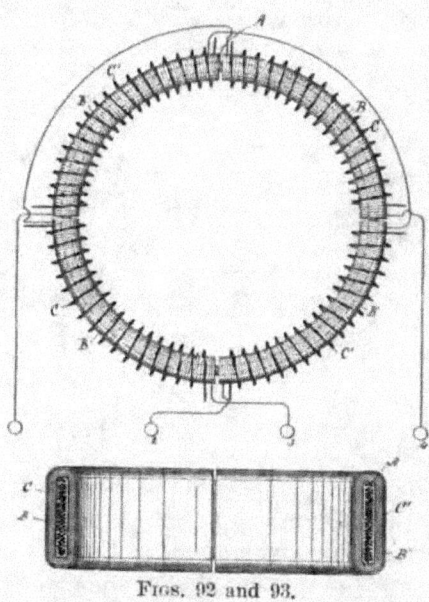

FIGS. 92 and 93.

the ends of the coils are led. The diametrically opposite coils в в and в′ в′ are connected, respectively, in series, and the four terminals are connected to the binding posts. The induced coils are connected together in any desired manner. For example, as shown in Fig. 94, c c may be connected in multiple arc when a quantity current is desired—as for running a group of incandescent lamps—while c′ c′ may be independently connected in series in a circuit including arc lamps or the like. The generator in this system will be adapted to the converter in the manner illustrated. For example, in the present case there are employed a pair of ordinary permanent or electro-magnets, E E, between which is mounted a cylindrical armature on a shaft, F, and wound with two coils, G G'. The terminals of these coils are connected, respectively, to four insulated contact or collecting rings, H H H' H', and the four line circuit wires L connect the brushes K, bearing on these rings, to the converter in the order shown. Noting the results of this combination, it will be observed that at a given point of time the coil G is in its neutral position and is generating little or no current, while the other coil, G', is in a position where it exerts its maximum effect. Assuming coil G to be connected in circuit with coils B B of the converter, and coil G' with coils B' B', it is evident that the poles

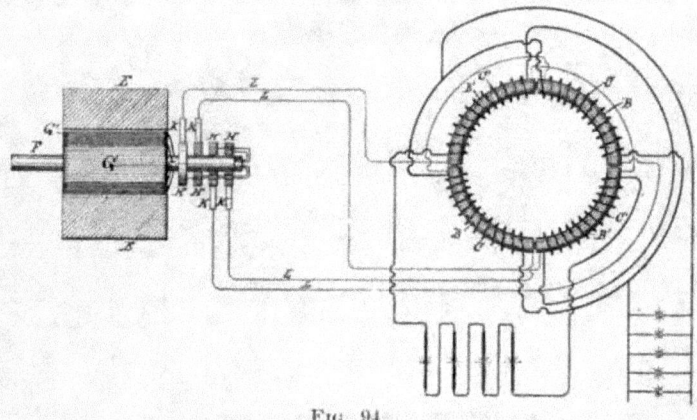

Fig. 94.

of the ring A will be determined by coils B' B' alone; but as the armature of the generator revolves, coil G develops more current and coil G' less, until G reaches its maximum and G' its neutral position. The obvious result will be to shift the poles of the ring A through one-quarter of its periphery. The movement of the coils through the next quarter of a turn—during which coil G' enters a field of opposite polarity and generates a current of opposite direction and increasing strength, while coil G, in passing from its maximum to its neutral position generates a current of decreasing strength and same direction as before—causes a further shifting of the poles through the second quarter of the ring. The second half-revolution will obviously be a repetition of the same action. By the shifting of the poles of the ring A, a powerful dynamic inductive effect on the coils c c′ is produced.   Besides the currents generated in the secondary coils by dynamo-magnetic induction, other currents will be set up in the same coils in consequence of many variations in the intensity of the poles in the ring A.   This should be avoided by maintaining the intensity of the poles constant, to accomplish which care should be taken in designing and proportioning the generator and in distributing the coils in the ring A, and balancing their effect. When this is done, the currents are produced by dynamo-magnetic induction only, the same result being obtained as though the poles were shifted by a commutator with an infinite number of segments.

The modifications which are applicable to other forms of converter are in many respects applicable to this, such as those pertaining more particularly to the form of the core, the relative lengths and resistances of the primary and secondary coils, and the arrangements for running or operating the same.

CHAPTER XXIV.

A Constant Current Transformer with Magnetic Shield Between Coils of Primary and Secondary.

Mr. Tesla has applied his principle of magnetic shielding of parts to the construction also of transformers, the shield being interposed between the primary and secondary coils. In transformers of the ordinary type it will be found that the wave of electromotive force of the secondary very nearly coincides with that of the primary, being, however, in opposite sign. At the same time the currents, both primary and secondary, lag behind their respective electromotive forces; but as this lag is practically or nearly the same in the case of each it follows that the maximum and minimum of the primary and secondary currents will nearly coincide, but differ in sign or direction, provided the secondary be not loaded or if it contain devices having the property of self-induction. On the other hand, the lag of the primary behind the impressed electromotive force may be diminished by loading the secondary with a non-inductive or dead resistance—such as incandescent lamps—whereby the time interval between the maximum or minimum periods of the primary and secondary currents is increased. This time interval, however, is limited, and the results obtained by phase difference in the operation of such devices as the Tesla alternating current motors can only be approximately realized by such means of producing or securing this difference, as above indicated, for it is desirable in such cases that there should exist between the primary and secondary currents, or those which, however produced, pass through the two circuits of the motor, a difference of phase of ninety degrees; or, in other words, the current in one circuit should be a maximum when that in the other circuit is a minimum. To attain to this condition more perfectly, an increased retardation of the secondary current is secured in the following manner: Instead of bringing the primary and secondary coils or circuits of a transformer into the closest possible relations, as has hitherto

been done, Mr. Tesla protects in a measure the secondary from
the inductive action or effect of the primary by surrounding
either the primary or the secondary with a comparatively thin
magnetic shield or screen. Under these modified conditions,
as long as the primary current has a small value, the shield
protects the secondary; but as soon as the primary current
has reached a certain strength, which is arbitrarily determined,
the protecting magnetic shield becomes saturated and the inductive action upon the secondary begins. It results, therefore, that
the secondary current begins to flow at a certain fraction of a
period later than it would without the interposed shield, and
since this retardation may be obtained without necessarily retarding the primary current also, an additional lag is secured, and
the time interval between the maximum or minimum periods of
the primary and secondary currents is increased. Such a trans-

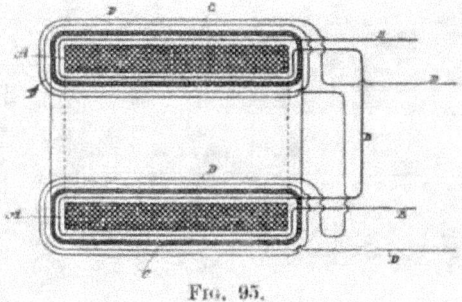

FIG. 95.

former may, by properly proportioning its several elements and
determining the proper relations between the primary and
secondary windings, the thickness of the magnetic shield, and
other conditions, be constructed to yield a constant current at all
loads.

Fig. 95 is a cross-section of a transformer embodying this improvement. Fig. 96 is a similar view of a modified form of
transformer, showing diagrammatically the manner of using the
same.

A A is the main core of the transformer, composed of a ring
of soft annealed and insulated or oxidized iron wire. Upon this
core is wound the secondary circuit or coil B B. This latter is
then covered with a layer or layers of annealed and insulated
iron wires C C, wound in a direction at right angles to the secondary

coil. Over the whole is then wound the primary coil or wire D D. From the nature of this construction it will be obvious that as long as the shield formed by the wires c is below magnetic saturation the secondary coil or circuit is effectually protected or shielded from the inductive influence of the primary, although on open circuit it may exhibit some electromotive force. When the strength of the primary reaches a certain value, the shield c, becoming saturated, ceases to protect the secondary from inductive action, and current is in consequence developed therein. For similar reasons, when the primary current weakens, the weakening of the secondary is retarded to the same or approximately the same extent.

The specific construction of the transformer is largely imma-

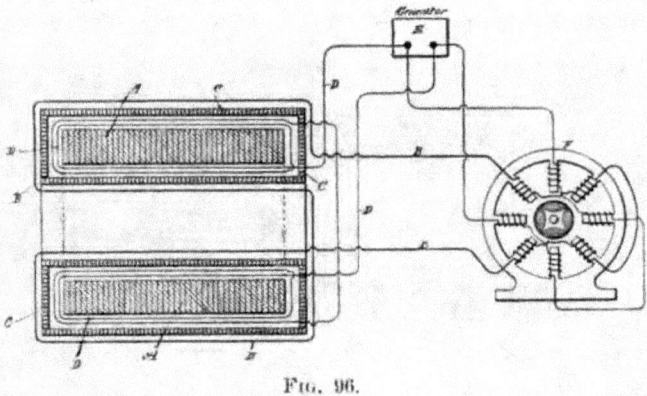

FIG. 96.

terial. In Fig. 96, for example, the core A is built up of thin insulated iron plates or discs. The primary circuit D is wound next the core A. Over this is applied the shield c, which in this case is made up of thin strips or plates of iron properly insulated and surrounding the primary, forming a closed magnetic circuit. The secondary B is wound over the shield c. In Fig. 96, also, E is a source of alternating or rapidly changing currents. The primary of the transformer is connected with the circuit of the generator. F is a two-circuit alternating current motor, one of the circuits being connected with the main circuit from the source E, and the other being supplied with currents from the secondary of the transformer.

PART II.

——

THE TESLA EFFECTS WITH HIGH FREQUENCY AND HIGH POTENTIAL CURRENTS.

CHAPTER XXV.

Introduction.—The Scope of the Tesla Lectures.

Before proceeding to study the three Tesla lectures here presented, the reader may find it of some assistance to have his attention directed to the main points of interest and significance therein. The first of these lectures was delivered in New York, at Columbia College, before the American Institute of Electrical Engineers, May 20, 1891. The urgent desire expressed immediately from all parts of Europe for an opportunity to witness the brilliant and unusual experiments with which the lecture was accompanied, induced Mr. Tesla to go to England early in 1892, when he appeared before the Institution of Electrical Engineers, and a day later, by special request, before the Royal Institution. His reception was of the most enthusiastic and flattering nature on both occasions. He then went, by invitation, to France, and repeated his novel demonstrations before the Société Internationale des Electriciens, and the Société Française de Physique. Mr. Tesla returned to America in the fall of 1892, and in February, 1893, delivered his third lecture before the Franklin Institute of Philadelphia, in fulfilment of a long standing promise to Prof. Houston. The following week, at the request of President James I. Ayer, of the National Electric Light Association, the same lecture was re-delivered in St. Louis. It had been intended to limit the invitations to members, but the appeals from residents in the city were so numerous and pressing that it became necessary to secure a very large hall. Hence it came about that the lecture was listened to by an audience of over 5,000 people, and was in some parts of a more popular nature than either of its predecessors. Despite this concession to the need of the hour and occasion, Mr. Tesla did not hesitate to show many new and brilliant experiments, and to advance the frontier of discovery far beyond any point he had theretofore marked publicly.

We may now proceed to a running review of the lectures themselves. The ground covered by them is so vast that only the leading ideas and experiments can here be touched upon; besides, it is preferable that the lectures should be carefully gone over for their own sake, it being more than likely that each student will discover a new beauty or stimulus in them. Taking up the course of reasoning followed by Mr. Tesla in his first lecture, it will be noted that he started out with the recognition of the fact, which he has now experimentally demonstrated, that for the production of light waves, primarily, electrostatic effects must be brought into play, and continued study has led him to the opinion that all electrical and magnetic effects may be referred to electrostatic molecular forces. This opinion finds a singular confirmation in one of the most striking experiments which he describes, namely, the production of a veritable flame by the agitation of electrostatically charged molecules. It is of the highest interest to observe that this result points out a way of obtaining a flame which consumes no material and in which no chemical action whatever takes place. It also throws a light on the nature of the ordinary flame, which Mr. Tesla believes to be due to electrostatic molecular actions, which, if true, would lead directly to the idea that even chemical affinities might be electrostatic in their nature and that, as has already been suggested, molecular forces in general may be referable to one and the same cause. This singular phenomenon accounts in a plausible manner for the unexplained fact that buildings are frequently set on fire during thunder storms without having been at all struck by lightning. It may also explain the total disappearance of ships at sea.

One of the striking proofs of the correctness of the ideas advanced by Mr. Tesla is the fact that, notwithstanding the employment of the most powerful electromagnetic inductive effects, but feeble luminosity is obtainable, and this only in close proximity to the source of disturbance; whereas, when the electrostatic effects are intensified, the same initial energy suffices to excite luminosity at considerable distances from the source. That there are only electrostatic effects active seems to be clearly proved by Mr. Tesla's experiments with an induction coil operated with alternating currents of very high frequency. He shows how tubes may be made to glow brilliantly at considerable distances from any object when placed in a powerful, rapidly alternating, electrostatic field, and he describes many interesting phenomena observed in such a field. His experiments open up the possibility of lighting an apartment by simply creating in it such an electro-static field, and this, in a certain way, would appear to be the ideal method of lighting a room, as it would allow the illuminating device to be freely moved about. The power with which these exhausted tubes, devoid of any electrodes, light up is certainly remarkable.

That the principle propounded by Mr. Tesla is a broad one is evident from the many ways in which it may be practically applied. We need only refer to the variety of the devices shown or described, all of which are novel in character and will, without doubt, lead to further important results at the hands of Mr. Tesla and other investigators. The experiment, for instance, of lighting up a single filament or block of refractory material with a single wire, is in itself sufficient to give Mr. Tesla's work the stamp of originality, and the numerous other experiments and effects which may be varied at will, are equally new and interesting. Thus, the incandescent filament spinning in an unexhausted globe, the well-known Crookes experiment on open circuit, and the many others suggested, will not fail to interest the reader. Mr. Tesla has made an exhaustive study of the various forms of the discharge presented by an induction coil when operated with these rapidly alternating currents, starting from the thread-like discharge and passing through various stages to the true electric flame.

A point of great importance in the introduction of high tension alternating current which Mr. Tesla brings out is the necessity of carefully avoiding all gaseous matter in the high tension apparatus. He shows that, at least with very rapidly alternating currents of high potential, the discharge may work through almost any practicable thickness of the best insulators, if air is present. In such cases the air included within the apparatus is violently agitated and by molecular bombardment the parts may be so greatly heated as to cause a rupture of the insulation. The practical outcome of this is, that, whereas with steady currents, any kind of insulation may be used, with rapidly alternating currents oils will probably be the best to employ, a fact which has been observed, but not until now satisfactorily explained. The recognition of the above fact is of special importance in the construction of the costly commercial induction coils which are often rendered useless in an unaccountable manner. The truth of these views of Mr. Tesla is made evident by the interesting experiments illustrative of the behavior of the air between charged surfaces, the luminous streams formed by the charged molecules appearing even when great thicknesses of the best insulators are interposed between the charged surfaces. These luminous streams afford in themselves a very interesting study for the experimenter. With these rapidly alternating currents they become far more powerful and produce beautiful light effects when they issue from a wire, pinwheel or other object attached to a terminal of the coil; and it is interesting to note that they issue from a ball almost as freely as from a point, when the frequency is very high.

From these experiments we also obtain a better idea of the importance of taking into account the capacity and self-induction in the apparatus employed and the possibilities offered by the use of condensers in conjunction with alternate currents, the employment of currents of high frequency, among other things, making it possible to reduce the condenser to practicable dimensions. Another point of interest and practical bearing is the fact, proved by Mr. Tesla, that for alternate currents, especially those of high frequency, insulators are required possessing a small specific inductive capacity, which at the same time have a high insulating power.

Mr. Tesla also makes interesting and valuable suggestion in regard to the economical utilization of iron in machines and transformers. He shows how, by maintaining by continuous magnetization a flow of lines through the iron, the latter may be kept near its maximum permeability and a higher output and economy may be secured in such apparatus. This principle may prove of considerable commercial importance in the development of alternating systems. Mr. Tesla's suggestion that the same result can be secured by heating the iron by hysteresis and eddy currents, and increasing the permeability in this manner, while it may appear less practical, nevertheless opens another direction for investigation and improvement.

The demonstration of the fact that with alternating currents of high frequency, sufficient energy may be transmitted under practicable conditions through the glass of an incandescent lamp by electrostatic or electromagnetic induction may lead to a departure in the construction of such devices. Another important experimental result achieved is the operation of lamps, and even motors, with the discharges of condensers, this method affording a means of converting direct or alternating currents. In this connection Mr. Tesla advocates the perfecting of apparatus capable of generating electricity of high tension from heat energy, believing this to be a better way of obtaining electrical energy for practical purposes, particularly for the production of light.

While many were probably prepared to encounter curious phenomena of impedance in the use of a condenser discharged disruptively, the experiments shown were extremely interesting on account of their paradoxical character. The burning of an incandescent lamp at any candle power when connected across a heavy metal bar, the existence of nodes on the bar and the possibility of exploring the bar by means of an ordinary Cardew voltmeter, are all peculiar developments, but perhaps the most interesting observation is the phenomenon of impedance observed in the lamp with a straight filament, which remains dark while the bulb glows.

Mr. Tesla's manner of operating an induction coil by means of the disruptive discharge, and thus obtaining enormous differences of potential from comparatively small and inexpensive coils, will be appreciated by experimenters and will find valuable application in laboratories. Indeed, his many suggestions and hints in regard to the construction and use of apparatus in these investigations will be highly valued and will aid materially in future research.

The London lecture was delivered twice. In its first form, before the Institution of Electrical Engineers, it was in some respects an amplification of several points not specially enlarged upon in the New York lecture, but brought forward many additional discoveries and new investigations. Its repetition, in another form, at the Royal Institution, was due to Prof. Dewar, who with Lord Rayleigh, manifested a most lively interest in Mr. Tesla's work, and whose kindness illustrated once more the strong English love of scientific truth and appreciation of its votaries. As an indefatigable experimenter, Mr. Tesla was certainly nowhere more at home than in the haunts of Faraday, and as the guest of Faraday's successor. This Royal Institution lecture summed up the leading points of Mr. Tesla's work, in the high potential, high frequency field, and we may here avail ourselves of so valuable a summarization, in a simple form, of a subject by no means easy of comprehension until it has been thoroughly studied.

In these London lectures, among the many notable points made was first, the difficulty of constructing the alternators to obtain the very high frequencies needed. To obtain the high frequencies it was necessary to provide several hundred polar projections, which were necessarily small and offered many drawbacks, and this the more as exceedingly high peripheral speeds had to be resorted to. In some of the first machines both armature and field had polar projections. These machines produced a curious noise, especially when the armature was started from the state of rest, the field being charged. The most efficient machine was found to be one with a drum armature, the iron body of which consisted of very thin wire annealed with special care. It was, of course, desirable to avoid the employment of iron in the armature, and several machines of this kind, with moving or stationary conductors were constructed, but the results obtained were not quite satisfactory, on account of the great mechanical and other difficulties encountered.

The study of the properties of the high frequency currents obtained from these machines is very interesting, as nearly every experiment discloses something new. Two coils traversed by such a current attract or repel each other with a force which, owing to the imperfection of our sense of touch, seems continuous. An interesting observation, already noted under another form, is that a piece of iron, surrounded by a coil through which the current is passing appears to be continuously magnetized. This apparent continuity might be ascribed to the deficiency of the sense of touch, but there is evidence that in currents of such high frequencies one of the impulses preponderates over the other.

As might be expected, conductors traversed by such currents are rapidly heated, owing to the increase of the resistance, and the heating effects are relatively much greater in the iron. The hysteresis losses in iron are so great that an iron core, even if finely subdivided, is heated in an incredibly short time. To give an idea of this, an ordinary iron wire $\frac{1}{6}$ inch in diameter inserted within a coil having 250 turns, with a current estimated to be five amperes passing through the coil, becomes within two seconds' time so hot as to scorch wood. Beyond a certain frequency, an iron core, no matter how finely subdivided, exercises a dampening effect, and it was easy to find a point at which the impedance of a coil was not affected by the presence of a core consisting of a bundle of very thin well annealed and varnished iron wires.

Experiments with a telephone, a conductor in a strong magnetic field, or with a condenser or arc, seem to afford certain proof that sounds far above the usually accepted limit of hearing would be perceived if produced with sufficient power. The arc produced by these currents possesses several interesting features. Usually it emits a note the pitch of which corresponds to twice the frequency of the current, but if the frequency be sufficiently high it becomes noiseless, the limit of audition being determined principally by the linear dimensions of the arc. A curious feature of the arc is its persistency, which is due partly to the inability of the gaseous column to cool and increase considerably in resistance, as is the case with low frequencies, and partly to the tendency of such a high frequency machine to maintain a constant current.

In connection with these machines the condenser affords a particularly interesting study. Striking effects are produced by proper adjustments of capacity and self-induction. It is easy to raise the electromotive force of the machine to many times the original value by simply adjusting the capacity of a condenser connected in the induced circuit. If the condenser be at some distance from the machine, the difference of potential on the terminals of the latter may be only a small fraction of that on the condenser.

But the most interesting experiences are gained when the tension of the currents from the machine is raised by means of an induction coil. In consequence of the enormous rate of change obtainable in the primary current, much higher potential differences are obtained than with coils operated in the usual ways, and, owing to the high frequency, the secondary discharge possesses many striking peculiarities. Both the electrodes behave generally alike, though it appears from some observations that one current impulse preponderates over the other, as before mentioned.

The physiological effects of the high tension discharge are found to be so small that the shock of the coil can be supported without any inconvenience, except perhaps a small burn produced by the discharge upon approaching the hand to one of the terminals. The decidedly smaller physiological effects of these currents are thought to be due either to a different distribution through the body or to the tissues acting as condensers. But in the case of an induction coil with a great many turns the harmlessness is principally due to the fact that but little energy is available in the external circuit when the same is closed through the experimenter's body, on account of the great impedance of the coil.

In varying the frequency and strenth of the currents through the primary of the coil, the character of the secondary discharge is greatly varied, and no less than five distincts forms are observed:—A weak, sensitive thread discharge, a powerful flaming discharge, and three forms of brush or streaming discharges. Each of these possesses certain noteworthy features, but the most interesting to study are the latter.

Under certain conditions the streams, which are presumably due to the violent agitation of the air molecules, issue freely from all points of the coil, even through a thick insulation. If there is the smallest air space between the primary and secondary, they will form there and surely injure the coil by slowly warming the insulation. As they form even with ordinary frequencies when the potential is excessive, the air-space must be most carefully avoided. These high frequency streamers differ in aspect and properties from those produced by a static machine. The wind produced by them is small and should altogether cease if still considerably higher frequencies could be obtained. A peculiarity is that they issue as freely from surfaces as from points. Owing to this, a metallic vane, mounted in one of the terminals of the coil so as to rotate freely, and having one of its sides covered with insulation, is spun rapidly around. Such a vane would not rotate with a steady potential, but with a high frequency coil it will spin, even if it be entirely covered with insulation, provided the insulation on one side be either thicker or of a higher specific inductive capacity. A Crookes electric radiometer is also spun around when connected to one of the terminals of the coil, but only at very high exhaustion or at ordinary pressures.

There is still another and more striking peculiarity of such a high frequency streamer, namely, it is hot. The heat is easily perceptible with frequencies of about 10,000, even if the potential is not excessively high. The heating effect is, of course, due to the molecular impacts and collisions. Could the frequency and potential be pushed far enough, then a brush could be produced resembling in every particular a flame and giving light and heat, yet without a chemical process taking place.

The hot brush, when properly produced, resembles a jet of burning gas escaping under great pressure, and it emits an extraordinary strong smell of ozone. The great ozonizing action is ascribed to the fact that the agitation of the molecules of the air is more violent in such a brush than in the ordinary streamer of a static machine. But the most powerful brush discharges were produced by employing currents of much higher frequencies than it was possible to obtain by means of the alternators. These currents were obtained by disruptively discharging a condenser and setting up oscillations. In this manner currents of a frequency of several hundred thousand were obtained.

Currents of this kind, Mr. Tesla pointed out, produce striking effects. At these frequencies, the impedance of a copper bar is so great that a potential difference of several hundred volts can be maintained between two points of a short and thick bar, and it is possible to keep an ordinary incandescent lamp burning at full candle power by attaching the terminals of the lamp to two points of the bar no more than a few inches apart. When the frequency is extremely high, nodes are found to exist on such a bar, and it is easy to locate them by means of a lamp.

By converting the high tension discharges of a low frequency coil in this manner, it was found practicable to keep a few lamps burning on the ordinary circuit in the laboratory, and by bringing the undulation to a low pitch, it was possible to operate small motors.

This plan likewise allows of converting high tension discharges of one direction into low tension unidirectional currents, by adjusting the circuit so that there are no oscillations. In passing the oscillating discharges through the primary of a specially constructed coil, it is easy to obtain enormous potential differences with only few turns of the secondary.

Great difficulties were at first experienced in producing a successful coil on this plan. It was found necessary to keep all air, or gaseous matter in general, away from the charged surfaces, and oil immersion was resorted to. The wires used were heavily covered with gutta-percha and wound in oil, or the air was pumped out by means of a Sprengel pump. The general arrangement was the following:—An ordinary induction coil, operated from a low frequency alternator, was used to charge Leyden jars. The jars were made to discharge over a single or multiple gap through the primary of the second coil. To insure the action of the gap, the arc was blown out by a magnet or air blast. To adjust the potential in the secondary a small oil condenser was used, or polished brass spheres of different sizes were screwed on the terminals and their distance adjusted.

When the conditions were carefully determined to suit each experiment, magnificent effects were obtained. Two wires, stretched through the room, each being connected to one of the terminals of the coil, emitted streams so powerful that the light from them allowed distinguishing the objects in the room; the wires became luminous even though covered with thick and most excellent insulation. When two straight wires, or two concentric circles of wire, are connected to the terminals, and set at the proper distance, a uniform luminous sheet is produced between them. It was possible in this way to cover an area of more than one meter square completely with the streams. By attaching to one terminal a large circle of wire and to the other terminal a small sphere, the streams are focused upon the sphere, produce a strongly lighted spot upon the same, and present the appearance of a luminous cone. A very thin wire glued upon a plate of hard rubber of great thickness, on the opposite side of which is fastened a tinfoil coating, is rendered intensely luminous when the coating is connected to the other terminal of the coil. Such an experiment can be performed also with low frequency currents, but much less satisfactorily.

When the terminals of such a coil, even of a very small one, are separated by a rubber or glass plate, the discharge spreads over the plate in the form of streams, threads or brilliant sparks, and affords a magnificent display, which cannot be equaled by the largest coil operated in the usual ways. By a simple adjustment it is possible to produce with the coil a succession of brilliant sparks, exactly as with a Holtz machine.

Under certain conditions, when the frequency of the oscillation is very great, white, phantom-like streams are seen to break forth from the terminals of the coil. The chief interesting feature about them is, that they stream freely against the outstretched hand or other conducting object without producing any sensation, and the hand may be approached very near to the terminal without a spark being induced to jump. This is due presumably to the fact that a considerable portion of the energy is carried away or dissipated in the streamers, and the difference of potential between the terminal and the hand is diminished.

It is found in such experiments that the frequency of the vibration and the quickness of succession of the sparks between the knobs affect to a marked degree the appearance of the streams. When the frequency is very low, the air gives way in more or less the same manner as by a steady difference of potential, and the streams consist of distinct threads, generally mingled with thin sparks, which probably correspond to the successive discharges occurring between the knobs. But when the frequency is very high, and the arc of the discharge produces a sound which is loud and smooth (which indicates both that oscillation takes place and that the sparks succeed each other with great rapidity), then the luminous streams formed are perfectly uniform. They are generally of a purplish hue, but when the molecular vibration is increased by raising the potential, they assume a white color.

The luminous intensity of the streams increases rapidly when the potential is increased; and with frequencies of only a few hundred thousand, could the coil be made to withstand a sufficiently high potential difference, there is no doubt that the space around a wire could be made to emit a strong light, merely by the agitation of the molecules of the air at ordinary pressure.

Such discharges of very high frequency which render luminous the air at ordinary pressure we have very likely occasion to witness in the aurora borealis. From many of these experiments it seems reasonable to infer that sudden cosmic disturbances, such as eruptions on the sun, set the electrostatic charge of the earth in an extremely rapid vibration, and produce the glow by the violent agitation of the air in the upper and even in the lower strata. It is thought that if the frequency were low, or even more so if the charge were not at all vibrating, the lower dense strata would break down as in a lightning discharge. Indications of such breaking down have been repeatedly observed, but they can be attributed to the fundamental disturbances, which are few in number, for the superimposed vibration would be so rapid as not to allow a disruptive break.

The study of these discharge phenomena has led Mr. Tesla to the recognition of some important facts. It was found, as already stated, that gaseous matter must be most carefully excluded from any dielectric which is subjected to great, rapidly changing electrostatic stresses. Since it is difficult to exclude the gas perfectly when solid insulators are used, it is necessary to resort to liquid dielectrics. When a solid dielectric is used, it matters little how thick and how good it is; if air be present, streamers form, which gradually heat the dielectric and impair its insulating power, and the discharge finally breaks through. Under ordinary conditions the best insulators are those which possess the highest specific inductive capacity, but such insulators are not the best to employ when working with these high frequency currents, for in most cases the higher specific inductive capacity is rather a disadvantage. The prime quality of the insulating medium for these currents is continuity. For this reason principally it is necessary to employ liquid insulators, such as oils. If two metal plates, connected to the terminals of the coil, are immersed in oil and set a distance apart, the coil may be kept working for any length of time without a break occurring, or without the oil being warmed, but if air bubbles are introduced, they become luminous; the air molecules, by their impact against the oil, heat it, and after some time cause the insulation to give way. If, instead of the oil, a solid plate of the best dielectric, even several times thicker than the oil intervening between the metal plates, is inserted between the latter, the air having free access to the charged surfaces, the dielectric ivariably is warmed and breaks down.

The employment of oil is advisable or necessary even with low frequencies, if the potentials are such that streamers form, but only in such cases, as is evident from the theory of the action. If the potentials are so low that streamers do not form, then it is even disadvantageous to employ oil, for it may, principally by confining the heat, be the cause of the breaking down of the insulation.

The exclusion of gaseous matter is not only desirable on account of the safety of the apparatus, but also on account of economy, especially in a condenser, in which considerable waste of power may occur merely owing to the presence of air, if the electric density on the charged surfaces is great.

In the course of these investigations a phenomenon of special scientific interest was observed. It may be ranked among the brush phenomena, in fact it is a kind of brush which forms at, or near, a single terminal in high vacuum. In a bulb with a conducting electrode, even if the latter be of aluminum, the brush has only a very short existence, but it can be preserved for a considerable length of time in a bulb devoid of any conducting electrode. To observe the phenomenon it is found best to employ a large spherical bulb having in its centre a small bulb supported on a tube sealed to the neck of the former. The large bulb being exhausted to a high degree, and the inside of the small bulb being connected to one of the terminals of the coil, under certain conditions there appears a misty haze around the small bulb, which, after passing through some stages, assumes the form of a brush, generally at right angles to the tube supporting the small bulb. When the brush assumes this form it may be brought to a state of extreme sensitiveness to electrostatic and magnetic influence. The bulb hanging straight down, and all objects being remote from it, the approach of the observer within a few paces will cause the brush to fly to the opposite side, and if he walks around the bulb it will always keep on the opposite side. It may begin to spin around the terminal long before it reaches that sensitive stage. When it begins to turn around, principally, but also before, it is affected by a magnet, and at a certain stage it is susceptible to magnetic influence to an astonishing degree. A small permanent magnet, with its poles at a distance of no more than two centimetres will affect it visibly at a distance of two metres, slowing down or accelerating the rotation according to how it is held relatively to the brush.

When the bulb hangs with the globe down, the rotation is always clockwise. In the southern hemisphere it would occur in the opposite direction, and on the (magnetic) equator the brush should not turn at all. The rotation may be reversed by a magnet kept at some distance. The brush rotates best, seemingly, when it is at right angles to the lines of force of the earth. It very likely rotates, when at its maximum speed, in synchronism with the alternations, say, 10,000 times a second. The rotation can be slowed down or accelerated by the approach or recession of the observer, or any conducting body, but it cannot be reversed by putting the bulb in any position. Very curious experiments may be performed with the brush when in its most sensitive state. For instance, the brush resting in one position, the experimenter may, by selecting a proper position, approach the hand at a certain considerable distance to the bulb, and he may cause the brush to pass off by merely stiffening the muscles of the arm, the mere change of configuration of the arm and the consequent imperceptible displacement being sufficient to disturb the delicate balance. When it begins to rotate slowly, and the hands are held at a proper distance, it is impossible to make even the slightest motion without producing a visible effect upon the brush. A metal plate connected to the other terminal of the coil affects it at a great distance, slowing down the rotation often to one turn a second.

Mr. Tesla hopes that this phenomenon will prove a valuable aid in the investigation of the nature of the forces acting in an electrostatic or magnetic field. If there is any motion which is measurable going on in the space, such a brush would be apt to reveal it. It is, so to speak, a beam of light, frictionless, devoid of inertia. On account of its marvellous sensitiveness to electrostatic or magnetic disturbances it may be the means of sending signals through submarine cables with any speed, and even of transmitting intelligence to a distance without wires.

In operating an induction coil with these rapidly alternating currents, it is astonishing to note, for the first time, the great importance of the relation of capacity, self-induction, and frequency as bearing upon the general result. The combined effect of these elements produces many curious effects. For instance, two metal plates are connected to the terminals and set at a small distance, so that an arc is formed between them. This arc *prevents* a strong current from flowing through the coil. If the arc be interrupted by the interposition of a glass plate, the capacity of the condenser obtained counteracts the self-induction, and a stronger current is made to pass. The effects of capacity are the most striking, for in these experiments, since the self-induction and frequency both are high, the critical capacity is very small, and need be but slightly varied to produce a very considerable change. The experimenter brings his body in contact with the terminals of the secondary of the coil, or attaches to one or both terminals insulated bodies of very small bulk, such as exhausted bulbs, and he produces a considerable rise or fall of potential on the secondary, and greatly affects the flow of the current through the primary coil.

In many of the phenomena observed, the presence of the air, or, generally speaking, of a medium of a gaseous nature (using this term not to imply specific properties, but in contradistinction to homogeneity or perfect continuity) plays an important part, as it allows energy to be dissipated by molecular impact or bombardment. The action is thus explained:—When an insulated body connected to a terminal of the coil is suddenly charged to high potential, it acts inductively upon the surrounding air, or whatever gaseous medium there might be. The molecules or atoms which are near it are, of course, more attracted, and move through a greater distance than the further ones. When the nearest molecules strike the body they are repelled, and collisions occur at all distances within the inductive distance. It is now clear that, if the potential be steady, but little loss of energy can be caused in this way, for the molecules which are nearest to the body having had an additional charge imparted to them by contact, are not attracted until they have parted, if not with all, at least with most of the additional charge, which can be accomplished only after a great many collisions. This is inferred from the fact that with a steady potential there is but little loss in dry air. When the potential, instead of being steady, is alternating, the conditions are entirely different. In this case a rhythmical bombardment occurs, no matter whether the molecules after coming in contact with the body lose the imparted charge or not, and, what is more, if the charge is not lost, the impacts are all the more violent. Still, if the frequency of the impulses be very small, the loss caused by the impacts and collisions would not be serious unless the potential was excessive. But when extremely high frequencies and more or less high potentials are used, the loss may be very great. The total energy lost per unit of time is proportionate to the product of the number of impacts per second, or the frequency and the energy lost in each impact. But the energy of an impact must be proportionate to the square of the electric density of the body, on the assumption that the charge imparted to the molecule is proportionate to that density. It is concluded from this that the total energy lost must be proportionate to the product of the frequency and the square of the electric density; but this law needs experimental confirmation. Assuming the preceding considerations to be true, then, by rapidly alternating the potential of a body immersed in an insulating gaseous medium, any amount of energy may be dissipated into space. Most of that energy, then, is not dissipated in the form of long ether waves, propagated to considerable distance, as is thought most generally, but is consumed in impact and collisional losses—that is, heat vibrations—on the surface and in the vicinity of the body. To reduce the dissipation it is necessary to work with a small electric density—the smaller, the higher the frequency.

The behavior of a gaseous medium to such rapid alternations of potential makes it appear plausible that electrostatic disturbances of the earth, produced by cosmic events, may have great influence upon the meteorological conditions. When such disturbances occur both the frequency of the vibrations of the charge and the potential are in all probability excessive, and the energy converted into heat may be considerable. Since the density must be unevenly distributed, either in consequence of the irregularity of the earth's surface, or on account of the condition of the atmosphere in various places, the effect produced would accordingly vary from place to place. Considerable variations in the temperature and pressure of the atmosphere may in this manner be caused at any point of the surface of the earth. The variations may be gradual or very sudden, according to the nature of the original disturbance, and may produce rain and storms, or locally modify the weather in any way.

From many experiences gathered in the course of these investigations it appears certain that in lightning discharges the air is an element of importance. For instance, during a storm a stream may form on a nail or pointed projection of a building. If lightning strikes somewhere in the neighborhood, the harmless static discharge may, in consequence of the oscillations set up, assume the character of a high-frequency streamer, and the nail or projection may be brought to a high temperature by the violent impact of the air molecules. Thus, it is thought, a building may be set on fire without the lightning striking it. In like manner small metallic objects may be fused and volatilized —as frequently occurs in lightning discharges—merely because they are surrounded by air. Were they immersed in a practically continuous medium, such as oil, they would probably be safe, as the energy would have to spend itself elsewhere.

An instructive experience having a bearing on this subject is the following:—A glass tube of an inch or so in diameter and several inches long is taken, and a platnium wire sealed into it, the wire running through the center of the tube from end to end. The tube is exhausted to a moderate degree. If a steady current is passed through the wire it is heated uniformly in all parts and the gas in the tube is of no consequence. But if high frequency discharges are directed through the wire, it is heated more on the ends than in the middle portion, and if the frequency, or rate of charge, is high enough, the wire might as well be cut in the middle as not, for most of the heating on the ends is due to the rarefied gas. Here the gas might only act as a conductor of no impedance, diverting the current from the wire as the impedance of the latter is enormously increased, and merely heating the ends of the wire by reason of their resistance to the passage of the discharge. But it is not at all necessary that the gas in the tube should be conducting; it might be at an extremely low pressure, still the ends of the wire would be heated : however, as is ascertained by experience, only the two ends would in such case not be electrically connected through the gaseous medium. Now, what with these frequencies and potentials occurs in an exhausted tube, occurs in the lightning discharge at ordinary pressure.

From the facility with which any amount of energy may be carried off through a gas, Mr. Tesla infers that the best way to render harmless a lightning discharge is to afford it in some way a passage through a volume of gas.

The recognition of some of the above facts has a bearing upon far-reaching scientific investigations in which extremely high frequencies and potentials are used. In such cases the air is an important factor to be considered. So, for instance, if two wires are attached to the terminals of the coil, and the streamers issue from them, there is dissipation of energy in the form of heat and light, and the wires behave like a condenser of larger capacity. If the wires be immersed in oil, the dissipation of energy is prevented, or at least reduced, and the apparent capacity is diminished. The action of the air would seem to make it very difficult to tell, from the measured or computed capacity of a condenser in which the air is acted upon, its actual capacity or vibration period, especially if the condenser is of very small surface and is charged to a very high potential. As many important results are dependant upon the correctness of the estimation of the vibration period, this subject demands the most careful scrutiny of investigators.

In Leyden jars the loss due to the presence of air is comparatively small, principally on account of the great surface of the coatings and the small external action, but if there are streamers on the top, the loss may be considerable, and the period of vibration is affected. In a resonator, the density is small, but the frequency is extreme, and may introduce a considerable error. It appears certain, at any rate, that the periods of vibration of a charged body in a gaseous and in a continuous medium, such as oil, are different, on account of the action of the former, as explained.

Another fact recognized, which is of some consequence, is, that in similar investigations the general considerations of static screening are not applicable when a gaseous medium is present. This is evident from the following experiment :—A short and wide glass tube is taken and covered with a substantial coating of bronze powder, barely allowing the light to shine a little through. The tube is highly exhausted and suspended on a metallic clasp from the end of a wire. When the wire is connected with one of the terminals of the coil, the gas inside of the tube is lighted in spite of the metal coating. Here the metal evidently does not screen the gas inside as it ought to, even if it be very thin and poorly conducting. Yet, in a condition of rest the metal coating, however thin, screens the inside perfectly.

One of the most interesting results arrived at in pursuing these experiments, is the demonstration of the fact that a gaseous medium, upon which vibration is impressed by rapid changes of electrostatic potential, is rigid. In illustration of this result an experiment made by Mr. Tesla may by cited :—A glass tube about one inch in diameter and three feet long, with outside condenser coatings on the ends, was exhausted to a certain point, when, the tube being suspended freely from a wire connecting the upper coating to one of the terminals of the coil, the discharge appeared in the form of a luminous thread passing through the axis of the tube. Usually the thread was sharply defined in the upper part of the tube and lost itself in the lower part. When a magnet or the finger was quickly passed near the upper part of the luminous thread, it was brought out of position by magnetic or electrostatic influence, and a transversal vibration like that of a suspended cord, with one or more distinct nodes, was set up, which lasted for a few minutes and gradually died out. By suspending from the lower condenser coating metal plates of different sizes, the speed of the vibration was varied. This vibration would seem to show beyond doubt that the thread possessed rigidity, at least to transversal displacements.

Many experiments were tried to demonstrate this property in

air at ordinary pressure. Though no positive evidence has been obtained, it is thought, nevertheless, that a high frequency brush or streamer, if the frequency could be pushed far enough, would be decidedly rigid. A small sphere might then be moved within it quite freely, but if thrown against it the sphere would rebound. An ordinary flame cannot possess rigidity to a marked degree because the vibration is directionless; but an electric arc, it is believed, must possess that property more or less. A luminous band excited in a bulb by repeated discharges of a Leyden jar must also possess rigidity, and if deformed and suddenly released should vibrate.

From like considerations other conclusions of interest are reached. The most probable medium filling the space is one consisting of independent carriers immersed in an insulating fluid. If through this medium enormous electrostatic stresses are assumed to act, which vary rapidly in intensity, it would allow the motion of a body through it, yet it would be rigid and elastic, although the fluid itself might be devoid of these properties. Furthermore, on the assumption that the independent carriers are of any configuration such that the fluid resistance to motion in one direction is greater than in another, a stress of that nature would cause the carriers to arrange themselves in groups, since they would turn to each other their sides of the greatest electric density, in which position the fluid resistance to approach would be smaller than to receding. If in a medium of the above characteristics a brush would be formed by a steady potential, an exchange of the carriers would go on continually, and there would be less carriers per unit of volume in the brush than in the space at some distance from the electrode, this corresponding to rarefaction. If the potential were rapidly changing, the result would be very different; the higher the freqency of the pulses, the slower would be the exchange of the carriers; finally, the motion of translation through measurable space would cease, and, with a sufficiently high frequency and intensity of the stress, the carriers would be drawn towards the electrode, and compression would result.

An interesting feature of these high frequency currents is that they allow of operating all kinds of devices by connecting the device with only one leading wire to the electric source. In fact, under certain conditions it may be more economical to supply the electrical energy with one lead than with two.

An experiment of special interest shown by Mr. Tesla, is the running, by the use of only one insulated line, of a motor operating on the principle of the rotating magnetic field enunciated by Mr. Tesla. A simple form of such a motor is obtained by winding upon a laminated iron core a primary and close to it a secondary coil, closing the ends of the latter and placing a freely movable metal disc within the influence of the moving field. The secondary coil may, however, be omitted. When one of the ends of the primary coil of the motor is connected to one of the terminals of the high frequency coil and the other end to an insulated metal plate, which, it should be stated, is not absolutely necessary for the success of the experiment, the disc is set in rotation.

Experiments of this kind seem to bring it within possibility to operate a motor at any point of the earth's surface from a central source, without any connection to the same except through the earth. If, by means of powerful machinery, rapid variations of the earth's potential were produced, a grounded wire reaching up to some height would be traversed by a current which could be increased by connecting the free end of the wire to a body of some size. The current might be converted to low tension and used to operate a motor or other device. The experiment, which would be one of great scientific interest, would probably best succeed on a ship at sea. In this manner, even if it were not possible to operate machinery, intelligence might be transmitted quite certainly.

In the course of this experimental study special attention was devoted to the heating effects produced by these currents, which are not only striking, but open up the possibility of producing a more efficient illuminant. It is sufficient to attach to the coil terminal a thin wire or filament, to have the temperature of the latter perceptibly raised. If the wire or filament be enclosed in a bulb, the heating effect is increased by preventing the circulation of the air. If the air in the bulb be strongly compressed, the displacements are smaller, the impacts less violent, and the heating effect is diminished. On the contrary, if the air in the bulb be exhausted, an inclosed lamp filament is brought to incandescence, and any amount of light may thus be produced.

The heating of the inclosed lamp filament depends on so many things of a different nature, that it is difficult to give a generally applicable rule under which the maximum heating occurs. As regards the size of the bulb, it is ascertained that at ordinary or only slightly differing atmospheric pressures, when air is a good insulator, the filament is heated more in a small bulb, because of the better confinement of heat in this case. At lower pressures, when air becomes conducting, the heating effect is greater in a large bulb, but at excessively high degrees of exhaustion there seems to be, beyond a certain and rather small size of the vessel, no perceptible difference in the heating.

The shape of the vessel is also of some importance, and it has been found of advantage for reasons of economy to employ a spherical bulb with the electrode mounted in its centre, where the rebounding molecules collide.

It is desirable on account of economy that all the energy supplied to the bulb from the source should reach without loss the body to be heated. The loss in conveying the energy from the source to the body may be reduced by employing thin wires heavily coated with insulation, and by the use of electrostatic screens. It is to be remarked, that the screen cannot be connected to the ground as under ordinary conditions.

In the bulb itself a large portion of the energy supplied may be lost by molecular bombardment against the wire connecting the body to be heated with the source. Considerable improvement was effected by covering the glass stem containing the wire with a closely fitting conducting tube. This tube is made to project a little above the glass, and prevents the cracking of the latter near the heated body. The effectiveness of the conducting tube is limited to very high degrees of exhaustion. It diminishes the energy lost in bombardment for two reasons; first, the charge given up by the atoms spreads over a greater area, and hence the electric density at any point is small, and the atoms are repelled with less energy than if they would strike against a good insulator; secondly, as the tube is electrified by the atoms which first come in contact with it, the progress of the following atoms against the tube is more or less checked by the repulsion which the electrified tube must exert upon the similarly electrified atoms. This, it is thought, explains why the discharge through a bulb is established with much greater facility when an insulator, than when a conductor, is present.

During the investigations a great many bulbs of different construction, with electrodes of different material, were experimented upon, and a number of observations of interest were made. Mr.

Tesla has found that the deterioration of the electrode is the less, the higher the frequency. This was to be expected, as then the heating is effected by many small impacts, instead by fewer and more violent ones, which quickly shatter the structure. The deterioration is also smaller when the vibration is harmonic. Thus an electrode, maintained at a certain degree of heat, lasts much longer with currents obtained from an alternator, than with those obtained by means of a disruptive discharge. One of the most durable electrodes was obtained from strongly compressed carborundum, which is a kind of carbon recently produced by Mr. E. G. Acheson, of Monongahela City, Pa. From experience, it is inferred, that to be most durable, the electrode should be in the form of a sphere with a highly polished surface.

In some bulbs refractory bodies were mounted in a carbon cup and put under the molecular impact. It was observed in such experiments that the carbon cup was heated at first, until a higher temperature was reached; then most of the bombardment was directed against the refractory body, and the carbon was relieved. In general, when different bodies were mounted in the bulb, the hardest fusible would be relieved, and would remain at a considerably lower temperature. This was necessitated by the fact that most of the energy supplied would find its way through the body which was more easily fused or "evaporated."

Curiously enough it appeared in some of the experiments made, that a body was fused in a bulb under the molecular impact by evolution of less light than when fused by the application of heat in ordinary ways. This may be ascribed to a loosening of the structure of the body under the violent impacts and changing stresses.

Some experiments seem to indicate that under certain conditions a body, conducting or nonconducting, may, when bombarded, emit light, which to all appearances is due to phosphorescence, but may in reality be caused by the incandescence of an infinitesimal layer, the mean temperature of the body being comparatively small. Such might be the case if each single rhythmical impact were capable of instantaneously exciting the retina, and the rhythm were just high enough to cause a continuous impression in the eye. According to this view, a coil operated by disruptive discharge would be eminently adapted to produce such a result, and it is found by experience that its power of exciting phosphorescence is extraordinarily great. It is capable of exciting phosphorescence at comparatively low degrees of exhaustion, and also projects shadows at pressures far greater than those at which the mean free path is comparable to the dimensions of the vessel. The latter observation is of some importance, inasmuch as it may modify the generally accepted views in regard to the "radiant state" phenomena.

A thought which early and naturally suggested itself to Mr. Tesla, was to utilize the great inductive effects of high frequency currents to produce light in a sealed glass vessel without the use of leading in wires. Accordingly, many bulbs were constructed in which the energy necessary to maintain a button or filament at high incandescence, was supplied through the glass by either electrostatic or electrodynamic induction. It was easy to regulate the intensity of the light emitted by means of an externally applied condenser coating connected to an insulated plate, or simply by means of a plate attached to the bulb which at the same time performed the function of a shade.

A subject of experiment, which has been exhaustively treated in England by Prof. J. J. Thomson, has been followed up independently by Mr. Tesla from the beginning of this study, namely, to excite by electrodynamic induction a luminous band in a closed tube or bulb. In observing the behavior of gases, and the luminous phenomena obtained, the importance of the electrostatic effects was noted and it appeared desirable to produce enormous potential differences, alternating with extreme rapidity. Experiments in this direction led to some of the most interesting results arrived at in the course of these investigations. It was found that by rapid alternations of a high electrostatic potential, exhausted tubes could be lighted at considerable distances from a conductor connected to a properly constructed coil, and that it was practicable to establish with the coil an alternating electrostatic field, acting through the whole room and lighting a tube wherever it was placed within the four walls. Phosphorescent bulbs may be excited in such a field, and it is easy to regulate the effect by connecting to the bulb a small insulated metal plate. It was likewise possible to maintain a filament or button mounted in a tube at bright incandescence, and, in one experiment, a mica vane was spun by the incandescence of a platinum wire.

Coming now to the lecture delivered in Philadelphia and St.

Louis, it may be remarked that to the superficial reader, Mr. Tesla's introduction, dealing with the importance of the eye, might appear as a digression, but the thoughtful reader will find therein much food for meditation and speculation. Throughout his discourse one can trace Mr. Tesla's effort to present in a popular way thoughts and views on the electrical phenomena which have in recent years captivated the scientific world, but of which the general public has even yet merely received an inkling. Mr. Tesla also dwells rather extensively on his well-known method of high-frequency conversion; and the large amount of detail information will be gratefully received by students and experimenters in this virgin field. The employment of apt analogies in explaining the fundamental principles involved makes it easy for all to gain a clear idea of their nature. Again, the ease with which, thanks to Mr. Tesla's efforts, these high-frequency currents may now be obtained from circuits carrying almost any kind of current, cannot fail to result in an extensive broadening of this field of research, which offers so many possibilities. Mr. Tesla, true philosopher as he is, does not hesitate to point out defects in some of his methods, and indicates the lines which to him seem the most promising. Particular stress is laid by him upon the employment of a medium in which the discharge electrodes should be immersed in order that this method of conversion may be brought to the highest perfection. He has evidently taken pains to give as much useful information as possible to those who wish to follow in his path, as he shows in detail the circuit arrangements to be adopted in all ordinary cases met with in practice, and although some of these methods were described by him two years before, the additional information is still timely and welcome.

In his experiments he dwells first on some phenomena produced by electrostatic force, which he considers in the light of modern theories to be the most important force in nature for us to investigate. At the very outset he shows a strikingly novel experiment illustrating the effect of a rapidly varying electrostatic force in a gaseous medium, by touching with one hand one of the terminals of a 200,000 volt transformer and bringing the other hand to the opposite terminal. The powerful streamers which issued from his hand and astonished his audiences formed a capital illustration of some of the views advanced, and afforded Mr. Tesla an opportunity of pointing out the true reasons why, with these currents, such an amount of energy can be passed through the body with impunity. He then showed by experiment the difference between a steady and a rapidly varying force upon the dielectric. This difference is most strikingly illustrated in the experiment in which a bulb attached to the end of a wire in connection with one of the terminals of the transformer is ruptured, although all extraneous bodies are remote from the bulb. He next illustrates how mechanical motions are produced by a varying electrostatic force acting through a gaseous medium. The importance of the action of the air is particularly illustrated by an interesting experiment.

Taking up another class of phenomena, namely, those of dynamic electricity, Mr. Tesla produced in a number of experiments a variety of effects by the employment of only a single wire with the evident intent of impressing upon his audience the idea that electric vibration or current can be transmitted with ease, without any return circuit; also how currents so transmitted can be converted and used for many practical purposes. A number of experiments are then shown, illustrating the effects of frequency, self-induction and capacity; then a number of ways of operating motive and other devices by the use of a single lead. A number of novel impedance phenomena are also shown which cannot fail to arouse interest.

Mr. Tesla next dwelt upon a subject which he thinks of great importance, that is, electrical resonance, which he explained in a popular way. He expressed his firm conviction that by observing proper conditions, intelligence, and possibly even power, can be transmitted through the medium or through the earth; and he considers this problem worthy of serious and immediate consideration.

Coming now to the light phenomena in particular, he illustrated the four distinct kinds of these phenomena in an original way, which to many must have been a revelation. Mr. Tesla attributes these light effects to molecular or atomic impacts produced by a varying electrostatic stress in a gaseous medium. He illustrated in a series of novel experiments the effect of the gas surrounding the conductor and shows beyond a doubt that with high frequency and high potential currents, the surrounding gas is of paramount importance in the heating of the conductor. He attributes the heating partially to a conduction current and partially to bombardment, and demonstrates that in many cases the

heating may be practically due to the bombardment alone.   He
pointed out also that the skin effect is largely modified by the
presence of the gas or of an atomic medium in general.   He
showed also some interesting experiments in which the effect of
convection is illustrated.   Probably one of the most curious experiments in this connection is that in which a thin platinum wire
stretched along the axis of an exhausted tube is brought to incandescence at certain points corresponding to the position of
the striæ, while at others it remains dark.   This experiment
throws an interesting light upon the nature of the striæ and may
lead to important revelations.

Mr. Tesla also demonstrated the dissipation of energy through
an atomic medium and dwelt upon the behavior of vacuous
space in conveying heat, and in this connection showed the curious
behavior of an electrode stream, from which he concludes that
the molecules of a gas probably cannot be acted upon directly
at measurable distances.

Mr. Tesla summarized the chief results arrived at in pursuing
his investigations in a manner which will serve as a valuable
guide to all who may engage in this work.   Perhaps most interest will centre on his general statements regarding the phenomena
of phosphorescence, the most important fact revealed in this direction being that when exciting a phosphorescent bulb a certain
definite potential gives the most economical result.

The lectures will now be presented in the order of their date
of delivery.

CHAPTER XXVI.

EXPERIMENTS WITH ALTERNATE CURRENTS OF VERY HIGH FREQUENCY AND THEIR APPLICATION TO METHODS OF ARTIFICIAL ILLUMINATION. [1]

THERE is no subject more captivating, more worthy of study, than nature. To understand this great mechanism, to discover the forces which are active, and the laws which govern them, is the highest aim of the intellect of man.

Nature has stored up in the universe infinite energy. The eternal recipient and transmitter of this infinite energy is the ether. The recognition of the existence of ether, and of the functions it performs, is one of the most important results of modern scientific research. The mere abandoning of the idea of action at a distance, the assumption of a medium pervading all space and connecting all gross matter, has freed the minds of thinkers of an ever present doubt, and, by opening a new horizon—new and unforeseen possibilities—has given fresh interest to phenomena with which we are familiar of old. It has been a great step towards the understanding of the forces of nature and their multifold manifestations to our senses. It has been for the enlightened student of physics what the understanding of the mechanism of the firearm or of the steam engine is for the barbarian. Phenomena upon which we used to look as wonders baffling explanation, we now see in a different light. The spark of an induction coil, the glow of an incandescent lamp, the manifestations of the mechanical forces of currents and magnets are no longer beyond our grasp; instead of the incomprehensible, as before, their observation suggests now in our minds a simple mechanism, and although as to its precise nature all is still conjecture, yet we know that the truth cannot be much longer hidden, and instinctively we feel that the understanding is dawning upon us. We still admire these beautiful phenomena, these

---

1. A lecture delivered before the American Institute of Electrical Engineers, at Columbia College, N. Y., May 20, 1891.

strange forces, but we are helpless no longer; we can in a certain measure explain them, account for them, and we are hopeful of finally succeeding in unraveling the mystery which surrounds them.

In how far we can understand the world around us is the ultimate thought of every student of nature. The coarseness of our senses prevents us from recognizing the ulterior construction of matter, and astronomy, this grandest and most positive of natural sciences, can only teach us something that happens, as it were, in our immediate neighborhood; of the remoter portions of the boundless universe, with its numberless stars and suns, we know nothing. But far beyond the limit of perception of our senses the spirit still can guide us, and so we may hope that even these unknown worlds—infinitely small and great—may in a measure become known to us. Still, even if this knowledge should reach us, the searching mind will find a barrier, perhaps forever unsurpassable, to the *true* recognition of that which *seems* to be, the mere *appearance* of which is the only and slender basis of all our philosophy.

Of all the forms of nature's immeasurable, all-pervading energy, which ever and ever changing and moving, like a soul animates the inert universe, electricity and magnetism are perhaps the most fascinating. The effects of gravitation, of heat and light we observe daily, and soon we get accustomed to them, and soon they lose for us the character of the marvelous and wonderful; but electricity and magnetism, with their singular relationship, with their seemingly dual character, unique among the forces in nature, with their phenomena of attractions, repulsions and rotations, strange manifestations of mysterious agents, stimulate and excite the mind to thought and research. What is electricity, and what is magnetism? These questions have been asked again and again. The most able intellects have ceaselessly wrestled with the problem; still the question has not as yet been fully answered. But while we cannot even to-day state what these singular forces are, we have made good headway towards the solution of the problem. We are now confident that electric and magnetic phenomena are attributable to ether, and we are perhaps justified in saying that the effects of static electricity are effects of ether under strain, and those of dynamic electricity and electro-magnetism effects of ether in motion. But this still leaves the question, as to what electricity and magnetism are, unanswered.

First, we naturally inquire, What is electricity, and is there such a thing as electricity? In interpreting electric phenomena, we may speak of electricity or of an electric condition, state or effect. If we speak of electric effects we must distinguish two such effects, opposite in character and neutralizing each other, as observation shows that two such opposite effects exist. This is unavoidable, for in a medium of the properties of ether, we cannot possibly exert a strain, or produce a displacement or motion of any kind, without causing in the surrounding medium an equivalent and opposite effect. But if we speak of electricity, meaning a *thing*, we must, I think, abandon the idea of two electricities, as the existence of two such things is highly improbable. For how can we imagine that there should be two things, equivalent in amount, alike in their properties, but of opposite character, both clinging to matter, both attracting and completely neutralizing each other? Such an assumption, though suggested by many phenomena, though most convenient for explaining them, has little to commend it. If there *is* such a thing as electricity, there can be only *one* such thing, and, excess and want of that one thing, possibly; but more probably its condition determines the positive and negative character. The old theory of Franklin, though falling short in some respects, is, from a certain point of view, after all, the most plausible one. Still, in spite of this, the theory of the two electricities is generally accepted, as it apparently explains electric phenomena in a more satisfactor manner. But a theory which better explains the facts is not necessarily true. Ingenious minds will invent theories to suit observation, and almost every independent thinker has his own views on the subject.

It is not with the object of advancing an opinion, but with the desire of acquainting you better with some of the results, which I will describe, to show you the reasoning I have followed, the departures I have made—that I venture to express, in a few words, the views and convictions which have led me to these results.

I adhere to the idea that there is a thing which we have been in the habit of calling electricity. The question is, What is that thing? or, What, of all things, the existence of which we know, have we the best reason to call electricity? We know that it acts like an incompressible fluid; that there must be a constant quantity of it in nature; that it can be neither produced nor destroyed;

and, what is more important, the electro-magnetic theory of light
and all facts observed teach us that electric and ether phenomena
are identical.   The idea at once suggests itself, therefore, that
electricity might be called ether.   In fact, this view has in a certain sense been advanced by Dr. Lodge.   His interesting work
has been read by everyone and many have been convinced by
his arguments.   His great ability and the interesting nature of
the subject, keep the reader spellbound; but when the impressions fade, one realizes that he has to deal only with ingenious
explanations.   I must confess, that I cannot believe in two electricities, much less in a doubly-constituted ether.   The puzzling
behavior of the ether as a solid to waves of light and heat, and
as a fluid to the motion of bodies through it, is certainly explained in the most natural and satisfactory manner by assuming
it to be in motion, as Sir William Thomson has suggested; but
regardless of this, there is nothing which would enable us to
conclude with certainty that, while a fluid is not capable of transmitting transverse vibrations of a few hundred or thousand per
second, it might not be capable of transmitting such vibrations
when they range into hundreds of million millions per second.
Nor can anyone prove that there are transverse ether waves
emitted from an alternate current machine, giving a small number of alternations per second; to such slow disturbances, the ether,
if at rest, may behave as a true fluid.

Returning to the subject, and bearing in mind that the existence of two electricities is, to say the least, highly improbable,
we must remember, that we have no evidence of electricity, nor
can we hope to get it, unless gross matter is present.   Electricity,
therefore, cannot be called ether in the broad sense of the term;
but nothing would seem to stand in the way of calling electricity
ether associated with matter, or bound ether; or, in other words,
that the so-called static charge of the molecule is ether associated
in some way with the molecule.   Looking at it in that light, we
would be justified in saying, that electricity is concerned in all
molecular actions.

Now, precisely what the ether surrounding the molecules is,
wherein it differs from ether in general, can only be conjectured.   It cannot differ in density, ether being incompressible;
it must, therefore, be under some strain or in motion, and the
latter is the most probable.   To understand its functions, it
would be necessary to have an exact idea of the physical con-

struction of matter, of which, of course, we can only form a mental picture.

But of all the views on nature, the one which assumes one matter and one force, and a perfect uniformity throughout, is the most scientific and most likely to be true. An infinitesimal world, with the molecules and their atoms spinning and moving in orbits, in much the same manner as celestial bodies, carrying with them and probably spinning with them ether, or in other words, carrying with them static charges, seems to my mind the most probable view, and one which, in a plausible manner, accounts for most of the phenomena observed. The spinning of the molecules and their ether sets up the ether tensions or electrostatic strains; the equalization of ether tensions sets up ether motions or electric currents, and the orbital movements produce the effects of electro and permanent magnetism.

About fifteen years ago, Prof. Rowland demonstrated a most interesting and important fact, namely, that a static charge carried around produces the effects of an electric current. Leaving out of consideration the precise nature of the mechanism, which produces the attraction and repulsion of currents, and conceiving the electrostatically charged molecules in motion, this experimental fact gives us a fair idea of magnetism. We can conceive lines or tubes of force which physically exist, being formed of rows of directed moving molecules; we can see that these lines must be closed, that they must tend to shorten and expand, etc. It likewise explains in a reasonable way, the most puzzling phenomenon of all, permanent magnetism, and, in general, has all the beauties of the Ampere theory without possessing the vital defect of the same, namely, the assumption of molecular currents. Without enlarging further upon the subject, I would say, that I look upon all electrostatic, current and magnetic phenomena as being due to electrostatic molecular forces.

The preceding remarks I have deemed necessary to a full understanding of the subject as it presents itself to my mind.

Of all these phenomena the most important to study are the current phenomena, on account of the already extensive and ever-growing use of currents for industrial purposes. It is now a century since the first practical source of current was produced, and, ever since, the phenomena which accompany the flow of currents have been diligently studied, and through the untiring efforts of scientific men the simple laws which govern them have been discovered. But these laws are found to hold good only when the currents are of a steady character. When the currents are rapidly varying in strength, quite different phenomena, often unexpected, present themselves, and quite different laws hold good, which even now have not been determined as fully as is desirable, though through the work, principally, of English scientists, enough knowledge has been gained on the subject to enable us to treat simple cases which now present themselves in daily practice.

The phenomena which are peculiar to the changing character of the currents are greatly exalted when the rate of change is increased, hence the study of these currents is considerably facilitated by the employment of properly constructed apparatus. It was with this and other objects in view that I constructed alternate current machines capable of giving more than two million reversals of current per minute, and to this circumstance it is principally due, that I am able to bring to your attention some of the results thus far reached, which I hope will prove to be a step in advance on account of their direct bearing upon one of the most important problems, namely, the production of a practical and efficient source of light.

The study of such rapidly alternating currents is very interesting. Nearly every experiment discloses something new. Many results may, of course, be predicted, but many more are unforeseen. The experimenter makes many interesting observations. For instance, we take a piece of iron and hold it against a magnet. Starting from low alternations and running up higher and higher we feel the impulses succeed each other faster and faster, get weaker and weaker, and finally disappear. We then observe a continuous pull; the pull, of course, is not continuous; it only appears so to us; our sense of touch is imperfect.

We may next establish an arc between the electrodes and observe, as the alternations rise, that the note which accompanies alternating arcs gets shriller and shriller, gradually weakens, and finally ceases. The air vibrations, of course, continue, but they are too weak to be perceived; our sense of hearing fails us.

We observe the small physiological effects, the rapid heating of the iron cores and conductors, curious inductive effects, interesting condenser phenomena, and still more interesting light phenomena with a high tension induction coil. All these experiments and observations would be of the greatest interest to the student, but their description would lead me too far from the principal subject. Partly for this reason, and partly on account of their vastly greater importance, I will confine myself to the description of the light effects produced by these currents.

In the experiments to this end a high tension induction coil or equivalent apparatus for converting currents of comparatively low into currents of high tension is used.

If you will be sufficiently interested in the results I shall describe as to enter into an experimental study of this subject; if you will be convinced of the truth of the arguments I shall advance— your aim will be to produce high frequencies and high potentials; in other words, powerful electrostatic effects. You will then encounter many difficulties, which, if completely overcome, would allow us to produce truly wonderful results.

First will be met the difficulty of obtaining the required frequencies by means of mechanical apparatus, and, if they be obtained otherwise, obstacles of a different nature will present themselves. Next it will be found difficult to provide the requisite insulation without considerably increasing the size of the apparatus, for the potentials required are high, and, owing to the rapidity of the alternations, the insulation presents peculiar difficulties. So, for instance, when a gas is present, the discharge may work, by the molecular bombardment of the gas and consequent heating, through as much as an inch of the best solid insulating material, such as glass, hard rubber, porcelain, sealing wax, etc.; in fact, through any known insulating substance. The chief requisite in the insulation of the apparatus is, therefore, the exclusion of any gaseous matter.

In general my experience tends to show that bodies which possess the highest specific inductive capacity, such as glass, afford a rather inferior insulation to others, which, while they are good insulators, have a much smaller specific inductive capacity, such as oils, for instance, the dielectric losses being no doubt greater in the former. The difficulty of insulating, of course, only exists when the potentials are excessively high, for with potentials such as a few thousand volts there is no particular difficulty encountered in conveying currents from a machine giving, say, 20,000 alternations per second, to quite a distance. This number of alternations, however, is by far too small for many purposes, though quite sufficient for some practical applications. This difficulty of insulating is fortunately not a vital drawback;

it affects mostly the size of the apparatus, for, when excessively high potentials would be used, the light-giving devices would be located not far from the apparatus, and often they would be quite close to it. As the air-bombardment of the insulated wire is dependent on condenser action, the loss may be reduced to a trifle by using excessively thin wires heavily insulated.

Another difficulty will be encountered in the capacity and self-induction necessarily possessed by the coil. If the coil be large, that is, if it contain a great length of wire, it will be generally unsuited for excessively high frequencies; if it be small, it may be well adapted for such frequencies, but the potential might then not be as high as desired. A good insulator, and preferably one possessing a small specific inductive capacity, would afford a two-fold advantage. First, it would enable us to construct a very small coil capable of withstanding enormous differences of potential; and secondly, such a small coil, by reason of its smaller capacity and self-induction, would be capable of a quicker and more vigorous vibration. The problem then of constructing a coil or induction apparatus of any kind possessing the requisite qualities I regard as one of no small importance, and it has occupied me for a considerable time.

The investigator who desires to repeat the experiments which I will describe, with an alternate current machine, capable of supplying currents of the desired frequency, and an induction coil, will do well to take the primary coil out and mount the secondary in such a manner as to be able to look through the tube upon which the secondary is wound. He will then be able to observe the streams which pass from the primary to the insulating tube, and from their intensity he will know how far he can strain the coil. Without this precaution he is sure to injure the insulation. This arrangment permits, however, an easy exchange of the primaries, which is desirable in these experiments.

The selection of the type of machine best suited for the purpose must be left to the judgment of the experimenter. There are here illustrated three distinct types of machines, which, besides others, I have used in my experiments.

Fig. 97 represents the machine used in my experiments before this Institute. The field magnet consists of a ring of wrought iron with 384 pole projections. The armature comprises a steel disc to which is fastened a thin, carefully welded rim of wrought iron. Upon the rim are wound several layers of fine, well annealed iron wire, which, when wound, is passed through shellac. The armature wires are wound around brass pins, wrapped with silk thread. The diameter of the armature wire in this type of machine should not be more than $\frac{1}{8}$ of the thickness of the pole projections, else the local action will be considerable.

Fig. 98 represents a larger machine of a different type. The field magnet of this machine consists of two like parts which either enclose an exciting coil, or else are independently wound.

FIG. 97.

Each part has 480 pole projections, the projections of one facing those of the other. The armature consists of a wheel of hard bronze, carrying the conductors which revolve between the projections of the field magnet. To wind the armature conductors, I have found it most convenient to proceed in the following manner. I construct a ring of hard bronze of the required size. This ring and the rim of the wheel are provided with the proper number of pins, and both fastened upon a plate. The armature conductors being wound, the pins are cut off and the ends of the conductors fastened by two rings which screw to the bronze ring and the rim of the wheel, respectively. The whole may then be taken off and forms a solid structure. The conductors in such a type of machine should consist of sheet copper, the thickness of which, of course, depends on the thickness of the pole projections; or else twisted thin wires should be employed.

Fig. 99 is a smaller machine, in many respects similar to the former, only here the armature conductors and the exciting coil are kept stationary, while only a block of wrought iron is revolved.

It would be uselessly lengthening this description were I to

Fig. 98.

dwell more on the details of construction of these machines. Besides, they have been described somewhat more elaborately in *The Electrical Engineer*, of March 18, 1891. I deem it well, however, to call the attention of the investigator to two things, the importance of which, though self evident, he is nevertheless apt to underestimate; namely, to the local action in the conductors which must be carefully avoided, and to the clearance, which must be small. I may add, that since it is desirable to use very high peripheral speeds, the armature should be of very large diameter in order to avoid impracticable belt speeds. Of the several types of these machines which have been constructed by me, I have found that the type illustrated in Fig. 97 caused me the least trouble in construction, as well as in maintenance, and on the whole, it has been a good experimental machine.

In operating an induction coil with very rapidly alternating currents, among the first luminous phenomena noticed are naturally those presented by the high-tension discharge. As the number of alternations per second is increased, or as—the number being high—the current through the primary is varied, the discharge gradually changes in appearance. It would be difficult to describe the minor changes which occur, and the conditions which

Fig. 99.

bring them about, but one may note five distinct forms of the discharge.

First, one may observe a weak, sensitive discharge in the form of a thin, feeble-colored thread. (Fig. 100a.) It always occurs when, the number of alternations per second being high, the current through the primary is very small. In spite of the excessively small current, the rate of change is great, and the difference of potential at the terminals of the secondary is therefore considerable, so that the arc is established at great distances; but the quantity of "electricity" set in motion is insignificant, barely sufficient to maintain a thin, threadlike arc. It is excessively sensitive and may be made so to such a degree that the mere act of breathing near the coil will affect it, and unless it is perfectly

well protected from currents of air, it wriggles around constantly.
Nevertheless, it is in this form excessively persistent, and when
the terminals are approached to, say, one-third of the striking
distance, it can be blown out only with difficulty. This exceptional persistency, when short, is largely due to the arc being
excessively thin; presenting, therefore, a very small surface
to the blast. Its great sensitiveness, when very long, is probably
due to the motion of the particles of dust suspended in the air.

When the current through the primary is increased, the discharge gets broader and stronger, and the effect of the capacity
of the coil becomes visible until, finally, under proper conditions,
a white flaming arc, Fig. 100 B, often as thick as one's finger, and
striking across the whole coil, is produced. It develops remarkable heat, and may be further characterized by the absence of
the high note which accompanies the less powerful discharges.
To take a shock from the coil under these conditions would not

FIG. 100a.   FIG. 100b.

be advisable, although under different conditions, the potential
being much higher, a shock from the coil may be taken with
impunity. To produce this kind of discharge the number of
alternations per second must not be too great for the coil used;
and, generally speaking, certain relations between capacity, self-
induction and frequency must be observed.

The importance of these elements in an alternate current circuit is now well-known, and under ordinary conditions, the general rules are applicable. But in an induction coil exceptional
conditions prevail. First, the self-induction is of little importance
before the arc is established, when it asserts itself, but perhaps
never as prominently as in ordinary alternate current circuits,
because the capacity is distributed all along the coil, and by reason
of the fact that the coil usually discharges through very great
resistances; hence the currents are exceptionally small. Secondly,

the capacity goes on increasing continually as the potential rises, in consequence of absorption which takes place to a considerable extent. Owing to this there exists no critical relationship between these quantities, and ordinary rules would not seem to be applicable. As the potential is increased either in consequence of the increased frequency or of the increased current through the primary, the amount of the energy stored becomes greater and greater, and the capacity gains more and more in importance. Up to a certain point the capacity is beneficial, but after that it begins to be an enormous drawback. It follows from this that each coil gives the best result with a given frequency and primary current. A very large coil, when operated with currents of very high frequency, may not give as much as $\frac{1}{8}$ inch spark. By adding capacity to the terminals, the condition may be improved, but what the coil really wants is a lower frequency.

When the flaming discharge occurs, the conditions are evidently such that the greatest current is made to flow through the circuit. These conditions may be attained by varying the frequency within wide limits, but the highest frequency at which the flaming arc can still be produced, determines, for a given primary current, the maximum striking distance of the coil. In the flaming discharge the *eclat* effect of the capacity is not perceptible; the rate at which the energy is being stored then just equals the rate at which it can be disposed of through the circuit. This kind of discharge is the severest test for a coil; the break, when it occurs, is of the nature of that in an overcharged Leyden jar. To give a rough approximation I would state that, with an ordinary coil of, say 10,000 ohms resistance, the most powerful arc would be produced with about 12,000 alternations per second.

When the frequency is increased beyond that rate, the potential, of course, rises, but the striking distance may, nevertheless, diminish, paradoxical as it may seem. As the potential rises the coil attains more and more the properties of a static machine until, finally, one may observe the beautiful phenomenon of the streaming discharge, Fig. 101, which may be produced across the whole length of the coil. At that stage streams begin to issue freely from all points and projections. These streams will also be seen to pass in abundance in the space between the primary and the insulating tube. When the potential is excessively high they will always appear, even if the frequency be low, and even if the primary be surrounded by as much as an inch of wax, hard rub-

ber, glass, or any other insulating substance. This limits greatly
the output of the coil, but I will later show how I have been able
to overcome to a considerable extent this disadvantage in the
ordinary coil.

Besides the potential, the intensity of the streams depends on
the frequency; but if the coil be very large they show themselves, no matter how low the frequencies used. For instance,
in a very large coil of a resistance of 67,000 ohms, constructed
by me some time ago, they appear with as low as 100 alternations
per second and less, the insulation of the secondary being ¾ inch
of ebonite. When very intense they produce a noise similar to
that produced by the charging of a Holtz machine, but much
more powerful, and they emit a strong smell of ozone. The
lower the frequency, the more apt they are to suddenly injure
the coil. With excessively high frequencies they may pass freely

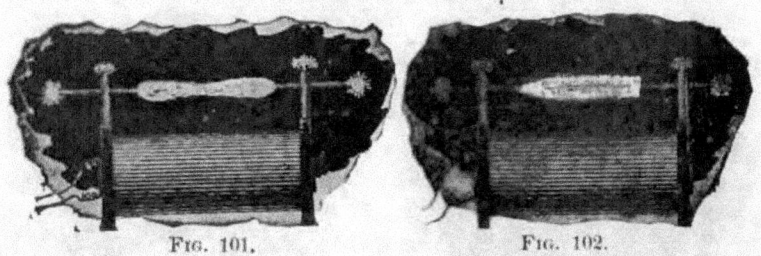

FIG. 101.   FIG. 102.

without producing any other effect than to heat the insulation
slowly and uniformly.

The existence of these streams shows the importance of constructing an expensive coil so as to permit of one's seeing
through the tube surrounding the primary, and the latter should
be easily exchangeable; or else the space between the primary
and secondary should be completely filled up with insulating
material so as to exclude all air. The non-observance of this
simple rule in the construction of commercial coils is responsible
for the destruction of many an expensive coil.

At the stage when the streaming discharge occurs, or with
somewhat higher frequencies, one may, by approaching the terminals quite nearly, and regulating properly the effect of capacity, produce a veritable spray of small silver-white sparks, or a
bunch of excessively thin silvery threads (Fig. 102) amidst a
powerful brush—each spark or thread possibly corresponding

to one alternation. This, when produced under proper conditions, is probably the most beautiful discharge, and when an air blast is directed against it, it presents a singular appearance. The spray of sparks, when received through the body, causes some inconvenience, whereas, when the discharge simply streams, nothing at all is likely to be felt if large conducting objects are held in the hands to protect them from receiving small burns.

If the frequency is still more increased, then the coil refuses to give any spark unless at comparatively small distances, and the fifth typical form of discharge may be observed (Fig. 103). The tendency to stream out and dissipate is then so great that when the brush is produced at one terminal no sparking occurs, even if, as I have repeatedly tried, the hand, or any conducting object, is held within the stream; and, what is more singular, the lumi-

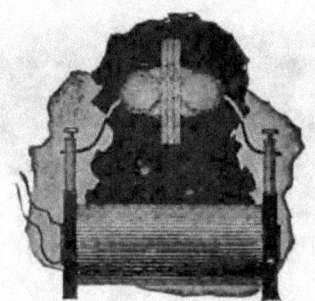

Fig. 103.   Fig. 104.

nous stream is not at all easily deflected by the approach of a conducting body.

At this stage the streams seemingly pass with the greatest freedom through considerable thicknesses of insulators, and it is particularly interesting to study their behavior. For this purpose it is convenient to connect to the terminals of the coil two metallic spheres which may be placed at any desired distance, Fig. 104. Spheres are preferable to plates, as the discharge can be better observed. By inserting dielectric bodies between the spheres, beautiful discharge phenomena may be observed. If the spheres be quite close and a spark be playing between them, by interposing a thin plate of ebonite between the spheres the spark instantly ceases and the discharge spreads into an intensely luminous circle several inches in diameter, provided the spheres are sufficiently large.   The passage of the streams heats, and, after a while, softens, the rubber so much that two plates may be made to stick together in this manner. If the spheres are so far apart that no spark occurs, even if they are far beyond the striking distance, by inserting a thick plate of glass the discharge is instantly induced to pass from the spheres to the glass in the form of luminous streams.   It appears almost as though these streams pass *through* the dielectric.   In reality this is not the case, as the streams are due to the molecules of the air which are violently agitated in the space between the oppositely charged surfaces of the spheres.   When no dielectric other than air is present, the bombardment goes on, but is too weak to be visible; by inserting a dielectric the inductive effect is much increased, and besides, the projected air molecules find an obstacle and the bombardment becomes so intense that the streams become luminous.   If by any mechanical means we could effect such a violent agitation of the molecules we could produce the same phenomenon.   A jet of air escaping through a small hole under enormous pressure and striking against an insulating substance, such as glass, may be luminous in the dark, and it might be possible to produce a phosphorescence of the glass or other insulators in this manner.

The greater the specific inductive capacity of the interposed dielectric, the more powerful the effect produced.   Owing to this, the streams show themselves with excessively high potentials even if the glass be as much as one and one-half to two inches thick.   But besides the heating due to bombardment, some heating goes on undoubtedly in the dielectric, being apparently greater in glass than in ebonite.   I attribute this to the greater specific inductive capacity of the glass, in consequence of which, with the same potential difference, a greater amount of energy is taken up in it than in rubber.   It is like connecting to a battery a copper and a brass wire of the same dimensions.   The copper wire, though a more perfect conductor, would heat more by reason of its taking more current.   Thus what is otherwise considered a virtue of the glass is here a defect.   Glass usually gives way much quicker than ebonite; when it is heated to a certain degree, the discharge suddenly breaks through at one point, assuming then the ordinary form of an arc.

The heating effect produced by molecular bombardment of the dielectric would, of course, diminish as the pressure of the air is increased, and at enormous pressure it would be negligible, unless the frequency would increase correspondingly.

It will be often observed in these experiments that when the spheres are beyond the striking distance, the approach of a glass plate, for instance, may induce the spark to jump between the spheres. This occurs when the capacity of the spheres is somewhat below the critical value which gives the greatest difference of potential at the terminals of the coil. By approaching a dielectric, the specific inductive capacity of the space between the spheres is increased, producing the same effect as if the capacity of the spheres were increased. The potential at the terminals may then rise so high that the air space is cracked. The experiment is best performed with dense glass or mica.

Another interesting observation is that a plate of insulating material, when the discharge is passing through it, is strongly attracted by either of the spheres, that is by the nearer one, this being obviously due to the smaller mechanical effect of the bombardment on that side, and perhaps also to the greater electrification.

From the behavior of the dielectrics in these experiments, we may conclude that the best insulator for these rapidly alternating currents would be the one possessing the smallest specific inductive capacity and at the same time one capable of withstanding the greatest differences of potential; and thus two diametrically opposite ways of securing the required insulation are indicated, namely, to use either a perfect vacuum or a gas under great pressure; but the former would be preferable. Unfortunately neither of these two ways is easily carried out in practice.

It is especially interesting to note the behavior of an excessively high vacuum in these experiments. If a test tube, provided with external electrodes and exhausted to the highest possible degree, be connected to the terminals of the coil, Fig. 105, the electrodes of the tube are instantly brought to a high temperature and the glass at each end of the tube is rendered intensely phosphorescent, but the middle appears comparatively dark, and for a while remains cool.

When the frequency is so high that the discharge shown in Fig. 103 is observed, considerable dissipation no doubt occurs in the coil. Nevertheless the coil may be worked for a long time, as the heating is gradual.

In spite of the fact that the difference of potential may be enormous, little is felt when the discharge is passed through the body, provided the hands are armed. This is to some extent due to the higher frequency, but principally to the fact that less energy is available externally, when the difference of potential reaches an enormous value, owing to the circumstance that, with the rise of potential, the energy absorbed in the coil increases as the square of the potential. Up to a certain point the energy available externally increases with the rise of potential, then it begins to fall off rapidly. Thus, with the ordinary high tension induction coil, the curious paradox exists, that, while with a given current through the primary the shock might be fatal, with many times that current it might be perfectly harmless, even if the frequency be the same. With high frequencies and excessively high potentials when the terminals are not connected to bodies of some size, practically all the energy supplied to the primary is

FIG. 105.   FIG. 106.

taken up by the coil. There is no breaking through, no local injury, but all the material, insulating and conducting, is uniformly heated.

To avoid misunderstanding in regard to the physiological effect of alternating currents of very high frequency, I think it necessary to state that, while it is an undeniable fact that they are incomparably less dangerous than currents of low frequencies, it should not be thought that they are altogether harmless. What has just been said refers only to currents from an ordinary high tension induction coil, which currents are necessarily very small ; if received directly from a machine or from a secondary of low resistance, they produce more or less powerful effects, and may cause serious injury, especially when used in conjunction with condensers.

The streaming discharge of a high tension induction coil differs in many respects from that of a powerful static machine. In color it has neither the violet of the positive, nor the brightness of the negative, static discharge, but lies somewhere between, being, of course, alternatively positive and negative. But since the streaming is more powerful when the point or terminal is electrified positively, than when electrified negatively, it follows that the point of the brush is more like the positive, and the root more like the negative, static discharge. In the dark, when the brush is very powerful, the root may appear almost white. The wind produced by the escaping streams, though it may be very strong—often indeed to such a degree that it may be felt quite a distance from the coil—is, nevertheless, considering the quantity of the discharge, smaller than that produced by the positive

FIG. 107.   FIG. 108.

brush of a static machine, and it affects the flame much less powerfully. From the nature of the phenomenon we can conclude that the higher the frequency, the smaller must, of course, be the wind produced by the streams, and with sufficiently high frequencies no wind at all would be produced at the ordinary atmospheric pressures. With frequencies obtainable by means of a machine, the mechanical effect is sufficiently great to revolve, with considerable speed, large pin-wheels, which in the dark present a beautiful appearance owing to the abundance of the streams (Fig. 106).

In general, most of the experiments usually performed with a static machine can be performed with an induction coil when operated with very rapidly alternating currents. The effects produced, however, are much more striking, being of incomparably greater power. When a small length of ordinary cotton covered wire, Fig. 107, is attached to one terminal of the coil, the streams issuing from all points of the wire may be so intense as to produce a considerable light effect. When the potentials and frequencies are very high, a wire insulated with gutta percha or rubber and attached to one of the terminals, appears to be covered with a luminous film. A very thin bare wire when attached to a terminal emits powerful streams and vibrates continually to and fro or spins in a circle, producing a singular effect (Fig. 108). Some of these experiments have been described by me in *The Electrical World*, of February 21, 1891.

Another peculiarity of the rapidly alternating discharge of the induction coil is its radically different behavior with respect to points and rounded surfaces.

If a thick wire, provided with a ball at one end and with a point at the other, be attached to the positive terminal of a static machine, practically all the charge will be lost through the point, on account of the enormously greater tension, dependent on the radius of curvature. But if such a wire is attached to one of the terminals of the induction coil, it will be observed that with very high frequencies streams issue from the ball almost as copiously as from the point (Fig. 109).

It is hardly conceivable that we could produce such a condition to an equal degree in a static machine, for the simple reason, that the tension increases as the square of the density, which in turn is proportional to the radius of curvature; hence, with a steady potential an enormous charge would be required to make streams issue from a polished ball while it is connected with a point. But with an induction coil the discharge of which alternates with great rapidity it is different. Here we have to deal with two distinct tendencies. First, there is the tendency to escape which exists in a condition of rest, and which depends on the radius of curvature; second, there is the tendency to dissipate into the surrounding air by condenser action, which depends on the surface. When one of these tendencies is a maximum, the other is at a minimum. At the point the luminous stream is principally due to the air molecules coming bodily in contact with the point; they are attracted and repelled, charged and discharged, and, their atomic charges being thus disturbed, vibrate and emit light waves. At the ball, on the contrary, there is no doubt that the effect is to a great extent produced inductively, the air molecules not *necessarily* coming in contact with the ball, though they undoubtedly do so. To convince ourselves of this we only need to exalt the condenser action, for instance, by enveloping the ball, at some distance, by a better conductor than the surrounding medium, the conductor being, of course, insulated; or else by surrounding it with a better dielectric and approaching an insulated conductor; in both cases the streams will break forth more copiously. Also, the larger the ball with a given frequency, or the higher the frequency, the more will the ball have the advantage over the point. But, since a certain intensity of action is required to render the streams visible, it is obvious that in the experiment described the ball should not be taken too large.

In consequence of this two-fold tendency, it is possible to produce by means of points, effects identical to those produced by

Fig. 109.   Fig. 110.

capacity. Thus, for instance, by attaching to one terminal of the coil a small length of soiled wire, presenting many points and offering great facility to escape, the potential of the coil may be raised to the same value as by attaching to the terminal a polished ball of a surface many times greater than that of the wire.

An interesting experiment, showing the effect of the points, may be performed in the following manner: Attach to one of the terminals of the coil a cotton covered wire about two feet in length, and adjust the conditions so that streams issue from the wire. In this experiment the primary coil should be preferably placed so that it extends only about half way into the secondary coil. Now touch the free terminal of the secondary with a conducting object held in the hand, or else connect it to an insulated body of some size. In this manner the potential on the wire may be enormously raised. The effect of this will be either to increase, or to diminish, the streams. If they increase, the wire is too short; if they diminish, it is too long. By adjusting the length of the wire, a point is found where the touching of the other terminal does not at all affect the streams. In this case the rise of potential is exactly counteracted by the drop through the coil. It will be observed that small lengths of wire produce considerable difference in the magnitude and luminosity of the streams. The primary coil is placed sidewise for two reasons: First, to increase the potential at the wire; and, second, to increase the drop through the coil. The sensitiveness is thus augmented.

There is still another and far more striking peculiarity of the brush discharge produced by very rapidly alternating currents. To observe this it is best to replace the usual terminals of the coil by two metal columns insulated with a good thickness of ebonite. It is also well to close all fissures and cracks with wax so that the brushes cannot form anywhere except at the tops of the columns. If the conditions are carefully adjusted—which, of course, must be left to the skill of the experimenter—so that the potential rises to an enormous value, one may produce two powerful brushes several inches long, nearly white at their roots, which in the dark bear a striking resemblance to two flames of a gas escaping under pressure (Fig. 110). But they do not only *resemble*, they *are* veritable flames, for they are hot. Certainly they are not as hot as a gas burner, *but they would be so if the frequency and the potential would be sufficiently high.* Produced with, say, twenty thousand alternations per second, the heat is easily perceptible even if the potential is not excessively high. The heat devoloped is, of course, due to the impact of the air molecules against the terminals and against each other. As, at the ordinary pressures, the mean free path is excessively small, it is possible that in spite of the enormous initial speed imparted to each molecule upon coming in contact with the terminal, its progress—by collision with other molecules—is retarded to such an extent, that it does not get away far from the terminal, but may strike the same many times in succession. The higher the frequency, the less the molecule is able to get away, and this the more so, as for a given effect the potential required is smaller; and a frequency is conceivable—perhaps even obtainable—at which practically the same molecules would strike the terminal. Under such conditions the exchange of the molecules would be very slow, and the heat produced at, and very near, the terminal would be excessive. But if the frequency would go on increasing constantly, the heat produced would begin to diminish for obvious reasons. In the positive brush of a static machine the exchange of the molecules is very rapid, the stream is constantly of one direction, and there are fewer collisions; hence the heating effect must be very small. Anything that impairs the facility of exchange tends to increase the local heat produced. Thus, if a bulb be held over the terminal of the coil so as to enclose the brush, the air contained in the bulb is very quickly brought to a high temperature. If a glass tube be held over the brush so as to allow the draught to carry the brush upwards, scorching hot air escapes at the top of the tube. Anything held within the brush is, of course, rapidly heated, and the possibility of using such heating effects for some purpose or other suggests itself.

When contemplating this singular phenomenon of the hot brush, we cannot help being convinced that a similar process must take place in the ordinary flame, and it seems strange that after all these centuries past of familiarity with the flame, now, in this era of electric lighting and heating, we are finally led to recognize, that since time immemorial we have, after all, always had "electric light and heat" at our disposal. It is also of no little interest to contemplate, that we have a possible way of producing—by other than chemical means—a veritable flame, which would give light and heat without any material being consumed, without any chemical process taking place, and to accomplish this, we only need to perfect methods of producing enormous frequencies and potentials. I have no doubt that if the potential could be made to alternate with sufficient rapidity and power, the brush formed at the end of a wire would lose its electrical characteristics and would become flamelike. The flame must be due to electrostatic molecular action.

This phenomenon now explains in a manner which can hardly be doubted the frequent accidents occurring in storms. It is well known that objects are often set on fire without the lightning striking them. We shall presently see how this can happen. On a nail in a roof, for instance, or on a projection of any kind, more or less conducting, or rendered so by dampness, a powerful brush may appear. If the lightning strikes somewhere in the

neighborhood the enormous potential may be made to alternate
or fluctuate perhaps many million times a second. The air
molecules are violently attracted and repelled, and by their impact produce such a powerful heating effect that a fire is started.
It is conceivable that a ship at sea may, in this manner, catch fire
at many points at once. When we consider, that even with the
comparatively low frequencies obtained from a dynamo machine,
and with potentials of no more than one or two hundred thous-
and volts, the heating effects are considerable, we may imagine
how much more powerful they must be with frequencies and potentials many times greater; and the above explanation seems, to
say the least, very probable. Similar explanations may have been
suggested, but I am not aware that, up to the present, the heating effects of a brush produced by a rapidly alternating potential

FIG. 111.

have been experimentally demonstrated, at least not to such a
remarkable degree.

By preventing completely the exchange of the air molecules,
the local heating effect may be so exalted as to bring a body to
incandescence. Thus, for instance, if a small button, or preferably a very thin wire or filament be enclosed in an unexhausted
globe and connected with the terminal of the coil, it may be
rendered incandescent. The phenomenon is made much more
interesting by the rapid spinning round in a circle of the top of
the filament, thus presenting the appearance of a luminous funnel, Fig. 111, which widens when the potential is increased.
When the potential is small the end of the filament may perform
irregular motions, suddenly changing from one to the other, or
it may describe an ellipse; but when the potential is very
high it always spins in a circle; and so does generally a thin

straight wire attached freely to the terminal of the coil. These motions are, of course, due to the impact of the molecules, and the irregularity in the distribution of the potential, owing to the roughness and dissymmetry of the wire or filament. With a perfectly symmetrical and polished wire such motions would probably not occur. That the motion is not likely to be due to others causes is evident from the fact that it is not of a definite direction, and that in a very highly exhausted globe it ceases altogether. The possibility of bringing a body to incandescence in an exhausted globe, or even when not at all enclosed, would seem to afford a possible way of obtaining light effects, which, in perfecting methods of producing rapidly alternating potentials, might be rendered available for useful purposes.

In employing a commercial coil, the production of very powerful brush effects is attended with considerable difficulties, for

Fig. 112a.

when these high frequencies and enormous potentials are used, the best insulation is apt to give way. Usually the coil is insulated well enough to stand the strain from convolution to convolution, since two double silk covered paraffined wires will withstand a pressure of several thousand volts; the difficulty lies principally in preventing the breaking through from the secondary to the primary, which is greatly facilitated by the streams issuing from the latter. In the coil, of course, the strain is greatest from section to section, but usually in a larger coil there are so many sections that the danger of a sudden giving way is not very great. No difficulty will generally be encountered in that direction, and besides, the liability of injuring the coil internally is very much reduced by the fact that the effect most likely to be produced is simply a gradual heating, which, when far enough advanced, could not fail to be observed. The principal necessity is then to prevent the streams between the primary and the tube, not only on account of the heating and possible injury, but also because the streams may diminish very considerably the potential difference available at the terminals. A few hints as to how this may be accomplished will probably be found useful in most of these experiments with the ordinary induction coil.

One of the ways is to wind a short primary, Fig. 112a, so that the difference of potential is not at that length great enough to cause the breaking forth of the streams through the insulating tube. The length of the primary should be determined by experiment. Both the ends of the coil should be brought out on one end through a plug of insulating material fitting in the tube as illustrated. In such a disposition one terminal of the secondary is attached to a body, the surface of which is determined with the

Fig. 112b.

greatest care so as to produce the greatest rise in the potential. At the other terminal a powerful brush appears, which may be experimented upon.

The above plan necessitates the employment of a primary of comparatively small size, and it is apt to heat when powerful effects are desirable for a certain length of time. In such a case it is better to employ a larger coil, Fig. 112b, and introduce it from one side of the tube, until the streams begin to appear. In this case the nearest terminal of the secondary may be connected to the primary or to the ground, which is practically the same thing, if the primary is connected directly to the machine. In the case of ground connections it is well to determine experimentally the frequency which is best suited under the conditions of the test. Another way of obviating the streams, more or less, is to make the primary in sections and supply it from separate, well insulated sources.

In many of these experiments, when powerful effects are wanted for a short time, it is advantageous to use iron cores with the primaries. In such case a very large primary coil may be wound and placed side by side with the secondary, and, the nearest terminal of the latter being connected to the primary, a laminated iron core is introduced through the primary into the secondary as far as the streams will permit. Under these conditions an excessively powerful brush, several inches long, which may be appropriately called "St. Elmo's hot fire," may be caused to appear at the other terminal of the secondary, producing striking effects. It is a most powerful ozonizer, so powerful indeed, that only a few minutes are sufficient to fill the whole room with the smell of ozone, and it undoubtedly possesses the quality of exciting chemical affinities.

For the production of ozone, alternating currents of very high frequency are eminently suited, not only on account of the advantages they offer in the way of conversion but also because of the fact, that the ozonizing action of a discharge is dependent on the frequency as well as on the potential, this being undoubtedly confirmed by observation.

In these experiments if an iron core is used it should be carefully watched, as it is apt to get excessively hot in an incredibly short time. To give an idea of the rapidity of the heating, I will state, that by passing a powerful current through a coil with many turns, the inserting within the same of a thin iron wire for no more than one second's time is sufficient to heat the wire to something like 100° C.

But this rapid heating need not discourage us in the use of iron cores in connection with rapidly alternating currents. I have for a long time been convinced that in the industrial distribution by means of transformers, some such plan as the following might be practicable. We may use a comparatively small iron core, subdivided, or perhaps not even subdivided. We may surround this core with a considerable thickness of material which is fire-proof and conducts the heat poorly, and on top of that we may place the primary and secondary windings. By using either higher frequencies or greater magnetizing forces, we may by hysteresis and eddy currents heat the iron core so far as to bring it nearly to its maximum permeability, which, as Hopkinson has shown, may be as much as sixteen times greater than that at ordinary temperatures. If the iron core were perfectly enclosed, it would not be deteriorated by the heat, and, if the enclosure of fire-proof material would be sufficiently thick, only a limited amount of energy could be radiated in spite of the high temperature. Transformers have been constructed by me on that plan, but for lack of time, no thorough tests have as yet been made.

Another way of adapting the iron core to rapid alternations, or, generally speaking, reducing the frictional losses, is to produce by continuous magnetization a flow of something like seven thousand or eight thousand lines per square centimetre through the core, and then work with weak magnetizing forces and preferably high frequencies around the point of greatest permeability. A higher efficiency of conversion and greater output are obtainable in this manner. I have also employed this principle in connection with machines in which there is no reversal of polarity. In these types of machines, as long as there are only few pole projections, there is no great gain, as the maxima and minima of magnetization are far from the point of maximum permeability: but when the number of the pole projections is very great, the required rate of change may be obtained, without the magnetization varying so far as to depart greatly from the point of maximum permeability, and the gain is considerable.

The above described arrangements refer only to the use of commercial coils as ordinarily constructed. If it is desired to construct a coil for the express purpose of performing with it such experiments as I have described, or, generally, rendering it capable of withstanding the greatest possible difference of potential, then a construction as indicated in Fig. 113 will be found of advantage. The coil in this case is formed of two independent parts which are wound oppositely, the connection between both being made near the primary. The potential in the middle being zero, there is not much tendency to jump to the primary and not much insulation is required. In some cases the middle point may, however, be connected to the primary or to the ground. In such a coil the places of greatest difference of potential are far apart and the coil is capable of withstanding an enormous strain. The two parts may be movable so as to allow a slight adjustment of the capacity effect.

As to the manner of insulating the coil, it will be found convenient to proceed in the following way: First, the wire should be boiled in paraffine until all the air is out; then the coil is wound by running the wire through melted paraffine, merely for the purpose of fixing the wire. The coil is then taken off from the spool, immersed in a cylindrical vessel filled with pure melted wax and boiled for a long time until the bubbles cease to appear. The whole is then left to cool down thoroughly, and then the mass is taken out of the vessel and turned up in a lathe. A coil made in this manner and with care is capable of withstanding enormous potential differences.

It may be found convenient to immerse the coil in paraffine oil or some other kind of oil; it is a most effective way of insulating, principally on account of the perfect exclusion of air, but it may

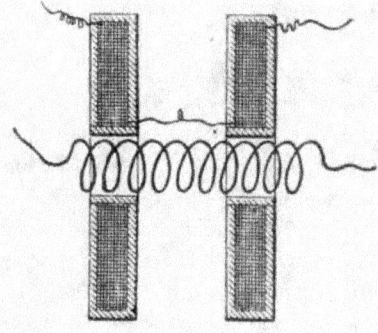

Fig. 113.

be found that, after all, a vessel filled with oil is not a very convenient thing to handle in a laboratory.

If an ordinary coil can be dismounted, the primary may be taken out of the tube and the latter plugged up at one end, filled with oil, and the primary reinserted. This affords an excellent insulation and prevents the formation of the streams.

Of all the experiments which may be performed with rapidly alternating currents the most interesting are those which concern the production of a practical illuminant. It cannot be denied that the present methods, though they were brilliant advances, are very wasteful. Some better methods must be invented, some more perfect apparatus devised. Modern research has opened new possibilities for the production of an efficient source of light, and the attention of all has been turned in the direction indicated by able pioneers. Many have been carried away by the enthusiasm and passion to discover, but in their zeal to reach results, some have been misled. Starting with the idea of producing electromagnetic waves, they turned their attention, perhaps, too much to the study of electro-magnetic effects, and neglected the study of electrostatic phenomena. Naturally, nearly every investigator availed himself of an apparatus similar to that used in earlier experiments. But in those forms of apparatus, while the electromagnetic inductive effects are enormous, the electrostatic effects are excessively small.

In the Hertz experiments, for instance, a high tension induction coil is short circuited by an arc, the resistance of which is very small, the smaller, the more capacity is attached to the terminals; and the difference of potential at these is enormously diminished. On the other hand, when the discharge is not passing between the terminals, the static effects may be considerable, but only qualitatively so, not quantitatively, since their rise and fall is very sudden, and since their frequency is small. In neither case, therefore, are powerful electrostatic effects perceivable. Similar conditions exist when, as in some interesting experiments of Dr. Lodge, Leyden jars are discharged disruptively. It has been thought — and I believe asserted — that in such cases most of the energy is radiated into space. In the light of the experiments which I have described above, it will now not be thought so. I feel safe in asserting that in such cases most of the energy is partly taken up and converted into heat in the arc of the discharge and in the conducting and insulating material of the jar, some energy being, of course, given off by electrification of the air; but the amount of the directly radiated energy is very small.

When a high tension induction coil, operated by currents alternating only 20,000 times a second, has its terminals closed through even a very small jar, practically all the energy passes through the dielectric of the jar, which is heated, and the electrostatic effects manifest themselves outwardly only to a very weak degree. Now the external circuit of a Leyden jar, that is, the arc and the connections of the coatings, may be looked upon as a circuit generating alternating currents of excessively high frequency and fairly high potential, which is closed through the coatings and the dielectric between them, and from the above it is evident that the external electrostatic effects must be very small, even if a recoil circuit be used. These conditions make it appear that with the apparatus usually at hand, the observation of powerful electrostatic effects was impossible, and what experience has been gained in that direction is only due to the great ability of the investigators.

But powerful electrostatic effects are a *sine qua non* of light production on the lines indicated by theory. Electro-magnetic effects are primarily unavailable, for the reason that to produce the required effects we would have to pass current impulses through a conductor, which, long before the required frequency of the impulses could be reached, would cease to transmit them. On the other hand, electro-magnetic waves many times longer than those of light, and producible by sudden discharge of a condenser, could not be utilized, it would seem, except we avail ourselves of their effect upon conductors as in the present methods, which are wasteful. We could not affect by means of such waves the static molecular or atomic charges of a gas, cause them to vibrate and to emit light. Long transverse waves cannot, apparently, produce such effects, since excessively small electro-magnetic disturbances may pass readily through miles of air. Such dark waves, unless they are of the length of true light waves, cannot, it would seem, excite luminous radiation in a Geissler tube, and the luminous effects, which are producible by induction in a tube devoid of electrodes, I am inclined to consider as being of an electrostatic nature.

To produce such luminous effects, straight electrostatic thrusts are required; these, whatever be their frequency, may disturb the molecular charges and produce light. Since current impulses of the required frequency cannot pass through a conductor of measurable dimensions, we must work with a gas, and then the production of powerful electrostatic effects becomes an imperative necessity.

It has occurred to me, however, that electrostatic effects are in many ways available for the production of light. For instance, we may place a body of some refractory material in a closed, and preferably more or less exhausted, globe, connect it to a source of high, rapidly alternating potential, causing the molecules of the gas to strike it many times a second at enormous speeds, and in this manner, with trillions of invisible hammers, pound it until it gets incandescent; or we may place a body in a very highly exhausted globe, in a non-striking vacuum, and, by employing very high frequencies and potentials, transfer sufficient energy from it to other bodies in the vicinity, or in general to the surroundings, to maintain it at any degree of incandescence; or we may, by means of such rapidly alternating high potentials, disturb the ether carried by the molecules of a gas or their static charges, causing them to vibrate and to emit light.

But, electrostatic effects being dependent upon the potential and frequency, to produce the most powerful action it is desirable to increase both as far as practicable. It may be possible to obtain quite fair results by keeping either of these factors small, provided the other is sufficiently great; but we are limited in both directions. My experience demonstrates that we cannot go below a certain frequency, for, first, the potential then becomes so great that it is dangerous; and, secondly, the light production is less efficient.

I have found that, by using the ordinary low frequencies, the physiological effect of the current required to maintain at a certain degree of brightness a tube four feet long, provided at the ends with outside and inside condenser coatings, is so powerful that, I think, it might produce serious injury to those not accustomed to such shocks; whereas, with twenty thousand alternations per second, the tube may be maintained at the same degree of brightness without any effect being felt. This is due principally to the fact that a much smaller potential is required to produce the same light effect, and also to the higher efficiency in the light production. It is evident that the efficiency in such cases is the greater, the higher the frequency, for the quicker the process of charging and discharging the molecules, the less energy will be lost in the form of dark radiation. But, unfortunately, we cannot go beyond a certain frequency on account of the difficulty of producing and conveying the effects.

I have stated above that a body inclosed in an unexhausted bulb may be intensely heated by simply connecting it with a source of rapidly alternating potential. The heating in such a case is, in all probability, due mostly to the bombardment of the molecules of the gas contained in the bulb. When the bulb is exhausted, the heating of the body is much more rapid, and there is no difficulty whatever in bringing a wire or filament to any degree of incandescence by simply connecting it to one terminal of a coil of the proper dimensions. Thus, if the well-known apparatus of Prof. Crookes, consisting of a bent platinum wire with vanes mounted over it (Fig. 114), be connected to one terminal of the coil—either one or both ends of the platinum wire being connected—the wire is rendered almost instantly incandescent, and the mica vanes are rotated as though a current from a battery were used. A thin carbon filament, or, preferably, a button of some refractory material (Fig. 115), even if it be a comparatively poor conductor, inclosed in an exhausted globe, may be rendered highly incandescent; and in this manner a simple lamp capable of giving any desired candle power is provided.

The success of lamps of this kind would depend largely on the selection of the light-giving bodies contained within the bulb. Since, under the conditions described, refractory bodies—which are very poor conductors and capable of withstanding for a long time excessively high degrees of temperature—may be used, such illuminating devices may be rendered successful.

It might be thought at first that if the bulb, containing the

FIG. 114.   FIG. 115.

filament or button of refractory material, be perfectly well exhausted—that is, as far as it can be done by the use of the best apparatus—the heating would be much less intense, and that in a perfect vacuum it could not occur at all. This is not confirmed by my experience; quite the contrary, the better the vacuum the more easily the bodies are brought to incandescence. This result is interesting for many reasons.

At the outset of this work the idea presented itself to me, whether two bodies of refractory material enclosed in a bulb exhausted to such a degree that the discharge of a large induction coil, operated in the usual manner, cannot pass through, could be rendered incandescent by mere condenser action. Obviously, to reach this result enormous potential differences and very high frequencies are required, as is evident from a simple calculation.

But such a lamp would possess a vast advantage over an ordinary incandescent lamp in regard to efficiency. It is well-known that the efficiency of a lamp is to some extent a function of the degree of incandescence, and that, could we but work a filament at many times higher degrees of incandescence, the efficiency would be much greater. In an ordinary lamp this is impracticable on account of the destruction of the filament, and it has been determined by experience how far it is advisable to push the incandescence. It is impossible to tell how much higher efficiency could be obtained if the filament could withstand indefinitely, as the investigation to this end obviously cannot be carried beyond a certain stage; but there are reasons for believing that it would be very considerably higher. An improvement might be made in the ordinary lamp by employing a short and thick carbon; but then the leading-in wires would have to be thick, and, besides, there are many other considerations which render such a modification entirely impracticable. But in a lamp as above described, the leading in wires may be very small, the incandescent refractory material may be in the shape of blocks offering a very small radiating surface, so that less energy would be required to keep them at the desired incandescence; and in addition to this, the refractory material need not be carbon, but may be manufactured from mixtures of oxides, for instance, with carbon or other material, or may be selected from bodies which are practically non-conductors, and capable of withstanding enormous degrees of temperature.

All this would point to the possibility of obtaining a much higher efficiency with such a lamp than is obtainable in ordinary lamps. In my experience it has been demonstrated that the blocks are brought to high degrees of incandescence with much lower potentials than those determined by calculation, and the blocks may be set at greater distances from each other. We may freely assume, and it is probable, that the molecular bombardment is an important element in the heating, even if the globe be exhausted with the utmost care, as I have done; for although the number of the molecules is, comparatively speaking, insignificant, yet on account of the mean free path being very great, there are fewer collisions, and the molecules may reach much higher speeds, so that the heating effect due to this cause may be considerable, as in the Crookes experiments with radiant matter.

But it is likewise possible that we have to deal here with an increased facility of losing the charge in very high vacuum, when the potential is rapidly alternating, in which case most of the heating would be directly due to the surging of the charges in the heated bodies. Or else the observed fact may be largely attributable to the effect of the points which I have mentioned above, in consequence of which the blocks or filaments contained in the vacuum are equivalent to condensers of many times greater surface than that calculated from their geometrical dimensions. Scientific men still differ in opinion as to whether a charge should, or should not, be lost in a perfect vacuum, or in other words, whether ether is, or is not, a conductor. If the

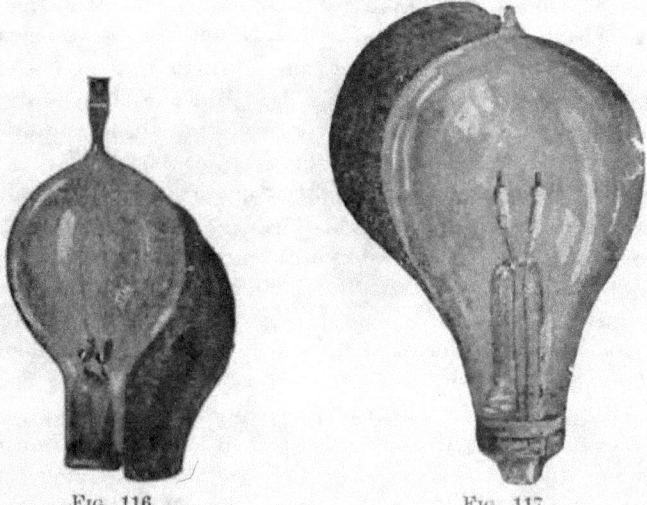

FIG. 116.   FIG. 117.

former were the case, then a thin filament enclosed in a perfectly exhausted globe, and connected to a source of enormous, steady potential, would be brought to incandescence.

Various forms of lamps on the above described principle, with the refractory bodies in the form of filaments, Fig. 116, or blocks, Fig. 117, have been constructed and operated by me, and investigations are being carried on in this line. There is no difficulty in reaching such high degrees of incandescence that ordinary carbon is to all appearance melted and volatilized. If the vacuum could be made absolutely perfect, such a lamp, although inoperative with apparatus ordinarily used, would, if operated with currents of the required character, afford an illuminant which would never be destroyed, and which would be far more efficient than an ordinary incandescent lamp. This perfection can, of course, never be reached, and a very slow destruction and gradual diminution in size always occurs, as in incandescent filaments; but there is no possibility of a sudden and premature disabling which occurs in the latter by the breaking of the filament, especially when the incandescent bodies are in the shape of blocks.

With these rapidly alternating potentials there is, however, no necessity of enclosing two blocks in a globe, but a single block, as in Fig. 115, or filament, Fig. 118, may be used. The potential in this case must of course be higher, but is easily obtainable, and besides it is not necessarily dangerous.

The facility with which the button or filament in such a lamp

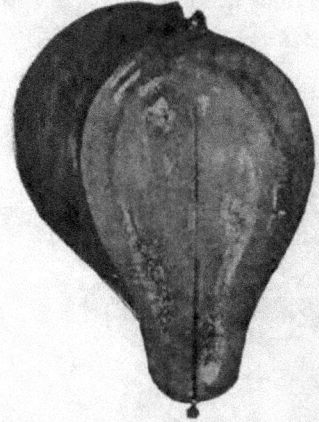

FIG. 118.

is brought to incandescence, other things being equal, depends on the size of the globe. If a perfect vacuum could be obtained, the size of the globe would not be of importance, for then the heating would be wholly due to the surging of the charges, and all the energy would be given off to the surroundings by radiation. But this can never occur in practice. There is always some gas left in the globe, and although the exhaustion may be carried to the highest degree, still the space inside of the bulb must be considered as conducting when such high potentials are used, and I assume that, in estimating the energy that may be given off from the filament to the surroundings, we may consider the inside surface of the bulb as one coating of a condenser, the air and other objects surrounding the bulb forming the other coating. When the alternations are very low there is no doubt that a considerable portion of the energy is given off by the electrification of the surrounding air.

In order to study this subject better, I carried on some experiments with excessively high potentials and low frequencies. I then observed that when the hand is approached to the bulb,— the filament being connected with one terminal of the coil,—a powerful vibration is felt, being due to the attraction and repulsion of the molecules of the air which are electrified by induction through the glass. In some cases when the action is very intense I have been able to hear a sound, which must be due to the same cause.

When the alternations are low, one is apt to get an excessively

 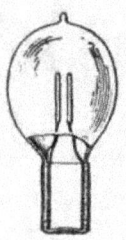

FIG. 119.   FIG. 120.

powerful shock from the bulb. In general, when one attaches bulbs or objects of some size to the terminals of the coil, one should look out for the rise of potential, for it may happen that by merely connecting a bulb or plate to the terminal, the potential may rise to many times its original value. When lamps are attached to the terminals, as illustrated in Fig. 119, then the capacity of the bulbs should be such as to give the maximum rise of potential under the existing conditions. In this manner one may obtain the required potential with fewer turns of wire.

The life of such lamps as described above depends, of course, largely on the degree of exhaustion, but to some extent also on the shape of the block of refractory material. Theoretically it would seem that a small sphere of carbon enclosed in a sphere of glass would not suffer deterioration from molecular bombardment, for, the matter in the globe being radiant, the molecules would move in straight lines, and would seldom strike the sphere obliquely. An interesting thought in connection with such a lamp is, that in it "electricity" and electrical energy apparently must move in the same lines.

The use of alternating currents of very high frequency makes it possible to transfer, by electrostatic or electromagnetic induction through the glass of a lamp, sufficient energy to keep a fila-

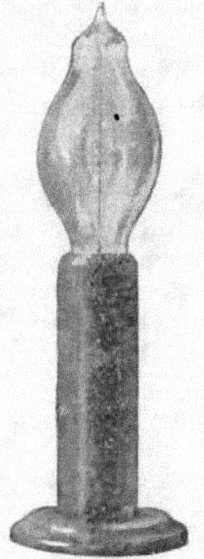

Fig. 121a.   Fig. 121b.

ment at incandescence and so do away with the leading-in wires. Such lamps have been proposed, but for want of proper apparatus they have not been successfully operated. Many forms of lamps on this principle with continuous and broken filaments have been constructed by me and experimented upon. When using a secondary enclosed within the lamp, a condenser is advantageously combined with the secondary. When the transference is effected by electrostatic induction, the potentials used are, of course, very high with frequencies obtainable from a machine. For instance, with a condenser surface of forty square centimetres, which is not impracticably large, and with glass of good quality 1 mm. thick, using currents alternating twenty thousand times a second, the potential required is approximately 9,000 volts. This may seem large, but since each lamp may be included in the secondary of a transformer of very small dimensions, it would not be inconvenient, and, moreover, it would not produce fatal injury. The transformers would all be preferably in series. The regulation would offer no difficulties, as with currents of such frequencies it is very easy to maintain a constant current.

In the accompanying engravings some of the types of lamps of this kind are shown. Fig. 120 is such a lamp with a broken filament, and Figs. 121 A and 121 B one with a single outside and inside coating and a single filament. I have also made lamps with two outside and inside coatings and a continuous loop connecting the latter. Such lamps have been operated by me with current impulses of the enormous frequencies obtainable by the disruptive discharge of condensers.

The disruptive discharge of a condenser is especially suited for operating such lamps—with no outward electrical connections—by means of electromagnetic induction, the electromagnetic inductive effects being excessively high; and I have been able to produce the desired incandescence with only a few short turns of wire. Incandescence may also be produced in this manner in a simple closed filament.

Leaving now out of consideration the practicability of such lamps, I would only say that they possess a beautiful and desirable feature, namely, that they can be rendered, at will, more or less brilliant simply by altering the relative position of the outside and inside condenser coatings, or inducing and induced circuits.

When a lamp is lighted by connecting it to one terminal only of the source, this may be facilitated by providing the globe with an outside condenser coating, which serves at the same time as a reflector, and connecting this to an insulated body of some size. Lamps of this kind are illustrated in Fig. 122 and Fig. 123. Fig. 124 shows the plan of connection. The brilliancy of the lamp may, in this case, be regulated within wide limits by varying the size of the insulated metal plate to which the coating is connected.

It is likewise practicable to light with one leading wire lamps such as illustrated in Fig. 116 and Fig. 117, by connecting one

terminal of the lamp to one terminal of the source, and the
other to an insulated body of the required size. In all cases
the insulated body serves to give off the energy into the surrounding space, and is equivalent to a return wire. Obviously,
in the two last-named cases, instead of connecting the wires to
an insulated body, connections may be made to the ground.

The experiments which will prove most suggestive and of
most interest to the investigator are probably those performed
with exhausted tubes. As might be anticipated, a source of such
rapidly alternating potentials is capable of exciting the tubes at
a considerable distance, and the light effects produced are remarkable.

During my investigations in this line I endeavored to excite

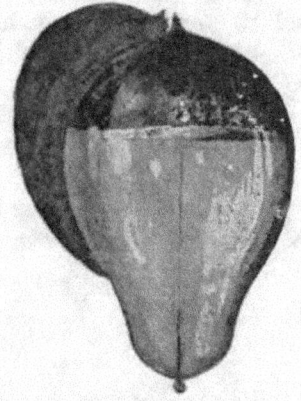

Fig. 122.

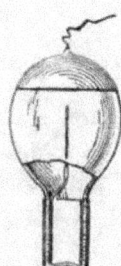

Fig. 123.

tubes, devoid of any electrodes, by electromagnetic induction,
making the tube the secondary of the induction device, and
passing through the primary the discharges of a Leyden jar.
These tubes were made of many shapes, and I was able to
obtain luminous effects which I then thought were due wholly
to electromagnetic induction. But on carefully investigating
the phenomena I found that the effects produced were more
of an electrostatic nature. It may be attributed to this circumstance that this mode of exciting tubes is very wasteful,
namely, the primary circuit being closed, the potential, and
consequently the electrostatic inductive effect, is much diminished.

When an induction coil, operated as above described, is used, there is no doubt that the tubes are excited by electrostatic induction, and that electromagnetic induction has little, if anything, to do with the phenomena.

This is evident from many experiments. For instance, if a tube be taken in one hand, the observer being near the coil, it is brilliantly lighted and remains so no matter in what position it is held relatively to the observer's body. Were the action electromagnetic, the tube could not be lighted when the observer's body is interposed between it and the coil, or at least its luminosity should be considerably diminished. When the tube is held exactly over the centre of the coil—the latter being wound in sections and the primary placed symmetrically to the secondary—it may remain completely dark, whereas it is rendered intensely luminous by moving it slightly to the right or left from the centre of the coil. It does not light because in the

Fig. 124.

middle both halves of the coil neutralize each other, and the electric potential is zero. If the action were electromagnetic, the tube should light best in the plane through the centre of the coil, since the electromagnetic effect there should be a maximum. When an arc is established between the terminals, the tubes and lamps in the vicinity of the coil go out, but light up again when the arc is broken, on account of the rise of potential. Yet the electromagnetic effect should be practically the same in both cases.

By placing a tube at some distance from the coil, and nearer to one terminal—preferably at a point on the axis of the coil—one may light it by touching the remote terminal with an insulated body of some size or with the hand, thereby raising the potential at that terminal nearer to the tube. If the tube is shifted nearer to the coil so that it is lighted by the action of the nearer terminal, it may be made to go out by holding, on an insulated support, the end of a wire connected to the remote terminal, in the vicinity of the nearer terminal, by this means counteracting the action of the latter upon the tube. These effects are evidently electrostatic. Likewise, when a tube is placed at a considerable distance from the coil, the observer may, standing upon an insulated support between coil and tube, light the latter by approaching the hand to it ; or he may even render it luminous by simply stepping between it and the coil. This would be impossible with electro-magnetic induction, for the body of the observer would act as a screen.

When the coil is energized by excessively weak currents, the experimenter may, by touching one terminal of the coil with the tube, extinguish the latter, and may again light it by bringing it out of contact with the terminal and allowing a small arc to form. This is clearly due to the respective lowering and raising of the potential at that terminal. In the above experiment, when the tube is lighted through a small arc, it may go out when the arc is broken, because the electrostatic inductive effect alone is too weak, though the potential may be much higher ; but when the arc is established, the electrification of the end of the tube is much greater, and it consequently lights.

If a tube is lighted by holding it near to the coil, and in the hand which is remote, by grasping the tube anywhere with the other hand, the part between the hands is rendered dark, and the singular effect of wiping out the light of the tube may be produced by passing the hand quickly along the tube and at the same time withdrawing it gently from the coil, judging properly the distance so that the tube remains dark afterwards.

If the primary coil is placed sidewise, as in Fig. 112 B for instance, and an exhausted tube be introduced from the other side in the hollow space, the tube is lighted most intensely because of the increased condenser action, and in this position the striæ are most sharply defined. In all these experiments described, and in many others, the action is clearly electrostatic.

The effects of screening also indicate the electrostatic nature of the phenomena and show something of the nature of electrification through the air. For instance, if a tube is placed in the direction of the axis of the coil, and an insulated metal plate be interposed, the tube will generally increase in brilliancy, or if it be too far from the coil to light, it may even be rendered luminous by interposing an insulated metal plate. The magnitude of the effects depends to some extent on the size of the plate. But if the metal plate be connected by a wire to the ground, its interposition will always make the tube go out even if it be very near the coil. In general, the interposition of a body between the coil and tube, increases or diminishes the brilliancy of the tube, or its facility to light up, according to whether it increases or diminishes the electrification. When experimenting with an insulated plate, the plate should not be taken too large, else it will generally produce a weakening effect by reason of its great facility for giving off energy to the surroundings.

If a tube be lighted at some distance from the coil, and a plate of hard rubber or other insulating substance be interposed, the tube may be made to go out. The interposition of the dielectric in this case only slightly increases the inductive effect, but diminishes considerably the electrification through the air.

In all cases, then, when we excite luminosity in exhausted tubes by means of such a coil, the effect is due to the rapidly alternating electrostatic potential; and, furthermore, it must be attributed to the harmonic alternation produced directly by the machine, and not to any superimposed vibration which might be thought to exist. Such superimposed vibrations are impossible when we work with an alternate current machine. If a spring be gradually tightened and released, it does not perform independent vibrations; for this a sudden release is necessary. So with the alternate currents from a dynamo machine; the medium is harmonically strained and released, this giving rise to only one kind of waves; a sudden contact or break, or a sudden giving way of the dielectric, as in the disruptive discharge of a Leyden jar, are essential for the production of superimposed waves.

In all the last described experiments, tubes devoid of any electrodes may be used, and there is no difficulty in producing by their means sufficient light to read by. The light effect is, however, considerably increased by the use of phosphorescent bodies such as yttria, uranium glass, etc. A difficulty will be found when the phosphorescent material is used, for with these powerful effects, it is carried gradually away, and it is preferable to use material in the form of a solid.

Instead of depending on induction at a distance to light the tube, the same may be provided with an external—and, if desired, also with an internal—condenser coating, and it may then

be suspended anywhere in the room from a conductor connected
to one terminal of the coil, and in this manner a soft illumination
may be provided.

The ideal way of lighting a hall or room would, however, be

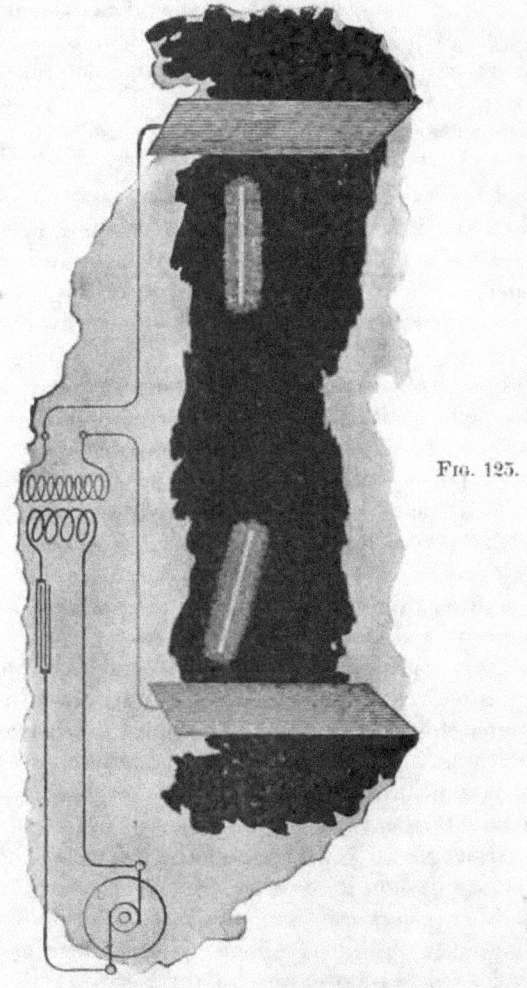

FIG. 125.

to produce such a condition in it that an illuminating device
could be moved and put anywhere, and that it is lighted, no matter where it is put and without being electrically connected to

anything. I have been able to produce such a condition by creating in the room a powerful, rapidly alternating electrostatic field. For this purpose I suspend a sheet of metal a distance from the ceiling on insulating cords and connect it to one terminal of the induction coil, the other terminal being preferably connected to the ground. Or else I suspend two sheets as illustrated in Fig. 125, each sheet being connected with one of the terminals of the coil, and their size being carefully determined. An exhausted tube may then be carried in the hand anywhere between the sheets or placed anywhere, even a certain distance beyond them; it remains always luminous.

In such an electrostatic field interesting phenomena may be observed, especially if the alternations are kept low and the potentials excessively high. In addition to the luminous phenomena mentioned, one may observe that any insulated conductor gives sparks when the hand or another object is approached to it, and the sparks may often be powerful. When a large conducting object is fastened on an insulating support, and the hand approached to it, a vibration, due to the rythmical motion of the air molecules is felt, and luminous streams may be perceived when the hand is held near a pointed projection. When a telephone receiver is made to touch with one or both of its terminals an insulated conductor of some size, the telephone emits a loud sound; it also emits a sound when a length of wire is attached to one or both terminals, and with very powerful fields a sound may be perceived even without any wire.

How far this principle is capable of practical application, the future will tell. It might be thought that electrostatic effects are unsuited for such action at a distance. Electromagnetic inductive effects, if available for the production of light, might be thought better suited. It is true the electrostatic effects diminish nearly with the cube of the distance from the coil, whereas the electromagnetic inductive effects diminish simply with the distance. But when we establish an electrostatic field of force, the condition is very different, for then, instead of the differential effect of both the terminals, we get their conjoint effect. Besides, I would call attention to the effect, that in an alternating electrostatic field, a conductor, such as an exhausted tube, for instance, tends to take up most of the energy, whereas in an electromagnetic alternating field the conductor tends to take up the least energy, the waves being reflected with but little loss.

This is one reason why it is difficult to excite an exhausted tube, at a distance, by electromagnetic induction. I have wound coils of very large diameter and of many turns of wire, and connected a Geissler tube to the ends of the coil with the object of exciting the tube at a distance; but even with the powerful inductive effects producible by Leyden jar discharges, the tube could not be excited unless at a very small distance, although some judgment was used as to the dimensions of the coil. I have also found that even the most powerful Leyden jar discharges are capable of exciting only feeble luminous effects in a closed exhausted tube, and even these effects upon thorough examination I have been forced to consider of an electrostatic nature.

How then can we hope to produce the required effects at a distance by means of electromagnetic action, when even in the closest proximity to the source of disturbance, under the most advantageous conditions, we can excite but faint luminosity? It is true that when acting at a distance we have the resonance to help us out. We can connect an exhausted tube, or whatever the illuminating device may be, with an insulated system of the proper capacity, and so it may be possible to increase the effect qualitatively, and only qualitatively, for we would not get *more* energy through the device. So we may, by resonance effect, obtain the required electromotive force in an exhausted tube, and excite faint luminous effects, but we cannot get enough energy to render the light practically available, and a simple calculation, based on experimental results, shows that even if all the energy which a tube would receive at a certain distance from the source should be wholly converted into light, it would hardly satisfy the practical requirements. Hence the necessity of directing, by means of a conducting circuit, the energy to the place of transformation. But in so doing we cannot very sensibly depart from present methods, and all we could do would be to improve the apparatus.

From these considerations it would seem that if this ideal way of lighting is to be rendered practicable it will be only by the use of electrostatic effects. In such a case the most powerful electrostatic inductive effects are needed; the apparatus employed must, therefore, be capable of producing high electrostatic potentials changing in value with extreme rapidity. High frequencies are especially wanted, for practical considerations make it desirable to keep down the potential. By the employment of machines, or, generally speaking, of any mechanical apparatus, but low frequencies can be reached; recourse must, therefore, be had to some other means. The discharge of a condenser affords us a means of obtaining frequencies by far higher than are obtainable mechanically, and I have accordingly employed condensers in the experiments to the above end.

When the terminals of a high tension induction coil, Fig. 126, are connected to a Leyden jar, and the latter is discharging disruptively into a circuit, we may look upon the arc playing between the knobs as being a source of alternating, or generally speaking, undulating currents, and then we have to deal with the familiar system of a generator of such currents, a circuit connected to it, and a condenser bridging the circuit. The condenser in such case is a veritable transformer, and since the frequency is excessive, almost any ratio in the strength of the currents in both the branches may be obtained. In reality the analogy is not quite complete, for in the disruptive discharge we have most generally a fundamental instantaneous variation of comparatively low frequency, and a superimposed harmonic vibration, and the laws governing the flow of currents are not the same for both.

In converting in this manner, the ratio of conversion should not be too great, for the loss in the arc between the knobs increases with the square of the current, and if the jar be discharged through very thick and short conductors, with the view of obtaining a very rapid oscillation, a very considerable portion of the energy stored is lost. On the other hand, too small ratios are not practicable for many obvious reasons.

As the converted currents flow in a practically closed circuit, the electrostatic effects are necessarily small, and I therefore convert them into currents or effects of the required character. I have effected such conversions in several ways. The preferred plan of connections is illustrated in Fig. 127. The manner of operating renders it easy to obtain by means of a small and inexpensive apparatus enormous differences of potential which have been usually obtained by means of large and expensive coils. For this it is only necessary to take an ordinary small coil, adjust to it a condenser and discharging circuit, forming the primary of an auxiliary small coil, and convert upward. As the inductive effect of the primary currents is excessively great, the second coil need have comparatively but very few turns. By properly adjusting the elements, remarkable results may be secured.

In endeavoring to obtain the required electrostatic effects in this manner, I have, as might be expected, encountered many difficulties which I have been gradually overcoming, but I am not as yet prepared to dwell upon my experiences in this direction.

I believe that the disruptive discharge of a condenser will play an important part in the future, for it offers vast possibilities, not only in the way of producing light in a more efficient manner and in the line indicated by theory, but also in many other respects.

For years the efforts of inventors have been directed towards obtaining electrical energy from heat by means of the thermopile. It might seem invidious to remark that but few know what is the real trouble with the thermopile. It is not the inefficiency or small output—though these are great drawbacks—but the fact that the thermopile has its phylloxera, that is, that by constant use it is deteriorated, which has thus far prevented its

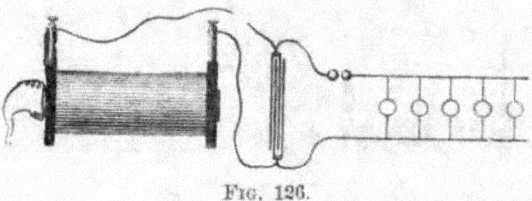

FIG. 126.

introduction on an industrial scale. Now that all modern research seems to point with certainty to the use of electricity of excessively high tension, the question must present itself to many whether it is not possible to obtain in a practicable manner this form of energy from heat. We have been used to look upon an electrostatic machine as a plaything, and somehow we couple with it the idea of the inefficient and impractical. But now we must think differently, for now we know that everywhere we have to deal with the same forces, and that it is a mere question of inventing proper methods or apparatus for rendering them available.

In the present systems of electrical distribution, the employment of the iron with its wonderful magnetic properties allows us to reduce considerably the size of the apparatus; but, in spite of this, it is still very cumbersome. The more we progress in the study of electric and magnetic phenomena, the more we become convinced that the present methods will be short-lived. For the production of light, at least, such heavy machinery would seem to be unnecessary. The energy required is very small, and if light can be obtained as efficiently as, theoretically, it appears possible, the apparatus need have but a very small output. There being a strong probability that the illuminating methods of the future will involve the use of very high potentials, it seems very desirable to perfect a contrivance capable of converting the energy of heat into energy of the requisite form. Nothing to speak of has been done towards this end, for the thought that electricity of some 50,000 or 100,000 volts pressure or more, even if obtained, would be unavailable for practical purposes, has deterred inventors from working in this direction.

In Fig. 126 a plan of connections is shown for converting currents of high, into currents of low, tension by means of the disruptive discharge of a condenser. This plan has been used by

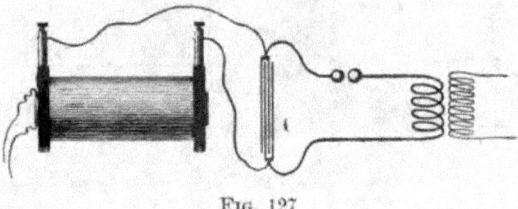

Fig. 127.

me frequently for operating a few incandescent lamps required in the laboratory. Some difficulties have been encountered in the arc of the discharge which I have been able to overcome to a great extent; besides this, and the adjustment necessary for the proper working, no other difficulties have been met with, and it was easy to operate ordinary lamps, and even motors, in this manner. The line being connected to the ground, all the wires could be handled with perfect impunity, no matter how high the potential at the terminals of the condenser. In these experiments a high tension induction coil, operated from a battery or from an alternate current machine, was employed to charge the condenser; but the induction coil might be replaced by an apparatus of a different kind, capable of giving electricity of such high tension. In this manner, direct or alternating currents may be converted, and in both cases the current-impulses may be of any desired frequency. When the currents charging the condenser are of the

same direction, and it is desired that the converted currents
should also be of one direction, the resistance of the discharging circuit should, of course, be so chosen that there are no
oscillations.

In operating devices on the above plan I have observed curious phenomena of impedance which are of interest.   For instance
if a thick copper bar be bent, as indicated in Fig. 128, and shunted
by ordinary incandescent lamps, then, by passing the discharge
between the knobs, the lamps may be brought to incandescence
although they are short-circuited.   When a large induction coil

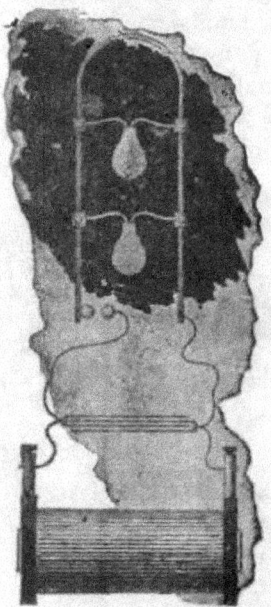

Fig. 128.

is employed it is easy to obtain nodes on the bar, which are
rendered evident by the different degree of brilliancy of the
lamps, as shown roughly in Fig. 128.   The nodes are never clearly
defined, but they are simply maxima and minima of potentials
along the bar.   This is probably due to the irregularity of the arc
between the knobs.   In general when the above-described plan
of conversion from high to low tension is used, the behavior of
the disruptive discharge may be closely studied.   The nodes may
also be investigated by means of an ordinary Cardew voltmeter

which should be well insulated. Geissler tubes may also be lighted across the points of the bent bar; in this case, of course, it is better to employ smaller capacities. I have found it practicable to light up in this manner a lamp, and even a Geissler tube, shunted by a short, heavy block of metal, and this result seems at first very curious. In fact, the thicker the copper bar in Fig. 128, the better it is for the success of the experiments, as they appear more striking. When lamps with long slender filaments are used it will be often noted that the filaments are from time to time violently vibrated, the vibration being smallest at the nodal points. This vibration seems to be due to an electrostatic action between the filament and the glass of the bulb.

In some of the above experiments it is preferable to use special lamps having a straight filament as shown in Fig. 129. When such a lamp is used a still more curious phenomenon than those

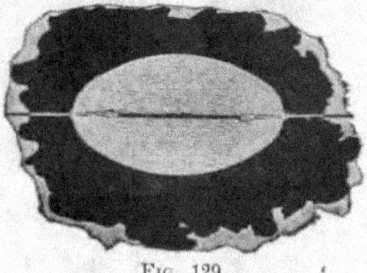

FIG. 129.

described may be observed. The lamp may be placed across the copper bar and lighted, and by using somewhat larger capacities, or, in other words, smaller frequencies or smaller impulsive impedances, the filament may be brought to any desired degree of incandescence. But when the impedance is increased, a point is reached when comparatively little current passes through the carbon, and most of it through the rarefied gas; or perhaps it may be more correct to state that the current divides nearly evenly through both, in spite of the enormous difference in the resistance, and this would be true unless the gas and the filament behave differently. It is then noted that the whole bulb is brilliantly illuminated, and the ends of the leading-in wires become incandescent and often throw off sparks in consequence of the violent bombardment, but the carbon filament remains dark. This is illustrated in Fig. 129. Instead of the filament a single wire extending through the whole bulb may be used, and in this case the phenomenon would seem to be still more interesting.

From the above experiment it will be evident, that when ordinary lamps are operated by the converted currents, those should be preferably taken in which the platinum wires are far apart, and the frequencies used should not be too great, else the discharge will occur at the ends of the filament or in the base of the lamp between the leading-in wires, and the lamp might then be damaged.

In presenting to you these results of my investigation on the subject under consideration, I have paid only a passing notice to facts upon which I could have dwelt at length, and among many observations I have selected only those which I thought most likely to interest you. The field is wide and completely unexplored, and at every step a new truth is gleaned, a novel fact observed.

How far the results here borne out are capable of practical applications will be decided in the future. As regards the production of light, some results already reached are encouraging and make me confident in asserting that the practical solution of the problem lies in the direction I have endeavored to indicate. Still, whatever may be the immediate outcome of these experiments I am hopeful that they will only prove a step in further development towards the ideal and final perfection. The possibilities which are opened by modern research are so vast that even the most reserved must feel sanguine of the future. Eminent scientists consider the problem of utilizing one kind of radiation without the others a rational one. In an apparatus designed for the production of light by conversion from any form of energy into that of light, such a result can never be reached, for no matter what the process of producing the required vibrations, be it electrical, chemical or any other, it will not be possible to obtain the higher light vibrations without going through the lower heat vibrations. It is the problem of imparting to a body a certain velocity without passing through all lower velocities. But there is a possibility of obtaining energy not only in the form of light, but motive power, and energy of any other form, in some more direct way from the medium. The time will be when this will be accomplished, and the time has come when one may utter such words before an enlightened audience without being considered a visionary. We are whirling through endless space with an inconceivable speed, all around us every-thing is spinning, everything is moving, everywhere is energy. There *must* be some way of availing ourselves of this energy more directly. Then, with the light obtained from the medium, with the power derived from it, with every form of energy obtained without effort, from the store forever inexhaustible, humanity will advance with giant strides. The mere contempla-tion of these magnificent possibilities expands our minds, strength-ens our hopes and fills our hearts with supreme delight.

CHAPTER XXVII.

EXPERIMENTS WITH ALTERNATE CURRENTS OF HIGH POTENTIAL AND HIGH FREQUENCY.[1]

I CANNOT find words to express how deeply I feel the honor of addressing some of the foremost thinkers of the present time, and so many able scientific men, engineers and electricians, of the country greatest in scientific achievements.

The results which I have the honor to present before such a gathering I cannot call my own. There are among you not a few who can lay better claim than myself on any feature of merit which this work may contain. I need not mention many names which are world-known—names of those among you who are recognized as the leaders in this enchanting science; but one, at least, I must mention—a name which could not be omitted in a demonstration of this kind. It is a name associated with the most beautiful invention ever made: it is Crookes!

When I was at college, a good while ago, I read, in a translation (for then I was not familiar with your magnificent language), the description of his experiments on radiant matter. I read it only once in my life—that time—yet every detail about that charming work I can remember to this day. Few are the books, let me say, which can make such an impression upon the mind of a student.

But if, on the present occasion, I mention this name as one of many your Institution can boast of, it is because I have more than one reason to do so. For what I have to tell you and to show you this evening concerns, in a large measure, that same vague world which Professor Crookes has so ably explored; and, more than this, when I trace back the mental process which led me to these advances—which even by myself cannot be considered trifling, since they are so appreciated by you—I believe that their real origin, that which started me to work in this

1. Lecture delivered before the Institution of Electrical Engineers, London, February, 1892.

direction, and brought me to them, after a long period of constant thought, was that fascinating little book which I read many years ago.

And now that I have made a feeble effort to express my homage and acknowledge my indebedness to him and others among you, I will make a second effort, which I hope you will not find so feeble as the first, to entertain you.

Give me leave to introduce the subject in a few words.

A short time ago I had the honor to bring before our American Institute of Electrical Engineers some results then arrived at by me in a novel line of work. I need not assure you that the many evidences which I have received that English scientific men and engineers were interested in this work have been for me a great reward and encouragement. I will not dwell upon the experiments already described, except with the view of completing, or more clearly expressing, some ideas advanced by me before, and also with the view of rendering the study here presented self-contained, and my remarks on the subject of this evening's lecture consistent.

This investigation, then, it goes without saying, deals with alternating currents, and to be more precise, with alternating currents of high potential and high frequency. Just in how much a very high frequency is essential for the production of the results presented is a question which, even with my present experience, would embarrass me to answer. Some of the experiments may be performed with low frequencies; but very high frequencies are desirable, not only on account of the many effects secured by their use, but also as a convenient means of obtaining, in the induction apparatus employed, the high potentials, which in their turn are necessary to the demonstration of most of the experiments here contemplated.

Of the various branches of electrical investigation, perhaps the most interesting and the most immediately promising is that dealing with alternating currents. The progress in this branch of applied science has been so great in recent years that it justifies the most sanguine hopes. Hardly have we become familiar with one fact, when novel experiences are met and new avenues of research are opened. Even at this hour possibilities not dreamed of before are, by the use of these currents, partly realized. As in nature all is ebb and tide, all is wave motion, so it seems that in all branches of industry alternating currents—electric wave motion—will have the sway.

One reason, perhaps, why this branch of science is being so rapidly developed is to be found in the interest which is attached to its experimental study. We wind a simple ring of iron with coils; we establish the connections to the generator, and with wonder and delight we note the effects of strange forces which we bring into play, which allow us to transform, to transmit and direct energy at will. We arrange the circuits properly, and we see the mass of iron and wires behave as though it were endowed with life, spinning a heavy armature, through invisible connections, with great speed and power—with the energy possibly conveyed from a great distance. We observe how the energy of an alternating current traversing the wire manifests itself—not so much in the wire as in the surrounding space—in the most surprising manner, taking the forms of heat, light, mechanical energy, and, most surprising of all, even chemical affinity. All these observations fascinate us, and fill us with an intense desire to know more about the nature of these phenomena. Each day we go to our work in the hope of discovering,—in the hope that some one, no matter who, may find a solution of one of the pending great problems,—and each succeeding day we return to our task with renewed ardor; and even if we *are* unsuccessful, our work has not been in vain, for in these strivings, in these efforts, we have found hours of untold pleasure, and we have directed our energies to the benefit of mankind.

We may take—at random, if you choose—any of the many experiments which may be performed with alternating currents; a few of which only, and by no means the most striking, form the subject of this evening's demonstration; they are all equally interesting, equally inciting to thought.

Here is a simple glass tube from which the air has been partially exhausted. I take hold of it; I bring my body in contact with a wire conveying alternating currents of high potential, and the tube in my hand is brilliantly lighted. In whatever position I may put it, wherever I move it in space, as far as I can reach, its soft, pleasing light persists with undiminished brightness.

Here is an exhausted bulb suspended from a single wire. Standing on an insulated support, I grasp it, and a platinum button mounted in it is brought to vivid incandescence.

Here, attached to a leading wire, is another bulb, which, as I touch its metallic socket, is filled with magnificent colors of phosphorescent light.

Here still another, which by my fingers' touch casts a shadow —the Crookes shadow—of the stem inside of it.

Here, again, insulated as I stand on this platform, I bring my body in contact with one of the terminals of the secondary of this induction coil—with the end of a wire many miles long—and you see streams of light break forth from its distant end, which is set in violent vibration.

Here, once more, I attach these two plates of wire gauze to the terminals of the coil; I set them a distance apart, and I set the coil to work. You may see a small spark pass between the plates. I insert a thick plate of one of the best dielectrics between them, and instead of rendering altogether impossible, as we are used to expect, I *aid* the passage of the discharge, which, as I insert the plate, merely changes in appearance and assumes the form of luminous streams.

Is there, I ask, can there be, a more interesting study than that of alternating currents?

In all these investigations, in all these experiments, which are so very, very interesting, for many years past—ever since the greatest experimenter who lectured in this hall discovered its principle—we have had a steady companion, an appliance familiar to every one, a plaything once, a thing of momentous importance now—the induction coil. There is no dearer appliance to the electrician. From the ablest among you, I dare say, down to the inexperienced student, to your lecturer, we all have passed many delightful hours in experimenting with the induction coil. We have watched its play, and thought and pondered over the beautiful phenomena which it disclosed to our ravished eyes. So well known is this apparatus, so familiar are these phenomena to every one, that my courage nearly fails me when I think that I have ventured to address so able an audience, that I have ventured to entertain you with that same old subject. Here in reality is the same apparatus, and here are the same phenomena, only the apparatus is operated somewhat differently, the phenomena are presented in a different aspect. Some of the results we find as expected, others surprise us, but all captivate our attention, for in scientific investigation each novel result achieved may be the centre of a new departure, each novel fact learned may lead to important developments.

Usually in operating an induction coil we have set up a vibration of moderate frequency in the primary, either by means of an interrupter or break, or by the use of an alternator. Earlier English investigators, to mention only Spottiswoode and J. E. H. Gordon, have used a rapid break in connection with the coil. Our knowledge and experience of to-day enables us to see clearly why these coils under the conditions of the test did not disclose any remarkable phenomena, and why able experimenters failed to perceive many of the curious effects which have since been observed.

In the experiments such as performed this evening, we operate the coil either from a specially constructed alternator capable of giving many thousands of reversals of current per second, or, by disruptively discharging a condenser through the primary, we set up a vibration in the secondary circuit of a frequency of many hundred thousand or millions per second, if we so desire; and in using either of these means we enter a field as yet unexplored.

It is impossible to pursue an investigation in any novel line without finally making some interesting observation or learning some useful fact. That this statement is applicable to the subject of this lecture the many curious and unexpected phenomena which we observe afford a convincing proof. By way of illustration, take for instance the most obvious phenomena, those of the discharge of the induction coil.

Here is a coil which is operated by currents vibrating with extreme rapidity, obtained by disruptively discharging a Leyden jar. It would not surprise a student were the lecturer to say that the secondary of this coil consists of a small length of comparatively stout wire; it would not surprise him were the lecturer to state that, in spite of this, the coil is capable of giving any potential which the best insulation of the turns is able to withstand; but although he may be prepared, and even be indifferent as to the anticipated result, yet the aspect of the discharge of the coil will surprise and interest him. Every one is familiar with the discharge of an ordinary coil; it need not be reproduced here. But, by way of contrast, here is a form of discharge of a coil, the primary current of which is vibrating several hundred thousand times per second. The discharge of an ordinary coil appears as a simple line or band of light. The discharge of this coil appears in the form of powerful brushes and luminous streams issuing from all points of the two straight wires attached to the terminals of the secondary. (Fig. 130.)

Now compare this phenomenon which you have just witnessed

with the discharge of a Holtz or Wimshurst machine—that other
interesting appliance so dear to the experimenter. What a difference there is between these phenomena! And yet, had I made
the necessary arrangements—which could have been made easily,
were it not that they would interfere with other experiments—I
could have produced with this coil sparks which, had I the coil

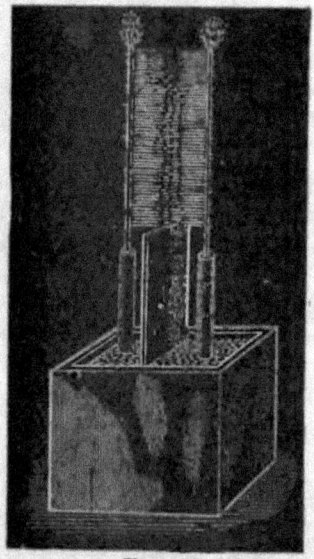

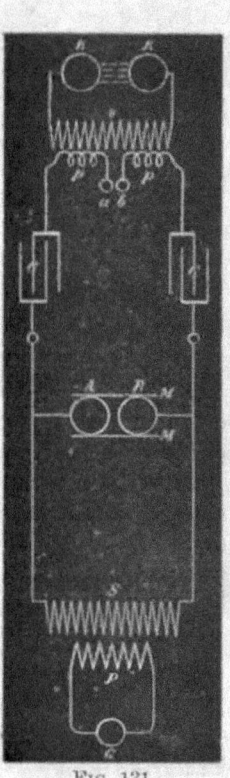

FIG. 130.   FIG. 131.

hidden from your view and only two knobs exposed, even the
keenest observer among you would find it difficult, if not impossible, to distinguish from those of an influence or friction machine. This may be done in many ways—for instance, by operating the induction coil which charges the condenser from an
alternating-current machine of very low frequency, and preferably adjusting the discharge circuit so that there are no oscillations
set up in it. We then obtain in the secondary circuit, if the
knobs are of the required size and properly set, a more or less

rapid succession of sparks of great intensity and small quantity, which possess the same brilliancy, and are accompanied by the same sharp crackling sound, as those obtained from a friction or influence machine.

Another way is to pass through two primary circuits, having a common secondary, two currents of a slightly different period, which produce in the secondary circuit sparks occurring at comparatively long intervals. But, even with the means at hand this evening, I may succeed in imitating the spark of a Holtz machine. For this purpose I establish between the terminals of the coil which charges the condenser a long, unsteady arc, which is periodically interrupted by the upward current of air produced by it. To increase the current of air I place on each side of the arc, and close to it, a large plate of mica. The condenser charged from this coil discharges into the primary circuit of a second coil through a small air gap, which is necessary to produce a sudden rush of current through the primary. The scheme of connections in the present experiment is indicated in Fig. 131.

G is an ordinarily constructed alternator, supplying the primary P of an induction coil, the secondary s of which charges the condensers or jars c c. The terminals of the secondary are connected to the inside coatings of the jars, the outer coatings being connected to the ends of the primary p p of a second induction coil. This primary p p has a small air gap a b.

The secondary s of this coil is provided with knobs or spheres κ κ of the proper size and set at a distance suitable for the experiment.

A long arc is established between the terminals A B of the first induction coil. M M are the mica plates.

Each time the arc is broken between A and B the jars are quickly charged and discharged through the primary p p, producing a snapping spark between the knobs κ κ. Upon the arc forming between A and B the potential falls, and the jars cannot be charged to such high potential as to break through the air gap a b until the arc is again broken by the draught.

In this manner sudden impulses, at long intervals, are produced in the primary p p, which in the secondary s give a corresponding number of impulses of great intensity. If the secondary knobs or spheres, κ κ, are of the proper size, the sparks show much resemblance to those of a Holtz machine.

But these two effects, which to the eye appear so very different, are only two of the many discharge phenomena. We only need to change the conditions of the test, and again we make other observations of interest.

When, instead of operating the induction coil as in the last two experiments, we operate it from a high frequency alternator, as in the next experiment, a systematic study of the phenomena is rendered much more easy. In such case, in varying the strength and frequency of the currents through the primary, we may observe five distinct forms of discharge, which I have described in my former paper on the subject before the American Institute of Electrical Engineers, May 20, 1891.

It would take too much time, and it would lead us too far from the subject presented this evening, to reproduce all these forms, but it seems to me desirable to show you one of them. It is a brush discharge, which is interesting in more than one respect. Viewed from a near position it resembles much a jet of gas escaping under great pressure. We know that the phenomenon is due to the agitation of the molecules near the terminal, and we anticipate that some heat must be developed by the impact of the molecules against the terminal or against each other. Indeed, we find that the brush is hot, and only a little thought leads us to the conclusion that, could we but reach sufficiently high frequencies, we could produce a brush which would give intense light and heat, and which would resemble in every particular an ordinary flame, save, perhaps, that both phenomena might not be due to the same agent—save, perhaps, that chemical affinity might not be *electrical* in its nature.

As the production of heat and light is here due to the impact of the molecules, or atoms of air, or something else besides, and, as we can augment the energy simply by raising the potential, we might, even with frequencies obtained from a dynamo machine, intensify the action to such a degree as to bring the terminal to melting heat. But with such low frequencies we would have to deal always with something of the nature of an electric current. If I approach a conducting object to the brush, a thin little spark passes, yet, even with the frequencies used this evening, the tendency to spark is not very great. So, for instance, if I hold a metallic sphere at some distance above the terminal, you may see the whole space between the terminal and sphere illuminated by the streams without the spark passing; and with the much higher frequencies obtainable by the disruptive discharge of a condenser, were it not for the sudden impulses, which are comparatively few in number, sparking would not occur even at very small distances. However, with incomparably higher frequencies, which we may yet find means to produce efficiently, and provided that electric impulses of such high frequencies could be transmitted through a conductor, the electrical characteristics of the brush discharge would completely vanish—no spark would pass, no shock would be felt—yet we would still have to deal with an *electric* phenomenon, but in the broad, modern interpretation of the word. In my first paper, before referred to, I have pointed out the curious properties of the brush, and described the best manner of producing it, but I have thought it worth while to endeavor to express myself more clearly in regard to this phenomenon, because of its absorbing interest.

When a coil is operated with currents of very high frequency, beautiful brush effects may be produced, even if the coil be of comparatively small dimensions. The experimenter may vary them in many ways, and, if it were for nothing else, they afford a pleasing sight. What adds to their interest is that they may be produced with one single terminal as well as with two—in fact, often better with one than with two.

But of all the discharge phenomena observed, the most pleasing to the eye, and the most instructive, are those observed with a coil which is operated by means of the disruptive discharge of a condenser. The power of the brushes, the abundance of the sparks, when the conditions are patiently adjusted, is often amazing. With even a very small coil, if it be so well insulated as to stand a difference of potential of several thousand volts per turn, the sparks may be so abundant that the whole coil may appear a complete mass of fire.

Curiously enough the sparks, when the terminals of the coil are set at a considerable distance, seem to dart in every possible direction as though the terminals were perfectly independent of each other. As the sparks would soon destroy the insulation, it is necessary to prevent them. This is best done by immersing the coil in a good liquid insulator, such as boiled-out oil. Immersion in a liquid may be considered almost an absolute necessity for the continued and successful working of such a coil.

It is, of course, out of the question, in an experimental lecture, with only a few minutes at disposal for the performance of each experiment, to show these discharge phenomena to advantage, as, to produce each phenomenon at its best, a very careful adjustment is required. But even if imperfectly produced, as they are likely to be this evening, they are sufficiently striking to interest an intelligent audience.

Before showing some of these curious effects I must, for the sake of completeness, give a short description of the coil and other apparatus used in the experiments with the disruptive discharge this evening.

It is contained in a box B (Fig. 132) of thick boards of hard

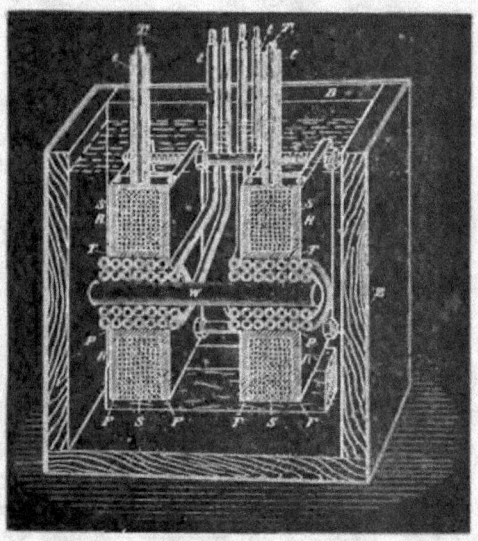

wood, covered on the outside with a zinc sheet z, which is carefully soldered all around. It might be advisable, in a strictly scientific investigation, when accuracy is of great importance, to do away with the metal cover, as it might introduce many errors, principally on account of its complex action upon the coil, as a condenser of very small capacity and as an electrostatic and electromagnetic screen. When the coil is used for such experiments as are here contemplated, the employment of the metal cover offers some practical advantages, but these are not of sufficient importance to be dwelt upon.

The coil should be placed symmetrically to the metal cover,

and the space between should, of course, not be too small, certainly not less than, say, five centimetres, but much more if possible; especially the two sides of the zinc box, which are at right angles to the axis of the coil, should be sufficiently remote from the latter, as otherwise they might impair its action and be a source of loss.

The coil consists of two spools of hard rubber R R, held apart at a distance of 10 centimetres by bolts c and nuts n, likewise of hard rubber. Each spool comprises a tube T of approximately 8 centimetres inside diameter, and 3 millimetres thick, upon which are screwed two flanges F F, 24 centimetres square, the space between the flanges being about 3 centimetres. The secondary, s s, of the best gutta percha-covered wire, has 26 layers, 10 turns in each, giving for each half a total of 260 turns. The two halves are wound oppositely and connected in series, the connection between both being made over the primary. This disposition, besides being convenient, has the advantage that when the coil is well balanced—that is, when both of its terminals T_1, T_1, are connected to bodies or devices of equal capacity—there is not much danger of breaking through to the primary, and the insulation between the primary and the secondary need not be thick. In using the coil it is advisable to attach to *both* terminals devices of nearly equal capacity, as, when the capacity of the terminals is not equal, sparks will be apt to pass to the primary. To avoid this, the middle point of the secondary may be connected to the primary, but this is not always practicable.

The primary P P is wound in two parts, and oppositely, upon a wooden spool w, and the four ends are led out of the oil through hard rubber tubes t t. The ends of the secondary T_1 T_1 are also led out of the oil through rubber tubes t_1 t_1 of great thickness. The primary and secondary layers are insulated by cotton cloth, the thickness of the insulation, of course, bearing some proportion to the difference of potential between the turns of the different layers. Each half of the primary has four layers, 24 turns in each, this giving a total of 96 turns. When both the parts are connected in series, this gives a ratio of conversion of about 1 : 2.7, and with the primaries in multiple, 1 : 5.4; but in operating with very rapidly alternating currents this ratio does not convey even an approximate idea of the ratio of the E. M. F's. in the primary and secondary circuits. The coil is held in position in the oil on wooden supports, there being about 5 centimetres thickness of oil all round. Where the oil is not specially needed, the space is filled with pieces of wood, and for this purpose principally the wooden box B surrounding the whole is used.

The construction here shown is, of course, not the best on general principles, but I believe it is a good and convenient one for the production of effects in which an excessive potential and a very small current are needed.

In connection with the coil I use either the ordinary form of discharger or a modified form. In the former I have introduced two changes which secure some advantages, and which are obvious. If they are mentioned, it is only in the hope that some experimenter may find them of use.

One of the changes is that the adjustable knobs A and B (Fig. 133), of the discharger are held in jaws of brass, J J, by spring pressure, this allowing of turning them successively into different

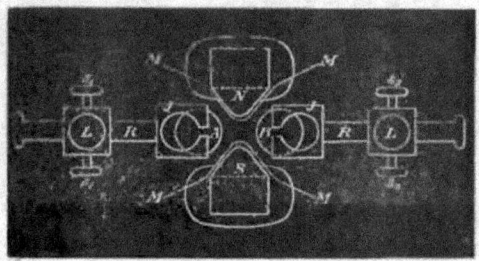

Fig. 133.

positions, and so doing away with the tedious process of frequent polishing up.

The other change consists in the employment of a strong electromagnet N S, which is placed with its axis at right angles to the line joining the knobs A and B, and produces a strong magnetic field between them. The pole pieces of the magnet are movable and properly formed so as to protrude between the brass knobs, in order to make the field as intense as possible; but to prevent the discharge from jumping to the magnet the pole pieces are protected by a layer of mica, M M, of sufficient thickness; $s_1 s_1$ and $s_2 s_2$ are screws for fastening the wires. On each side one of the screws is for large and the other for small wires. L L are screws for fixing in position the rods R R, which support the knobs.

In another arrangement with the magnet I take the discharge between the rounded pole pieces themselves, which in such case are insulated and preferably provided with polished brass caps.

The employment of an intense magnetic field is of advantage principally when the induction coil or transformer which charges the condenser is operated by currents of very low frequency. In such a case the number of the fundamental discharges between the knobs may be so small as to render the currents produced in the secondary unsuitable for many experiments. The intense magnetic field then serves to blow out the arc between the knobs as soon as it is formed, and the fundamental discharges occur in quicker succession.

Instead of the magnet, a draught or blast of air may be employed with some advantage. In this case the arc is preferably

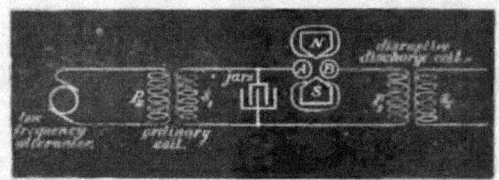

FIG. 134.

established between the knobs A B, in Fig. 131 (the knobs *a b* being generally joined, or entirely done away with), as in this disposition the arc is long and unsteady, and is easily affected by the draught.

When a magnet is employed to break the arc, it is better to choose the connection indicated diagrammatically in Fig. 134, as in this case the currents forming the arc are much more powerful, and the magnetic field exercises a greater influence. The use of the magnet permits, however, of the arc being replaced by a vacuum tube, but I have encountered great difficulties in working with an exhausted tube.

The other form of discharger used in these and similar experiments is indicated in Figs. 135 and 136. It consists of a number of brass pieces *c c* (Fig. 135), each of which comprises a spherical middle portion *m* with an extension *e* below—which is merely used to fasten the piece in a lathe when polishing up the discharging surface—and a column above, which consists of a knurled flange *f* surmounted by a threaded stem *l* carrying a nut *n*, by means of which a wire is fastened to the column. The flange *f* conveniently serves for holding the brass piece when fastening the

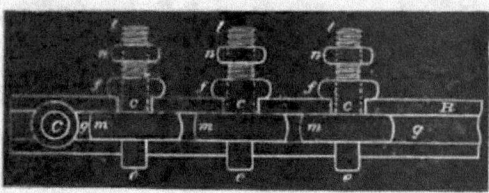

Fig. 135.

wire, and also for turning it in any position when it becomes necessary to present a fresh discharging surface. Two stout strips of hard rubber ʀ ʀ, with planed grooves *g g* (Fig. 136) to fit the middle portion of the pieces *c c*, serve to clamp the latter and hold them firmly in position by means of two bolts c c (of which only one is shown) passing through the ends of the strips.

In the use of this kind of discharger I have found three principal advantages over the ordinary form. First, the dielectric strength of a given total width of air space is greater when a great many small air gaps are used instead of one, which permits

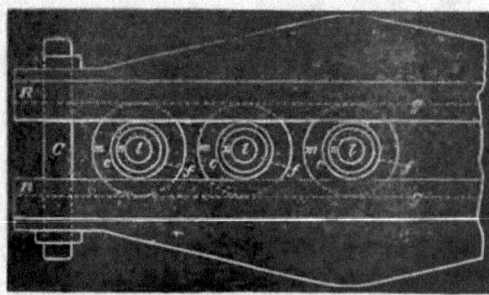

Fig. 136.

of working with a smaller length of air gap, and that means smaller loss and less deterioration of the metal; secondly, by reason of splitting the arc up into smaller arcs, the polished surfaces are made to last much longer; and, thirdly, the apparatus affords some gauge in the experiments. I usually set the pieces by putting between them sheets of uniform thickness at a certain very small distance which is known from the experiments of Sir William Thomson to require a certain electromotive force to be bridged by the spark.

It should, of course, be remembered that the sparking distance is much diminished as the frequency is increased. By taking any number of spaces the experimenter has a rough idea of the electromotive force, and he finds it easier to repeat an experiment, as he has not the trouble of setting the knobs again and again. With this kind of discharger I have been able to maintain an oscillating motion without any spark being visible with the naked eye between the knobs, and they would not show a very appreciable rise in temperature. This form of discharge also lends itself to many arrangements of condensers and circuits which are often very convenient and time-saving. I have used it preferably in a disposition similar to that indicated in Fig. 131, when the currents forming the arc are small.

I may here mention that I have also used dischargers with single or multiple air gaps, in which the discharge surfaces were rotated with great speed. No particular advantage was, however, gained by this method, except in cases where the currents from the condenser were large and the keeping cool of the surfaces was necessary, and in cases when, the discharge not being oscillating of itself, the arc as soon as established was broken by the air current, thus starting the vibration at intervals in rapid succession. I have also used mechanical interrupters in many ways. To avoid the difficulties with frictional contacts, the preferred plan adopted was to establish the arc and rotate through it at great speed a rim of mica provided with many holes and fastened to a steel plate. It is understood, of course, that the employment of a magnet, air current, or other interrupter, produces no effect worth noticing, unless the self-induction, capacity and resistance are so related that there are oscillations set up upon each interruption.

I will now endeavor to show you some of the most noteworthy of these discharge phenomena.

I have stretched across the room two ordinary cotton covered wires, each about seven metres in length. They are supported on insulating cords at a distance of about thirty centimetres. I attach now to each of the terminals of the coil one of the wires, and set the coil in action. Upon turning the lights off in the room you see the wires strongly illuminated by the streams issuing abundantly from their whole surface in spite of the cotton covering, which may even be very thick. When the experiment is performed under good conditions, the light from the wires is sufficiently intense to allow distinguishing the objects in a room. To produce the best result it is, of course, necessary to adjust carefully the capacity of the jars, the arc between the knobs and the length of the wires. My experience is that calculation of the length of the wires leads, in such case, to no result whatever. The experimenter will do best to take the wires at the start very long, and then adjust by cutting off first long pieces, and then smaller and smaller ones as he approaches the right length.

A convenient way is to use an oil condenser of very small capacity, consisting of two small adjustable metal plates, in connection with this and similar experiments. In such case I take wires rather short and at the beginning set the condenser plates at maximum distance. If the streams from the wires increase by approach of the plates, the length of the wires is about right; if they diminish, the wires are too long for that frequency and potential. When a condenser is used in connection with experiments with such a coil, it should be an oil condenser by all means, as in using an air condenser considerable energy might be wasted. The wires leading to the plates in the oil should be very thin, heavily coated with some insulating compound, and provided with a conducting covering—this preferably extending under the surface of the oil. The conducting cover should not be too near the terminals, or ends, of the wire, as a spark would be apt to jump from the wire to it. The conducting coating is used to diminish the air losses, in virtue of its action as an electrostatic screen. As to the size of the vessel containing the oil, and the size of the plates, the experimenter gains at once an idea from a rough trial. The size of the plates *in oil* is, however, calculable, as the dielectric losses are very small.

In the preceding experiment it is of considerable interest to know what relation the quantity of the light emitted bears to the frequency and potential of the electric impulses. My opinion is that the heat as well as light effects produced should be proportionate, under otherwise equal conditions of test, to the product of frequency and square of potential, but the experimental verification of the law, whatever it may be, would be exceedingly difficult. One thing is certain, at any rate, and that is, that in augmenting the potential and frequency we rapidly intensify the streams ; and, though it may be very sanguine, it is surely not altogether hopeless to expect that we may succeed in producing a practical illuminant on these lines. We would then be simply using burners or flames, in which there would be no chemical process, no consumption of material, but merely a transfer of energy, and which would, in all probability, emit more light and less heat than ordinary flames.

The luminous intensity of the streams is, of course, considerably

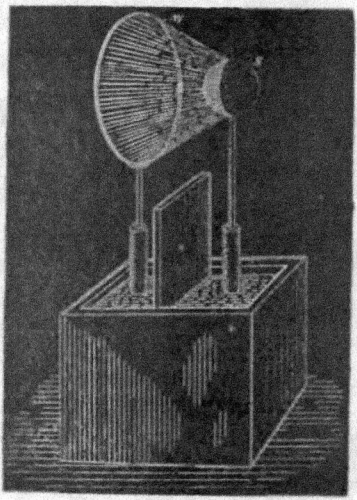

Fig. 137.

increased when they are focused upon a small surface. This may be shown by the following experiment :

I attach to one of the terminals of the coil a wire w (Fig. 137), bent in a circle of about 30 centimetres in diameter, and to the other terminal I fasten a small brass sphere s, the surface of the wire being preferably equal to the surface of the sphere, and the centre of the latter being in a line at right angles to the plane of the wire circle and passing through its centre. When the discharge is established under proper conditions, a luminous hollow cone is formed, and in the dark one-half of the brass sphere is strongly illuminated, as shown in the cut.

By some artifice or other it is easy to concentrate the streams upon small surfaces and to produce very strong light effects. Two thin wires may thus be rendered intensely luminous.

In order to intensify the streams the wires should be very thin and short; but as in this case their capacity would be generally too small for the coil—at least for such a one as the present—it is necessary to augment the capacity to the required value, while, at the same time, the surface of the wires remains very small. This may be done in many ways.

Here, for instance, I have two plates, R R, of hard rubber (Fig. 138), upon which I have glued two very thin wires *w w*, so as to form a name. The wires may be bare or covered with the best insulation—it is immaterial for the success of the experiment. Well insulated wires, if anything, are preferable. On the back

FIG. 138.

of each plate, indicated by the shaded portion, is a tinfoil coating *t t*. The plates are placed in line at a sufficient distance to prevent a spark passing from one wire to the other. The two tinfoil coatings I have joined by a conductor c, and the two wires I presently connect to the terminals of the coil. It is now easy, by varying the strength and frequency of the currents through the primary, to find a point at which the capacity of the system is best suited to the conditions, and the wires become so strongly luminous that, when the light in the room is turned off the name formed by them appears in brilliant letters.

It is perhaps preferable to perform this experiment with a coil operated from an alternator of high frequency, as then, owing to the harmonic rise and fall, the streams are very uniform, though they are less abundant than when produced with such a coil as the present one. This experiment, however, may be performed with low frequencies, but much less satisfactorily.

When two wires, attached to the terminals of the coil, are set at the proper distance, the streams between them may be so intense as to produce a continuous luminous sheet. To show this phenomenon I have here two circles, c and c (Fig. 139), of rather stout wire, one being about 80 centimetres and the other 30 centimetres in diameter. To each of the terminals of the coil I attach one of the circles. The supporting wires are so bent that

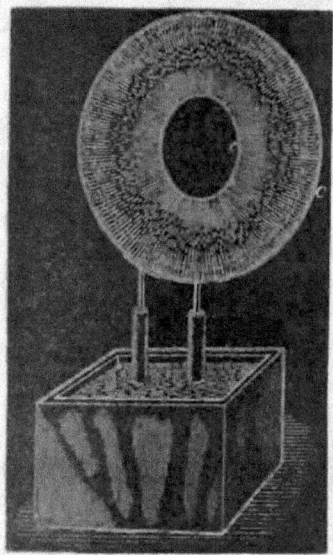

Fig. 139.

the circles may be placed in the same plane, coinciding as nearly as possible. When the light in the room is turned off and the coil set to work, you see the whole space between the wires uniformly filled with streams, forming a luminous disc, which could be seen from a considerable distance, such is the intensity of the streams. The outer circle could have been much larger than the present one; in fact, with this coil I have used much larger circles, and I have been able to produce a strongly luminous sheet, covering an area of more than one square metre, which is a remarkable effect with this very small coil. To avoid uncertainty, the circle has been taken smaller, and the area is now about 0.43 square metre.

The frequency of the vibration, and the quickness of succession of the sparks between the knobs, affect to a marked degree the appearance of the streams. When the frequency is very low, the air gives way in more or less the same manner, as by a steady difference of potential, and the streams consist of distinct threads, generally mingled with thin sparks, which probably correspond to the successive discharges occurring between the knobs. But when the frequency is extremely high, and the arc of the discharge produces a very *loud* and *smooth* sound—showing both that oscillation takes place and that the sparks succeed each other with great rapidity—then the luminous streams formed are perfectly uniform. To reach this result very small coils and jars of small capacity should be used. I take two tubes of thick Bohemian glass, about 5 centimetres in diameter and 20 centimetres long. In each of the tubes I slip a primary of very thick copper wire. On the top of each tube I wind a secondary of much thinner gutta-percha covered wire. The two secondaries I connect in series, the primaries preferably in multiple arc. The tubes are then placed in a large glass vessel, at a distance of 10 to 15 centimetres from each other, on insulating supports, and the vessel is filled with boiled-out oil, the oil reaching about an inch above the tubes. The free ends of the secondary are lifted out of the coil and placed parallel to each other at a distance of about ten centimetres. The ends which are scraped should be dipped in the oil. Two four-pint jars joined in series may be used to discharge through the primary. When the necessary adjustments in the length and distance of the wires above the oil and in the arc of discharge are made, a luminous sheet is produced between the wires which is perfectly smooth and textureless, like the ordinary discharge through a moderately exhausted tube.

I have purposely dwelt upon this apparently insignificant experiment. In trials of this kind the experimenter arrives at the startling conclusion that, to pass ordinary luminous discharges through gases, no particular degree of exhaustion is needed, but that the gas may be at ordinary or even greater pressure. To accomplish this, a very high frequency is essential; a high potential is likewise required, but this is merely an incidental necessity. These experiments teach us that, in endeavoring to discover novel methods of producing light by the agitation of atoms, or molecules, of a gas, we need not limit our research to the vacuum tube, but may look forward quite seriously to the possibility of obtaining the light effects without the use of any vessel whatever, with air at ordinary pressure.

Such discharges of very high frequency, which render luminous the air at ordinary pressures, we have probably occasion often to witness in Nature. I have no doubt that if, as many believe, the aurora borealis is produced by sudden cosmic disturbances, such as eruptions at the sun's surface, which set the electrostatic charge of the earth in an extremely rapid vibration, the red glow observed is not confined to the upper rarefied strata of the air, but the discharge traverses, by reason of its very high frequency, also the dense atmosphere in the form of a *glow*, such as we ordinarily produce in a slightly exhausted tube. If the frequency were very low, or even more so, if the charge were not at all vibrating, the dense air would break down as in a lightning discharge. Indications of such breaking down of the lower dense strata of the air have been repeatedly observed at the occurence of this marvelous phenomenon; but if it does occur, it can only be attributed to the fundamental disturbances, which are few in number, for the vibration produced by them would be far too rapid to allow a disruptive break. It is the original and irregular impulses which affect the instruments; the superimposed vibrations probably pass unnoticed.

When an ordinary low frequency discharge is passed through moderately rarefied air, the air assumes a purplish hue. If by some means or other we increase the intensity of the molecular, or atomic, vibration, the gas changes to a white color. A similar change occurs at ordinary pressures with electric impulses of very high frequency. If the molecules of the air around a wire are moderately agitated, the brush formed is reddish or violet; if the vibration is rendered sufficiently intense, the streams become white. We may accomplish this in various ways. In the experiment before shown with the two wires across the room, I have endeavored to secure the result by pushing to a high value both the frequency and potential; in the experiment with the thin wires glued on the rubber plate I have concentrated the action upon a very small surface—in other words, I have worked with a great electric density.

A most curious form of discharge is observed with such a coil

when the frequency and potential are pushed to the extreme
limit. To perform the experiment, every part of the coil should
be heavily insulated, and only two small spheres—or, better still,
two sharp-edged metal discs (*d d*, Fig. 140) of no more than
a few centimetres in diameter—should be exposed to the air.
The coil here used is immersed in oil, and the ends of the
secondary reaching out of the oil are covered with an air-tight
cover of hard rubber of great thickness. All cracks, if there
are any, should be carefully stopped up, so that the brush discharge cannot form anywhere except on the small spheres or
plates which are exposed to the air. In this case, since there
are no large plates or other bodies of capacity attached to the
terminals, the coil is capable of an extremely rapid vibration.

Fig. 140.

The potential may be raised by increasing, as far as the experimenter judges proper, the rate of change of the primary current. With a coil not widely differing from the present, it is
best to connect the two primaries in multiple arc; but if the
secondary should have a much greater number of turns the
primaries should preferably be used in series, as otherwise the
vibration might be too fast for the secondary. It occurs under
these conditions that misty white streams break forth from the
edges of the discs and spread out phantom-like into space.
With this coil, when fairly well produced, they are about 25 to
30 centimetres long. When the hand is held against them no
sensation is produced, and a spark, causing a shock, jumps from

the terminal only upon the hand being brought much nearer. If the oscillation of the primary current is rendered intermittent by some means or other, there is a corresponding throbbing of the streams, and now the hand or other conducting object may be brought in still greater proximity to the terminal without a spark being caused to jump.

Among the many beautiful phenomena which may be produced with such a coil, I have here selected only those which appear to possess some features of novelty, and lead us to some conclusions of interest. One will not find it at all difficult to produce in the laboratory, by means of it, many other phenomena which appeal to the eye even more than these here shown, but present no particular feature of novelty.

Early experimenters describe the display of sparks produced by an ordinary large induction coil upon an insulating plate separating the terminals. Quite recently Siemens performed some experiments in which fine effects were obtained, which were seen by many with interest. No doubt large coils, even if operated with currents of low frequencies, are capable of producing beautiful effects. But the largest coil ever made could not, by far, equal the magnificent display of streams and sparks obtained from such a disruptive discharge coil when properly adjusted. To give an idea, a coil such as the present one will cover easily a plate of one metre in diameter completely with the streams. The best way to perform such experiments is to take a very thin rubber or a glass plate and glue on one side of it a narrow ring of tinfoil of very large diameter, and on the other a circular washer, the centre of the latter coinciding with that of the ring, and the surfaces of both being preferably equal, so as to keep the coil well balanced. The washer and ring should be connected to the terminals by heavily insulated thin wires. It is easy in observing the effect of the capacity to produce a sheet of uniform streams, or a fine network of thin silvery threads, or a mass of loud brilliant sparks, which completely cover the plate.

Since I have advanced the idea of the conversion by means of the disruptive discharge, in my paper before the American Institute of Electrical Engineers at the beginning of the past year, the interest excited in it has been considerable. It affords us a means for producing any potentials by the aid of inexpensive coils operated from ordinary systems of distribution, and—what is perhaps more appreciated—it enables us to convert currents of any frequency into currents of any other lower or higher frequency. But its chief value will perhaps be found in the help which it will afford us in the investigations of the phenomena of phosphorescence, which a disruptive discharge coil is capable of exciting in innumerable cases where ordinary coils, even the largest, would utterly fail.

Considering its probable uses for many practical purposes, and its possible introduction into laboratories for scientific research, a few additional remarks as to the construction of such a coil will perhaps not be found superfluous.

It is, of course, absolutely necessary to employ in such a coil wires provided with the best insulation.

Good coils may be produced by employing wires covered with several layers of cotton, boiling the coil a long time in pure wax, and cooling under moderate pressure. The advantage of such a coil is that it can be easily handled, but it cannot probably give as satisfactory results as a coil immersed in pure oil. Besides, it seems that the presence of a large body of wax affects the coil disadvantageously, whereas this does not seem to be the case with oil. Perhaps it is because the dielectric losses in the liquid are smaller.

I have tried at first silk and cotton covered wires with oil immersions, but I have been gradually led to use gutta-percha covered wires, which proved most satisfactory. Gutta-percha insulation adds, of course, to the capacity of the coil, and this, especially if the coil be large, is a great disadvantage when extreme frequencies are desired; but, on the other hand, gutta-percha will withstand much more than an equal thickness of oil, and this advantage should be secured at any price. Once the coil has been immersed, it should never be taken out of the oil for more than a few hours, else the gutta-percha will crack up and the coil will not be worth half as much as before. Gutta-percha is probably slowly attacked by the oil, but after an immersion of eight to nine months I have found no ill effects.

I have obtained two kinds of gutta-percha wire known in commerce: in one the insulation sticks tightly to the metal, in the other it does not. Unless a special method is followed to expel all air, it is much safer to use the first kind. I wind the coil within an oil tank so that all interstices are filled up with the oil. Between the layers I use cloth boiled out thoroughly in oil, calculating the thickness according to the difference of potential between the turns. There seems not to be a very great difference whatever kind of oil is used; I use paraffine or linseed oil.

To exclude more perfectly the air, an excellent way to proceed, and easily practicable with small coils, is the following: Construct a box of hardwood of very thick boards which have been for a long time boiled in oil. The boards should be so joined as to safely withstand the external air pressure. The coil being placed and fastened in position within the box, the latter is closed with a strong lid, and covered with closely fitting metal sheets, the joints of which are soldered very carefully. On the top two small holes are drilled, passing through the metal sheet and the wood, and in these holes two small glass tubes are inserted and the joints made air-tight. One of the tubes is connected to a vacuum pump, and the other with a vessel containing a sufficient quantity of boiled-out oil. The latter tube has a very small hole at the bottom, and is provided with a stopcock. When a fairly good vacuum has been obtained, the stopcock is opened and the oil slowly fed in. Proceeding in this manner, it is impossible that any big bubbles, which are the principal danger, should remain between the turns. The air is most completely excluded, probably better than by boiling out, which, however, when gutta-percha coated wires are used, is not practicable.

For the primaries I use ordinary line wire with a thick cotton coating. Strands of very thin insulated wires properly interlaced would, of course, be the best to employ for the primaries, but they are not to be had.

In an experimental coil the size of the wires is not of great importance. In the coil here used the primary is No. 12 and the secondary No. 24 Brown & Sharpe gauge wire; but the sections may be varied considerably. It would only imply different adjustments; the results aimed at would not be materially affected.

I have dwelt at some length upon the various forms of brush discharge because, in studying them, we not only observe phenomena which please our eye, but also afford us food for thought, and lead us to conclusions of practical importance. In the use of alternating currents of very high tension, too much precaution cannot be taken to prevent the brush discharge. In a main conveying such currents, in an induction coil or transformer, or in a condenser, the brush discharge is a source of great danger to the insulation. In a condenser, especially, the gaseous matter must be most carefully expelled, for in it the charged surfaces are near each other, and if the potentials are high, just as sure as a weight will fall if let go, so the insulation will give way if a single gaseous bubble of some size be present, whereas, if all gaseous matter were carefully excluded, the condenser would safely withstand a much higher difference of potential. A main conveying alternating currents of very high tension may be injured merely by a blow hole or small crack in the insulation, the more so as a blowhole is apt to contain gas at low pressure; and as it appears almost impossible to completely obviate such little imperfections, I am led to believe that in our future distribution of electrical energy by currents of very high tension, liquid insulation will be used. The cost is a great drawback, but if we employ an oil as an insulator the distribution of electrical energy with something like 100,000 volts, and even more, becomes, at least with higher frequencies, so easy that it could be hardly called an engineering feat. With oil insulation and alternate current motors, transmissions of power can be affected with safety and upon an industrial basis at distances of as much as a thousand miles.

A peculiar property of oils, and liquid insulation in general, when subjected to rapidly changing electric stresses, is to disperse any gaseous bubbles which may be present, and diffuse them through its mass, generally long before any injurious break can occur. This feature may be easily observed with an ordinary induction coil by taking the primary out, plugging up the end of the tube upon which the secondary is wound, and filling it with some fairly transparent insulator, such as paraffine oil. A primary of a diameter something like six millimetres smaller than the inside of the tube may be inserted in the oil. When the coil is set to work one may see, looking from the top through the oil, many luminous points—air bubbles which are caught by inserting the primary, and which are rendered luminous in consequence of the violent bombardment. The occluded air, by its impact against the oil, heats it; the oil begins to circulate, carrying some of the air along with it, until the bubbles are dispersed and the luminous points disappear. In this manner, unless large bubbles are occluded in such way that circulation is rendered impossible, a damaging break is averted, the only effect being a moderate warming up of the oil. If, instead of the liquid, a solid insulation, no matter how thick, were used, a breaking through and injury of the apparatus would be inevitable.

The exclusion of gaseous matter from any apparatus in which the dielectric is subjected to more or less rapidly changing electric forces is, however, not only desirable in order to avoid a possible injury of the apparatus, but also on account of economy. In a condenser, for instance, as long as only a solid or only a liquid dielectric is used, the loss is small; but if a gas under ordinary or small pressure be present the loss may be very great. Whatever the nature of the force acting in the dielectric may be, it seems that in a solid or liquid the molecular displacement produced by the force is small: hence the product of force and displacement is insignificant, unless the force be very great; but in a gas the displacement, and therefore this product, is considerable; the molecules are free to move, they reach high speeds, and the energy of their impact is lost in heat or otherwise. If the gas be strongly compressed, the displacement due to the force is made smaller, and the losses are reduced.

In most of the succeeding experiments I prefer, chiefly on account of the regular and positive action, to employ the alternator before referred to. This is one of the several machines constructed by me for the purpose of these investigations. It has 384 pole projections, and is capable of giving currents of a frequency of about 10,000 per second. This machine has been illustrated and briefly described in my first paper before the American Institute of Electrical Engineers, May 20th, 1891, to which I have already referred. A more detailed description, sufficient to enable any engineer to build a similar machine, will be found in several electrical journals of that period.

The induction coils operated from the machine are rather small, containing from 5,000 to 15,000 turns in the secondary. They are immersed in boiled-out linseed oil, contained in wooden boxes covered with zinc sheet.

I have found it advantageous to reverse the usual position of the wires, and to wind, in these coils, the primaries on the top; thus allowing the use of a much larger primary, which, of course, reduces the danger of overheating and increases the output of the coil. I make the primary on each side at least one centimetre shorter than the secondary, to prevent the breaking through on the ends, which would surely occur unless the insulation on the top of the secondary be very thick, and this, of course, would be disadvantageous.

When the primary is made movable, which is necessary in

some experiments, and many times convenient for the purposes of adjustment, I cover the secondary with wax, and turn it off in a lathe to a diameter slightly smaller than the inside of the primary coil. The latter I provide with a handle reaching out of the oil, which serves to shift it in any position along the secondary.

I will now venture to make, in regard to the general manipulation of induction coils, a few observations bearing upon points which have not been fully appreciated in earlier experiments with such coils, and are even now often overlooked.

The secondary of the coil possesses usually such a high self-induction that the current through the wire is inappreciable, and may be so even when the terminals are joined by a conductor of small resistance. If capacity is added to the terminals, the self-induction is counteracted, and a stronger current is made to flow through the secondary, though its terminals are insulated from each other. To one entirely unacquainted with the properties of alternating currents nothing will look more puzzling. This feature was illustrated in the experiment performed at the beginning with the top plates of wire gauze attached to the terminals and the rubber plate. When the plates of wire gauze were close together, and a small arc passed between them, the arc *prevented* a strong current from passing through the secondary, because it did away with the capacity on the terminals; when the rubber plate was inserted between, the capacity of the condenser formed counteracted the self-induction of the secondary, a stronger current passed now, the coil performed more work, and the discharge was by far more powerful.

The first thing, then, in operating the induction coil is to combine capacity with the secondary to overcome the self-induction. If the frequencies and potentials are very high, gaseous matter should be carefully kept away from the charged surfaces. If Leyden jars are used, they should be immersed in oil, as otherwise considerable dissipation may occur if the jars are greatly strained. When high frequencies are used, it is of equal importance to combine a condenser with the primary. One may use a condenser connected to the ends of the primary or to the terminals of the alternator, but the latter is not to be recommended, as the machine might be injured. The best way is undoubtedly to use the condenser in series with the primary and with the alternator, and to adjust its capacity so as to annul the self-induction of both the latter. The condenser should be adjustable by very small steps, and for a finer adjustment a small oil condenser with movable plates may be used conveniently.

I think it best at this juncture to bring before you a phenomenon, observed by me some time ago, which to the purely scientific investigator may perhaps appear more interesting than any of the results which I have the privilege to present to you this evening.

It may be quite properly ranked among the brush phenomena—in fact, it is a brush, formed at, or near, a single terminal in high vacuum.

In bulbs provided with a conducting terminal, though it be of

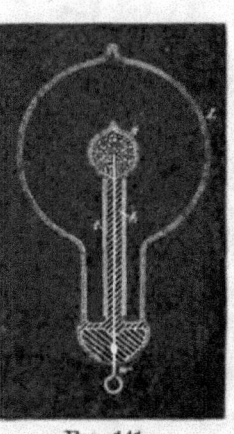

Fig. 141.

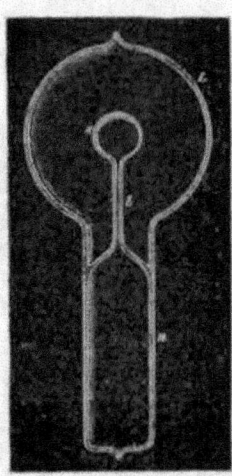

Fig. 142.

aluminum, the brush has but an ephemeral existence, and cannot, unfortunately, be indefinitely preserved in its most sensitive state, even in a bulb devoid of any conducting electrode. In studying the phenomenon, by all means a bulb having no leading-in wire should be used. I have found it best to use bulbs constructed as indicated in Figs. 141 and 142.

In Fig. 141 the bulb comprises an incandescent lamp globe *L*, in the neck of which is sealed a barometer tube *b*, the end of which is blown out to form a small sphere *s*. This sphere should be sealed as closely as possible in the centre of the large globe. Before sealing, a thin tube *t*, of aluminum sheet, may be slipped in the barometer tube, but it is not important to employ it.

The small hollow sphere *s* is filled with some conducting powder, and a wire *w* is cemented in the neck for the purpose of connecting the conducting powder with the generator.

The construction shown in Fig. 142 was chosen in order to remove from the brush any conducting body which might possibly affect it. The bulb consists in this case of a lamp globe *L*, which has a neck *n*, provided with a tube *b* and small sphere *s*, sealed to it, so that two entirely independent compartments are formed, as indicated in the drawing. When the bulb is in use the neck *n* is provided with a tinfoil coating, which is connected to the generator and acts inductively upon the moderately rarefied and highly conducted gas inclosed in the neck. From there the current passes through the tube *b* into the small sphere *s*, to act by induction upon the gas contained in the globe *L*.

It is of advantage to make the tube *b* very thick, the hole

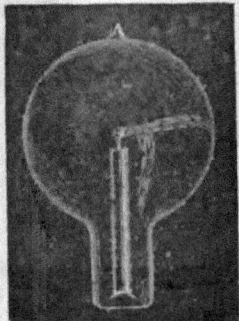

Fig. 143.

through it very small, and to blow the sphere *s* very thin. It is of the greatest importance that the sphere *s* be placed in the centre of the globe *L*.

Figs. 143, 144 and 145 indicate different forms, or stages, of the brush. Fig. 143 shows the brush as it first appears in a bulb provided with a conducting terminal; but, as in such a bulb it very soon disappears—often after a few minutes—I will confine myself to the description of the phenomenon as seen in a bulb without conducting electrode. It is observed under the following conditions:

When the globe *L* (Figs. 141 and 142) is exhausted to a very high degree, generally the bulb is not excited upon connecting the wire *w* (Fig. 141) or the tinfoil coating of the bulb (Fig.

142) to the terminal of the induction coil. To excite it, it is
usually sufficient to grasp the globe L with the hand. An intense phosphorescence then spreads at first over the globe, but
soon gives place to a white, misty light. Shortly afterward one
may notice that the luminosity is unevenly distributed in the
globe, and after passing the current for some time the bulb appears as in Fig. 144. From this stage the phenomenon will
gradually pass to that indicated in Fig. 145, after some minutes,
hours, days or weeks, according as the bulb is worked. Warming the bulb or increasing the potential hastens the transit.

When the brush assumes the form indicated in Fig. 145, it may
be brought to a state of extreme sensitiveness to electrostatic

Fig. 144.   Fig. 145.

and magnetic influence. The bulb hanging straight down from
a wire, and all objects being remote from it, the approach of the
observer at a few paces from the bulb will cause the brush to fly
to the opposite side, and if he walks around the bulb it will
always keep on the opposite side. It may begin to spin around
the terminal long before it reaches that sensitive stage. When
it begins to turn around, principally, but also before, it is affected
by a magnet, and at a certain stage it is susceptible to magnetic
influence to an astonishing degree. A small permanent magnet,
with its poles at a distance of no more than two centimetres, will
affect it visibly at a distance of two metres, slowing down or accelerating the rotation according to how it is held relatively to

the brush. I think I have observed that at the stage when it is most sensitive to magnetic, it is not most sensitive to electrostatic, influence. My explanation is, that the electrostatic attraction between the brush and the glass of the bulb, which retards the rotation, grows much quicker than the magnetic influence when the intensity of the stream is increased.

When the bulb hangs with the globe L down, the rotation is always clockwise. In the southern hemisphere it would occur in the opposite direction and on the equator the brush should not turn at all. The rotation may be reversed by a magnet kept at some distance. The brush rotates best, seemingly, when it is at right angles to the lines of force of the earth. It very likely rotates, when at its maximum speed, in synchronism with the alternations, say, 10,000 times a second. The rotation can be slowed down or accelerated by the approach or receding of the observer, or any conducting body, but it cannot be reversed by putting the bulb in any position. When it is in the state of the highest sensitiveness and the potential or frequency be varied, the sensitiveness is rapidly diminished. Changing either of these but little will generally stop the rotation. The sensitiveness is likewise affected by the variations of temperature. To attain great sensitiveness it is necessary to have the small sphere s in the centre of the globe L, as otherwise the electrostatic action of the glass of the globe will tend to stop the rotation. The sphere s should be small and of uniform thickness; any dissymmetry of course has the effect to diminish the sensitiveness.

The fact that the brush rotates in a definite direction in a permanent magnetic field seems to show that in alternating currents of very high frequency the positive and negative impulses are not equal, but that one always preponderates over the other.

Of course, this rotation in one direction may be due to the action of the two elements of the same current upon each other, or to the action of the field produced by one of the elements upon the other, as in a series motor, without necessarily one impulse being stronger than the other. The fact that the brush turns, as far as I could observe, in any position, would speak for this view. In such case it would turn at any point of the earth's surface. But, on the other hand, it is then hard to explain why a permanent magnet should reverse the rotation, and one must assume the preponderance of impulses of one kind.

As to the causes of the formation of the brush or stream, I

think it is due to the electrostatic action of the globe and the dissymmetry of the parts. If the small bulb *s* and the globe *L* were perfect concentric spheres, and the glass throughout of the same thickness and quality, I think the brush would not form, as the tendency to pass would be equal on all sides. That the formation of the stream is due to an irregularity is apparent from the fact that it has the tendency to remain in one position, and rotation occurs most generally only when it is brought out of this position by electrostatic or magnetic influence. When in an extremely sensitive state it rests in one position, most curious experiments may be performed with it. For instance, the experimenter may, by selecting a proper position, approach the hand at a certain considerable distance to the bulb, and he may cause the brush to pass off by merely stiffening the muscles of the arm. When it begins to rotate slowly, and the hands are held at a proper distance, it is impossible to make even the slightest motion without producing a visible effect upon the brush. A metal plate connected to the other terminal of the coil affects it at a great distance, slowing down the rotation often to one turn a second.

I am firmly convinced that such a brush, when we learn how to produce it properly, will prove a valuable aid in the investigation of the nature of the forces acting in an electrostatic or magnetic field. If there is any motion which is measurable going on in the space, such a brush ought to reveal it. It is, so to speak, a beam of light, frictionless, devoid of inertia.

I think that it may find practical applications in telegraphy. With such a brush it would be possible to send dispatches across the Atlantic, for instance, with any speed, since its sensitiveness may be so great that the slightest changes will affect it. If it were possible to make the stream more intense and very narrow, its deflections could be easily photographed.

I have been interested to find whether there is a rotation of the stream itself, or whether there is simply a stress traveling around the bulb. For this purpose I mounted a light mica fan so that its vanes were in the path of the brush. If the stream itself was rotating the fan would be spun around. I could produce no distinct rotation of the fan, although I tried the experiment repeatedly ; but as the fan exerted a noticeable influence on the stream, and the apparent rotation of the latter was, in this case, never quite satisfactory, the experiment did not appear to be conclusive.

I have been unable to produce the phenomenon with the disruptive discharge coil, although every other of these phenomena can be well produced by it—many, in fact, much better than with coils operated from an alternator.

It may be possible to produce the brush by impulses of one direction, or even by a steady potential, in which case it would be still more sensitive to magnetic influence.

In operating an induction coil with rapidly alternating currents, we realize with astonishment, for the first time, the great importance of the relation of capacity, self-induction and frequency as regards the general results. The effects of capacity are the most striking, for in these experiments, since the self-induction and frequency both are high, the critical capacity is very small, and need be but slightly varied to produce a very considerable change. The experimenter may bring his body in contact with the terminals of the secondary of the coil, or attach to one or both terminals insulated bodies of very small bulk, such as bulbs, and he may produce a considerable rise or fall of potential, and greatly affect the flow of the current through the primary. In the experiment before shown, in which a brush appears at a wire attached to one terminal, and the wire is vibrated when the experimenter brings his insulated body in contact with the other terminal of the coil, the sudden rise of potential was made evident.

I may show you the behavior of the coil in another manner which possesses a feature of some interest. I have here a little light fan of aluminum sheet, fastened to a needle and arranged to rotate freely in a metal piece screwed to one of the terminals of the coil. When the coil is set to work, the molecules of the air are rhythmically attracted and repelled. As the force with which they are repelled is greater than that with which they are attracted, it results that there is a repulsion exerted on the surfaces of the fan. If the fan were made simply of a metal sheet, the repulsion would be equal on the opposite sides, and would produce no effect. But if one of the opposing surfaces is screened, or if, generally speaking, the bombardment on this side is weakened in some way or other, there remains the repulsion exerted upon the other, and the fan is set in rotation. The screening is best effected by fastening upon one of the opposing sides of the fan insulated conducting coatings, or, if the fan is made in the shape of an ordinary propeller screw, by fastening on one side, and close to it, an insulated metal plate. The static screen may, however, be omitted, and simply a thickness of insulating material fastened to one of the sides of the fan.

To show the behavior of the coil, the fan may be placed upon the terminal and it will readily rotate when the coil is operated by currents of very high frequency. With a steady potential, of course, and even with alternating currents of very low frequency, it would not turn, because of the very slow exchange of air and, consequently, smaller bombardment; but in the latter case it might turn if the potential were excessive. With a pin wheel, quite the opposite rule holds good; it rotates best with a steady potential, and the effort is the smaller the higher the frequency. Now, it is very easy to adjust the conditions so that the potential is normally not sufficient to turn the fan, but that by connecting the other terminal of the coil with an insulated body it rises to a much greater value, so as to rotate the fan, and it is likewise possible to stop the rotation by connecting to the terminal a body of different size, thereby diminishing the potential.

Instead of using the fan in this experiment, we may use the " electric " radiometer with similar effect. But in this case it will be found that the vanes will rotate only at high exhaustion or at ordinary pressures; they will not rotate at moderate pressures, when the air is highly conducting. This curious observation was made conjointly by Professor Crookes and myself. I attribute the result to the high conductivity of the air, the molecules of which then do not act as independent carriers of electric charges, but act all together as a single conducting body. In such case, of course, if there is any repulsion at all of the molecules from the vanes, it must be very small. It is possible, however, that the result is in part due to the fact that the greater part of the discharge passes from the leading-in wire through the highly conducting gas, instead of passing off from the conducting vanes.

In trying the preceding experiment with the electric radiometer the potential should not exceed a certain limit, as then the electrostatic attraction between the vanes and the glass of the bulb may be so great as to stop the rotation.

A most curious feature of alternate currents of high frequencies and potentials is that they enable us to perform many experiments by the use of one wire only. In many respects this feature is of great interest.

In a type of alternate current motor invented by me some years ago I produced rotation by inducing, by means of a single alternating current passed through a motor circuit, in the mass or other circuits of the motor, secondary currents, which, jointly with the primary or inducing current, created a moving field of force. A simple but crude form of such a motor is obtained by winding upon an iron core a primary, and close to it a secondary coil, joining the ends of the latter and placing a freely movable metal disc within the influence of the field produced by both. The iron core is employed for obvious reasons, but it is not essential to the operation. To improve the motor, the iron core is made to encircle the armature. Again to improve, the secondary coil is made to partly overlap the primary, so that it cannot free itself from a strong inductive action of the latter, repel its lines as it may. Once more to improve, the proper difference of phase is obtained between the primary and secondary currents by a condenser, self-induction, resistance or equivalent windings.

I had discovered, however, that rotation is produced by means of a single coil and core; my explanation of the phenomenon, and leading thought in trying the experiment, being that there must be a true time lag in the magnetization of the core. I remember the pleasure I had when, in the writings of Professor Ayrton, which came later to my hand, I found the idea of the time lag advocated. Whether there is a true time lag, or whether the retardation is due to eddy currents circulating in minute paths, must remain an open question, but the fact is that a coil wound upon an iron core and traversed by an alternating current creates a moving field of force, capable of setting an armature in rotation. It is of some interest, in conjunction with the historical Arago experiment, to mention that in lag or phase motors I have produced rotation in the opposite direction to the moving field, which means that in that experiment the magnet may not rotate, or may even rotate in the opposite direction to the moving disc. Here, then, is a motor (diagrammatically illustrated in Fig. 146), comprising a coil and iron core, and a freely movable copper disc in proximity to the latter.

To demonstrate a novel and interesting feature, I have, for a reason which I will explain, selected this type of motor. When the ends of the coil are connected to the terminals of an alternator the disc is set in rotation. But it is not this experiment, now well known, which I desire to perform. What I wish to

show you is that this motor rotates with *one single* connection between it and the generator; that is to say, one terminal of the
motor is connected to one terminal of the generator—in this case
the secondary of a high-tension induction coil—the other terminals of motor and generator being insulated in space. To produce rotation it is generally (but not absolutely) necessary to
connect the free end of the motor coil to an insulated body of
some size. The experimenter's body is more than sufficient. If
he touches the free terminal with an object held in the hand, a
current passes through the coil and the copper disc is set in rotation. If an exhausted tube is put in series with the coil, the tube
lights brilliantly, showing the passage of a strong current. In-

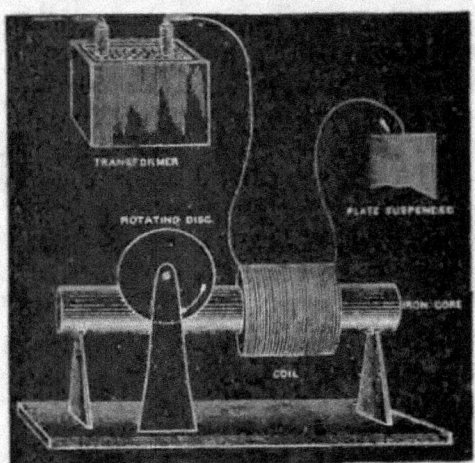

Fig. 146.

stead of the experimenter's body, a small metal sheet suspended
on a cord may be used with the same result. In this case the
plate acts as a condenser in series with the coil. It counteracts
the self-induction of the latter and allows a strong current to
pass. In such a combination, the greater the self-induction of
the coil the smaller need be the plate, and this means that a lower
frequency, or eventually a lower potential, is required to operate
the motor. A single coil wound upon a core has a high self-
induction; for this reason, principally, this type of motor was
chosen to perform the experiment. Were a secondary closed
coil wound upon the core, it would tend to diminish the self-

induction, and then it would be necessary to employ a much higher frequency and potential. Neither would be advisable, for a higher potential would endanger the insulation of the small primary coil, and a higher frequency would result in a materially diminished torque.

It should be remarked that when such a motor with a closed secondary is used, it is not at all easy to obtain rotation with excessive frequencies, as the secondary cuts off almost completely the lines of the primary—and this, of course, the more, the higher the frequency—and allows the passage of but a minute current. In such a case, unless the secondary is closed through a condenser, it is almost essential, in order to produce rotation, to make the primary and secondary coils overlap each other more or less.

But there is an additional feature of interest about this motor, namely, it is not necessary to have even a single connection between the motor and generator, except, perhaps, through the ground; for not only is an insulated plate capable of giving off energy into space, but it is likewise capable of deriving it from an alternating electrostatic field, though in the latter case the available energy is much smaller. In this instance one of the motor terminals is connected to the insulated plate or body located within the alternating electrostatic field, and the other terminal preferably to the ground.

It is quite possible, however, that such "no wire" motors, as they might be called, could be operated by conduction through the rarefied air at considerable distances. Alternate currents, especially of high frequencies, pass with astonishing freedom through even slightly rarefied gases. The upper strata of the air are rarefied. To reach a number of miles out into space requires the overcoming of difficulties of a merely mechanical nature. There is no doubt that with the enormous potentials obtainable by the use of high frequencies and oil insulation, luminous discharges might be passed through many miles of rarefied air, and that, by thus directing the energy of many hundreds or thousands of horsepower, motors or lamps might be operated at considerable distances from stationary sources. But such schemes are mentioned merely as possibilities. We shall have no need to transmit power in this way. We shall have no need to *transmit* power at all. Ere many generations pass, our machinery will be driven by a power obtainable at any point of the universe. This idea is not novel. Men have been led to it long ago by instinct or reason. It has been expressed in many ways, and in many places, in the history of old and new. We find it in the delightful myth of Antheus, who derives power from the earth; we find it among the subtle speculations of one of your splendid mathematicians, and in many hints and statements of thinkers of the present time. Throughout space there is energy. Is this energy static or kinetic? If static our hopes are in vain; if kinetic—and this we know it is, for certain—then it is a mere question of time when men will succeed in attaching their machinery to the very wheelwork of nature. Of all, living or dead, Crookes came nearest to doing it. His radiometer will turn in the light of day and in the darkness of the night; it will turn everywhere where there is heat, and heat is everywhere. But, unfortunately, this beautiful little machine, while it goes down to posterity as the most interesting, must likewise be put on record as the most inefficient machine ever invented!

The preceding experiment is only one of many equally interesting experiments which may be performed by the use of only one wire with alternations of high potential and frequency. We may connect an insulated line to a source of such currents, we may pass an inappreciable current over the line, and on any point of the same we are able to obtain a heavy current, capable of fusing a thick copper wire. Or we may, by the help of some artifice, decompose a solution in any electrolytic cell by connecting only one pole of the cell to the line or source of energy. Or we may, by attaching to the line, or only bringing into its vicinity, light up an incandescent lamp, an exhausted tube, or a phosphorescent bulb.

However impracticable this plan of working may appear in many cases, it certainly seems practicable, and even recommendable, in the production of light. A perfected lamp would require but little energy, and if wires were used at all we ought to be able to supply that energy without a return wire.

It is now a fact that a body may be rendered incandescent or phosphorescent by bringing it either in single contact or merely in the vicinity of a source of electric impulses of the proper character, and that in this manner a quantity of light sufficient to afford a practical illuminant may be produced. It is, therefore, to say the least, worth while to attempt to determine the best conditions and to invent the best appliances for attaining this object.

Some experiences have already been gained in this direction, and I will dwell on them briefly, in the hope that they might prove useful.

The heating of a conducting body inclosed in a bulb, and connected to a source of rapidly alternating electric impulses, is dependent on so many things of a different nature, that it would be difficult to give a generally applicable rule under which the maximum heating occurs. As regards the size of the vessel, I have lately found that at ordinary or only slightly differing atmospheric pressures, when air is a good insulator, and hence practically the same amount of energy by a certain potential and frequency is given off from the body, whether the bulb be small or large, the body is brought to a higher temperature if enclosed in a small bulb, because of the better confinement of heat in this case.

At lower pressures, when air becomes more or less conducting, or if the air be sufficiently warmed to become conducting, the body is rendered more intensely incandescent in a large bulb, obviously because, under otherwise equal conditions of test, more energy may be given off from the body when the bulb is large.

At very high degrees of exhaustion, when the matter in the bulb becomes "radiant," a large bulb has still an advantage, but a comparatively slight one, over the small bulb.

Finally, at excessively high degrees of exhaustion, which cannot be reached except by the employment of special means, there seems to be, beyond a certain and rather small size of vessel, no perceptible difference in the heating.

These observations were the result of a number of experiments, of which one, showing the effect of the size of the bulb at a high degree of exhaustion, may be described and shown here, as it presents a feature of interest. Three spherical bulbs of 2 inches, 3 inches and 4 inches diameter were taken, and in the centre of each was mounted an equal length of an ordinary incandescent lamp filament of uniform thickness. In each bulb the piece of filament was fastened to the leading-in wire of platinum, contained in a glass stem sealed in the bulb; care being taken, of course, to make everything as nearly alike as possible. On each glass stem in the inside of the bulb was slipped a highly polished tube made of aluminum sheet, which fitted the stem and was held on it by spring pressure. The function of this aluminum tube will be explained subsequently. In each bulb an equal length of filament protruded above the metal tube. It is sufficient to say now that under these conditions equal lengths of filament of the same thickness—in other words, bodies of equal bulk—were brought to incandescence. The three bulbs were sealed to a glass tube, which was connected to a Sprengel pump. When a high vacuum had been reached, the glass tube carrying the bulbs was sealed off. A current was then turned on successively on each bulb, and it was found that the filaments came to about the same brightness, and, if anything, the smallest bulb, which was placed midway between the two larger ones, may have been slightly brighter. This result was expected, for when either of the bulbs was connected to the coil the luminosity spread through the other two, hence the three bulbs constituted really one vessel. When all the three bulbs were connected in multiple arc to the coil, in the largest of them the filament glowed brightest, in the next smaller it was a little less bright, and in the smallest it only came to redness. The bulbs were then sealed off and separately tried. The brightness of the filaments was now such as would have been expected on the supposition that the energy given off was proportionate to the surface of the bulb, this surface in each case representing one of the coatings of a condenser. Accordingly, there was less difference between the largest and the middle sized than between the latter and the smallest bulb.

An interesting observation was made in this experiment. The three bulbs were suspended from a straight bare wire connected to a terminal of a coil, the largest bulb being placed at the end of the wire, at some distance from it the smallest bulb, and at an equal distance from the latter the middle-sized one. The carbons glowed then in both the larger bulbs about as expected, but the smallest did not get its share by far. This observation led me to exchange the position of the bulbs, and I then observed that whichever of the bulbs was in the middle was by far less bright than it was in any other position. This mystifying result was, of course, found to be due to the electrostatic action between the bulbs. When they were placed at a considerable distance, or when they were attached to the corners of an equilateral triangle of copper wire, they glowed in about the order determined by their surfaces.

As to the shape of the vessel, it is also of some importance, especially at high degrees of exhaustion. Of all the possible constructions, it seems that a spherical globe with the refractory body mounted in its centre is the best to employ. By experience it has been demonstrated that in such a globe a refractory body of a given bulk is more easily brought to incandescence than when differently shaped bulbs are used. There is also an advantage in giving to the incandescent body the shape of a sphere, for self-evident reasons. In any case the body should be mounted in the centre, where the atoms rebounding from the glass collide. This object is best attained in the spherical bulb; but it is also attained in a cylindrical vessel with one or two straight filaments coinciding with its axis, and possibly also in parabolical or spherical bulbs with refractory body or bodies placed in the focus or foci of the same; though the latter is not probable, as the electrified atoms should in all cases rebound normally from the surface they strike, unless the speed were excessive, in which case they *would* probably follow the general law of reflection. No matter what shape the vessel may have, if the exhaustion be low, a filament mounted in the globe is brought to the same degree of incandescence in all parts; but if the exhaustion be high and the bulb be spherical or pear-shaped, as usual, focal points form and the filament is heated to a higher degree at or near such points.

To illustrate the effect, I have here two small bulbs which are alike, only one is exhausted to a low and the other to a very high degree. When connected to the coil, the filament in the former glows uniformly throughout all its length; whereas in the latter, that portion of the filament which is in the centre of the bulb glows far more intensely than the rest. A curious point is that the phenomenon occurs even if two filaments are mounted in a bulb, each being connected to one terminal of the coil, and, what is still more curious, if they be very near together, provided the vacuum be very high. I noted in experiments with such bulbs that the filaments would give way usually at a certain point, and in the first trials I attributed it to a defect in the carbon. But when the phenomenon occurred many times in succession I recognized its real cause.

In order to bring a refractory body inclosed in a bulb to incandescence, it is desirable, on account of economy, that all the energy supplied to the bulb from the source should reach without loss the body to be heated; from there, and from nowhere else, it should be radiated. It is, of course, out of the question to reach this theoretical result, but it is possible by a proper construction of the illuminating device to approximate it more or less.

For many reasons, the refractory body is placed in the centre of the bulb, and it is usually supported on a glass stem containing the leading-in wire. As the potential of this wire is alternated, the rarefied gas surrounding the stem is acted upon inductively, and the glass stem is violently bombarded and heated. In this manner by far the greater portion of the energy supplied to the bulb—especially when exceedingly high frequencies are used— may be lost for the purpose contemplated. To obviate this loss, or at least to reduce it to a minimum, I usually screen the rarefied gas surrounding the stem from the inductive action of the leading-in wire by providing the stem with a tube or coating of conducting material. It seems beyond doubt that the best among metals to employ for this purpose is aluminum, on account of its many remarkable properties. Its only fault is that it is easily fusible, and, therefore, its distance from the incandescing body should be properly estimated. Usually, a thin tube, of a diameter somewhat smaller than that of the glass stem, is made of the finest aluminum sheet, and slipped on the stem. The tube is conveniently prepared by wrapping around a rod fastened in a lathe a piece of aluminum sheet of proper size, grasping the sheet firmly with clean chamois leather or blotting paper, and spinning the rod very fast. The sheet is wound tightly around the rod, and a highly polished tube of one or three layers of the sheet is obtained. When slipped on the stem, the pressure is generally sufficient to prevent it from slipping off, but, for safety, the lower edge of the sheet may be turned inside. The upper inside corner of the sheet—that is, the one which is nearest to the refractory incandescent body—should be cut out diagonally, as it often happens that, in consequence of the intense heat, this corner turns toward the inside and comes very near to, or in contact with, the wire, or filament, supporting the refractory body. The greater part of the energy supplied to the bulb is then used up in heating the metal tube, and the bulb is rendered useless for the purpose. The aluminum sheet should project above the glass stem more or less—one inch or so—or else, if the glass be too close to the incandescing body, it may be strongly heated and become more or less conducting, whereupon it may be ruptured, or may, by its conductivity, establish a good electrical connection between the metal tube and the leading-in wire, in which case, again, most of the energy will be lost in heating the former. Perhaps the best way is to make the top of the glass tube, for about an inch, of a much smaller diameter. To still further reduce the danger arising from the heating of the glass stem, and also with the view of preventing an electrical connection between the metal tube and the electrode, I preferably wrap the stem with several layers of thin mica, which extends at least as far as the metal tube. In some bulbs I have also used an outside insulating cover.

The preceding remarks are only made to aid the experimenter in the first trials, for the difficulties which he encounters he may soon find means to overcome in his own way.

To illustrate the effect of the screen, and the advantage of using it, I have here two bulbs of the same size, with their stems, leading-in wires and incandescent lamp filaments tied to the latter, as nearly alike as possible. The stem of one bulb is provided with an aluminum tube, the stem of the other has none. Originally the two bulbs were joined by a tube which was connected to a Sprengel pump. When a high vacuum had been reached, first the connecting tube, and then the bulbs, were sealed off; they are therefore of the same degree of exhaustion. When they are separately connected to the coil giving a certain potential, the carbon filament in the bulb provided with the aluminum screen is rendered highly incandescent, while the filament in the other bulb may, with the same potential, not even come to redness, although in reality the latter bulb takes generally more energy than the former. When they are both connected together to the terminal, the difference is even more apparent, showing the importance of the screening. The metal tube placed on the stem containing the leading-in wire performs really two distinct functions: First, it acts more or less as an electrostatic screen, thus economizing the energy supplied to the bulb; and, second, to whatever extent it may fail to act electrostatically, it acts mechanically, preventing the bombardment, and consequently intense heating and possible deterioration of the slender support of the refractory incandescent body, or of the glass stem containing the leading-in wire. I say *slender* support, for it is evident that in order to confine the heat more completely to the incandescing body its support should be very thin, so as to carry away the smallest possible amount of heat by conduction. Of all the supports used I have found an ordinary incandescent lamp filament to be the best, principally because among conductors it can withstand the highest degree of heat.

The effectiveness of the metal tube as an electrostatic screen depends largely on the degree of exhaustion.

At excessively high degrees of exhaustion—which are reached by using great care and special means in connection with the Sprengel pump—when the matter in the globe is in the ultra-radiant state, it acts most perfectly. The shadow of the upper edge of the tube is then sharply defined upon the bulb.

At a somewhat lower degree of exhaustion, which is about the ordinary "non-striking" vacuum, and generally as long as the matter moves predominantly in straight lines, the screen still does well. In elucidation of the preceding remark it is necessary to state that what is a "non-striking" vacuum for a coil operated as ordinarily, by impulses, or currents, of low frequency, is not so, by far, when the coil is operated by currents of very high frequency. In such case the discharge may pass with great freedom through the rarefied gas through which a low frequency discharge may not pass, even though the potential be much higher. At ordinary atmospheric pressures just the reverse rule holds good: the higher the frequency, the less the spark discharge is able to jump between the terminals, especially if they are knobs or spheres of some size.

Finally, at very low degrees of exhaustion, when the gas is well conducting, the metal tube not only does not act as an electro-static screen, but even is a drawback, aiding to a considerable extent the dissipation of the energy laterally from the leading-in wire. This, of course, is to be expected. In this case, namely, the metal tube is in good electrical connection with the leading-in wire, and most of the bombardment is directed upon the tube. As long as the electrical connection is not good, the conducting tube is always of some advantage, for although it may not greatly economize energy, still it protects the support of the refractory button, and is the means of concentrating more energy upon the same.

To whatever extent the aluminum tube performs the function of a screen, its usefulness is therefore limited to very high degrees of exhaustion when it is insulated from the electrode—that is, when the gas as a whole is non-conducting, and the molecules, or atoms, act as independent carriers of electric charges.

In addition to acting as a more or less effective screen, in the true meaning of the word, the conducting tube or coating may also act, by reason of its conductivity, as a sort of equalizer or dampener of the bombardment against the stem. To be explicit, I assume the action to be as follows: Suppose a rhythmical bombardment to occur against the conducting tube by reason of its imperfect action as a screen, it certainly must happen that some molecules, or atoms, strike the tube sooner than others. Those which come first in contact with it give up their superfluous charge, and the tube is electrified, the electrification instantly spreading over its surface. But this must diminish the energy lost in the bombardment, for two reasons: first, the charge given up by the atoms spreads over a great area, and hence the electric density at any point is small, and the atoms are repelled with less energy than they would be if they struck against a good insulator; secondly, as the tube is electrified by the atoms which first come in contact with it, the progress of the following atoms against the tube is more or less checked by the repulsion which

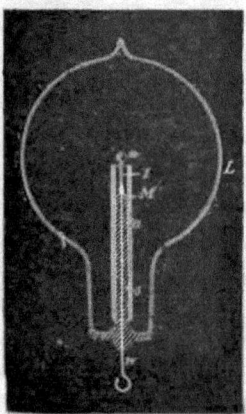

Fig. 147.   Fig. 148.

the electrified tube must exert upon the similarly electrified atoms. This repulsion may perhaps be sufficient to prevent a large portion of the atoms from striking the tube, but at any rate it must diminish the energy of their impact. It is clear that when the exhaustion is very low, and the rarefied gas well conducting, neither of the above effects can occur, and, on the other hand, the fewer the atoms, with the greater freedom they move; in other words, the higher the degree of exhaustion, up to a limit, the more telling will be both the effects.

What I have just said may afford an explanation of the phenomenon observed by Prof. Crookes, namely, that a discharge through a bulb is established with much greater facility when an insulator than when a conductor is present in the same. In my opinion, the conductor acts as a dampener of the motion of the atoms in the two ways pointed out; hence, to cause a visible discharge to pass through the bulb, a much higher potential is needed if a conductor, especially of much surface, be present.

For the sake of elucidating of some of the remarks before made, I must now refer to Figs. 147, 148 and 149, which illustrate various arrangements with a type of bulb most generally used.

Fig. 147 is a section through a spherical bulb L, with the glass stem *s*, contains the leading-in wire *w*, which has a lamp filament *l* fastened to it, serving to support the refractory button *m* in the centre. M is a sheet of thin mica wound in several layers around the stem *s*, and *a* is the aluminum tube.

Fig. 148 illustrates such a bulb in a somewhat more advanced stage of perfection. A metallic tube s is fastened by means of some cement to the neck of the tube. In the tube is screwed a plug P, of insulating material, in the centre of which is fastened a metallic terminal *t*, for the connection to the leading-in wire *w*. This terminal must be well insulated from the metal tube s; therefore, if the cement used is conducting—and most generally it is sufficiently so—the space between the plug P and the neck of the bulb should be filled with some good insulating material, such as mica powder.

Fig. 149 shows a bulb made for experimental purposes. In this bulb the aluminum tube is provided with an external connection, which serves to investigate the effect of the tube under various conditions. It is referred to chiefly to suggest a line of experiment followed.

Since the bombardment against the stem containing the leading-in wire is due to the inductive action of the latter upon the rarefied gas, it is of advantage to reduce this action as far as practicable by employing a very thin wire, surrounded by a very thick insulation of glass or other material, and by making the wire passing through the rarefied gas as short as practicable. To combine these features I employ a large tube T (Fig. 150), which protrudes into the bulb to some distance, and carries on the top a very short glass stem *s*, into which is sealed the leading-in wire *w*, and I protect the top of the glass stem against the heat by a small aluminum tube *a* and a layer of mica underneath the same, as usual. The wire *w*, passing through the large tube to the outside of the bulb, should be well insulated—with a glass tube, for instance—and the space between ought to be filled out with some excellent insulator. Among many insulating powders I have found that mica powder is the best to employ. If this precaution is not taken, the tube T, protruding into the bulb, will surely be cracked in consequence of the heating by the brushes which are apt to form in the upper part of the tube, near the exhausted globe, especially if the vacuum be excellent, and therefore the potential necessary to operate the lamp be very high.

Fig. 151 illustrates a similar arrangement, with a large tube T protruding into the part of the bulb containing the refractory button m. In this case the wire leading from the outside into the bulb is omitted, the energy required being supplied through

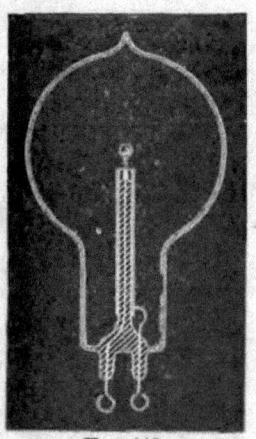

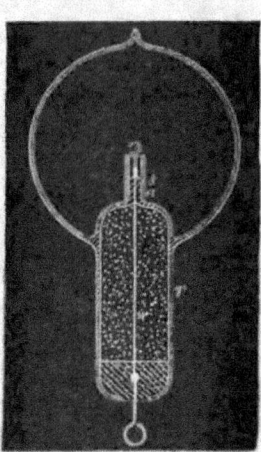

FIG. 149.   FIG. 150,

condenser coatings c c. The insulating packing P should in this construction be tightly fitting to the glass, and rather wide, or otherwise the discharge might avoid passing through the wire w, which connects the inside condenser coating to the incandescent button m.

The molecular bombardment against the glass stem in the bulb is a source of great trouble. As an illustration I will cite a phenomenon only too frequently and unwillingly observed. A bulb, preferably a large one, may be taken, and a good conducting body, such as a piece of carbon, may be mounted in it upon a platinum wire sealed in the glass stem. The bulb may be exhausted to a fairly high degree, nearly to the point when phosphorescence begins to appear. When the bulb is connected with the coil, the piece of carbon, if small, may become highly incandescent at first, but its brightness immediately diminishes, and then the discharge may break through the glass somewhere in the middle of the stem, in the form of bright sparks, in spite of the fact that the platinum wire is in good electrical connection with the rarefied gas through the piece of carbon or metal at the top. The first sparks are singularly bright, recalling those drawn from a clear surface of mercury. But, as they heat the glass rapidly, they, of course, lose their brightness, and cease when the glass at the ruptured place becomes incandescent, or generally sufficiently hot to conduct. When observed for the first time the phenomenon must appear very curious, and shows in a striking manner how radically different alternate currents, or impulses, of high frequency behave, as compared with steady currents, or currents of low frequency. With such currents—namely, the latter—the phenomenon would of course not occur. When frequencies such as are obtained by mechanical means are used, I think that the rupture of the glass is more or less the consequence of the bombard. ment, which warms it up and impairs its insulating power ; but with frequencies obtainable with condensers I have no doubt that the glass may give way without previous heating. Although this appears most singular at first, it is in reality what we might expect to occur. The energy supplied to the wire leading into the bulb is given off partly by direct action through the carbon button, and partly by inductive action through the glass surrounding the wire. The case is thus analogous to that in which a condenser shunted by a conductor of low resistance is connected to a source of alternating current. As long as the frequencies are low, the conductor gets the most and the condenser is perfectly safe ; but when the frequency becomes excessive, the *role* of the conductor may become quite insignificant. In the latter case the difference of potential at the terminals of the condenser may become so great as to rupture the dielectric, notwithstanding the fact that the terminals are joined by a conductor of low resis tance.

It is, of course, not necessary, when it is desired to produce the incandescence of a body inclosed in a bulb by means of these currents, that the body should be a conductor, for even a perfect non-conductor may be quite as readily heated. For this purpose it is sufficient to surround a conducting electrode with a non-conducting material, as, for instance, in the bulb described before in Fig. 150, in which a thin incandescent lamp filament is coated with a non-conductor, and supports a button of the same material on the top. At the start the bombardment goes on by inductive action through the non-conductor, until the same is sufficiently heated to become conducting, when the bombardment continues in the ordinary way.

A different arrangement used in some of the bulbs constructed is illustrated in Fig. 152. In this instance a non-conductor *m* is mounted in a piece of common arc light carbon so as to project some small distance above the latter. The carbon piece is connected to the leading-in wire passing through a glass stem, which

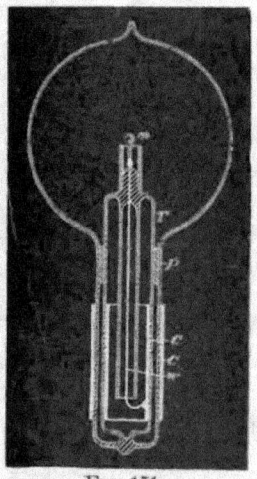

Fig. 151.   Fig. 152.

is wrapped with several layers of mica. An aluminum tube *a* is employed as usual for screening. It is so arranged that it reaches very nearly as high as the carbon and only the non-conductor *m* projects a little above it. The bombardment goes at first against the upper surface of carbon, the lower parts being protected by the aluminum tube. As soon, however, as the non-conductor *m* is heated it is rendered good conducting, and then it becomes the centre of the bombardment, being most exposed to the same.

I have also constructed during these experiments many such single-wire bulbs with or without internal electrode, in which the radiant matter was projected against, or focused upon, the body to be rendered incandescent.   Fig. 153 (page 263) illustrates one of the bulbs used.   It consists of a spherical globe L, provided with a long neck *n*, on top, for increasing the action in some cases by the application of an external conducting coating.   The globe L is blown out on the bottom into a very small bulb *b*, which serves to hold it firmly in a socket s of insulating material into which it is cemented.   A fine lamp filament *f*, supported on a wire *w*, passes through the centre of the globe L.   The filament is rendered incandescent in the middle portion, where the bombardment proceeding from the lower inside surface of the globe is most intense.   The lower portion of the globe, as far as the socket s reaches, is rendered conducting, either by a tinfoil coating or otherwise, and the external electrode is connected to a terminal of the coil.

The arrangement diagrammatically indicated in Fig. 153 was found to be an inferior one when it was desired to render incandescent a filament or button supported in the centre of the globe, but it was convenient when the object was to excite phosphorescence.

In many experiments in which bodies of different kind were mounted in the bulb as, for instance, indicated in Fig. 152, some observations of interest were made.

It was found, among other things, that in such cases, no matter where the bombardment began, just as soon as a high temperature was reached there was generally one of the bodies which seemed to take most of the bombardment upon itself, the other, or others, being thereby relieved.   The quality appeared to depend principally on the point of fusion, and on the facility with which the body was "evaporated," or, generally speaking, disintegrated—meaning by the latter term not only the throwing off of atoms, but likewise of large lumps.   The observation made was in accordance with generally accepted notions.   In a highly exhausted bulb, electricity is carried off from the electrode by independent carriers, which are partly the atoms, or molecules, of the residual atmosphere, and partly the atoms, molecules, or lumps thrown off from the electrode.   If the electrode is composed of bodies of different character, and if one of these is more easily disentegrated than the other, most of the electricity supplied is carried off from that body, which is then brought to a higher temperature than the others, and this the more, as upon an increase of the temperature the body is still more easily disintregrated.

It seems to me quite probable that a similar process takes place in the bulb even with a homogeneous electrode, and I think it to be the principal cause of the disintegration. There is bound to be some irregularity, even if the surface is highly polished, which, of course, is impossible with most of the refractory bodies employed as electrodes. Assume that a point of the electrode gets hotter; instantly most of the discharge passes through that point, and a minute patch it probably fused and evaporated. It is now possible that in consequence of the violent disintegration the spot attacked sinks in temperature, or that a counter force is created, as in an arc; at any rate, the local tearing off meets with the limitations incident to the experiment, whereupon the same process occurs on another place. To the eye the electrode appears uniformly brilliant, but there are upon it points constantly shifting and wandering around, of a temperature far above the mean, and this materially hastens the process of deterioration. That some such thing occurs, at least when the electrode is at a lower temperature, sufficient experimental evidence can be obtained in the following manner: Exhaust a bulb to a very high degree, so that with a fairly high potential the discharge cannot pass—that is, not a *luminous* one, for a weak invisible discharge occurs always, in all probability. Now raise slowly and carefully the potential, leaving the primary current on no more than for an instant. At a certain point, two, three, or half a dozen phosphorescent spots will appear on the globe. These places of the glass are evidently more violently bombarded than others, this being due to the unevenly distributed electric density, necessitated, of course, by sharp projections, or, generally speaking, irregularities of the electrode. But the luminous patches are constantly changing in position, which is especially well observable if one manages to produce very few, and this indicates that the configuration of the electrode is rapidly changing.

From experiences of this kind I am led to infer that, in order to be most durable, the refractory button in the bulb should be in the form of a sphere with a highly polished surface. Such a small sphere could be manufactured from a diamond or some other crystal, but a better way would be to fuse, by the employment of extreme degrees of temperature, some oxide—as, fo instance, zirconia—into a small drop, and then keep it in the bulb at a temperature somewhat below its point of fusion.

Interesting and useful results can, no doubt, be reached in the direction of extreme degrees of heat. How can such high temperatures be arrived at? How are the highest degrees of heat reached in nature? By the impact of stars, by high speeds and collisions. In a collision any rate of heat generation may be attained. In a chemical process we are limited. When oxygen and hydrogen combine, they fall, metaphorically speaking, from a definite height. We cannot go very far with a blast, nor by confining heat in a furnace, but in an exhausted bulb we can concentrate any amount of energy upon a minute button. Leaving practicability out of consideration, this, then, would be the means which, in my opinion, would enable us to reach the highest temperature. But a great difficulty when proceeding in this way is encountered, namely, in most cases the body is carried off before it can fuse and form a drop. This difficulty exists principally with an oxide, such as zirconia, because it cannot be compressed in so hard a cake that it would not be carried off quickly. I have endeavored repeatedly to fuse zirconia, placing it in a cup of arc light carbon, as indicated in Fig. 152. It glowed with a most intense light, and the stream of the particles projected out of the carbon cup was of a vivid white; but whether it was compressed in a cake or made into a paste with carbon, it was carried off before it could be fused. The carbon cup, containing zirconia, had to be mounted very low in the neck of a large bulb, as the heating of the glass by the projected particles of the oxide was so rapid that in the first trial the bulb was cracked almost in an instant, when the current was turned on. The heating of the glass by the projected particles was found to be always greater when the carbon cup contained a body which was rapidly carried off—I presume, because in such cases, with the same potential, higher speeds were reached, and also because, per unit of time, more matter was projected—that is, more particles would strike the glass.

The before-mentioned difficulty did not exist, however, when the body mounted in the carbon cup offered great resistance to deterioration. For instance, when an oxide was first fused in an oxygen blast, and then mounted in the bulb, it melted very readily into a drop.

Generally, during the process of fusion, magnificent light effects were noted, of which it would be difficult to give an adequate idea. Fig. 152 is intended to illustrate the effect observed with a ruby drop. At first one may see a narrow funnel of white light projected against the top of the globe, where it produces an irregularly outlined phosphorescent patch. When the point of the ruby fuses, the phosphorescence becomes very powerful; but as the atoms are projected with much greater speed from the surface of the drop, soon the glass gets hot and "tired," and now only the outer edge of the patch glows. In this manner an intensely phosphorescent, sharply defined line, *l*, corresponding to the outline of the drop, is produced, which spreads slowly over the globe as the drop gets larger. When the mass begins to boil, small bubbles and cavities are formed, which cause dark colored spots to sweep across the globe. The bulb may be turned downward without fear of the drop falling off, as the mass possesses considerable viscosity.

I may mention here another feature of some interest, which I believe to have noted in the course of these experiments, though the observations do not amount to a certitude. It *appeared* that under the molecular impact caused by the rapidly alternating potential, the body was fused and maintained in that state at a lower temperature in a highly exhausted bulb than was the case at normal pressure and application of heat in the ordinary way—that is, at least, judging from the quantity of the light emitted. One of the experiments performed may be mentioned here by way of illustration. A small piece of pumice stone was stuck on a platinum wire, and first melted to it in a gas burner. The wire was next placed between two pieces of charcoal, and a burner applied, so as to produce an intense heat, sufficient to melt down the pumice stone into a small glass-like button. The platinum wire had to be taken of sufficient thickness, to prevent its melting in the fire. While in the charcoal fire, or when held in a burner to get a better idea of the degree of heat, the button glowed with great brilliancy. The wire with the button was then mounted in a bulb, and upon exhausting the same to a high degree, the current was turned on slowly, so as to prevent the cracking of the button. The button was heated to the point of fusion, and when it melted, it did not, apparently, glow with the same brilliancy as before, and this would indicate a lower temperature. Leaving out of consideration the observer's possible, and even probable, error, the question is, can a body under these conditions be brought from a solid to a liquid state with the evolution of *less* light?

When the potential of a body is rapidly alternated, it is certain that the structure is jarred. When the potential is very high, although the vibrations may be few—say 20,000 per second—the effect upon the structure may be considerable. Suppose, for example, that a ruby is melted into a drop by a steady application of energy. When it forms a drop, it will emit visible and invisible waves, which will be in a definite ratio, and to the eye the drop will appear to be of a certain brilliancy. Next, suppose we diminish to any degree we choose the energy steadily supplied, and, instead, supply energy which rises and falls according to a certain law. Now, when the drop is formed, there will be emitted from it three different kinds of vibrations—the ordinary visible, and two kinds of invisible waves: that is, the ordinary dark waves of all lengths, and, in addition, waves of a well defined character. The latter would not exist by a steady supply of the energy; still they help to jar and loosen the structure. If this really be the case, then the ruby drop will emit relatively less visible and more invisible waves than before. Thus it would seem that when a platinum wire, for instance, is fused by currents alternating with extreme rapidity, it emits at the point of fusion less light and more visible radiation than it does when melted by a steady current, though the total energy used up in the process of fusion is the same in both cases. Or, to cite another example, a lamp filament is not capable of withstanding as long with currents of extreme frequency as it does with steady currents, assuming that it be worked at the same luminous intensity. This means that for rapidly alternating currents the filament should be shorter and thicker. The higher the frequency—that is, the greater the departure from the steady flow—the worse it would be for the filament. But if the truth of this remark were demonstrated, it would be erroneous to conclude that such a refractory button as used in these bulbs would be deteriorated quicker by currents of extremely high frequency than by steady or low frequency currents. From experience I may say that just the opposite holds good: the button withstands the bombardment better with currents of very high frequency. But this is due to the fact that a high frequency discharge passes through a rarefied gas with much greater freedom than a steady or low frequency discharge, and this will mean that with the former we can work with a lower potential or with a less violent impact. As long, then, as the gas is of no consequence, a steady or low frequency current is better; but as soon as the action of the gas is desired and important, high frequencies are preferable.

In the course of these experiments a great many trials were made with all kinds of carbon buttons. Electrodes made of ordinary carbon buttons were decidedly more durable when the buttons were obtained by the application of enormous pressure. Electrodes prepared by depositing carbon in well known ways did not show up well; they blackened the globe very quickly. From many experiences I conclude that lamp filaments obtained in this manner can be advantageously used only with low potentials and low frequency currents. Some kinds of carbon withstand so well that, in order to bring them to the point of fusion, it is necessary to employ very small buttons. In this case the observation is rendered very difficult on account of the intense heat produced. Nevertheless there can be no doubt that all kinds of carbon are fused under the molecular bombardment, but the liquid state must be one of great instability. Of all the bodies tried there were two which withstood best—diamond and carborundum. These two showed up about equally, but the latter was preferable for many reasons. As it is more than likely that this body is not yet generally known, I will venture to call your attention to it.

It has been recently produced by Mr. E. G. Acheson, of Monongahela City, Pa., U. S. A. It is intended to replace ordinary diamond powder for polishing precious stones, etc., and I have been informed that it accomplishes this object quite successfully. I do not know why the name "carborundum" has been given to it, unless there is something in the process of its manufacture which justifies this selection. Through the kindness of the inventor, I obtained a short while ago some samples which I desired to test in regard to their qualities of phosphorescence and capability of withstanding high degrees of heat.

Carborundum can be obtained in two forms—in the form of "crystals" and of powder. The former appear to the naked eye dark colored, but are very brilliant; the latter is of nearly the same color as ordinary diamond powder, but very much finer. When viewed under a microscope the samples of crystals given to me did not appear to have any definite form, but rather resembled pieces of broken up egg coal of fine quality. The majority were opaque, but there were some which were transparent and colored. The crystals are a kind of carbon containing some impurities; they are extremely hard, and withstand for a long time even an oxygen blast. When the blast is directed against them they at first form a cake of some compactness, probably in consequence of the fusion of impurities they contain. The mass withstands for a very long time the blast without further fusion; but a slow carrying off, or burning, occurs, and, finally, a small quantity of a glass-like residue is left, which, I suppose, is melted alumina. When compressed strongly they conduct very well, but not as well as ordinary carbon. The powder, which is obtained from the crystals in some way, is practically non-conducting. It affords a magnificent polishing material for stones.

The time has been too short to make a satisfactory study of the properties of this product, but enough experience has been gained in a few weeks I have experimented upon it to say that it does possess some remarkable properties in many respects. It withstands excessively high degrees of heat, it is little deteriorated by molecular bombardment, and it does not blacken the globe as ordinary carbon does. The only difficulty which I have experienced in its use in connection with these experiments was to find some binding material which would resist the heat and the effect of the bombardment as successfully as carborundum itself does.

I have here a number of bulbs which I have provided with buttons of carborundum. To make such a button of carborundum crystals I proceed in the following manner: I take an ordinary lamp filament and dip its point in tar, or some other thick substance or paint which may be readily carbonized. I next pass the point of the filament through the crystals, and then hold it vertically over a hot plate. The tar softens and forms a drop on the point of the filament, the crystals adhering to the surface of the drop. By regulating the distance from the plate the tar is slowly dried out and the button becomes solid. I then once more dip the button in tar and hold it again over a plate until the tar is evaporated, leaving only a hard mass which firmly binds the crystals. When a larger button is required I repeat the process several times, and I generally also cover the filament a certain distance below the button with crystals. The button being mounted in a bulb, when a good vacuum has been reached, first a weak and then a strong discharge is passed through the bulb to carbonize the tar and expel all gases, and later it is brought to a very intense incandescence.

When the powder is used I have found it best to proceed as follows: I make a thick paint of carborundum and tar, and pass a lamp filament through the paint. Taking then most of the paint off by rubbing the filament against a piece of chamois leather, I hold it over a hot plate until the tar evaporates and the coating becomes firm. I repeat this process as many times as it is necessary to obtain a certain thickness of coating. On the point of the coated filament I form a button in the same manner.

There is no doubt that such a button—properly prepared under great pressure—of carborundum, especially of powder of the best quality, will withstand the effect of the bombardment fully as well as anything we know. The difficulty is that the binding material gives way, and the carborundum is slowly thrown off after some time. As it does not seem to blacken the globe in the least, it might be found useful for coating the filaments of ordinary incandescent lamps, and I think that it is even possible to produce thin threads or sticks of carborundum which will replace the ordinary filaments in an incandescent lamp. A carborundum coating seems to be more durable than other coatings, not only because the carborundum can withstand high degrees of heat, but also because it seems to unite with the carbon better than any other material I have tried. A coating of zirconia or any other oxide, for instance, is far more quickly destroyed. I prepared buttons of diamond dust in the same manner as of carborundum, and these came in durability nearest to those prepared of carborundum, but the binding paste gave way much more quickly in the diamond buttons; this, however, I attributed to the size and irregularity of the grains of the diamond.

It was of interest to find whether carborundum possesses the quality of phosphorescence. One is, of course, prepared to encounter two difficulties: first, as regards the rough product, the "crystals," they are good conducting, and it is a fact that conductors do not phosphoresce; second, the powder, being exceedingly fine, would not be apt to exhibit very prominently this quality, since we know that when crystals, even such as diamond or ruby, are finely powdered, they lose the property of phosphorescence to a considerable degree.

The question presents itself here, can a conductor phosphoresce? What is there in such a body as a metal, for instance, that would deprive it of the quality of phosphorescence, unless it is that property which characterizes it as a conductor? For it is a fact that most of the phosphorescent bodies lose that quality when they are sufficiently heated to become more or less conducting.

Then, if a metal be in a large measure, or perhaps entirely, deprived of that property, it should be capable of phosphorescence. Therefore it is quite possible that at some extremely high frequency, when behaving practically as a non-conductor, a metal or any other conductor might exhibit the quality of phosphorescence, even though it be entirely incapable of phosphorescing under the impact of a low-frequency discharge. There is, however, another possible way how a conductor might at least *appear* to phosphoresce.

Considerable doubt still exists as to what really is phosphorescence, and as to whether the various phenomena comprised under this head are due to the same causes. Suppose that in an exhausted bulb, under the molecular impact, the surface of a piece of metal or other conductor is rendered strongly luminous, but at the same time it is found that it remains comparatively cool, would not this luminosity be called phosphorescence? Now such a result, theoretically at least, is possible, for it is a mere question of potential or speed. Assume the potential of the electrode, and consequently the speed of the projected atoms, to be sufficiently high, the surface of the metal piece, against which the atoms are projected, would be rendered highly incandescent, since the process of heat generation would be incomparably faster than that of radiating or conducting away from the surface of the collision. In the eye of the observer a single impact of the atoms would cause an instantaneous flash, but if the impacts were repeated with sufficient rapidity, they would produce a continuous impression upon his retina. To him then the surface of the metal would appear continuously incandescent and of constant luminous intensity, while in reality the light would be either intermittent, or at least changing periodically in intensity. The metal piece would rise in temperature until equilibrium was attained—that is, until the energy continuously radiated would equal that intermittently supplied. But the supplied energy might under such conditions not be sufficient to bring the body to any more than a very moderate mean temperature, especially if the frequency of the atomic impacts be very low—just enough that the fluctuation of the intensity of the light emitted could not be detected by the eye. The body would now, owing to the manner in which the energy is supplied, emit a strong light, and yet be at a comparatively very low mean temperature. How should the observer name the luminosity thus produced? Even if the analysis of the light would teach him something definite, still he would probably rank it under the phenomena of phosphorescence. It is conceivable that in such a way both conducting and non-conducting bodies may be maintained at a certain luminous intensity, but the energy required would very greatly vary with the nature and properties of the bodies.

These and some foregoing remarks of a speculative nature were made merely to bring out curious features of alternate currents or electric impulses. By their help we may cause a body to emit *more* light, while at a certain mean temperature, than it would emit if brought to that temperature by a steady supply; and, again, we may bring a body to the point of fusion, and cause it to emit *less* light than when fused by the application of energy in ordinary ways. It all depends on how we supply the energy, and what kind of vibrations we set up; in one case the vibrations are more, in the other less, adapted to affect our sense of vision.

Some effects, which I had not observed before, obtained with carborundum in the first trials, I attributed to phosphorescence, but in subsequent experiments it appeared that it was devoid of that quality. The crystals possess a noteworthy feature. In a bulb provided with a single electrode in the shape of a small circular metal disc, for instance, at a certain degree of exhaustion the electrode is covered with a milky film, which is separated by a dark space from the glow filling the bulb. When the metal disc is covered with carborundum crystals, the film is far more intense, and snow-white. This I found later to be merely an effect of the bright surface of the crystals, for when an aluminum electrode was highly polished, it exhibited more or less the same phenomenon. I made a number of experiments with the samples of crystals obtained, principally because it would have been of special interest to find that they are capable of phosphorescence, on account of their being conducting. I could not produce phosphorescence distinctly, but I must remark that a decisive opinion cannot be formed until other experimenters have gone over the same ground.

The powder behaved in some experiments as though it contained alumina, but it did not exhibit with sufficient distinctness the red of the latter. Its dead color brightens considerably under the molecular impact, but I am now convinced it does not phosphoresce. Still, the tests with the powder are not conclusive, because powdered carborundum probably does not behave like a phosphorescent sulphide, for example, which could be finely powdered without impairing the phosphorescence, but rather like powdered ruby or diamond, and therefore it would be necessary, in order to make a decisive test, to obtain it in a large lump and polish up the surface.

If the carborundum proves useful in connection with these and similar experiments, its chief value will be found in the production of coatings, thin conductors, buttons, or other electrodes capable of withstanding extremely high degrees of heat.

The production of a small electrode, capable of withstanding enormous temperatures, I regard as of the greatest importance in the manufacture of light. It would enable us to obtain, by means of currents of very high frequencies, certainly 20 times, if not more, the quantity of light which is obtained in the present incandescent lamp by the same expenditure of energy. This estimate may appear to many exaggerated, but in reality I think it is far from being so. As this statement might be misunderstood, I think it is necessary to expose clearly the problem with which, in this line of work, we are confronted, and the manner in which, in my opinion, a solution will be arrived at.

Any one who begins a study of the problem will be apt to think that what is wanted in a lamp with an electrode is a very high degree of incandescence of the electrode. There he will be mistaken. The high incandescence of the button is a necessary evil, but what is really wanted is the high incandescence of the gas surrounding the button. In other words, the problem in such a lamp is to bring a mass of gas to the highest possible incandescence. The higher the incandescence, the quicker the mean vibration, the greater is the economy of the light production. But to maintain a mass of gas at a high degree of incandescence in a glass vessel, it will always be necessary to keep the incandescent mass away from the glass; that is, to confine it as much as possible to the central portion of the globe.

In one of the experiments this evening a brush was produced at the end of a wire. The brush was a flame, a source of heat and light. It did not emit much perceptible heat, nor did it glow with an intense light; but is it the less a flame because it does not scorch my hand? Is it the less a flame because it does not hurt my eyes by its brilliancy? The problem is precisely to produce in the bulb such a flame, much smaller in size, but incomparably more powerful. Were there means at hand for producing electric impulses of a sufficiently high frequency, and for transmitting them, the bulb could be done away with, unless it were used to protect the electrode, or to economize the energy by confining the heat. But as such means are not at disposal, it becomes necessary to place the terminal in the bulb and rarefy the air in the same. This is done merely to enable the apparatus to perform the work which it is not capable of performing at ordinary air pressure. In the bulb we are able to intensify the action to any degree—so far that the brush emits a powerful light.

The intensity of the light emitted depends principally on the frequency and potential of the impulses, and on the electric density on the surface of the electrode. It is of the greatest importance to employ the smallest possible button, in order to push the density very far. Under the violent impact of the molecules of the gas surrounding it, the small electrode is of course brought to an extremely high temperature, but around it is a mass of highly incandescent gas, a flame photosphere, many hundred times the volume of the electrode. With a diamond, carborundum or zirconia button the photosphere can be as much as one thousand times the volume of the button. Without much reflection one would think that in pushing so far the incandescence of the electrode it would be instantly volatilized. But after a careful consideration one would find that, theoretically, it should not occur, and in this fact—which, moreover, is experimentally demonstrated—lies principally the future value of such a lamp.

At first, when the bombardment begins, most of the work is performed on the surface of the button, but when a highly conducting photosphere is formed the button is comparatively relieved. The higher the incandescence of the photosphere, the more it approaches in conductivity to that of the electrode, and the more, therefore, the solid and the gas form one conducting body. The consequence is that the further the incandescence is forced the more work, comparatively, is performed on the gas, and the less on the electrode. The formation of a powerful photosphere is consequently the very means for protecting the electrode. This protection, of course, is a relative one, and it should not be thought that by pushing the incandescence higher the electrode is actually less deteriorated. Still, theoretically, with extreme frequencies, this result must be reached, but probably at a temperature too high for most of the refractory bodies known. Given, then, an electrode which can withstand to a very high limit the effect of the bombardment and outward strain, it would be safe, no matter how much it was forced beyond that limit. In an incandescent lamp quite different considerations apply. There the gas is not at all concerned; the whole of the work is performed on the filament; and the the life of the lamp diminishes so rapidly with the increase of the degree of incandescence that economical reasons compel us to work it at a low incandescence. But if an incandescent lamp is operated with currents of very high frequency, the action of the gas cannot be neglected, and the rules for the most economical working must be considerably modified.

In order to bring such a lamp with one or two electrodes to a great perfection, it is necessary to employ impulses of very high frequency. The high frequency secures, among others, two chief advantages, which have a most important bearing upon the economy of the light production. First, the deterioration of the electrode is reduced by reason of the fact that we employ a great many small impacts, instead of a few violent ones, which quickly shatter the structure; secondly, the formation of a large photosphere is facilitated.

In order to reduce the deterioration of the electrode to the minimum, it is desirable that the vibration be harmonic, for any suddenness hastens the process of destruction. An electrode lasts much longer when kept at incandescence by currents, or impulses, obtained from a high frequency alternator, which rise and fall more or less harmonically, than by impulses obtained from a disruptive discharge coil. In the latter case there is no doubt that most of the damage is done by the fundamental sudden discharges.

One of the elements of loss in such a lamp is the bombardment of the globe. As the potential is very high, the molecules are projected with great speed; they strike the glass, and usually excite a strong phosphorescence. The effect produced is very pretty, but for economical reasons it would be perhaps preferable to prevent, or at least reduce to a minimum, the bombardment against the globe, as in such case it is, as a rule, not the object to excite phosphorescence, and as some loss of energy results from the bombardment. This loss in the bulb is principally dependent on the potential of the impulses and on the electric density on the surface of the electrode. In employing very high frequencies the loss of energy by the bombardment is greatly reduced, for, first, the potential needed to perform a given amount of work is much smaller; and, secondly, by producing a highly conducting photosphere around the electrode, the same result is obtained as though the electrode were much larger, which is equivalent to a smaller electric density. But be it by the diminution of the maximum potential or of the density, the gain is effected in the same manner, namely, by avoiding violent shocks, which strain the glass much beyond its limit of elasticity. If the frequency could be brought high enough, the loss due to the imperfect elasticity of the glass would be entirely negligible. The loss due to bombardment of the globe may, however, be reduced by using two electrodes instead of one. In such case each of the electrodes may be connected to one of the terminals; or else, if it is preferable to use only one wire, one electrode may be connected to one terminal and the other to the ground or to an insulated body of some surface, as, for instance, a shade on the lamp. In the latter case, unless some judgment is used, one of the electrodes might glow more intensely than the other.

But on the whole I find it preferable, when using such high frequencies, to employ only one electrode and one connecting wire. I am convinced that the illuminating device of the near future will not require for its operation more than one lead, and, at any rate, it will have no leading-in wire, since the energy required can be as well transmitted through the glass. In experimental bulbs the leading-in wire is not generally used on account of convenience, as in employing condenser coatings in the manner indicated in Fig. 151, for example, there is some difficulty in fitting the parts, but these difficulties would not exist if a great many bulbs were manufactured; otherwise the energy can be conveyed through the glass as well as through a wire, and with these high frequencies the losses are very small. Such illustrating devices will necessarily involve the use of very high potentials, and this, in the eyes of practical men, might be an objectionable feature. Yet, in reality, high potentials are not objectionable—certainly not in the least so far as the safety of the devices is concerned.

There are two ways of rendering an electric appliance safe. One is to use low potentials, the other is to determine the dimensions of the apparatus so that it is safe, no matter how high a potential is used. Of the two, the latter seems to me the better way, for then the safety is absolute, unaffected by any possible combination of circumstances which might render even a low-potential appliance dangerous to life and property. But the practical conditions require not only the judicious determination of the dimensions of the apparatus; they likewise necessitate the employment of energy of the proper kind. It is easy, for instance, to construct a transformer capable of giving, when operated from an ordinary alternate current machine of low tension, say 50,000 volts, which might be required to light a highly exhausted phosphorescent tube, so that, in spite of the high potential, it is perfectly safe, the shock from it producing no inconvenience. Still such a transformer would be expensive, and in itself inefficient; and, besides, what energy was obtained from it would not be economically used for the production of light. The economy demands the employment of energy in the form of extremely rapid vibrations. The problem of producing light has been likened to that of maintaining a certain high-pitch note by means of a bell. It should be said a *barely audible* note; and even these words would not express it, so wonderful is the sensitiveness of the eye. We may deliver powerful blows at long intervals, waste a good deal of energy, and still not get what we want; or we may keep up the note by delivering frequent taps, and get nearer to the object sought by the expenditure of much less energy. In the production of light, as far as the illuminating device is concerned, there can be only one rule—that is, to use as high frequencies as can be obtained; but the means for the production and conveyance of impulses of such character impose, at present at least, great limitations. Once it is decided to use very high frequencies, the return wire becomes unnecessary, and all the appliances are simplified. By the use of obvious means the same result is obtained as though the return wire were used. It is sufficient for this purpose to bring in contact with the bulb, or merely in the vicinity of the same, an insulated body of some surface. The surface need, of course, be the smaller, the higher the frequency and potential used, and necessarily, also, the higher the economy of the lamp or other device.

This plan of working has been resorted to on several occasions this evening. So, for instance, when the incandescence of a button was produced by grasping the bulb with the hand, the body of the experimenter merely served to intensify the action. The bulb used was similar to that illustrated in Fig. 148, and

the coil was excited to a small potential, not sufficient to bring
the button to incandescence when the bulb was hanging from
the wire; and incidentally, in order to perform the experiment
in a more suitable manner, the button was taken so large that a
perceptible time had to elapse before, upon grasping the bulb, it
could be rendered incandescent. The contact with the bulb was,
of course, quite unnecessary. It is easy, by using a rather large
bulb with an exceedingly small electrode, to adjust the conditions
so that the latter is brought to bright incandescence by the mere
approach of the experimenter within a few feet of the bulb, and
that the incandescence subsides upon his receding.

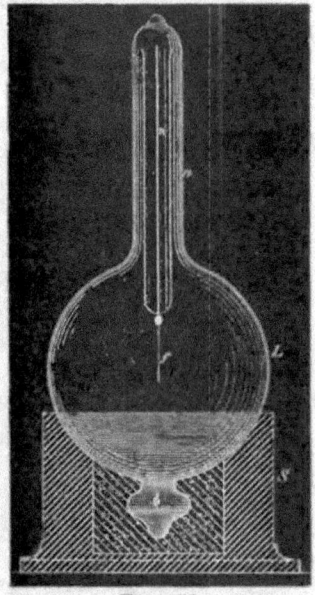

Fig. 153.

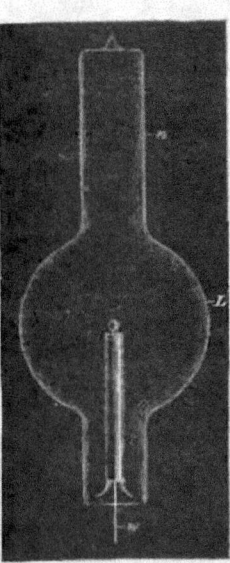

Fig. 154.

In another experiment, when phosphorescence was excited, a
similar bulb was used. Here again, originally, the potential was
not sufficient to excite phosphorescence until the action was intensified—in this case, however, to present a different feature, by
touching the socket with a metallic object held in the hand. The
electrode in the bulb was a carbon button so large that it could
not be brought to incandescence, and thereby spoil the effect
produced by phosphorescence.

Again, in another of the early experiments, a bulb was used, as illustrated in Fig. 141. In this instance, by touching the bulb with one or two fingers, one or two shadows of the stem inside were projected against the glass, the touch of the finger producing the same results as the application of an external negative electrode under ordinary circumstances.

In all these experiments the action was intensified by augmenting the capacity at the end of the lead connected to the terminal. As a rule, it is not necessary to resort to such means, and would be quite unnecessary with still higher frequencies; but when it *is* desired, the bulb, or tube, can be easily adapted to the purpose.

In Fig. 153, for example, an experimental bulb, L, is shown, which is provided with a neck, n, on the top, for the application of an external tinfoil coating, which may be connected to a body of larger surface. Such a lamp as illustrated in Fig. 154 may also be lighted by connecting the tinfoil coating on the neck n to the terminal, and the leading-in wire, w, to an insulated plate. If the bulb stands in a socket upright, as shown in the cut, a shade of conducting material may be slipped in the neck, n, and the action thus magnified.

A more perfected arrangement used in some of these bulbs is illustrated in Fig. 155. In this case the construction of the bulb is as shown and described before, when reference was made to Fig. 148. A zinc sheet, z, with a tubular extension, T, is applied over the metallic socket, s. The bulb hangs downward from the terminal, t, the zinc sheet, z, performing the double office of intensifier and reflector. The reflector is separated from the terminal, t, by an extension of the insulating plug, p.

A similar disposition with a phosphorescent tube is illustrated in Fig. 156. The tube, T, is prepared from two short tubes of different diameter, which are sealed on the ends. On the lower end is placed an inside conducting coating, c, which connects to the wire w. The wire has a hook on the upper end for suspension, and passes through the centre of the inside tube, which is filled with some good and tightly packed insulator. On the outside of the upper end of the tube, T, is another conducting coating, c_1, upon which is slipped a metallic reflector z, which should be separated by a thick insulation from the end of wire w.

The economical use of such a reflector or intensifier would require that all energy supplied to an air condenser should be recoverable, or, in other words, that there should not be any losses, neither in the gaseous medium nor through its action elsewhere. This is far from being so, but, fortunately, the losses may be reduced to anything desired. A few remarks are necessary on this subject, in order to make the experiences gathered in the course of these investigations perfectly clear.

Suppose a small helix with many well insulated turns, as in experiment Fig. 146, has one of its ends connected to one of the terminals of the induction coil, and the other to a metal plate, or, for the sake of simplicity, a sphere, insulated in space. When the coil is set to work, the potential of the sphere is alternated, and a small helix now behaves as though its free end were connected to the other terminal of the induction coil. If an iron rod be held within a small helix, it is quickly brought to a high

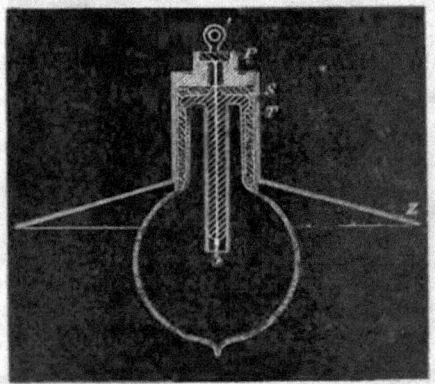

Fig. 155.

temperature, indicating the passage of a strong current through the helix. How does the insulated sphere act in this case? It can be a condenser, storing and returning the energy supplied to it, or it can be a mere sink of energy, and the conditions of the experiment determine whether it is rather one than the other. The sphere being charged to a high potential, it acts inductively upon the surrounding air, or whatever gaseous medium there might be. The molecules, or atoms, which are near the sphere, are of course more attracted, and move through a greater distance than the farther ones. When the nearest molecules strike the sphere, they are repelled, and collisions occur at all distances within the inductive action of the sphere. It is now clear that, if the potential be steady, but little loss of energy can be caused in this way, for the molecules which are nearest to the sphere, having had an additional charge imparted to them by contact, are not attracted until they have parted, if not with all, at least with most of the additional charge, which can be accomplished only after a great many collisions. From the fact, that with a steady potential there is but little loss in dry air, one must come to such a conclusion. When the potential of a sphere, instead of being steady, is alternating, the conditions are entirely different. In this case a rhythmical bombardment occurs, no matter whether the molecules, after coming in contact with the sphere, lose the imparted

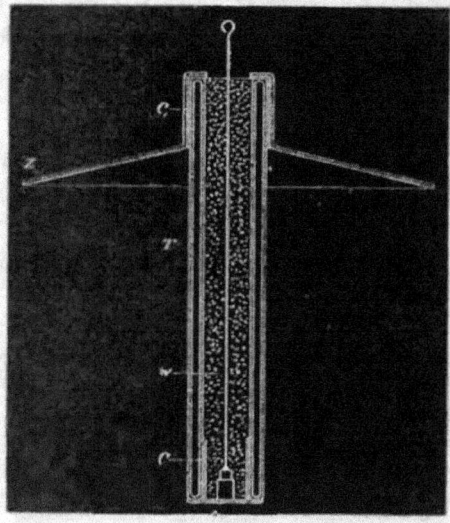

Fig. 156.

charge or not; what is more, if the charge is not lost, the impacts are only the more violent. Still, if the frequency of the impulses be very small, the loss caused by the impacts and collisions would not be serious, unless the potential were excessive. But when extremely high frequencies and more or less high potentials are used, the loss may very great. The total energy lost per unit of time is proportionate to the product of the number of impacts per second, or the frequency and the energy lost in each impact. But the energy of an impact must be proportionate to the square of the electric density of the sphere, since the charge imparted to the molecule is proportionate to that density. I conclude from this that the total energy lost must be proportionate to the product of the frequency and the square of the electric density ; but this law needs experimental confirmation. Assuming the preceding considerations to be true, then, by rapidly alternating the potential of a body immersed in an insulating gaseous medium, any amount of energy may be dissipated into space. Most of that energy then, I believe, is not dissipated in the form of long ether waves, propagated to considerable distance, as is thought most generally, but is consumed—in the case of an insulated sphere, for example—in impact and collisional losses—that is, heat vibrations—on the surface and in the vicinity of the sphere. To reduce the dissipation, it is necessary to work with a small electric density—the smaller, the higher the frequency.

But since, on the assumption before made, the loss is diminished with the square of the density, and since currents of very high frequencies involve considerable waste when transmitted through conductors, it follows that, on the whole, it is better to employ one wire than two. Therefore, if motors, lamps, or devices of any kind are perfected, capable of being advantageously operated by currents of extremely high frequency, economical reasons will make it advisable to use only one wire, especially if the distances are great.

When energy is absorbed in a condenser, the same behaves as though its capacity were increased. Absorption always exists more or less, but generally it is small and of no consequence as long as the frequencies are not very great. In using extremely high frequencies, and, necessarily in such case, also high potentials, the absorption—or, what is here meant more particularly by this term, the loss of energy due to the presence of a gaseous medium—is an important factor to be considered, as the energy absorbed in the air condenser may be any fraction of the supplied energy. This would seem to make it very difficult to tell from the measured or computed capacity of an air condenser its actual capacity or vibration period, especially if the condenser is of very small surface and is charged to a very high potential. As many important results are dependent upon the correctness of the estimation of the vibration period, this subject demands the most careful scrutiny of other investigators. To reduce the probable error as much as possible in experiments of the kind alluded to, it is advisable to use spheres or plates of large surface, so as to

make the density exceedingly small.   Otherwise, when it is
practicable, an oil condenser should be used in preference.   In
oil or other liquid dielectrics there are seemingly no such losses
as in gaseous media.   It being impossible to exclude entirely the
gas in condensers with solid dielectrics, such condensers should
be immersed in oil, for economical reasons, if nothing else ; they
can then be strained to the utmost, and will remain cool.   In
Leyden jars the loss due to air is comparatively small, as the tinfoil coatings are large, close together, and the charged surfaces
not directly exposed ; but when the potentials are very high, the
loss may be more or less considerable at, or near, the upper edge
of the foil, where the air is principally acted upon.   If the jar
be immersed in boiled-out oil, it will be capable of performing
four times the amount of work which it can for any length of
time when used in the ordinary way, and the loss will be inappreciable.

It should not be thought that the loss in heat in an air condenser is necessarily associated with the formation of *visible*
streams or brushes.   If a small electrode, inclosed in an unexhausted bulb, is connected to one of the terminals of the coil,
streams can be seen to issue from the electrode, and the air in the
bulb is heated ; if instead of a small electrode a large sphere is
inclosed in the bulb, no streams are observed, still the air is
heated.

Nor should it be thought that the temperature of an air condenser would give even an approximate idea of the loss in heat
incurred, as in such case heat must be given off much more
quickly, since there is, in addition to the ordinary radiation, a
very active carrying away of heat by independent carriers going
on, and since not only the apparatus, but the air at some distance
from it is heated in consequence of the collisions which must
occur.

Owing to this, in experiments with such a coil, a rise of temperature can be distinctly observed only when the body connected
to the coil is very small.   But with apparatus on a larger scale,
even a body of considerable bulk would be heated, as, for instance,
the body of a person ; and I think that skilled physicians might
make observations of utility in such experiments, which, if the
apparatus were judiciously designed, would not present the slightest danger.

A question of some interest, principally to meteorologists,

presents itself here. How does the earth behave? The earth is an air condenser, but is it a perfect or a very imperfect one—a mere sink of energy? There can be little doubt that to such small disturbance as might be caused in an experiment, the earth behaves as an almost perfect condenser. But it might be different when its charge is set in vibration by some sudden disturbance occurring in the heavens. In such case, as before stated, probably only little of the energy of the vibrations set up would be lost into space in the form of long ether radiations, but most of the energy, I think, would spend itself in molecular impacts and collisions, and pass off into space in the form of short heat, and possibly light, waves. As both the frequency of the vibrations of the charge and the potential are in all probability excessive, the energy converted into heat may be considerable. Since the density must be unevenly distributed, either in consequence of the irregularity of the earth's surface, or on account of the condition of the atmosphere in various places, the effect produced would accordingly vary from place to place. Considerable variations in the temperature and pressure of the atmosphere may in this manner be caused at any point of the surface of the earth. The variations may be gradual or very sudden, according to the nature of the general disturbance, and may produce rain and storms, or locally modify the weather in any way.

From the remarks before made, one may see what an important factor of loss the air in the neighborhood of a charged surface becomes when the electric density is great and the frequency of the impulses excessive. But the action, as explained, implies that the air is insulating—that is, that it is composed of independent carriers immersed in an insulating medium. This is the case only when the air is at something like ordinary or greater, or at extremely small, pressure. When the air is slightly rarefied and conducting, then true conduction losses occur also. In such case, of course, considerable energy may be dissipated into space even with a steady potential, or with impulses of low frequency, if the density is very great.

When the gas is at very low pressure, an electrode is heated more because higher speeds can be reached. If the gas around the electrode is strongly compressed, the displacements, and consequently the speeds, are very small, and the heating is insignificant. But if in such case the frequency could be sufficiently increased, the electrode would be brought to a high tem-

perature as well as if the gas were at very low pressure; in fact,
exhausting the bulb is only necessary because we cannot produce,
(and possibly not convey) currents of the required frequency.

Returning to the subject of electrode lamps, it is obviously of
advantage in such a lamp to confine as much as possible the heat
to the electrode by preventing the circulation of the gas in the
bulb. If a very small bulb be taken, it would confine the heat
better than a large one, but it might not be of sufficient capacity
to be operated from the coil, or, if so, the glass might get too
hot. A simple way to improve in this direction is to employ a
globe of the required size, but to place a small bulb, the diameter
of which is properly estimated, over the refractory button con-

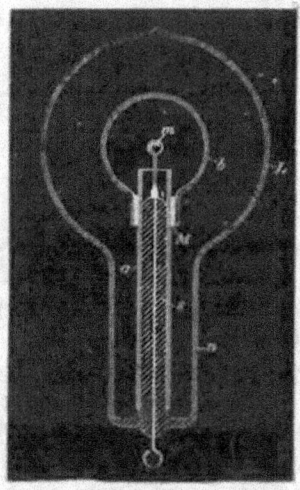

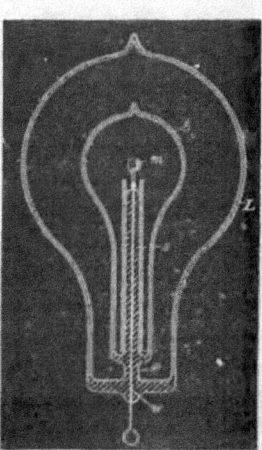

FIG. 157.   FIG. 158.

tained in the globe. This arrangement is illustrated in Fig. 157.

The globe L has in this case a large neck n, allowing the small
bulb b to slip through. Otherwise the construction is the same
as shown in Fig. 147, for example. The small bulb is conveniently supported upon the stem s, carrying the refractory button
m. It is separated from the aluminum tube a by several layers
of mica M, in order to prevent the cracking of the neck by the
rapid heating of the aluminum tube upon a sudden turning on
of the current. The inside bulb should be as small as possible
when it is desired to obtain light only by incandescence of the
electrode. If it is desired to produce phosphorescence, the bulb

should be larger, else it would be apt to get too hot, and the phosphorescence would cease. In this arrangement usually only the small bulb shows phosphorescence, as there is practically no bombardment against the outer globe. In some of these bulbs constructed as illustrated in Fig. 157, the small tube was coated with phosphorescent paint, and beautiful effects were obtained. Instead of making the inside bulb large, in order to avoid undue heating, it answers the purpose to make the electrode m larger. In this case the bombardment is weakened by reason of the smaller electric density.

Many bulbs were constructed on the plan illustrated in Fig. 158. Here a small bulb b, containing the refractory button m, upon being exhausted to a very high degree was sealed in a large globe L, which was then moderately exhausted and sealed off. The principal advantage of this construction was that it allowed of reaching extremely high vacua, and, at the same time of using a large bulb. It was found, in the course of experiments with bulbs such as illustrated in Fig. 158, that it was well to make the stem s, near the seal at e, very thick, and the leading-in wire w thin, as it occurred sometimes that the stem at e was heated and the bulb was cracked. Often the outer globe L was exhausted only just enough to allow the discharge to pass through, and the space between the bulbs appeared crimson, producing a curious effect. In some cases, when the exhaustion in globe L was very low, and the air good conducting, it was found necessary, in order to bring the button m to high incandescence, to place, preferably on the upper part of the neck of the globe, a tinfoil coating which was connected to an insulated body, to the ground, or to the other terminal of the coil, as the highly conducting air weakened the effect somewhat, probably by being acted upon inductively from the wire w, where it entered the bulb at e. Another difficulty—which, however, is always present when the refractory button is mounted in a very small bulb—existed in the construction illustrated in Fig. 158, namely, the vacuum in the bulb b would be impaired in a comparatively short time.

The chief idea in the two last described constructions was to confine the heat to the central portion of the globe by preventing the exchange of air. An advantage is secured, but owing to the heating of the inside bulb and slow evaporation of the glass, the vacuum is hard to maintain, even if the construction illustrated in Fig. 157 be chosen, in which both bulbs communicate.

But by far the better way—the ideal way—would be to reach sufficiently high frequencies. The higher the frequency, the slower would be the exchange of the air, and I think that a frequency may be reached, at which there would be no exchange whatever of the air molecules around the terminal. We would then produce a flame in which there would be no carrying away of material, and a queer flame it would be, for it would be rigid! With such high frequencies the inertia of the particles would come into play. As the brush, or flame, would gain rigidity in virtue of the inertia of the particles, the exchange of the latter would be prevented. This would necessarily occur, for, the number of impulses being augmented, the potential energy of each would diminish, so that finally only atomic vibrations could be set up, and the motion of translation through measurable space would cease. Thus an ordinary gas burner connected to a source of rapidly alternating potential might have its efficiency augmented to a certain limit, and this for two reasons—because of the additional vibration imparted, and because of a slowing down of the process of carrying off. But the renewal being rendered difficult, a renewal being necessary to maintain the *burner*, a continued increase of the frequency of the impulses, assuming they could be transmitted to and impressed upon the flame, would result in the "extinction" of the latter, meaning by this term only the cessation of the chemical process.

I think, however, that in the case of an electrode immersed in a fluid insulating medium, and surrounded by independent carriers of electric charges, which can be acted upon inductively, a sufficient high frequency of the impulses would probably result in a gravitation of the gas all around toward the electrode. For this it would be only necessary to assume that the independent bodies are irregularly shaped; they would then turn toward the electrode their side of the greatest electric density, and this would be a position in which the fluid resistance to approach would be smaller than that offered to the receding.

The general opinion, I do not doubt, is that it is out of the question to reach any such frequencies as might—assuming some of the views before expressed to be true—produce any of the results which I have pointed out as mere possibilities. This may be so, but in the course of these investigations, from the observation of many phenomena, I have gained the conviction that these frequencies would be much lower than one is apt to estimate at first. In a flame we set up light vibrations by causing molecules, or atoms, to collide. But what is the ratio of the frequency of the collisions and that of the vibrations set up? Certainly it must be incomparably smaller than that of the strokes of the bell and the sound vibrations, or that of the discharges and the oscillations of the condenser. We may cause the molecules of the gas to collide by the use of alternate electric impulses of high frequency, and so we may imitate the process in a flame; and from experiments with frequencies which we are now able to obtain, I think that the result is producible with impulses which are transmissible through a conductor.

In connection with thoughts of a similar nature, it appeared to me of great interest to demonstrate the rigidity of a vibrating gaseous column. Although with such low frequencies as, say 10,000 per second, which I was able to obtain without difficulty from a specially constructed alternator, the task looked discouraging at first, I made a series of experiments. The trials with air at ordinary pressure led to no result, but with air moderately rarefied I obtain what I think to be an unmistakable experimental evidence of the property sought for. As a result of this kind might lead able investigators to conclusions of importance, I will describe one of the experiments performed.

It is well known that when a tube is slightly exhausted, the discharge may be passed through it in the form of a thin luminous thread. When produced with currents of low frequency, obtained from a coil operated as usual, this thread is inert. If a magnet be approached to it, the part near the same is attracted or repelled, according to the direction of the lines of force of the magnet. It occurred to me that if such a thread would be produced with currents of very high frequency, it should be more or less rigid, and as it was visible it could be easily studied. Accordingly I prepared a tube about one inch in diameter and one metre long, with outside coating at each end. The tube was exhausted to a point at which, by a little working, the thread discharge could be obtained. It must be remarked here that the general aspect of the tube, and the degree of exhaustion, are quite other than when ordinary low frequency currents are used. As it was found preferable to work with one terminal, the tube prepared was suspended from the end of a wire connected to the terminal, the tinfoil coating being connected to the wire, and to the lower coating sometimes a small insulated plate was attached.   When the thread was formed, it extended through the upper part of the tube and lost itself in the lower end.   If it possessed rigidity it resembled, not exactly an elastic cord stretched tight between two supports, but a cord suspended from a height with a small weight attached at the end.   When the finger or a small magnet was approached to the upper end of the luminous thread, it could be brought locally out of position by electrostatic or magnetic action ; and when the disturbing object was very quickly removed, an analogous result was produced, as though a suspended cord would be displaced and quickly released near the point of suspension.   In doing this the luminous thread was set in vibration, and two very sharply marked nodes, and a third indistinct one, were formed.   The vibration, once set up, continued for fully eight minutes, dying gradually out.   The speed of the vibration often varied perceptibly, and it could be observed that the electrostatic attraction of the glass affected the vibrating thread ; but it was clear that the electrostatic action was not the cause of the vibration, for the thread was most generally stationary, and could always be set in vibration by passing the finger quickly near the upper part of the tube.   With a magnet the thread could be split in two and both parts vibrated. By approaching the hand to the lower coating of the tube, or insulation plate if attached, the vibration was quickened ; also, as far as I could see, by raising the potential or frequency.   Thus, either increasing the frequency or passing a stronger discharge of the same frequency corresponded to a tightening of the cord. I did not obtain any experimental evidence with condenser discharges.   A luminous band excited in the bulb by repeated discharges of a Leyden jar must possess rigidity, and if deformed and suddenly released, should vibrate.   But probably the amount of vibrating matter is so small that in spite of the extreme speed, the inertia cannot prominently assert itself.   Besides, the observation in such a case is rendered extremely difficult on account of the fundamental vibration.

The demonstration of the fact—which still needs better experimental confirmation—that a vibrating gaseous column possesses rigidity, might greatly modify the views of thinkers. When with low frequencies and insignificant potentials indications of that property may be noted, how must a gaseous medium behave under the influence of enormous electrostatic stresses which may be active in the interstellar space, and which may alternate with inconceivable rapidity ? The existence of such an electrostatic, rhythmically throbbing force—of a vibrating electrostatic field—would show a possible way how solids might have formed from the ultra-gaseous uterus, and how transverse and all kinds of vibrations may be transmitted through a gaseous medium filling all space. Then, ether might be a true fluid, devoid of rigidity, and at rest, it being merely necessary as a connecting link to enable interaction. What determines the rigidity of a body ? It must be the speed and the amount of motive matter. In a gas the speed may be considerable, but the density is exceedingly small; in a liquid the speed would be likely to be small, though the density may be considerable ; and in both cases the inertia resistance offered to displacement is practically *nil*. But place a gaseous (or liquid) column in an intense, rapidly alternating electrostatic field, set the particles vibrating with enormous speeds, then the inertia resistance asserts itself. A body might move with more or less freedom through the vibrating mass, but as a whole it would be rigid.

There is a subject which I must mention in connection with these experiments: it is that of high vacua. This is a subject, the study of which is not only interesting, but useful, for it may lead to results of great practical importance. In commercial apparatus, such as incandescent lamps, operated from ordinary systems of distribution, a much higher vacuum than is obtained at present would not secure a very great advantage. In such a case the work is performed on the filament, and the gas is little concerned ; the improvement, therefore, would be but trifling. But when we begin to use very high frequencies and potentials, the action of the gas becomes all important, and the degree of exhaustion materially modifies the results. As long as ordinary coils, even very large ones, were used, the study of the subject was limited, because just at a point when it became most interesting it had to be interrupted on account of the "non-striking" vacuum being reached. But at present we are able to obtain from a small disruptive discharge coil potentials much higher than even the largest coil was capable of giving, and, what is more, we can make the potential alternate with great rapidity. Both of these results enable us now to pass a luminous discharge through almost any vacua obtainable, and the field of our investigations is greatly extended. Think we as we may, of all the possible directions to develop a practical illuminant, the line of high vacua seems to be the most promising at present. But to reach extreme vacua the appliances must be much more improved, and ultimate perfection will not be attained until we shall have discharged the mechanical and perfected an *electrical* vacuum pump. Molecules and atoms can be thrown out of a bulb under the action of an enormous potential : *this* will be the principle of the vacuum pump of the future. For the present, we must secure the best results we can with mechanical appliances. In this respect, it might not be out of the way to say a few words about the method of, and apparatus for, producing excessively

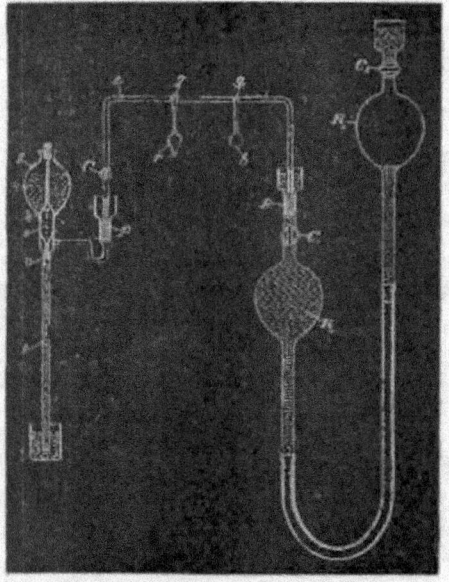

Fig. 159.

high degrees of exhaustion of which I have availed myself in the course of these investigations. It is very probable that other experimenters have used similar arrangements ; but as it is possible that there may be an item of interest in their description, a few remarks, which will render this investigation more complete, might be permitted.

The apparatus is illustrated in a drawing shown in Fig. 159. *s* represents a Sprengel pump, which has been specially constructed to better suit the work required. The stop-cock which is usually employed has been omitted, and instead of it a hollow stopper *s* has been fitted in the neck of the reservoir R. This stopper has a small hole *h*, through which the mercury descends; the size of the outlet *o* being properly determined with respect to the section of the fall tube *t*, which is sealed to the reservoir instead of being connected to it in the usual manner. This arrangement overcomes the imperfections and troubles which often arise from the use of the stopcock on the reservoir and the connections of the latter with the fall tube.

The pump is connected through a ∪-shaped tube *t* to a very large reservoir R_1. Especial care was taken in fitting the grinding surfaces of the stoppers *p* and p_1, and both of these and the mercury caps above them were made exceptionally long. After the ∪-shaped tube was fitted and put in place, it was heated, so as to soften and take off the strain resulting from imperfect fitting. The ∪-shaped tube was provided with a stopcock *c*, and two ground connections *g* and g_1,—one for a small bulb *b*, usually containing caustic potash, and the other for the receiver *r*, to be exhausted.

The reservoir R_1, was connected by means of a rubber tube to a slightly larger reservoir R_2, each of the two reservoirs being provided with a stopcock c_1 and c_2, respectively. The reservoir R_2 could be raised and lowered by a wheel and rack, and the range of its motion was so determined that when it was filled with mercury and the stopcock c_2 closed, so as to form a Torricellian vacuum in it when raised, it could be lifted so high that the reservoir R_1 would stand a little above stopcock c_1; and when this stopcock was closed and the reservoir R_2 descended, so as to form a Torricellian vacuum in reservoir R_1, it could be lowered so far as to completely empty the latter, the mercury filling the reservoir R_2 up to a little above stopcock c_2.

The capacity of the pump and of the connections was taken as small as possible relatively to the volume of reservoir, R_1, since, of course, the degree of exhaustion depended upon the ratio of these quantities.

With this apparatus I combined the usual means indicated by former experiments for the production of very high vacua. In most of the experiments it was most convenient to use caustic potash. I may venture to say, in regard to its use, that much time is saved and a more perfect action of the pump insured by fusing and boiling the potash as soon as, or even before, the pump settles down.   If this course is not followed, the sticks, as ordinarily employed, may give off moisture at a certain very slow rate, and the pump may work for many hours without reaching a very high vacuum.   The potash was heated either by a spirit lamp or by passing a discharge through it, or by passing a current through a wire contained in it.   The advantage in the latter case was that the heating could be more rapidly repeated.

Generally the process of exhaustion was the following :—At the start, the stop-cocks c and c_1 being open, and all other connections closed, the reservoir R_2 was raised so far that the mercury filled the reservoir R_1 and a part of the narrow connecting U-shaped tube.   When the pump was set to work, the mercury would, of course, quickly rise in the tube, and reservoir R_2 was lowered, the experimenter keeping the mercury at about the same level.   The reservoir R_2 was balanced by a long spring which facilitated the operation, and the friction of the parts was generally sufficient to keep it in almost any position.   When the Sprengel pump had done its work, the reservoir R_2 was further lowered and the mercury descended in R_1 and filled R_2, whereupon stopcock c_2 was closed.   The air adhering to the walls of R_1 and that absorbed by the mercury was carried off, and to free the mercury of all air the reservoir R_2 was for a long time worked up and down.   During this process some air, which would gather below stopcock c_2, was expelled from R_2 by lowering it far enough and opening the stopcock, closing the latter again before raising the reservoir.   When all the air had been expelled from the mercury, and no air would gather in R_2 when it was lowered, the caustic potash was resorted to.   The reservoir R_2 was now again raised until the mercury in R_1 stood above stopcock c_1.   The caustic potash was fused and boiled, and moisture partly carried off by the pump and partly re-absorbed ; and this process of heating and cooling was repeated many times, and each time, upon the moisture being absorbed or carried off, the reservoir R_2 was for a long time raised and lowered.   In this manner all the moisture was carried off from the mercury, and both the reservoirs were in proper condition to be used.   The reservoir R_2 was then again raised to the top, and the pump was kept working for a long time.   When the highest vacuum obtainable with the pump had been reached, the potash bulb was usually wrapped with cotton which was sprinkled with ether so as to keep the potash at a very low temperature, then the reservoir R_2 was lowered, and upon reservoir R_1 being emptied the receiver was quickly sealed up.

When a new bulb was put on, the mercury was always raised above stopcock c_1, which was closed, so as to always keep the mercury and both the reservoirs in fine condition, and the mercury was never withdrawn from R_1 except when the pump had reached the highest degree of exhaustion. It is necessary to observe this rule if it is desired to use the apparatus to advantage.

By means of this arrangement I was able to proceed very quickly, and when the apparatus was in perfect order it was possible to reach the phosphorescent stage in a small bulb in less than fifteen minutes, which is certainly very quick work for a small laboratory arrangement requiring all in all about 100 pounds of mercury. With ordinary small bulbs the ratio of the capacity of the pump, receiver, and connections, and that of reservoir R was about 1 to 20, and the degrees of exhaustion reached were necessarily very high, though I am unable to make a precise and reliable statement how far the exhaustion was carried.

What impresses the investigator most in the course of these experiences is the behavior of gases when subjected to great, rapidly alternating, electrostatic stresses. But he must remain in doubt as to whether the effects observed are due wholly to the molecules, or atoms, of the gas which chemical analysis discloses to us, or whether there enters into play another medium of a gaseous nature, comprising atoms, or molecules, immersed in a fluid pervading the space. Such a medium surely must exist, and I am convinced that, for instance, even if air were absent, the surface and neighborhood of a body in space would be heated by rapidly alternating the potential of the body; but no such heating of the surface or neighborhood could occur if all free atoms were removed and only a homogeneous, incompressible, and elastic fluid—such as ether is supposed to be—would remain, for then there would be no impacts, no collisions. In such a case, as far as the body itself is concerned, only frictional losses in the inside could occur.

It is a striking fact that the discharge through a gas is established with ever-increasing freedom as the frequency of the impulses is augmented. It behaves in this respect quite contrarily to a metallic conductor. In the latter the impedance enters prominently into play as the frequency is increased, but the gas acts much as a series of condensers would; the facility with which the discharge passes through, seems to depend on the rate of change of potential. If it acts so, then in a vacuum tube even of great length, and no matter how strong the current, self-induction could not assert itself to any appreciable degree. We have, then, as far as we can now see, in the gas a conductor which is capable of transmitting electric impulses of any frequency which we may be able to produce. Could the frequency be brought high enough, then a queer system of electric distribution, which would be likely to interest gas companies, might be realized: metal pipes filled with gas—the metal being the insulator, the gas the conductor—supplying phosphorescent bulbs, or perhaps devices as yet uninvented. It is certainly possible to take a hollow core of copper, rarefy the gas in the same, and by passing impulses of sufficiently high frequency through a circuit around it, bring the gas inside to a high degree of incandescence; but as to the nature of the forces there would be considerable uncertainty, for it would be doubtful whether with such impulses the copper core would act as a static screen. Such paradoxes and apparent impossibilities we encounter at every step in this line of work, and therein lies, to a great extent, the charm of the study.

I have here a short and wide tube which is exhausted to a high degree and covered with a substantial coating of bronze, the coating barely allowing the light to shine through. A metallic cap, with a hook for suspending the tube, is fastened around the middle portion of the latter, the clasp being in contact with the bronze coating. I now want to light the gas inside by suspending the tube on a wire connected to the coil. Any one who would try the experiment for the first time, not having any previous experience, would probably take care to be quite alone when making the trial, for fear that he might become the joke of his assistants. Still, the bulb lights in spite of the metal coating, and the light can be distinctly perceived through the latter. A long tube covered with aluminum bronze lights when held in one hand—the other touching the terminal of the coil—quite powerfully. It might be objected that the coatings are not sufficiently conducting; still, even if they were highly resistant, they ought to screen the gas. They certainly screen it perfectly in a condition of rest, but far from perfectly when the charge is surging in the coating. But the loss of energy which occurs within the tube, notwithstanding the screen, is occasioned principally by the presence of the gas. Were we to take a large hollow metallic sphere and fill it with a perfect, incompressible, fluid dielectric, there would be no loss inside of the sphere, and consequently the inside might be considered as perfectly screened, though the potential be very rapidly alternating. Even were the sphere filled with oil, the loss would be incomparably smaller than when the fluid is replaced by a gas, for in the latter case the force produces displacements; that means impact and collisions in the inside.

No matter what the pressure of the gas may be, it becomes an important factor in the heating of a conductor when the electric density is great and the frequency very high. That in the heating of conductors by lightning discharges, air is an element of great importance, is almost as certain as an experimental fact. I may illustrate the action of the air by the following experiment: I take a short tube which is exhausted to a moderate degree and has a platinum wire running through the middle from one end to the other. I pass a steady or low frequency current through the wire, and it is heated uniformly in all parts. The heating here is due to conduction, or frictional losses, and the gas around the wire has—as far as we can see—no function to perform. But now let me pass sudden discharges, or high frequency currents, through the wire. Again the wire is heated, this time principally on the ends and least in the middle portion; and if the frequency of the impulses, or the rate of change, is high enough, the wire might as well be cut in the middle as not, for practically all heating is due to the rarefied gas. Here the gas might only act as a conductor of no impedance diverting the current from the wire as the impedance of the latter is enormously increased, and merely heating the ends of the wire by reason of their resistance to the passage of the discharge. But it is not at all necessary that the gas in the tube should be conducting; it might be at an extremely low pressure, still the ends of the wire would be heated—as, however, is ascertained by experience— only the two ends would in such case not be electrically connected through the gaseous medium. Now what with these frequencies and potentials occurs in an exhausted tube, occurs in the lightning discharges at ordinary pressure. We only need remember one of the facts arrived at in the course of these investigations, namely, that to impulses of very high frequency the gas at ordinary pressure behaves much in the same manner as though it were at moderately low pressure. I think that in lightning discharges frequently wires or conducting objects are volatilized merely because air is present, and that, were the conductor immersed in an insulating liquid, it would be safe, for then the energy would have to spend itself somewhere else. From the behavior of gases under sudden impulses of high potential, I am led to conclude that there can be no surer way of diverting a lightning discharge than by affording it a passage through a volume of gas, if such a thing can be done in a practical manner.

There are two more features upon which I think it necessary to dwell in connection with these experiments—the " radiant state " and the " non-striking vacuum."

Any one who has studied Crookes' work must have received the impression that the " radiant state " is a property of the gas inseparably connected with an extremely high degree of exhaustion. But it should be remembered that the phenomena observed in an exhausted vessel are limited to the character and capacity of the apparatus which is made use of. I think that in a bulb a molecule, or atom, does not precisely move in a straight line because it meets no obstacle, but because the velocity imparted to it is sufficient to propel it in a sensibly straight line. The mean free path is one thing, but the velocity—the energy associated with the moving body—is another, and under ordinary circumstances I believe that it is a mere question of potential or speed. A disruptive discharge coil, when the potential is pushed very far, excites phosphorescence and projects shadows, at comparatively low degrees of exhaustion. In a lightning discharge, matter moves in straight lines at ordinary pressure when the mean free path is exceedingly small, and frequently images of wires or other metallic objects have been produced by the particles thrown off in straight lines.

I have prepared a bulb to illustrate by an experiment the correctness of these assertions. In a globe L, Fig. 160, I have mounted upon a lamp filament f a piece of lime l. The lamp filament is connected with a wire which leads into the bulb, and the general construction of the latter is as indicated in Fig. 148, before described. The bulb being suspended from a wire connected to the terminal of the coil, and the latter being set to work, the lime piece l and the projecting parts of the filament f are bombarded. The degree of exhaustion is just such that with the potential the coil is capable of giving, phosphorescence of the glass is produced, but disappears as soon as the vacuum is impaired. The lime containing moisture, and moisture being given off as soon as heating occurs, the phosphorescence lasts only for a few moments. When the lime has been sufficiently heated, enough moisture has been given off to impair materially the vacuum of the bulb. As the bombardment goes on, one point of the lime piece is more heated than other points, and the result is that finally practically all the discharge passes through that point which is intensely heated, and a white stream of lime particles (Fig. 160) then breaks forth from that point. This stream is composed of "radiant" matter, yet the degree of exhaustion is low. But the particles move in straight lines because the velocity imparted to them is great, and this is due to three causes—to the great electric density, the high temperature of the small point, and the fact that the particles of the lime are easily

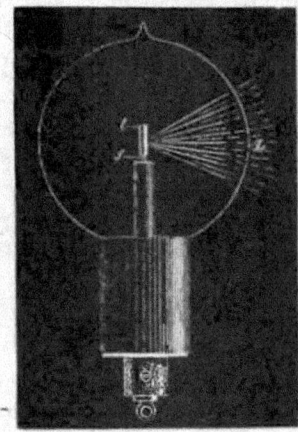

Fig. 160.

torn and thrown off—far more easily than those of carbon. With frequencies such as we are able to obtain, the particles are bodily thrown off and projected to a considerable distance; but with sufficiently high frequencies no such thing would occur; in such case only a stress would spread or a vibration would be propagated through the bulb. It would be out of the question to reach any such frequency on the assumption that the atoms move with the speed of light; but I believe that such a thing is impossible; for this an enormous potential would be required. With potentials which we are able to obtain, even with a disruptive discharge coil, the speed must be quite insignificant.

As to the "non-striking vacuum," the point to be noted is, that it can occur only with low frequency impulses, and it is

necessitated by the impossibility of carrying off enough energy
with such impulses in high vacuum, since the few atoms which
are around the terminal upon coming in contact with the same,
are repelled and kept at a distance for a comparatively long
period of time, and not enough work can be performed to render
the effect perceptible to the eye.   If the difference of potential
between the terminals is raised, the dielectric breaks down.   But
with very high frequency impulses there is no necessity for such
breaking down, since any amount of work can be performed by
continually agitating the atoms in the exhausted vessel, provided
the frequency is high enough.   It is easy to reach—even with

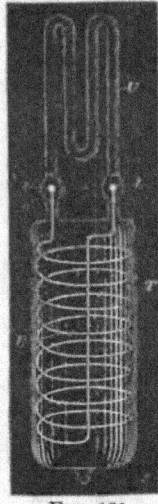

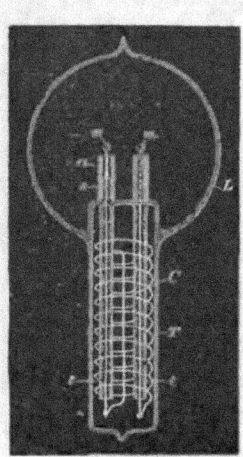

FIG. 161.   FIG. 162.

frequencies obtained from an alternator as here used—a stage at
which the discharge does not pass between two electrodes in a
narrow tube, each of these being connected to one of the terminals of the coil, but it is difficult to reach a point at which a
luminous discharge would not occur around each electrode.

A thought which naturally presents itself in connection with
high frequency currents, is to make use of their powerful electro-
dynamic inductive action to produce light effects in a sealed glass
globe.   The leading-in wire is one of the defects of the present
incandescent lamp, and if no other improvement were made,
that imperfection at least should be done away with.   Following

this thought, I have carried on experiments in various directions, of which some were indicated in my former paper. I may here mention one or two more lines of experiment which have been followed up.

Many bulbs were constructed as shown in Fig. 161 and Fig. 162.

In Fig. 161, a wide tube, T, was sealed to a smaller W shaped tube v, of phosphorescent glass. In the tube T, was placed a coil c, of aluminum wire, the ends of which were provided with small spheres, t and t_1, of aluminum, and reached into the v tube. The tube T was slipped into a socket containing a primary coil, through which usually the discharges of Leyden jars were directed, and the rarefied gas in the small v tube was excited to strong luminosity by the high-tension current induced in the coil c. When Leyden jar discharges were used to induce currents in the coil c, it was found necessary to pack the tube T tightly with insulating powder, as a discharge would occur frequently between the turns of the coil, especially when the primary was thick and the air gap, through which the jars discharged, large, and no little trouble was experienced in this way.

In Fig. 162 is illustrated another form of the bulb constructed. In this case a tube T is sealed to a globe L. The tube contains a coil c, the ends of which pass through two small glass tubes t and t_1, which are sealed to the tube T. Two refractory buttons m and m_1 are mounted on lamp filaments which are fastened to the ends of the wires passing through the glass tubes t and t_1. Generally in bulbs made on this plan the globe L communicated with the tube T. For this purpose the ends of the small tubes t and t_1 were heated just a trifle in the burner, merely to hold the wires, but not to interfere with the communication. The tube T, with the small tubes, wires through the same, and the refractory buttons m and m_1, were first prepared, and then sealed to globe L, whereupon the coil c was slipped in and the connections made to its ends. The tube was then packed with insulating powder, jamming the latter as tight as possible up to very nearly the end ; then it was closed and only a small hole left through which the remainder of the powder was introduced, and finally the end of the tube was closed. Usually in bulbs constructed as shown in Fig. 162 an aluminum tube a was fastened to the upper end s of each of the tubes t and t_1, in order to protect that end against the heat. The buttons m and m_1 could be brought to any degree of incandescence by passing the discharges of Leyden jars around the coil c. In such bulbs with two buttons a very curious effect is produced by the formation of the shadows of each of the two buttons.

Another line of experiment, which has been assiduously followed, was to induce by electro-dynamic induction a current or luminous discharge in an exhausted tube or bulb. This matter has received such able treatment at the hands of Prof. J. J. Thomson, that I could add but little to what he has made known, even had I made it the special subject of this lecture. Still, since experiments in this line have gradually led me to the present views and results, a few words must be devoted here to this subject.

It has occured, no doubt, to many that as a vacuum tube is made longer, the electromotive force per unit length of the tube, necessary to pass a luminous discharge through the latter, becomes continually smaller; therefore, if the exhausted tube be made long enough, even with low frequencies a luminous discharge could be induced in such a tube closed upon itself. Such a tube might be placed around a hall or on a ceiling, and at once a simple appliance capable of giving considerable light would be obtained. But this would be an appliance hard to manufacture and extremely unmanageable. It would not do to make the tube up of small lengths, because there would be with ordinary frequencies considerable loss in the coatings, and besides, if coatings were used, it would be better to supply the current directly to the tube by connecting the coatings to a transformer. But even if all objections of such nature were removed, with low frequencies the light conversion itself would be inefficient, as I have before stated. In using extremely high frequencies the length of the secondary—in other words, the size of the vessel—can be reduced as much as desired, and the efficiency of the light conversion is increased, provided that means are invented for efficiently obtaining such high frequencies. Thus one is led, from theoretical and practical considerations, to the use of high frequencies, and this means high electromotive forces and small currents in the primary. When one works with condenser charges—and they are the only means up to the present known for reaching these extreme frequencies—one gets to electromotive forces of several thousands of volts per turn of the primary. We cannot multiply the electro-dynamic inductive effect by taking more turns in the primary, for we arrive at the conclusion that the best way is to work with one single turn—though we must sometimes depart from this rule—and we must get along with whatever inductive effect we can obtain with one turn. But before one has long experimented with the extreme frequencies required to set up in a small bulb an electromotive force of several thousands of volts, one realizes the great importance of electrostatic effects, and these effects grow relatively to the electro-dynamic in significance as the frequency is increased.

Now, if anything is desirable in this case, it is to increase the frequency, and this would make it still worse for the electro-dynamic effects. On the other hand, it is easy to exalt the electrostatic action as far as one likes by taking more turns on the secondary, or combining self-induction and capacity to raise the potential. It should also be remembered that, in reducing the the current to the smallest value and increasing the potential, the electric impulses of high frequency can be more easily transmitted through a conductor.

These and similar thoughts determined me to devote more attention to the electrostatic phenomena, and to endeavor to produce potentials as high as possible, and alternating as fast as they could be made to alternate. I then found that I could excite vacuum tubes at considerable distance from a conductor connected to a properly constructed coil, and that I could, by converting the oscillatory current of a conductor to a higher potential, establish electrostatic alternating fields which acted through the whole extent of the room, lighting up a tube no matter where it was held in space. I thought I recognized that I had made a step in advance, and I have persevered in this line; but I wish to say that I share with all lovers of science and progress the one and only desire—to reach a result of utility to men in any direction to which thought or experiment may lead me. I think that this departure is the right one, for I cannot see, from the observation of the phenomena which manifest themselves as the frequency is increased, what there would remain to act between two circuits conveying, for instance, impulses of several hundred millions per second, except electrostatic forces. Even with such trifling frequencies the energy would be practically all potential, and my conviction has grown strong that, to whatever kind of motion light may be due, it is produced by tremendous electrostatic stresses vibrating with extreme rapidity.

Of all these phenomena observed with currents, or electric impulses, of high frequency, the most fascinating for an audience are certainly those which are noted in an electrostatic field acting through considerable distance; and the best an unskilled lecturer can do is to begin and finish with the exhibition of these singular effects. I take a tube in my hand and move it about, and it is lighted wherever I may hold it; throughout space the invisible forces act. But I may take another tube and it might not light, the vacuum being very high. I excite it by means of a disruptive discharge coil, and now it will light in the electrostatic

Fig. 163.   Fig. 164.

field. I may put it away for a few weeks or months, still it retains the faculty of being excited. What change have I produced in the tube in the act of exciting it? If a motion imparted to atoms, it is difficult to perceive how it can persist so long without being arrested by frictional losses; and if a strain exerted in the dielectric, such as a simple electrification would produce, it is easy to see how it may persist indefinitely, but very difficult to understand why such a condition should aid the excitation when we have to deal with potentials which are rapidly alternating.

Since I have exhibited these phenomena for the first time, I have obtained some other interesting effects. For instance, I have produced the incandescence of a button, filament, or wire enclosed in a tube. To get to this result it was necessary to economize the energy which is obtained from the field, and direct most of it on the small body to be rendered incandescent. At the beginning the task appeared difficult, but the experiences gathered permitted me to reach the result easily. In Fig. 163 and Fig. 164, two such tubes are illustrated, which are prepared for the occasion. In Fig. 163 a short tube T_1, sealed to another long tube T, is provided with a stem s, with a platinum wire sealed in the latter. A very thin lamp filament l, is fastened to this wire and connection to the outside is made through a thin copper wire w. The tube is provided with outside and inside coatings, c and c_1, respectively, and is filled as far as the coatings reach with conducting, and the space above with insulating, powder. These coatings are merely used to enable me to perform two experiments with the tube—namely, to produce the effect desired either by direct connection of the body of the experimenter or of another body to the wire w, or by acting inductively through the glass. The stem s is provided with an aluminum tube a, for purposes before explained, and only a small part of the filament reaches out of this tube. By holding the tube T_1 anywhere in the electrostatic field, the filament is rendered incandescent.

A more interesting piece of apparatus is illustrated in Fig. 164. The construction is the same as before, only instead of the lamp filament a small platinum wire p, sealed in a stem s, and bent above it in a circle, is connected to the copper wire w, which is joined to an inside coating c. A small stem s_1, is provided with a needle, on the point of which is arranged, to rotate very freely, a very light fan of mica v. To prevent the fan from falling out, a thin stem of glass g, is bent properly and fastened to the aluminum tube. When the glass tube is held anywhere in the electrostatic field the platinum wire becomes incandescent, and the mica vanes are rotated very fast.

Intense phosphorescence may be excited in a bulb by merely connecting it to a plate within the field, and the plate need not be any larger than an ordinary lamp shade. The phosphorescence excited with these currents is incomparably more powerful than with ordinary apparatus. A small phosphorescent bulb, when attached to a wire connected to a coil, emits sufficient light to allow reading ordinary print at a distance of five to six paces. It was of interest to see how some of the phosphorescent bulbs of Professor Crookes would behave with these currents, and he has had the kindness to lend me a few for the occasion. The effects produced are magnificent, especially by the sulphide of calcium and sulphide of zinc. With the disruptive discharge coil they glow intensely merely by holding them in the hand and connecting the body to the terminal of the coil.

To whatever results investigations of this kind may lead, the chief interest lies, for the present, in the possibilities they offer for the production of an efficient illuminating device. In no branch of electric industry is an advance more desired than in the manufacture of light. Every thinker, when considering the barbarous methods employed, the deplorable losses incurred in our best systems of light production, must have asked himself, What is likely to be the light of the future? Is it to be an incandescent solid, as in the present lamp, or an incandescent gas, or a phosphorescent body, or something like a burner, but incomparably more efficient?

There is little chance to perfect a gas burner; not, perhaps, because human ingenuity has been bent upon that problem for centuries without a radical departure having been made—though the argument is not devoid of force—but because in a burner the highest vibrations can never be reached, except by passing through all the low ones. For how is a flame to proceed unless by a fall of lifted weights? Such process cannot be maintained without renewal, and renewal is repeated passing from low to high vibrations. One way only seems to be open to improve a burner, and that is by trying to reach higher degrees of incandescence. Higher incandescence is equivalent to a quicker vibration: that means more light from the same material, and that again, means more economy. In this direction some improvements have been made, but the progress is hampered by many limitations. Discarding, then, the burner, there remains the three ways first mentioned, which are essentially electrical.

Suppose the light of the immediate future to be a solid, rendered incandescent by electricity. Would it not seem that it is better to employ a small button than a frail filament? From many considerations it certainly must be concluded that a button is capable of a higher economy, assuming, of course, the difficulties connected with the operation of such a lamp to be effectively overcome. But to light such a lamp we require a high potential; and to get this economically, we must use high frequencies.

Such considerations apply even more to the production of light by the incandescence of a gas, or by phosphorescence. In all cases we require high frequencies and high potentials. These thoughts occurred to me a long time ago.

Incidentally we gain, by the use of high frequencies, many advantages, such as higher economy in the light production, the possibility of working with one lead, the possibility of doing away with the leading-in wire, etc.

The question is, how far can we go with frequencies? Ordinary conductors rapidly lose the facility of transmitting electric impulses when the frequency is greatly increased. Assume the means for the production of impulses of very great frequency brought to the utmost perfection, every one will naturally ask how to transmit them when the necessity arises. In transmitting such impulses through conductors we must remember that we have to deal with *pressure* and *flow*, in the ordinary interpretation of these terms. Let the pressure increase to an enormous value, and let the flow correspondingly diminish, then such impulses—variations merely of pressure, as it were—can no doubt be transmitted through a wire even if their frequency be many hundreds of millions per second. It would, of course, be out of question to transmit such impulses through a wire immersed in a gaseous medium, even if the wire were provided with a thick and excellent insulation, for most of the energy would be lost in molecular bombardment and consequent heating. The end of the wire connected to the source would be heated, and the remote end would receive but a trifling part of the energy supplied. The prime necessity, then, if such electric impulses are to be used, is to find means to reduce as much as possible the dissipation.

The first thought is, to employ the thinnest possible wire surrounded by the thickest practicable insulation. The next thought is to employ electrostatic screens. The insulation of the wire may be covered with a thin conducting coating and the latter connected to the ground. But this would not do, as then all the energy would pass through the conducting coating to the ground and nothing would get to the end of the wire. If a ground connection is made it can only be made through a conductor offering an enormous impedance, or through a condenser of extremely small capacity. This, however, does not do away with other difficulties.

If the wave length of the impulses is much smaller than the length of the wire, then corresponding short waves will be set up in the conducting coating, and it will be more or less the same as though the coating were directly connected to earth. It is therefore necessary to cut up the coating in sections much shorter than the wave length. Such an arrangement does not still afford a perfect screen, but it is ten thousand times better than none. I think it preferable to cut up the conducting coating in small sections, even if the current waves be much longer than the coating.

If a wire were provided with a perfect electrostatic screen, it would be the same as though all objects were removed from it at infinite distance. The capacity would then be reduced to the capacity of the wire itself, which would be very small. It would then be possible to send over the wire current vibrations of very high frequencies at enormous distances, without affecting greatly the character of the vibrations. A perfect screen is of course out of the question, but I believe that with a screen such as I have just described telephony could be rendered practicable across the Atlantic. According to my ideas, the gutta-percha covered wire should be provided with a third conducting coating subdivided in sections. On the top of this should be again placed a layer of gutta-percha and other insulation, and on the top of the whole the armor. But such cables will not be constructed, for ere long intelligence—transmitted without wires—will throb through the earth like a pulse through a living organism. The wonder is that, with the present state of knowledge and the experiences gained, no attempt is being made to disturb the electrostatic or magnetic condition of the earth, and transmit, if nothing else, intelligence.

It has been my chief aim in presenting these results to point out phenomena or features of novelty, and to advance ideas which I am hopeful will serve as starting points of new departures. It has been my chief desire this evening to entertain you with some novel experiments. Your applause, so frequently and generously accorded, has told me that I have succeeded.

In conclusion, let me thank you most heartily for your kindness and attention, and assure you that the honor I have had in addressing such a distinguished audience, the pleasure I have had in presenting these results to a gathering of so many able men—and among them also some of those in whose work for many years past I have found enlightenment and constant pleasure—I shall never forget.

CHAPTER XXVIII.

On Light and Other High Frequency Phenomena.[1]

INTRODUCTORY.—SOME THOUGHTS ON THE EYE.

When we look at the world around us, on Nature, we are impressed with its beauty and grandeur. Each thing we perceive, though it may be vanishingly small, is in itself a world, that is, like the whole of the universe, matter and force governed by law,—a world, the contemplation of which fills us with feelings of wonder and irresistibly urges us to ceaseless thought and inquiry. But in all this vast world, of all objects our senses reveal to us, the most marvellous, the most appealing to our imagination, appears no doubt a highly developed organism, a thinking being. If there is anything fitted to make us admire Nature's handiwork, it is certainly this inconceivable structure, which performs its innumerable motions of obedience to external influence. To understand its workings, to get a deeper insight into this Nature's masterpiece, has ever been for thinkers a fascinating aim, and after many centuries of arduous research men have arrived at a fair understanding of the functions of its organs and senses. Again, in all the perfect harmony of its parts, of the parts which constitute the material or tangible of our being, of all its organs and senses, the eye is the most wonderful. It is the most precious, the most indispensable of our perceptive or directive organs, it is the great gateway through which all knowledge enters the mind. Of all our organs, it is the one, which is in the

---

1. A lecture delivered before the Franklin Institute, Philadelphia, February, 1893, and before the National Electric Light Association, St. Louis, March, 1893.

most intimate relation with that which we call intellect. So intimate is this relation, that it is often said, the very soul shows itself in the eye.

It can be taken as a fact, which the theory of the action of the eye implies, that for each external impression, that is, for each image produced upon the retina, the ends of the visual nerves, concerned in the conveyance of the impression to the mind, must be under a peculiar stress or in a vibratory state. It now does not seem improbable that, when by the power of thought an image is evoked, a distinct reflex action, no matter how weak, is exerted upon certain ends of the visual nerves, and therefore upon the retina. Will it ever be within human power to analyze the condition of the retina when disturbed by thought or reflex action, by the help of some optical or other means of such sensitiveness, that a clear idea of its state might be gained at any time? If this were possible, then the problem of reading one's thoughts with precision, like the characters of an open book, might be much easier to solve than many problems belonging to the domain of positive physical science, in the solution of which many, if not the majority, of scientific men implicitly believe. Helmholtz, has shown that the fundi of the eye are themselves, luminous, and he was able to *see*, in total darkness, the movement of his arm by the light of his own eyes. This is one of the most remarkable experiments recorded in the history of science, and probably only a few men could satisfactorily repeat it, for it is very likely, that the luminosity of the eyes is associated with uncommon activity of the brain and great imaginative power. It is fluorescence of brain action, as it were.

Another fact having a bearing on this subject which has probably been noted by many, since it is stated in popular expressions, but which I cannot recollect to have found chronicled as a positive result of observation is, that at times, when a sudden idea or image presents itself to the intellect, there is a distinct and sometimes painful sensation of luminosity produced in the eye, observable even in broad daylight.

The saying then, that the soul shows itself in the eye, is deeply founded, and we feel that it expresses a great truth. It has a profound meaning even for one who, like a poet or artist, only following his inborn instinct or love for Nature, finds delight in aimless thoughts and in the mere contemplation of natural phenomena, but a still more profound meaning for one who, in the spirit of positive scientific investigation, seeks to ascertain the causes of the effects. It is principally the natural philospher, the physicist, for whom the eye is the subject of the most intense admiration.

Two facts about the eye must forcibly impress the mind of the physicist, notwithstanding he may think or say that it is an imperfect optical instrument, forgetting, that the very conception of that which is perfect or seems so to him, has been gained through this same instrument. First, the eye is, as far as our positive knowledge goes, the only organ which is *directly* affected by that subtile medium, which as science teaches us, must fill all space ; secondly, it is the most sensitive of our organs, incomparably more sensitive to external impressions than any other.

The organ of hearing implies the impact of ponderable bodies, the organ of smell the transference of detached material particles, and the organs of taste, and of touch or force, the direct contact, or at least some interference of ponderable matter, and this is true even in those instances of animal organisms, in which some of these organs are developed to a degree of truly marvelous perfection. This being so, it seems wonderful that the organ of sight solely should be capable of being stirred by that, which all our other organs are powerless to detect, yet which plays an essential part in all natural phenomena, which transmits all energy and sustains all motion and, that most intricate of all, life, but which has properties such that even a scientifically trained mind cannot help drawing a distinction between it and all that is called matter. Considering merely this, and the fact that the eye, by its marvelous power, widens our otherwise very narrow range of perception far beyond the limits of the small world which is our own, to embrace myriads of other worlds, suns and stars in the infinite depths of the universe, would make it justifiable to assert, that it is an organ of a higher order. Its performances are beyond comprehension. Nature as far as we know never produced anything more wonderful. We can get barely a faint idea of its prodigious power by analyzing what it does and by comparing. When ether waves impinge upon the human body, they produce the sensations of warmth or cold, pleasure or pain, or perhaps other sensations of which we are not aware, and any degree or intensity of these sensations, which degrees are infinite in number, hence an infinite number of distinct sensations. But our sense of touch, or our sense of force, cannot reveal to us these differences in degree or intensity, unless they are very great. Now we can readily conceive how an organism, such as the human, in the eternal process of evolution, or more philosophically speaking, adaptation to Nature, being constrained to the use of only the sense of touch or force, for instance, might develop this sense to such a degree of senstiveness or perfection, that it would be capable of distinguishing the minutest differences in the temperature of a body even at some distance, to a hundredth, or thousandth, or millionth part of a degree. Yet, even this apparently impossible performance would not begin to compare with that of the eye, which is capable of distinguishing and conveying to the mind in a single instant innumerable peculiarities of the body, be it in form, or color, or other respects. This power of the eye rests upon two things, namely, the rectilinear propagation of the disturbance by which it is effected, and upon its sensitiveness. To say that the eye is sensitive is not saying anything. Compared with it, all other organs are monstrously crude. The organ of smell which guides a dog on the trail of a deer, the organ of touch or force which guides an insect in its wanderings, the organ of hearing, which is affected by the slightest disturbances of the air, are sensitive organs, to be sure, but what are they compared with the human eye! No doubt it responds to the faintest echoes or reverberations of the medium; no doubt, it brings us tidings from other worlds, infinitely remote, but in a language we cannot as yet always understand. And why not? Because we live in a medium filled with air and other gases, vapors and a dense mass of solid particles flying about. These play an important part in many phenomena; they fritter away the energy of the vibrations before they can reach the eye; they too, are the carriers of germs of destruction, they get into our lungs and other organs, clog up the channels and imperceptibly, yet inevitably, arrest the stream of life. Could we but do away with all ponderable matter in the line of sight of the telescope, it would reveal to us undreamt of marvels. Even the unaided eye, I think, would be capable of distinguishing in the pure medium, small objects at distances measured probably by hundreds or perhaps thousands of miles.

But there is something else about the eye which impresses us still more than these wonderful features which we observed, viewing it from the standpoint of a physicist, merely as an optical instrument,—something which appeals to us more than its marvelous faculty of being directly affected by the vibrations of the medium, without interference of gross matter, and more than its inconceivable sensitiveness and discerning power. It is its significance in the processes of life. No matter what one's views on nature and life may be, he must stand amazed when, for the first time in his thoughts, he realizes the importance of the eye in the physical processes and mental performances of the human organism. And how could it be otherwise, when he realizes, that the eye is the means through which the human race has acquired the entire knowledge it possesses, that it controls all our motions, more still, all our actions.

There is no way of acquiring knowledge except through the eye. What is the foundation of all philosophical systems of ancient and modern times, in fact, of all the philosophy of man? *I am, I think; I think, therefore I am.* But how could I think and how would I know that I exist, if I had not the eye? For knowledge involves consciousness; consciousness involves ideas, conceptions; conceptions involve pictures or images, and images the sense of vision, and therefore the organ of sight. But how about blind men, will be asked? Yes, a blind man may depict in magnificent poems, forms and scenes from real life, from a world he physically does not see. A blind man may touch the keys of an instrument with unerring precision, may model the fastest boat, may discover and invent, calculate and construct, may do still greater wonders— but all the blind men who have done such things have descended from those who had seeing eyes. Nature may reach the same result in many ways. Like a wave in the physical world, in the infinite ocean of the medium which pervades all, so in the world of organisms, in life, an impulse started proceeds onward, at times, may be, with the speed of light, at times, again, so slowly that for ages and ages it seems to stay, passing through processes of a complexity inconceivable to men, but in all its forms, in all its stages, its energy ever and ever integrally present. A single ray of light from a distant star falling upon the eye of a tyrant in bygone times, may have altered the course of his life, may have changed the destiny of nations, may have transformed the surface of the globe, so intricate, so inconceivably complex are the processes in Nature. In no way can we get such an overwhelming idea of the grandeur of Nature, as when we consider, that in accordance with the law of the conservation of energy, throughout the infinite, the forces are in a perfect balance, and hence the energy of a single thought may determine the motion of a Universe. It is not necessary that every individual, not even that every generation or many generations, should have the physical instrument of sight, in order to be able to form images and to think, that is, form ideas or conceptions; but sometime or other, during the process of evolution, the eye certainly must have existed, else thought, as we understand it, would be impossible; else conceptions, like spirit, intellect, mind, call it as you may, could not exist. It is conceivable, that in some other world, in some other beings, the eye is replaced by a different organ, equally or more perfect, but these beings cannot be men.

Now what prompts us all to voluntary motions and actions of any kind? Again the eye. If I am conscious of the motion, I must have an idea or conception, that is, an image, therefore the eye. If I am not precisely conscious of the motion, it is, because the images are vague or indistinct, being blurred by the superimposition of many. But when I perform the motion, does the impulse which prompts me to the action come from within or from without? The greatest physicists have not disdained to endeavor to answer this and similar questions and have at times abandoned themselves to the delights of pure and unrestrained thought. Such questions are generally considered not to belong to the realm of positive physical science, but will before long be annexed to its domain. Helmholtz has probably thought more on life than any modern scientist. Lord Kelvin expressed his belief that life's process is electrical and that there is a force inherent to the organism and determining its motions. Just as much as I am convinced of any physical truth I am convinced that the motive impulse must come from the outside. For, consider the lowest organism we know—and there are probably many lower ones—an aggregation of a few cells only. If it is capable of voluntary motion it can perform an infinite number of motions, all definite and precise. But now a mechanism consisting of a finite number of parts and few at that, cannot perform an infinite number of definite motions, hence the impulses which govern its movements must come from the environment. So, the atom, the ulterior element of the Universe's structure, is tossed about in space eternally, a play to external influences, like a boat in a troubled sea. Were it to stop its motion *it would die.* Matter at rest, if such a thing could exist, would be matter dead. Death of matter! Never has a sentence of deeper philosophical meaning been uttered. This is the way in which Prof. Dewar forcibly expresses it in the description of his admirable experiments, in which liquid oxygen is handled as one handles water, and air at ordinary pressure is made to condense and even to solidify by the intense cold. Experiments, which serve to illustrate, in his language, the last feeble manifestations of life, the last quiverings of matter about to die. But human eyes shall not witness such death. There is no death of matter, for throughout the infinite universe, all has to move, to vibrate, that is, to live.

I have made the preceding statements at the peril of treading upon metaphysical ground, in my desire to introduce the subject of this lecture in a manner not altogether uninteresting, I may hope, to an audience such as I have the honor to address. But now, then, returning to the subject, this divine organ of sight, this indispensable instrument for thought and all intellectual enjoyment, which lays open to us the marvels of this universe, through which we have acquired what knowledge we possess, and which prompts us to, and controls, all our physical and mental activity. By what is it affected? By light! What is light?

We have witnessed the great strides which have been made in all departments of science in recent years. So great have been the advances that we cannot refrain from asking ourselves, Is this all true, or is it but a dream? Centuries ago men have lived, have thought, discovered, invented, and have believed that they were soaring, while they were merely proceeding at a snail's pace. So we too may be mistaken. But taking the truth of the observed events as one of the implied facts of science, we must rejoice in the immense progress already made and still more in the anticipation of what must come, judging from the possibilities opened up by modern research. There is, however, an advance which we have been witnessing, which must be particularly gratifying to every lover of progress. It is not a discovery, or an invention, or an achievement in any particular direction. It is an advance in all directions of scientific thought and experiment. I mean the generalization of the natural forces and phenomena, the looming up of a certain broad idea on the scientific horizon. It is this idea which has, however, long ago taken possession of the most advanced minds, to which I desire to call your attention, and which I intend to illustrate in a general way, in these experiments, as the first step in answering the question "What is light?" and to realize the modern meaning of this word.

It is beyond the scope of my lecture to dwell upon the subject of light in general, my object being merely to bring presently to your notice a certain class of light effects and a number of phenomena observed in pursuing the study of these effects. But to be consistent in my remarks it is necessary to state that, according to that idea, now accepted by the majority of scientific men as a positive result of theoretical and experimental investigation, the various forms or manifestations of energy which were generally designated as "electric" or more precisely "electromagnetic" are energy manifestations of the same nature as those of radiant heat and light. Therefore the phenomena of light and heat and others besides these, may be called electrical phenomena. Thus electrical science has become the mother science of all and its study has become all important. The day when we shall know exactly what "electricity" is, will chronicle an event probably greater, more important than any other recorded in the history of the human race. The time will come when the comfort, the very existence, perhaps, of man will depend upon that wonderful agent. For our existence and comfort we require heat, light and mechanical power. How do we now get all these? We get them from fuel, we get them by consuming material. What will man do when the forests disappear, when the coal fields are exhausted? Only one thing, according to our present knowledge will remain; that is, to transmit power at great distances. Men will go to the waterfalls, to the tides, which are the stores of an infinitesimal part of Nature's immeasurable energy. There will they harness the energy and transmit the same to their settlements, to warm their homes by, to give them light, and to keep their obedient slaves, the machines, toiling. But how will they transmit this energy if not by electricity? Judge then, if the comfort, nay, the very existence, of man will not depend on electricity. I am aware that this view is not that of a practical engineer, but neither is it that of an illusionist, for it is certain, that power transmission, which at present is merely a stimulus to enterprise, will some day be a dire necessity.

It is more important for the student, who takes up the study of light phenomena, to make himself thoroughly acquainted with certain modern views, than to peruse entire books on the subject of light itself, as disconnected from these views. Were I therefore to make these demonstrations before students seeking information—and for the sake of the few of those who may be present, give me leave to so assume—it would be my principal endeavor to impress these views upon their minds in this series of experiments.

It might be sufficient for this purpose to perform a simple and well-known experiment. I might take a familiar appliance, a Leyden jar, charge it from a frictional machine, and then discharge it. In explaining to you its permanent state when charged, and its transitory condition when discharging, calling your attention to the forces which enter into play and to the various phenomena they produce, and pointing out the relation of the forces and phenomena, I might fully succeed in illustrating that modern idea. No doubt, to the thinker, this simple experiment would appeal as much as the most magnificent display. But this is to be an experimental demonstration, and one which should possess, besides instructive, also entertaining features and as such, a simple experiment, such as the one cited, would not go very far towards the attainment of the lecturer's aim. I must therefore choose another way of illustrating, more spectacular certainly, but perhaps also more instructive. Instead of the frictional machine and Leyden jar, I shall avail myself in these experiments, of an induction coil of peculiar properties, which was described in detail by me in a lecture before the London Institution of Electrical Engineers, in Feb., 1892. This induction coil is capable of yielding currents of enormous potential differences, alternating with extreme rapidity. With this apparatus I shall endeavor to show you three distinct classes of effects, or phenomena, and it is my desire that each experiment, while serving for the purposes of illustration, should at the same time teach us some novel truth, or show us some novel aspect of this fascinating science. But before doing this, it seems proper and useful to dwell upon the apparatus employed, and method of obtaining the high potentials and high-frequency currents which are made use of in these experiments.

ON THE APPARATUS AND METHOD OF CONVERSION.

These high-frequency currents are obtained in a peculiar manner. The method employed was advanced by me about two years ago in an experimental lecture before the American Institute of Electrical Engineers. A number of ways, as practiced in the laboratory, of obtaining these currents either from continuous or low frequency alternating currents, is diagramatically indicated in Fig. 165, which will be later described in detail. The general

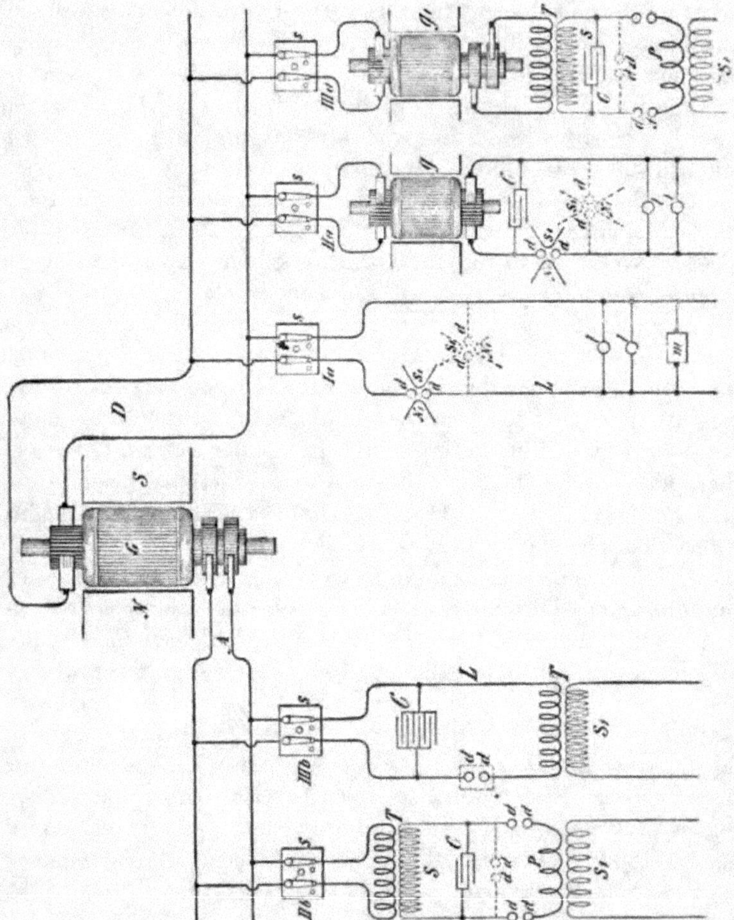

Fig. 165.

plan is to charge condensers, from a direct or alternate-current
source, preferably of high-tension, and to discharge them
disruptively while observing well-known conditions necessary to maintain the oscillations of the current.   In view of the
general interest taken in high-frequency currents and effects producible by them, it seems to me advisable to dwell at some length
upon this method of conversion.   In order to give you a clear
idea of the action, I will suppose that a continuous-current generator is employed, which is often very convenient. It is desirable
that the generator should possess such high tension as to be able
to break through a small air space. If this is not the case, then
auxiliary means have to be resorted to, some of which will be indicated subsequently. When the condensers are charged to a
certain potential, the air, or insulating space, gives way and a disruptive discharge occurs. There is then a sudden rush of current
and generally a large portion of accumulated electrical energy
spends itself. The condensers are thereupon quickly charged and
the same process is repeated in more or less rapid succession.
To produce such sudden rushes of current it is necessary to observe certain conditions. If the rate at which the condensers are
discharged is the same as that at which they are charged, then,
clearly, in the assumed case the condensers do not come into
play. If the rate of discharge be smaller than the rate of charging, then, again, the condensers cannot play an important part.
But if, on the contrary, the rate of discharging is greater than
that of charging, then a succession of rushes of current is obtained. It is evident that, if the rate at which the energy is
dissipated by the discharge is very much greater than the rate of
supply to the condensers, the sudden rushes will be comparatively few, with long-time intervals between. This always occurs
when a condenser of considerable capacity is charged by means
of a comparatively small machine. If the rates of supply and
dissipation are not widely different, then the rushes of current
will be in quicker succession, and this the more, the more nearly
equal both the rates are, until limitations incident to each case
and depending upon a number of causes are reached. Thus we
are able to obtain from a continuous-current generator as rapid a
succession of discharges as we like. Of course, the higher the
tension of the generator, the smaller need be the capacity of the
condensers, and for this reason, principally, it is of advantage to
employ a generator of very high tension. Besides, such a generator permits the attaining of greater rates of vibration.

The rushes of current may be of the same direction under the conditions before assumed, but most generally there is an oscillation superimposed upon the fundamental vibration of the current. When the conditions are so determined that there are no oscillations, the current impulses are unidirectional and thus a means is provided of transforming a continuous current of high tension, into a direct current of lower tension, which I think may find employment in the arts.

This method of conversion is exceedingly interesting and I was much impressed by its beauty when I first conceived it. It is ideal in certain respects. It involves the employment of no mechanical devices of any kind, and it allows of obtaining currents of any desired frequency from an ordinary circuit, direct or alternating. The frequency of the fundamental discharges depending on the relative rates of supply and dissipation can be readily varied within wide limits, by simple adjustments of these quantities, and the frequency of the superimposed vibration by the determination of the capacity, self-induction and resistance of the circuit. The potential of the currents, again, may be raised as high as any insulation is capable of withstanding safely by combining capacity and self-induction or by induction in a secondary, which need have but comparatively few turns.

As the conditions are often such that the intermittence or oscillation does not readily establish itself, especially when a direct current source is employed, it is of advantage to associate an interrupter with the arc, as I have, some time ago, indicated the use of an air-blast or magnet, or other such device readily at hand. The magnet is employed with special advantage in the conversion of direct currents, as it is then very effective. If the primary source is an alternate current generator, it is desirable, as I have stated on another occasion, that the frequency should be low, and that the current forming the arc be large, in order to render the magnet more effective.

A form of such discharger with a magnet which has been found convenient, and adopted after some trials, in the conversion of direct currents particularly, is illustrated in Fig. 166. N s are the pole pieces of a very strong magnet which is excited by a coil c. The pole pieces are slotted for adjustment and can be fastened in any position by screws $s\ s_1$. The discharge rods $d\ d_1$, thinned down on the ends in order to allow a closer approach of the magnetic pole pieces, pass through the columns of brass $b\ b_1$ and are fastened in position by screws $s_2\ s_2$. Springs $r\ r_1$ and collars $c\ c_1$

are slipped on the rods, the latter serving to set the points of the
rods at [a certain suitable distance by means of screws s_3 s_3, and
the former to draw the points apart.　When it is desired to start
the arc, one of the large rubber handles h h_1 is tapped quickly
with the hand, whereby the points of the rods are brought in
contact but are instantly separated by the springs r r_1.　Such an
arrangement has been found to be often necessary, namely in
cases when the E. M. F. was not large enough to cause the discharge
to break through the gap, and also when it was desirable to avoid
short circuiting of the generator by the metallic contact of the
rods.　The rapidity of the interruptions of the current with a
magnet depends on the intensity of the magnetic field and on the

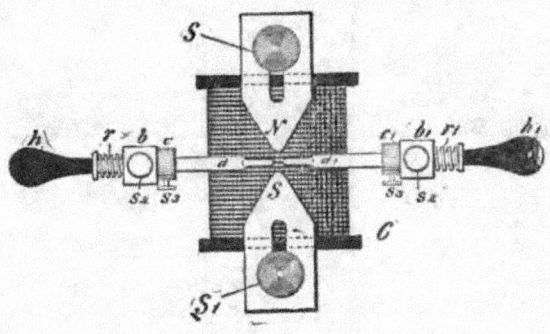

Fig. 166.

potential difference at the end of the arc.　The interruptions are
generally in such quick succession as to produce a musical sound.
Years ago it was observed that when a powerful induction coil
is discharged between the poles of a strong magnet, the discharge
produces a loud noise, not unlike a small pistol shot.　It was
vaguely stated that the spark was intensified by the presence of
the magnetic field.　It is now clear that the discharge current,
flowing for some time, was interrupted a great number of times
by the magnet, thus producing the sound.　The phenomenon is
especially marked when the field circuit of a large magnet or
dynamo is broken in a powerful magnetic field.

When the current through the gap is comparatively large, it is of advantage to slip on the points of the discharge rods pieces of very hard carbon and let the arc play between the carbon pieces. This preserves the rods, and besides has the advantage of keeping the air space hotter, as the heat is not conducted away as quickly through the carbons, and the result is that a smaller E. M. F. in the arc gap is required to maintain a succession of discharges.

Another form of discharger, which may be employed with advantage in some cases, is illustrated in Fig. 167. In this form the discharge rods d d_1 pass through perforations in a wooden

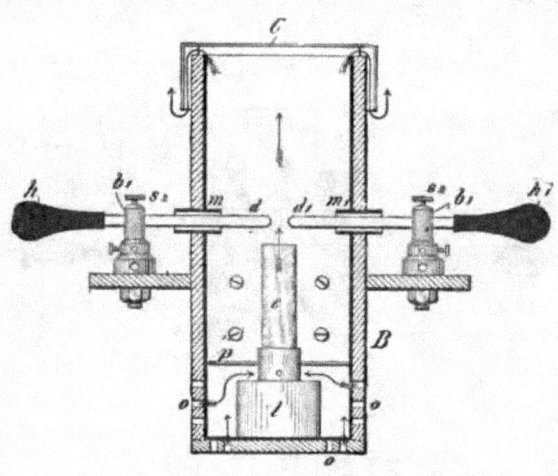

Fig. 167.

box B, which is thickly coated with mica on the inside, as indicated by the heavy lines. The perforations are provided with mica tubes m m_1 of some thickness, which are preferably not in contact with the rods d d_1. The box has a cover c which is a little larger and descends on the outside of the box. The spark gap is warmed by a small lamp l contained in the box. A plate p above the lamp allows the draught to pass only through the chimney e of the lamp, the air entering through holes o o in or near the bottom of the box and following the path indicated by the arrows. When the discharger is in operation, the door of the box is closed so that the light of the arc is not visible outside.

It is desirable to exclude the light as perfectly as possible, as it interferes with some experiments. This form of discharger is simple and very effective when properly manipulated. The air being warmed to a certain temperature, has its insulating power impaired; it becomes dielectrically weak, as it were, and the consequence is that the arc can be established at much greater distance. The arc should, of course, be sufficiently insulating to allow the discharge to pass through the gap *disruptively*. The arc formed under such conditions, when long, may be made extremely sensitive, and the weak draught through the lamp chimney *c* is quite sufficient to produce rapid interruptions. The adjustment is made by regulating the temperature and velocity of the draught. Instead of using the lamp, it answers the purpose to provide for a draught of warm air in other ways. A very simple way which has been practiced is to enclose the arc in a long vertical tube, with plates on the top and bottom for regulating the temperature and velocity of the air current. Some provision had to be made for deadening the sound.

The air may be rendered dielectrically weak also by rarefaction. Dischargers of this kind have likewise been used by me in connection with a magnet. A large tube is for this purpose provided with heavy electrodes of carbon or metal, between which the discharge is made to pass, the tube being placed in a powerful magnetic field. The exhaustion of the tube is carried to a point at which the discharge breaks through easily, but the pressure should be more than 75 millimetres, at which the ordinary thread discharge occurs. In another form of discharger, combining the features before mentioned, the discharge was made to pass between two adjustable magnetic pole pieces, the space between them being kept at an elevated temperature.

It should be remarked here that when such, or interrupting devices of any kind, are used and the currents are passed through the primary of a disruptive discharge coil, it is not, as a rule, of advantage to produce a number of interruptions of the current per second greater than the natural frequency of vibration of the dynamo supply circuit, which is ordinarily small. It should also be pointed out here, that while the devices mentioned in connection with the disruptive discharge are advantageous under certain conditions, they may be sometimes a source of trouble, as they produce intermittences and other irregularities in the vibration which it would be very desirable to overcome.

There is, I regret to say, in this beautiful method of conversion a defect, which fortunately is not vital, and which I have been gradually overcoming. I will best call attention to this defect and indicate a fruitful line of work, by comparing the electrical process with its mechanical analogue. The process may be illustrated in this manner. Imagine a tank with a wide opening at the bottom, which is kept closed by spring pressure, but so that it snaps off *suddenly* when the liquid in the tank has reached a certain height. Let the fluid be supplied to the tank by means of a pipe feeding at a certain rate. When the critical height of the liquid is reached, the spring gives way and the bottom of the tank drops out. Instantly the liquid falls through the wide opening, and the spring, reasserting itself, closes the bottom again. The tank is now filled, and after a certain time interval the same process is repeated. It is clear, that if the pipe feeds the fluid quicker than the bottom outlet is capable of letting it pass through, the bottom will remain off and the tank will still overflow. If the rates of supply are exactly equal, then the bottom lid will remain partially open and no vibration of the same and of the liquid column will generally occur, though it might, if started by some means. But if the inlet pipe does not feed the fluid fast enough for the outlet, then there will be always vibration. Again, in such case, each time the bottom flaps up or down, the spring and the liquid column, if the pliability of the spring and the inertia of the moving parts are properly chosen, will perform independent vibrations. In this analogue the fluid may be likened to electricity or electrical energy, the tank to the condenser, the spring to the dielectric, and the pipe to the conductor through which electricity is supplied to the condenser. To make this analogy quite complete it is necessary to make the assumption, that the bottom, each time it gives way, is knocked violently against a non-elastic stop, this impact involving some loss of energy; and that, besides, some dissipation of energy results due to frictional losses. In the preceding analogue the liquid is supposed to be under a steady pressure. If the presence of the fluid be assumed to vary rhythmically, this may be taken as corresponding to the case of an alternating current. The process is then not quite as simple to consider, but the action is the same in principle.

It is desirable, in order to maintain the vibration economically, to reduce the impact and frictional losses as much as possible.

As regards the latter, which in the electrical analogue correspond to the losses due to the resistance of the circuits, it is impossible to obviate them entirely, but they can be reduced to a minimum by a proper selection of the dimensions of the circuits and by the the employment of thin conductors in the form of strands. But the loss of energy caused by the first breaking through of the dielectric—which in the above example corresponds to the violent knock of the bottom against the inelastic stop—would be more important to overcome. At the moment of the breaking through, the air space has a very high resistance, which is probably reduced to a very small value when the current has reached some strength, and the space is brought to a high temperature. It would materially diminish the loss of energy if the space were always kept at an extremely high temperature, but then there would be no disruptive break. By warming the space moderately by means of a lamp or otherwise, the economy as far as the arc is concerned is sensibly increased. But the magnet or other interrupting device does not diminish the loss in the arc. Likewise, a jet of air only facilitates the carrying off of the energy. Air, or a gas in general, behaves curiously in this respect. When two bodies charged to a very high potential, discharge disruptively through an air space, any amount of energy may be carried off by the air. This energy is evidently dissipated by bodily carriers, in impact and collisional losses of the molecules. The exchange of the molecules in the space occurs with inconceivable rapidity. A powerful discharge taking place between two electrodes, they may remain entirely cool, and yet the loss in the air may represent any amount of energy. It is perfectly practicable, with very great potential differences in the gap, to dissipate several horse-power in the arc of the discharge without even noticing a small increase in the temperature of the electrodes. All the frictional losses occur then practically in the air. If the exchange of the air molecules is prevented, as by enclosing the air hermetically, the gas inside of the vessel is brought quickly to a high temperature, even with a very small discharge. It is difficult to estimate how much of the energy is lost in sound waves, audible or not, in a powerful discharge. When the currents through the gap are large, the electrodes may become rapidly heated, but this is not a reliable measure of the energy wasted in the arc, as the loss through the gap itself may be comparatively small. The air or a gas in general is, at ordinary pressure at least, clearly not the best medium through which a disruptive discharge should occur. Air or other gas under great pressure is of course a much more suitable medium for the discharge gap. I have carried on long-continued experiments in this direction, unfortunately less practicable on account of the difficulties and expense in getting air under great pressure. But even if the medium in the discharge space is solid or liquid, still the same losses take place, though they are generally smaller, for just as soon as the arc is established, the solid or liquid is volatilized. Indeed, there is no body known which would not be disintegrated by the arc, and it is an open question among scientific men, whether an arc discharge could occur at all in the air itself without the particles of the electrodes being torn off. When the current through the gap is very small and the arc very long, I believe that a relatively considerable amount of heat is taken up in the disintegration of the electrodes, which partially on this account may remain quite cold.

The ideal medium for a discharge gap should only *crack*, and the ideal electrode should be of some material which cannot be disintegrated. With small currents through the gap it is best to employ aluminum, but not when the currents are large. The disruptive break in the air, or more or less in any ordinary medium, is not of the nature of a crack, but it is rather comparable to the piercing of innumerable bullets through a mass offering great frictional resistances to the motion of the bullets, this involving considerable loss of energy. A medium which would merely crack when strained electrostatically—and this possibly might be the case with a perfect vacuum, that is, pure ether—would involve a very small loss in the gap, so small as to be entirely negligible, at least theoretically, because a crack may be produced by an infinitely small displacement. In exhausting an oblong bulb provided with two aluminum terminals, with the greatest care, I have succeeded in producing such a vacuum that the secondary discharge of a disruptive discharge coil would break disruptively through the bulb in the form of fine spark streams. The curious point was that the discharge would completely ignore the terminals and start far behind the two aluminum plates which served as electrodes. This extraordinary high vacuum could only be maintained for a very short while. To return to the ideal medium, think, for the sake of illustration, of a piece of glass or similar body clamped in a vice, and the latter tightened more and more. At a certain point a minute increase of the pressure will cause the glass to crack. The loss of energy involved in splitting the glass may be practically nothing, for though the force is great, the displacement need be but extremely small. Now imagine that the glass would possess the property of closing again perfectly the crack upon a minute diminution of the pressure. This is the way the dielectric in the discharge space should behave. But inasmuch as there would be always some loss in the gap, the medium, which should be continuous, should exchange through the gap at a rapid rate. In the preceding example, the glass being perfectly closed, it would mean that the dielectric in the discharge space possesses a great insulating power; the glass being cracked, it would signify that the medium in the space is a good conductor. The dielectric should vary enormously in resistance by minute variations of the E. M. F. across the discharge space. This condition is attained, but in an extremely imperfect manner, by warming the air space to a certain critical temperature, dependent on the E. M. F. across the gap, or by otherwise impairing the insulating power of the air. But as a matter of fact the air does never break down *disruptively*, if this term be rigorously interpreted, for before the sudden rush of the current occurs, there is always a weak current preceding it, which rises first gradually and then with comparative suddenness. That is the reason why the rate of change is very much greater when glass, for instance, is broken through, than when the break takes place through an air space of equivalent dielectric strength. As a medium for the discharge space, a solid, or even a liquid, would be preferable therefor. It is somewhat difficult to conceive of a solid body which would possess the property of closing instantly after it has been cracked. But a liquid, especially under great pressure, behaves practically like a solid, while it possesses the property of closing the crack. Hence it was thought that a liquid insulator might be more suitable as a dielectric than air. Following out this idea, a number of different forms of dischargers in which a variety of such insulators, sometimes under great pressure, were employed, have been experimented upon. It is thought sufficient to dwell in a few words upon one of the forms experimented upon. One of these dischargers is illustrated in Figs. 168*a* and 168*b*.

A hollow metal pulley P (Fig. 168*a*), was fastened upon an arbor *a*, which by suitable means was rotated at a considerable

speed. On the inside of the pulley, but disconnected from the
same, was supported a thin disc h (which is shown thick for the
sake of clearness), of hard rubber in which there were embedded
two metal segments s s with metallic extensions e e into which
were screwed conducting terminals t t covered with thick tubes
of hard rubber t t. The rubber disc h with its metallic segments
s s, was finished in a lathe, and its entire surface highly polished
so as to offer the smallest possible frictional resistance to the motion through a fluid. In the hollow of the pulley an insulating
liquid such as a thin oil was poured so as to reach very nearly to
the opening left in the flange f, which was screwed tightly on the
front side of the pulley. The terminals t t, were connected to the
opposite coatings of a battery of condensers so that the discharge
occurred through the liquid. When the pulley was rotated, the
liquid was forced against the rim of the pulley and considerable
fluid pressure resulted. In this simple way the discharge gap

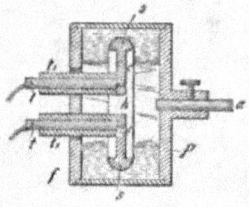 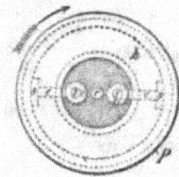

Fig. 168a.   Fig. 168b.

was filled with a medium which behaved practically like a solid,
which possessed the quality of closing instantly upon the occurrence of the break, and which moreover was circulating through
the gap at a rapid rate. Very powerful effects were produced by
discharges of this kind with liquid interrupters, of which a number of different forms were made. It was found that, as expected, a longer spark for a given length of wire was obtainable
in this way than by using air as an interrupting device. Generally the speed, and therefore also the fluid pressure, was limited
by reason of the fluid friction, in the form of discharger described,
but the practically obtainable speed was more than sufficient to
produce a number of breaks suitable for the circuits ordinarily
used. In such instances the metal pulley P was provided with a
few projections inwardly, and a definite number of breaks was
then produced which could be computed from the speed of

rotation of the pulley. Experiments were also carried on with liquids of different insulating power with the view of reducing the loss in the arc. When an insulating liquid is moderately warmed, the loss in the arc is diminished.

A point of some importance was noted in experiments with various discharges of this kind. It was found, for instance, that whereas the conditions maintained in these forms were favorable for the production of a great spark length, the current so obtained was not best suited to the production of light effects. Experience undoubtedly has shown, that for such purposes a harmonic rise and fall of the potential is preferable. Be it that a solid is rendered incandescent, or phosphorescent, or be it that energy is transmitted by condenser coating through the glass, it is quite certain that a harmonically rising and falling potential produces less destructive action, and that the vacuum is more permanently maintained. This would be easily explained if it were ascertained that the process going on in an exhausted vessel is of an electrolytic nature.

In the diagrammatical sketch, Fig. 165, which has been already referred to, the cases which are most likely to be met with in practice are illustrated. One has at his disposal either direct or alternating currents from a supply station. It is convenient for an experimenter in an isolated laboratory to employ a machine G, such as illustrated, capable of giving both kinds of currents. In such case it is also preferable to use a machine with multiple circuits, as in many experiments it is useful and convenient to have at one's disposal currents of different phases. In the sketch, D represents the direct and A the alternating circuit. In each of these, three branch circuits are shown, all of which are provided with double line switches s s s s s. Consider first the direct current conversion; 1a represents the simplest case. If the E. M. F. of the generator is sufficient to break through a small air space, at least when the latter is warmed or otherwise rendered poorly insulating, there is no difficulty in maintaining a vibration with fair economy by judicious adjustment of the capacity, self-induction and resistance of the circuit L containing the devices l l m. The magnet N, s, can be in this case advantageously combined with the air space. The discharger d d with the magnet may be placed either way, as indicated by the full or by the dotted lines. The circuit 1a with the connections and devices is supposed to possess dimensions such as are suitable for the maintenance of a vibration. But usually the E. M. F. on the circuit or branch 1*a* will be something like a 100 volts or so, and in this case it is not sufficient to break through the gap. Many different means may be used to remedy this by raising the E. M. F. across the gap. The simplest is probably to insert a large self-induction coil in series with the circuit L. When the arc is established, as by the discharger illustrated in Fig. 166, the magnet blows the arc out the instant it is formed. Now the extra current of the break, being of high E. M. F., breaks through the gap, and a path of low resistance for the dynamo current being again provided, there is a sudden rush of current from the dynamo upon the weakening or subsidence of the extra current. This process is repeated in rapid succession, and in this manner I have maintained oscillation with as low as 50 volts, or even less, across the gap. But conversion on this plan is not to be recommended on account of the too heavy currents through the gap and consequent heating of the electrodes; besides, the frequencies obtained in this way are low, owing to the high self-induction necessarily associated with the circuit. It is very desirable to have the E. M. F. as high as possible, first, in order to increase the economy of the conversion, and, secondly, to obtain high frequencies. The difference of potential in this electric oscillation is, of course, the equivalent of the stretching force in the mechanical vibration of the spring. To obtain very rapid vibration in a circuit of some inertia, a great stretching force or difference of potential is necessary. Incidentally, when the E. M. F. is very great, the condenser which is usually employed in connection with the circuit need but have a small capacity, and many other advantages are gained. With a view of raising the E. M. F. to a many times greater value than obtainable from ordinary distribution circuits, a rotating transformer *g* is used, as indicated at 11*a*, Fig. 165, or else a separate high potential machine is driven by means of a motor operated from the generator G. The latter plan is in fact preferable, as changes are easier made. The connections from the high tension winding are quite similar to those in branch 1*a* with the exception that a condenser c, which should be adjustable, is connected to the high tension circuit. Usually, also, an adjustable self-induction coil in series with the circuit has been employed in these experiments. When the tension of the currents is very high, the magnet ordinarily used in connection with the discharger is of comparatively small value, as it is quite easy to adjust the dimensions of the circuit so that oscillation is maintained. The employment of a steady E. M. F. in the high frequency conversion affords some advantages over the employment of alternating E. M. F., as the adjustments are much simpler and the action can be easier controlled. But unfortunately one is limited by the obtainable potential difference. The winding also breaks down easily in consequence of the sparks which form between the sections of the armature or commutator when a vigorous oscillation takes place. Besides, these transformers are expensive to build. It has been found by experience that it is best to follow the plan illustrated at IIIa. In this arrangement a rotating transformer g, is employed to convert the low tension direct currents into low frequency alternating currents, preferably also of small tension. The tension of the currents is then raised in a stationary transformer T. The secondary s of this transformer is connected to an adjustable condenser c which discharges through the gap or discharger $d\,d$, placed in either of the ways indicated, through the primary P of a disruptive discharge coil, the high frequency current being obtained from the secondary s of this coil, as described on previous occasions. This will undoubtedly be found the cheapest and most convenient way of converting direct currents.

The three branches of the circuit A represent the usual cases met in practice when alternating currents are converted. In Fig. 1b a condenser c., generally of large capacity, is connected to the circuit L containing the devices $l\,l$, $m\,m$. The devices mm are supposed to be of high self-induction so as to bring the frequency of the circuit more or less to that of the dynamo. In this instance the discharger $d\,d$ should best have a number of makes and breaks per second equal to twice the frequency of the dynamo. If not so, then it should have at least a number equal to a multiple or even fraction of the dynamo frequency. It should be observed, referring to 1b, that the conversion to a high potential is also effected when the discharger $d\,d$, which is shown in the sketch, is omitted. But the effects which are produced by currents which rise instantly to high values, as in a disruptive discharge, are entirely different from those produced by dynamo currents which rise and fall harmonically. So, for instance, there might be in a given case a number of makes and breaks at $d\,d$ equal to just twice the frequency of the dynamo, or in other words, there may be the same number of fundamental oscillations as would be produced without the discharge gap, and there might even not be any quicker superimposed vibration; yet the differences of potential at the various points of the circuit, the impedance and other phenomena, dependent upon the rate of change, will bear no similarity in the two cases. Thus, when working with currents discharging disruptively, the element chiefly to be considered is not the frequency, as a student might be apt to believe, but the rate of change per unit of time. With low frequencies in a certain measure the same effects may be obtained as with high frequencies, provided the rate of change is sufficiently great. So if a low frequency current is raised to a potential of, say, 75,000 volts, and the high tension current passed through a series of high resistance lamp filaments, the importance of the rarefied gas surrounding the filament is clearly noted, as will be seen later; or, if a low frequency current of several thousand amperes is passed through a metal bar, striking phenomena of impedance are observed, just as with currents of high frequencies. But it is, of course, evident that with low frequency currents it is impossible to obtain such rates of change per unit of time as with high frequencies, hence the effects produced by the latter are much more prominent. It is deemed advisable to make the preceding remarks, inasmuch as many more recently described effects have been unwittingly identified with high frequencies. Frequency alone in reality does not mean anything, except when an undisturbed harmonic oscillation is considered.

In the branch III*b* a similar disposition to that in I*b* is illustrated, with the difference that the currents discharging through the gap *d d* are used to induce currents in the secondary s of a transformer T. In such case the secondary should be provided with an adjustable condenser for the purpose of tuning it to the primary.

II*b* illustrates a plan of alternate current high frequency conversion which is most frequently used and which is found to be most convenient. This plan has been dwelt upon in detail on previous occasions and need not be described here.

Some of these results were obtained by the use of a high frequency alternator. A description of such machines will be found in my original paper before the American Institute of Electrical Engineers, and in periodicals of that period, notably in THE ELECTRICAL ENGINEER of March 18, 1891.

I will now proceed with the experiments.

ON PHENOMENA PRODUCED BY ELECTROSTATIC FORCE.

The first class of effects I intend to show you are effects produced by electrostatic force. It is the force which governs the the motion of the atoms, which causes them to collide and develop the life-sustaining energy of heat and light, and which causes them to aggregate in an infinite variety of ways, according to Nature's fanciful designs, and to form all these wondrous structures we perceive around us; it is, in fact, if our present views be true, the most important force for us to consider in Nature. As the term *electrostatic* might imply a steady electric condition, it should be remarked, that in these experiments the force is not constant, but varies at a rate which may be considered moderate, about one million times a second, or thereabouts. This enables me to produce many effects which are not producible with an unvarying force.

When two conducting bodies are insulated and electrified, we say that an electrostatic force is acting between them. This force manifests itself in attractions, repulsions and stresses in the bodies and space or medium without. So great may be the strain exerted in the air, or whatever separates the two conducting bodies, that it may break down, and we observe sparks or bundles of light or streamers, as they are called. These streamers form abundantly when the force through the air is rapidly varying. I will illustrate this action of electrostatic force in a novel experiment in which I will employ the induction coil before referred to. The coil is contained in a trough filled with oil, and placed under the table. The two ends of the secondary wire pass through the two thick columns of hard rubber which protrude to some height above the table. It is necessary to insulate the ends or terminals of the secondary heavily with hard rubber, because even dry wood is by far too poor an insulator for these currents of enormous potential differences. On one of the terminals of the coil, I have placed a large sphere of sheet brass, which is connected to a larger insulated brass plate, in order to enable me to perform the experiments under conditions, which, as you will see, are more suitable for this experiment. I now set the coil to work and approach the free terminal with a metallic object held in my hand, this simply to avoid burns. As I approach the metallic object to a distance of eight or ten inches, a torrent of furious sparks breaks forth from the end of the secondary wire, which passes through the rubber column. The sparks cease when the metal in my hand touches the wire. My arm is now traversed by a powerful electric current, vibrating at about the rate of one million times a second. All around me the electrostatic force makes itself felt, and the air molecules and particles of dust flying about are acted upon and are hammering violently against my body. So great is this agitation of the particles, that when the lights are turned out you may see streams of feeble light appear on some parts of my body. When such a streamer breaks out on any part of the body, it produces a sensation like the pricking of a needle. Were the potentials sufficiently high and the frequency of the vibration rather low, the skin would probably be ruptured under the tremendous strain, and the blood would rush out with great force in the form of fine spray or jet so thin as to be invisible, just as oil will when placed on the positive terminal of

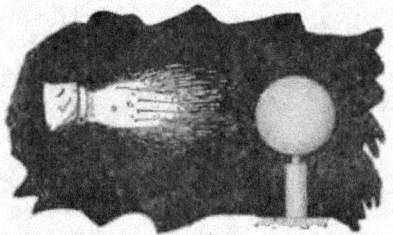

FIG. 169.

a Holtz machine. The breaking through of the skin though it may seem impossible at first, would perhaps occur, by reason of the tissues under the skin being incomparably better conducting. This, at least, appears plausible, judging from some observations.

I can make these streams of light visible to all, by touching with the metallic object one of the terminals as before, and approaching my free hand to the brass sphere, which is connected to the second terminal of the coil. As the hand is approached, the air between it and the sphere, or in the immediate neighborhood, is more violently agitated, and you see streams of light now break forth from my finger tips and from the whole hand (Fig. 169). Were I to approach the hand closer, powerful sparks would jump from the brass sphere to my hand, which might be injurious. The streamers offer no particular inconvenience, except that in the ends of the finger tips a burning sensation is felt. They should not be confounded with those produced by an influence machine, because in many respects they behave differently. I have attached the brass sphere and plate to one of the terminals in order to prevent the formation of visible streamers on that terminal, also in order to prevent sparks from jumping at a considerable distance. Besides, the attachment is favorable for the working of the coil.

The streams of light which you have observed issuing from my hand are due to a potential of about 200,000 volts, alternating in rather irregular intervals, sometimes like a million times a second. A vibration of the same amplitude, but four times as fast, to maintain which over 3,000,000 volts would be required, would be more than sufficient to envelop my body in a complete sheet of flame. But this flame would not burn me up; quite contrarily, the probability is that I would not be injured in the least. Yet a hundredth part of that energy, otherwise directed, would be amply sufficient to kill a person.

The amount of energy which may thus be passed into the body of a person depends on the frequency and potential of the currents, and by making both of these very great, a vast amount of energy may be passed into the body without causing any discomfort, except perhaps, in the arm, which is traversed by a true conduction current. The reason why no pain in the body is felt, and no injurious effect noted, is that everywhere, if a current be imagined to flow through the body, the direction of its flow would be at right angles to the surface; hence the body of the experimenter offers an enormous section to the current, and the density is very small, with the exception of the arm, perhaps, where the density may be considerable. But if only a small fraction of that energy would be applied in such a way that a current would traverse the body in the same manner as a low frequency current, a shock would be received which might be fatal. A direct or low frequency alternating current is fatal, I think, principally because its distribution through the body is not uniform, as it must divide itself in minute streamlets of great density, whereby some organs are vitally injured. That such a process occurs I have not the least doubt, though no evidence might apparently exist, or be found upon examination. The surest to injure and destroy life, is a continuous current, but the most painful is an alternating current of very low frequency. The expression of these views, which are the result of long continued experiment and observation, both with steady and varying currents, is elicited by the interest which is at present taken in this subject, and by the manifestly erroneous ideas which are daily propounded in journals on this subject.

I may illustrate an effect of the electrostatic force by another striking experiment, but before, I must call your attention to one or two facts. I have said before, that when the medium between two oppositely electrified bodies is strained beyond a certain limit it gives way and, stated in popular language, the opposite electric charges unite and neutralize each other. This breaking down of the medium occurs principally when the force acting between the bodies is steady, or varies at a moderate rate. Were the variation sufficiently rapid, such a destructive break would not occur, no matter how great the force, for all the energy would be spent in radiation, convection and mechanical and chemical action. Thus the *spark* length, or greatest distance which a *spark* will jump between the electrified bodies is the

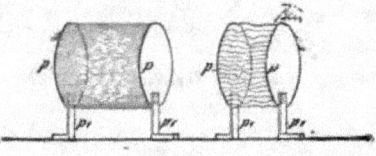

Fig. 170a.   Fig. 170b.

smaller, the greater the variation or time rate of change. But this rule may be taken to be true only in a general way, when comparing rates which are widely different.

I will show you by an experiment the difference in the effect produced by a rapidly varying and a steady or moderately varying force. I have here two large circular brass plates *p p* (Fig. 170a and Fig. 170b), supported on movable insulating stands on the table, connected to the ends of the secondary of a coil similar to the one used before. I place the plates ten or twelve inches apart and set the coil to work. You see the whole space between the plates, nearly two cubic feet, filled with uniform light, Fig. 170a. This light is due to the streamers you have seen in the first experiment, which are now much more intense. I have already pointed out the importance of these streamers in commercial apparatus and their still greater importance in some purely scientific investigations. Often they are too weak to be visible, but

they always exist, consuming energy and modifying the action
of the apparatus. When intense, as they are at present, they
produce ozone in great quantity, and also, as Professor Crookes
has pointed out, nitrous acid. So quick is the chemical action that
if a coil, such as this one, is worked for a very long time it will
make the atmosphere of a small room unbearable, for the eyes
and throat are attacked. But when moderately produced, the
streamers refresh the atmosphere wonderfully, like a thunder-
storm, and exercises unquestionably a beneficial effect.

In this experiment the force acting between the plates changes
in intensity and direction at a very rapid rate. I will now make
the rate of change per unit time much smaller. This I effect by
rendering the discharges through the primary of the induction
coil less frequent, and also by diminishing the rapidity of the vibration in the secondary. The former result is conveniently secured by lowering the E. M. F. over the air gap in the primary
circuit, the latter by approaching the two brass plates to a distance of about three or four inches. When the coil is set to work,
you see no streamers or light between the plates, yet the medium
between them is under a tremendous strain. I still further augment the strain by raising the E. M. F. in the primary circuit, and
soon you see the air give way and the hall is illuminated by a
shower of brilliant and noisy sparks, Fig. 170b. These sparks could
be produced also with unvarying force; they have been for many
years a familiar phenomenon, though they were usually obtained
from an entirely different apparatus. In describing these two
phenomena so radically different in appearance, I have advisedly
spoken of a " force " acting between the plates. It would be in
accordance with accepted views to say, that there was an " alternating E. M. F," acting between the plates. This term is quite
proper and applicable in all cases where there is evidence of at
least a possibility of an essential inter-dependence of the electric
state of the plates, or electric action in their neighborhood. But
if the plates were removed to an infinite distance, or if at a finite
distance, there is no probability or necessity whatever for such
dependence. I prefer to use the term " electrostatic force," and
to say that such a force is acting around each plate or electrified insulated body in general. There is an inconvenience in using this
expression as the term incidentally means a steady electric condition; but a proper nomenclature will eventually settle this difficulty.

I now return to the experiment to which I have already alluded, and with which I desire to illustrate a striking effect produced by a rapidly varying electrostatic force. I attach to the end of the wire, l (Fig. 171), which is in connection with one of the terminals of the secondary of the induction coil, an exhausted bulb b. This bulb contains a thin carbon filament f, which is fastened to a platinum wire w, sealed in the glass and leading outside of the bulb, where it connects to the wire l. The bulb may be exhausted to any degree attainable with ordinary apparatus. Just a moment before, you have witnessed the breaking down of the air between the charged brass plates. You know that a plate of glass, or any other insulating material, would break down in like manner. Had I therefore a metallic coating attached to the outside of the bulb, or placed near the same, and

FIG. 171. FIG. 172 a. FIG. 172 b.

were this coating connected to the other terminal of the coil, you would be prepared to see the glass give way if the strain were sufficiently increased. Even were the coating not connected to the other terminal, but to an insulated plate, still, if you have followed recent developments, you would naturally expect a rupture of the glass.

But it will certainly surprise you to note that under the action of the varying electrostatic force, the glass gives way when all other bodies are removed from the bulb. In fact, all the surrounding bodies we perceive might be removed to an infinite distance without affecting the result in the slightest. When the coil is set to work, the glass is invariably broken through at the seal, or other narrow channel, and the vacuum is quickly impaired.

Such a damaging break would not occur with a steady force, even if the same were many times greater. The break is due to the agitation of the molecules of the gas within the bulb, and outside of the same. This agitation, which is generally most violent in the narrow pointed channel near the seal, causes a heating and rupture of the glass. This rupture, would, however, not occur, not even with a varying force, if the medium filling the inside of the bulb, and that surrounding it, were perfectly homogeneous. The break occurs much quicker if the top of the bulb is drawn out into a fine fibre. In bulbs used with these coils such narrow, pointed channels must therefore be avoided.

When a conducting body is immersed in air, or similar insulating medium, consisting of, or containing, small freely movable particles capable of being electrified, and when the electrification of the body is made to undergo a very rapid change—which is equivalent to saying that the electrostatic force acting around the body is varying in intensity,—the small particles are attracted and repelled, and their violent impacts against the body may cause a mechanical motion of the latter. Phenomena of this kind are noteworthy, inasmuch as they have not been observed before with apparatus such as has been commonly in use. If a very light conducting sphere be suspended on an exceedingly fine wire, and charged to a steady potential, however high, the sphere will remain at rest. Even if the potential would be rapidly varying, provided that the small particles of matter, molecules or atoms, are evenly distributed, no motion of the sphere should result. But if one side of the conducting sphere is covered with a thick insulating layer, the impacts of the particles will cause the sphere to move about, generally in irregular curves, Fig. 172a. In like manner, as I have shown on a previous occasion, a fan of sheet metal, Fig. 172b, covered partially with insulating material as indicated, and placed upon the terminal of the coil so as to turn freely on it, is spun around.

All these phenomena you have witnessed and others which will be shown later, are due to the presence of a medium like air, and would not occur in a continuous medium. The action of the air may be illustrated still better by the following experiment. I take a glass tube t, Fig. 173, of about an inch in diameter, which has a platinum wire w sealed in the lower end, and to which is attached a thin lamp filament f. I connect the wire with the terminal of the coil and set the coil to work. The

platinum wire is now electrified positively and negatively
in rapid succession and the wire and air inside of the tube
is rapidly heated by the impacts of the particles, which may be
so violent as to render the filament incandescent. But if I pour
oil in the tube, just as soon as the wire is covered with the oil,
all action apparently ceases and there is no marked evidence of
heating. The reason of this is that the oil is a practically continuous medium. The displacements in such a continuous medium
are, with these frequencies, to all appearance incomparably
smaller than in air, hence the work performed in such a medium
is insignificant. But oil would behave very differently with frequencies many times as great, for even though the displacements

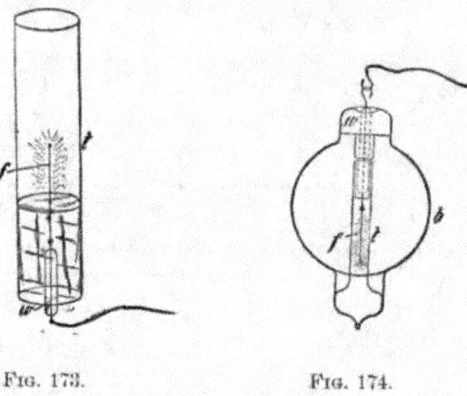

FIG. 173.   FIG. 174.

be small, if the frequency were much greater, considerable work
might be performed in the oil.

The electrostatic attractions and repulsions between bodies of
measurable dimensions are, of all the manifestations of this force,
the first so-called *electrical* phenomena noted. But though they
have been known to us for many centuries, the precise nature of
the mechanism concerned in these actions is still unknown to us,
and has not been even quite satisfactorily explained. What kind
of mechanism must that be ? We cannot help wondering when
we observe two magnets attracting and repelling each other with
a force of hundreds of pounds with apparently nothing between
them. We have in our commercial dynamos magnets capable of
sustaining in mid-air tons of weight. But what are even these

forces acting between magnets when compared with the tremendous attractions and repulsions produced by electrostatic force, to which there is apparently no limit as to intensity. In lightning discharges bodies are often charged to so high a potential that they are thrown away with inconceivable force and torn asunder or shattered into fragments. Still even such effects cannot compare with the attractions and repulsions which exist between charged molecules or atoms, and which are sufficient to project them with speeds of many kilometres a second, so that under their violent impact bodies are rendered highly incandescent and are volatilized. It is of special interest for the thinker who inquires into the nature of these forces to note that whereas the actions between individual molecules or atoms occur seemingly under any conditions, the attractions and repulsions of bodies of measurable dimensions imply a medium possessing insulating properties. So, if air, either by being rarefied or heated, is rendered more or less conducting, these actions between two electrified bodies practically cease, while the actions between the individual atoms continue to manifest themselves.

An experiment may serve as an illustration and as a means of bringing out other features of interest. Some time ago I showed that a lamp filament or wire mounted in a bulb and connected to one of the terminals of a high tension secondary coil is set spinning, the top of the filament generally describing a circle. This vibration was very energetic when the air in the bulb was at ordinary pressure and became less energetic when the air in the bulb was strongly compressed. It ceased altogether when the air was exhausted so as to become comparatively good conducting. I found at that time that no vibration took place when the bulb was very highly exhausted. But I conjectured that the vibration which I ascribed to the electrostatic action between the walls of the bulb and the filament should take place also in a highly exhausted bulb. To test this under conditions which were more favorable, a bulb like the one in Fig. 174, was constructed. It comprised a globe b, in the neck of which was sealed a platinum wire w carrying a thin lamp filament f. In the lower part of the globe a tube t was sealed so as to surround the filament. The exhaustion was carried as far as it was practicable with the apparatus employed.

This bulb verified my expectation, for the filament was set spinning when the current was turned on, and became incandescent. It also showed another interesting feature, bearing upon the preceding remarks, namely, when the filament had been kept incandescent some time, the narrow tube and the space inside were brought to an elevated temperature, and as the gas in the tube then became conducting, the electrostatic attraction between the glass and the filament became very weak or ceased, and the filament came to rest. When it came to rest it would glow far more intensely. This was probably due to its assuming the position in the centre of the tube where the molecular bombardment was most intense, and also partly to the fact that the individual impacts were more violent and that no part of the supplied energy was converted into mechanical movement. Since, in accordance with accepted views, in this experiment the incandescence must be attributed to the impacts of the particles, molecules or atoms in the heated space, these particles must therefore, in order to explain such action, be assumed to behave as independent carriers of electric charges immersed in an insulating medium; yet there is no attractive force between the glass tube and the filament because the space in the tube is, as a whole, conducting.

It is of some interest to observe in this connection that whereas the attraction between two electrified bodies may cease owing to the impairing of the insulating power of the medium in which they are immersed, the repulsion between the bodies may still be observed. This may be explained in a plausible way. When the bodies are placed at some distance in a poorly conducting medium, such as slightly warmed or rarefied air, and are suddenly electrified, opposite electric charges being imparted to them, these charges equalize more or less by leakage through the air. But if the bodies are similarly electrified, there is less opportunity afforded for such dissipation, hence the repulsion observed in such case is greater than the attraction. Repulsive actions in a gaseous medium are however, as Prof. Crookes has shown, enhanced by molecular bombardment.

ON CURRENT OR DYNAMIC ELECTRICITY PHENOMENA.

So far, I have considered principally effects produced by a varying electrostatic force in an insulating medium, such as air. When such a force is acting upon a conducting body of measurable dimensions, it causes within the same, or on its surface, displacements of the electricity and gives rise to electric currents, and these produce another kind of phenomena, some of which I

shall presently endeavor to illustrate.   In presenting this second
class of electrical effects, I will avail myself principally of such
as are producible without any return circuit, hoping to interest
you the more by presenting these phenomena in a more or less
novel aspect.

It has been a long time customary, owing to the limited
experience with vibratory currents, to consider an electric current as something circulating in a closed conducting path.   It
was astonishing at first to realize that a current may flow through
the conducting path even if the latter be interrupted, and it
was still more surprising to learn, that sometimes it may be
even easier to make a current flow under such conditions
than through a closed path.   But that old idea is gradually dis-
appearing, even among practical men, and will soon be entirely
forgotten.

If I connect an insulated metal plate P, Fig. 175, to one of the
terminals T of the induction coil by means of a wire, though this

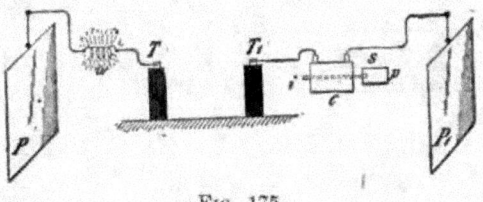

FIG. 175.

plate be very well insulated, a current passes through the
wire when the coil is set to work.   First I wish to give you
evidence that there *is* a current passing through the connecting
wire.   An obvious way of demonstrating this is to insert between
the terminal of the coil and the insulated plate a very thin platinum or german silver wire *w* and bring the latter to incandescence or fusion by the current.   This requires a rather large plate
or else current impulses of very high potential and frequency.
Another way is to take a coil c, Fig. 175, containing many turns of
thin insulated wire and to insert the same in the path of the current to the plate. When I connect one of the ends of the coil to the
wire leading to another insulated plate P_1, and its other end to the
terminal T_1 of the induction coil, and set the latter to work, a current passes through the inserted coil c and the existence of the
current may be made manifest in various ways.   For instance, I

insert an iron core i within the coil. The current being one of
very high frequency, will, if it be of some strength, soon bring the
iron core to a noticeably higher temperature, as the hysteresis and
current losses are great with such high frequencies. One might
take a core of some size, laminated or not, it would matter little ;
but ordinary iron wire $\frac{1}{16}$th or $\frac{1}{8}$th of an inch thick is suitable
for the purpose. While the induction coil is working, a current
traverses the inserted coil and only a few moments are sufficient
to bring the iron wire i to an elevated temperature sufficient to
soften the sealing-wax s, and cause a paper washer p fastened by
it to the iron wire to fall off. But with the apparatus such as I
have here, other, much more interesting, demonstrations of this
kind can be made. I have a secondary s, Fig 176, of coarse wire,
wound upon a coil similar to the first. In the preceding experiment the current through the coil c, Fig. 175, was very small, but
there being many turns a strong heating effect was, nevertheless,

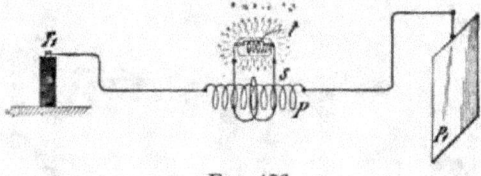

Fig. 176.

produced in the iron wire. Had I passed that current through a
conductor in order to show the heating of the latter, the current
might have been too small to produce the effect desired. But with
this coil provided with a secondary winding, I can now transform
the feeble current of high tension which passes through the primary P into a strong secondary current of low tension, and this
current will quite certainly do what I expect. In a small glass
tube (t, Fig. 176), I have enclosed a coiled platinum wire, w, this
merely in order to protect the wire. On each end of the glass
tube is sealed a terminal of stout wire to which one of the ends of
the platinum wire w, is connected. I join the terminals of the
secondary coil to these terminals and insert the primary p,
between the insulated plate P_1, and the terminal T_1, of the induction coil as before. The latter being set to work, instantly the
platinum wire w is rendered incandescent and can be fused, even
if it be very thick.

Instead of the platinum wire I now take an ordinary 50-volt 16 c. p. lamp. When I set the induction coil in operation the lamp filament is brought to high incandescence. It is, however, not necessary to use the insulated plate, for the lamp (*l*, Fig. 177) is rendered incandescent even if the plate r_1 be disconnected. The secondary may also be connected to the primary as indicated by the dotted line in Fig. 177, to do away more or less with the electrostatic induction or to modify the action otherwise.

I may here call attention to a number of interesting observations with the lamp. First, I disconnect one of the terminals of the lamp from the secondary s. When the induction coil plays, a glow is noted which fills the whole bulb. This glow is due to electrostatic induction. It increases when the bulb is grasped with the hand, and the capacity of the experimenter's body thus added to the secondary circuit. The secondary, in effect, is equivalent to a metallic coating, which would be placed near the pri-

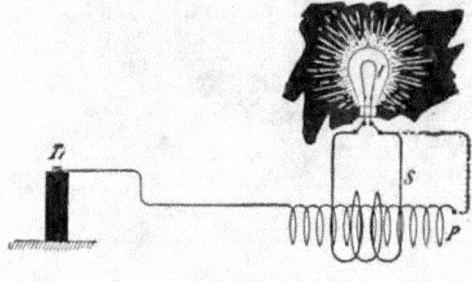

FIG. 177.

mary. If the secondary, or its equivalent, the coating, were placed symmetrically to the primary, the electrostatic induction would be nil under ordinary conditions, that is, when a primary return circuit is used, as both halves would neutralize each other. The secondary *is* in fact placed symmetrically to the primary, but the action of both halves of the latter, when only one of its ends is connected to the induction coil, is not exactly equal; hence electrostatic induction takes place, and hence the glow in the bulb. I can nearly equalize the action of both halves of the primary by connecting the other, free end of the same to the insulated plate, as in the preceding experiment. When the plate is connected, the glow disappears. With a smaller plate it would not entirely disappear and then it would contribute to the brightness of the filament when the secondary is closed, by warming the air in the bulb.

To demonstrate another interesting feature, I have adjusted
the coils used in a certain way. I first connect both the terminals
of the lamp to the secondary, one end of the primary being connected to the terminal T_1 of the induction coil and the other to
the insulated plate P_1 as before. When the current is turned on,
the lamp glows brightly, as shown in Fig. 178*b*, in which c is a fine
wire coil and s a coarse wire secondary wound upon it. If the
insulated plate P_1 is disconnected, leaving one of the ends *a* of the

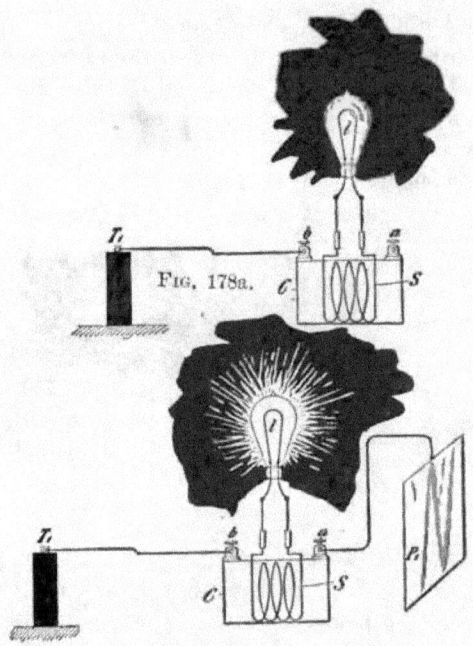

FIG. 178a.

FIG. 178b.

primary insulated, the filament becomes dark or generally it diminishes in brightness (Fig. 178*a*). Connecting again the plate P_1
and raising the frequency of the current, I make the filament
quite dark or barely red (Fig. 179*b*). Once more I will disconnect the plate. One will of course infer that when the plate is
disconnected, the current through the primary will be weakened,
that therefore the E. M. F. will fall in the secondary s, and that
the brightness of the lamp will diminish. This might be the
case and the result can be secured by an easy adjustment of the

coils; also by varying the frequency and potential of the currents. But it is perhaps of greater interest to note, that the lamp increases in brightness when the plate is disconnected (Fig. 179*a*). In this case all the energy the primary receives is now sunk into it, like the charge of a battery in an ocean cable, but most of that energy is recovered through the secondary and used to light the lamp. The current traversing the primary is strongest at the end *b* which is connected to the terminal T_1 of the induction coil, and

FIG. 179a.

FIG 179b.

diminishes in strength towards the remote end *a*. But the dynamic inductive effect exerted upon the secondary s is now greater than before, when the suspended plate was connected to the primary. These results might have been produced by a number of causes. For instance, the plate P_1 being connected, the reaction from the coil c may be such as to diminish the potential at the terminal T_1 of the induction coil, and therefore weaken the current through the primary of the coil c. Or the disconnecting of the plate may diminish the capacity effect with relation to the primary of the latter coil to such an extent that the current through it is diminished, though the potential at the terminal T_1 of the induction coil may be the same or even higher. Or the result might have been produced by the change of phase of the primary and secondary currents and consequent reaction. But the chief determining factor is the relation of the self-induction and capacity of coil c and plate P_1 and the frequency of the currents. The greater brightness of the filament in Fig. 179a, is, however, in part due to the heating of the rarefied gas in the lamp by electrostatic induction, which, as before remarked, is greater when the suspended plate is disconnected.

Still another feature of some interest I may here bring to your attention. When the insulated plate is disconnected and the secondary of the coil opened, by approaching a small object to the secondary, but very small sparks can be drawn from it, showing that the electrostatic induction is small in this case. But upon the secondary being closed upon itself or through the lamp, the filament glowing brightly, strong sparks are obtained from the secondary. The electrostatic induction is now much greater, because the closed secondary determines a greater flow of current through the primary and principally through that half of it which is connected to the induction coil. If now the bulb be grasped with the hand, the capacity of the secondary with reference to the primary is augmented by the experimenter's body and the luminosity of the filament is increased, the incandescence now being due partly to the flow of current through the filament and partly to the molecular bombardment of the rarefied gas in the bulb.

The preceding experiments will have prepared one for the next following results of interest, obtained in the course of these investigations. Since I can pass a current through an insulated wire merely by connecting one of its ends to the source of electrical energy, since I can induce by it another current, magnetize an iron core, and, in short, perform all operations as though a return circuit were used, clearly I can also drive a motor by the aid of only one wire. On a former occasion I have described a simple form of motor comprising a single exciting coil, an iron core and disc. Fig. 180 illustrates a modified way of operating such an alternate current motor by currents induced in a transformer connected to one lead, and several other arrangements of circuits for operating a certain class of alternating motors founded on the action of currents of differing phase. In view of the present state of the art it is thought sufficient to describe these arrangements in a few words only. The diagram, Fig. 180 II., shows a primary coil P, connected with one of its ends to the line L leading from a high tension transformer terminal T_1. In inductive relation to this primary P is a secondary s of coarse wire in the circuit of which is a coil c. The currents induced in the secondary energize the iron core i, which is preferably, but not necessarily, subdivided, and set the metal disc d in rotation. Such a motor M_2 as diagramatically shown in Fig. 180 II., has been called a "magnetic lag motor," but this expression may be objected to by those who attribute the rotation of the disc to eddy currents circulating in minute paths when the core i is finally subdivided. In order to operate such a motor effectively on the plan indicated, the frequencies should not be too high, not more than four or five thousand, though the rotation is produced even with ten thousand per second, or more.

In Fig. 180 I., a motor M_1 having two energizing circuits, A and B, is diagrammatically indicated. The circuit A is connected to the line L and in series with it is a primary P, which may have its free end connected to an insulated plate P_1, such connection being indicated by the dotted lines. The other motor circuit B is connected to the secondary s which is in inductive relation to the primary P. When the transformer terminal T_1 is alternately electrified, currents traverse the open line L and also circuit A and primary P. The currents through the latter induce secondary currents in the circuit s, which pass through the energizing coil B of the motor. The currents through the secondary s and those through the primary P differ in phase 90 degrees, or nearly so, and are capable of rotating an armature placed in inductive relation to the circuits A and B.

In Fig. 180 III., a similar motor M_3 with two energizing circuits A_1 and B_1 is illustrated. A primary P, connected with one of its ends to the line L has a secondary s, which is preferably wound for a tolerably high E. M. F., and to which the two energizing circuits of the motor are connected, one directly to the ends of the secondary and the other through a condenser c, by the action of which the currents traversing the circuit A_1 and B_1 are made to differ in phase.

In Fig. 180 IV., still another arrangement is shown. In this case two primaries P_1 and P_2 are connected to the line L, one

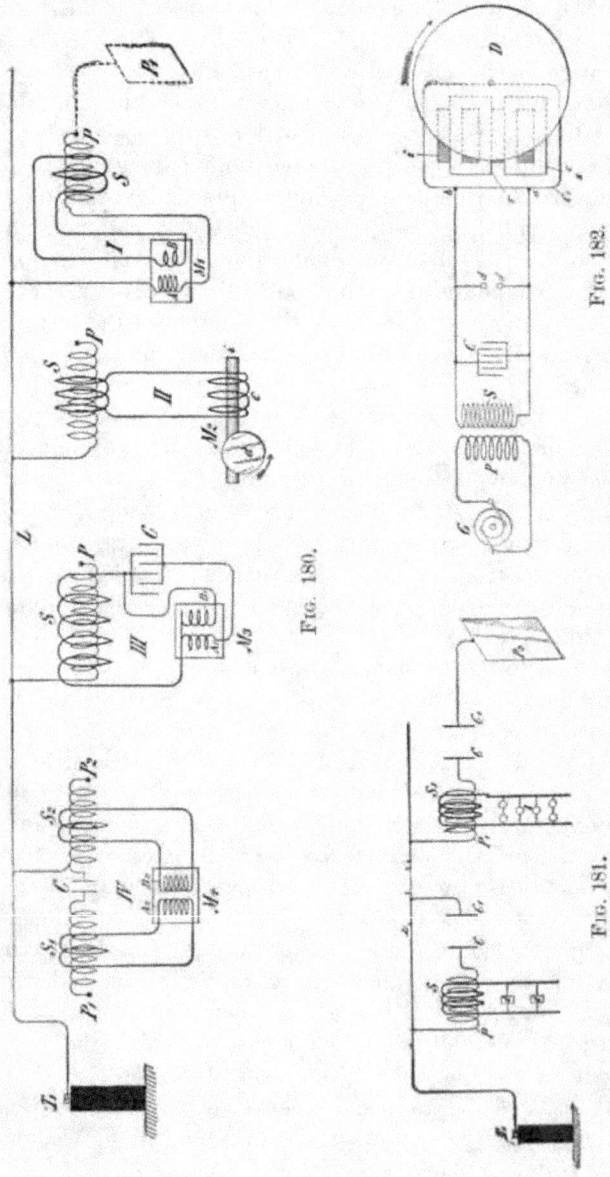

Fig. 180.

Fig. 181.

Fig. 182.

through a condenser c of small capacity, and the other directly. The primaries are provided with secondaries s_1 and s_2 which are in series with the energizing circuits, A_2 and B_2 and a motor M_3, the condenser c again serving to produce the requisite difference in the phase of the currents traversing the motor circuits. As such phase motors with two or more circuits are now well known in the art, they have been here illustrated diagrammatically. No difficulty whatever is found in operating a motor in the manner indicated, or in similar ways; and although such experiments up to this day present only scientific interest, they may at a period not far distant, be carried out with practical objects in view.

It is thought useful to devote here a few remarks to the subject of operating devices of all kinds by means of only one leading wire. It is quite obvious, that when high-frequency currents are made use of, ground connections are—at least when the E. M. F. of the currents is great—better than a return wire. Such ground connections are objectionable with steady or low frequency currents on account of destructive chemical actions of the former and disturbing influences exerted by both on the neighboring circuits; but with high frequencies these actions practically do not exist. Still, even ground connections become superfluous when the E. M. F. is very high, for soon a condition is reached, when the current may be passed more economically through open, than through closed, conductors. Remote as might seem an industrial application of such single wire transmission of energy to one not experienced in such lines of experiment, it will not seem so to anyone who for some time has carried on investigations of such nature. Indeed I cannot see why such a plan should not be practicable. Nor should it be thought that for carrying out such a plan currents of very high frequency are expressly required, for just as soon as potentials of say 30,000 volts are used, the single wire transmission may be effected with low frequencies, and experiments have been made by me from which these inferences are made.

When the frequencies are very high it has been found in laboratory practice quite easy to regulate the effects in the manner shown in diagram Fig. 181. Here two primaries P and P_1 are shown, each connected with one of its ends to the line L and with the other end to the condenser plates c and c, respectively. Near these are placed other condenser plates c_1 and c_1, the former being connected to the line L and the latter to an insulated larger plate p_2. On the primaries are wound secondaries s and s_1, of coarse wire, connected to the devices d and l respectively. By varying the distances of the condenser plates c and c_1, and c and c_1 the currents through the secondaries s and s_1 are varied in intensity. The curious feature is the great sensitiveness, the slightest change in the distance of the plates producing considerable variations in the intensity or strength of the currents. The sensitiveness may be rendered extreme by making the frequency such, that the primary itself, without any plate attached to its free end, satisfies, in conjunction with the closed secondary, the condition of resonance. In such condition an extremely small change in the capacity of the free terminal produces great variations. For instance, I have been able to adjust the conditions so that the mere approach of a person to the coil produces a considerable change in the brightness of the lamps attached to the secondary. Such observations and experiments possess, of course, at present, chiefly scientific interest, but they may soon become of practical importance.

Very high frequencies are of course not practicable with motors on account of the necessity of employing iron cores. But one may use sudden discharges of low frequency and thus obtain certain advantages of high-frequency currents without rendering the iron core entirely incapable of following the changes and without entailing a very great expenditure of energy in the core. I have found it quite practicable to operate with such low frequency disruptive discharges of condensers, alternating-current motors. A certain class of such motors which I advanced a few years ago, which contain closed secondary circuits, will rotate quite vigorously when the discharges are directed through the exciting coils. One reason that such a motor operates so well with these discharges is that the difference of phase between the primary and secondary currents is 90 degrees, which is generally not the case with harmonically rising and falling currents of low frequency. It might not be without interest to show an experiment with a simple motor of this kind, inasmuch as it is commonly thought that disruptive discharges are unsuitable for such purposes. The motor is illustrated in Fig. 182. It comprises a rather large iron core i with slots on the top into which are embedded thick copper washers c c. In proximity to the core is a freely-movable metal disc p. The core is provided with a primary exciting coil c_1 the ends a and b of which are connected to the terminals of the secondary s of an ordinary transformer, the primary p of the latter being connected to an alternating distribution circuit or generator g of low or moderate frequency. The terminals of the secondary s are attached to a condenser c which discharges through an air gap $d\,d$ which may be placed in series or shunt to the coil c_1. When the conditions are properly chosen the disc d rotates with considerable effort and the iron core i does not get very perceptibly hot. With currents from a high-frequency alternator, on the contrary, the core gets rapidly hot and the disc rotates with a much smaller effort. To perform the experiment properly it should be first ascertained that the disc d is not set in rotation when the discharge is not occurring at $d\,d$. It is preferable to use a large iron core and a condenser of large capacity so as to bring the superimposed quicker oscillation to a very low pitch or to do away with it entirely. By observing certain elementary rules I have also found it practicable to operate ordinary series or shunt direct-current motors with such disruptive discharges, and this can be done with or without a return wire.

IMPEDANCE PHENOMENA.

Among the various current phenomena observed, perhaps the most interesting are those of impedance presented by conductors to currents varying at a rapid rate. In my first paper before the American Institute of Electrical Engineers, I have described a few striking observations of this kind. Thus I showed that when such currents or sudden discharges are passed through a thick metal bar there may be points on the bar only a few inches apart, which have a sufficient potential difference between them to maintain at bright incandescence an ordinary filament lamp. I have also described the curious behavior of rarefied gas surrounding a conductor, due to such sudden rushes of current. These phenomena have since been more carefully studied and one or two novel experiments of this kind are deemed of sufficient interest to be described here.

Referring to Fig. 183a, b and b_1 are very stout copper bars connected at their lower ends to plates c and c_1, respectively, of a condenser, the opposite plates of the latter being connected to the terminals of the secondary s of a high-tension transformer, the primary p of which is supplied with alternating currents from an ordinary low-frequency dynamo g or distribution circuit. The condenser discharges through an adjustable gap *d d* as usual. By establishing a rapid vibration it was found quite easy to perform the following curious experiment. The bars B and B_1 were joined at the top by a low-voltage lamp l_3; a little lower was placed by means of clamps *c c*, a 50-volt lamp l_2; and still lower another 100-volt lamp l_1; and finally, at a certain distance below the latter lamp, an exhausted tube T. By carefully determining the positions of these devices it was found practicable to maintain them

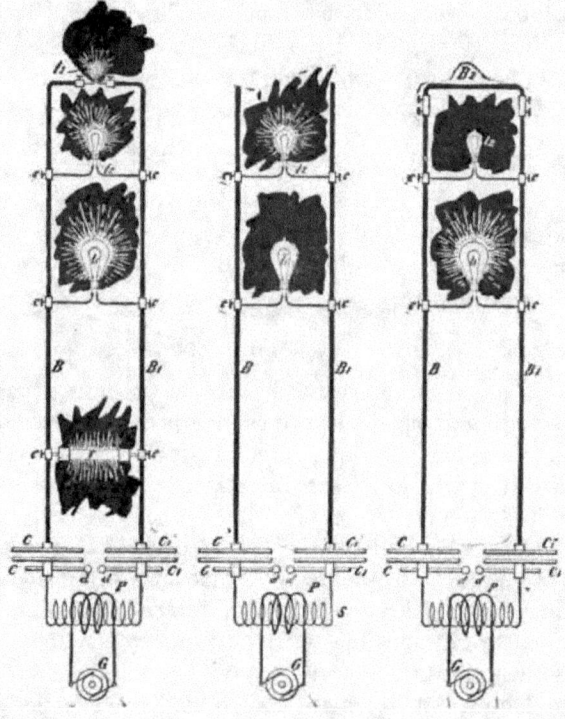

Figs. 183a, 183b and 183c.

all at their proper illuminating power. Yet they were all connected in multiple arc to the two stout copper bars and required widely different pressures. This experiment requires of course some time for adjustment but is quite easily performed.

In Figs. 183*b* and 183*c*, two other experiments are illustrated which, unlike the previous experiment, do not require very careful adjustments. In Fig. 183*b*, two lamps, l_1 and l_2, the former a

100-volt and the latter a 50-volt are placed in certain positions as indicated, the 100-volt lamp being below the 50-volt lamp.   When the arc is playing at $d\,d$ and the sudden discharges are passed through the bars B B₁, the 50-volt lamp will, as a rule, burn brightly, or at least this result is easily secured, while the 100-volt lamp will burn very low or remain quite dark, Fig. 183b.   Now the bars B B₁ may be joined at the top by a thick cross bar B₂ and it is quite easy to maintain the 100-volt lamp at full candle-power while the 50-volt lamp remains dark, Fig. 183c.   These results, as I have pointed out previously, should not be considered to be due exactly to frequency but rather to the time rate of change which may be great, even with low frequencies.   A great many other results of the same kind, equally interesting, especially to those who are only used to manipulate steady currents, may be obtained and they afford precious clues in investigating the nature of electric currents.

In the preceding experiments I have already had occasion to show some light phenomena and it would now be proper to study these in particular; but to make this investigation more complete I think it necessary to make first a few remarks on the subject of electrical resonance which has to be always observed in carrying out these experiments.

ON ELECTRICAL RESONANCE.

The effects of resonance are being more and more noted by engineers and are becoming of great importance in the practical operation of apparatus of all kinds with alternating currents.   A few general remarks may therefore be made concerning these effects. It is clear, that if we succeed in employing the effects of resonance practically in the operation of electric devices the return wire will, as a matter of course, become unnecessary, for the electric vibration may be conveyed with one wire just as well as, and sometimes even better than, with two.   The question first to answer is, then, whether pure resonance effects are producible.   Theory and experiment both show that such is impossible in Nature, for as the oscillation becomes more and more vigorous, the losses in the vibrating bodies and environing media rapidly increase and necessarily check the vibration which otherwise would go on increasing forever.   It is a fortunate circumstance that pure resonance is not producible, for if it were there is no telling what dangers might not lie in wait for the innocent experimenter.   But to a certain degree resonance is producible, the magnitude of the effects being limited by the imperfect conductivity and imperfect elasticity of the media or, generally stated, by frictional losses. The smaller these losses, the more striking are the effects. The same is the case in mechanical vibration. A stout steel bar may be set in vibration by drops of water falling upon it at proper intervals; and with glass, which is more perfectly elastic, the resonance effect is still more remarkable, for a goblet may be burst by singing into it a note of the proper pitch. The electrical resonance is the more perfectly attained, the smaller the resistance or the impedance of the conducting path and the more perfect the dielectric. In a Leyden jar discharging through a short stranded cable of thin wires these requirements are probably best fulfilled, and the resonance effects are therefore very prominent. Such is not the case with dynamo machines, transformers and their circuits, or with commercial apparatus in general in which the presence of iron cores complicates the action or renders it impossible. In regard to Leyden jars with which resonance effects are frequently demonstrated, I would say that the effects observed are often *attributed* but are seldom *due* to true resonance, for an error is quite easily made in this respect. This may be undoubtedly demonstrated by the following experiment. Take, for instance, two large insulated metallic plates or spheres which I shall designate A and B; place them at a certain small distance apart and charge them from a frictional or influence machine to a potential so high that just a slight increase of the difference of potential between them will cause the small air or insulating space to break down. This is easily reached by making a few preliminary trials. If now another plate—fastened on an insulating handle and connected by a wire to one of the terminals of a high tension secondary of an induction coil, which is maintained in action by an alternator (preferably high frequency)—is approached to one of the charged bodies A or B, so as to be nearer to either one of them, the discharge will invariably occur between them; at least it will, if the potential of the coil in connection with the plate is sufficiently high. But the explanation of this will soon be found in the fact that the approached plate acts inductively upon the bodies A and B and causes a spark to pass between them. When this spark occurs, the charges which were previously imparted to these bodies from the influence machine, must needs be lost, since the bodies are brought in electrical connection through the arc formed. Now this arc is formed whether there be resonance or not. But even if the spark would not be produced, still there is an alternating E. M. F. set up between the bodies when the plate is brought near one of them; therefore the approach of the plate, if it *does* not always actually, will, at any rate, *tend* to break down the air space by inductive action. Instead of the spheres or plates A and B we may take the coatings of a Leyden jar with the same result, and in place of the machine,—which is a high frequency alternator preferably, because it is more suitable for the experiment and also for the argument,—we may take another Leyden jar or battery of jars. When such jars are discharging through a circuit of low resistance the same is traversed by currents of very high frequency. The plate may now be connected to one of the coatings of the second jar, and when it is brought near to the first jar just previously charged to a high potential from an influence machine, the result is the same as before, and the first jar will discharge through a small air space upon the second being caused to discharge. But both jars and their circuits need not be tuned any closer than a basso profundo is to the note produced by a mosquito, as small sparks will be produced through the air space, or at least the latter will be considerably more strained owing to the setting up of an alternating E. M. F. by induction, which takes place when one of the jars begins to discharge. Again another error of a similar nature is quite easily made. If the circuits of the two jars are run parallel and close together, and the experiment has been performed of discharging one by the other, and now a coil of wire be added to one of the circuits whereupon the experiment does not succeed, the conclusion that this is due to the fact that the circuits are now not tuned, would be far from being safe. For the two circuits act as condenser coatings and the addition of the coil to one of them is equivalent to bridging them, at the point where the coil is placed, by a small condenser, and the effect of the latter might be to prevent the spark from jumping through the discharge space by diminishing the alternating E. M. F. acting across the same. All these remarks, and many more which might be added but for fear of wandering too far from the subject, are made with the pardonable intention of cautioning the unsuspecting student, who might gain an entirely unwarranted opinion of his skill at seeing every experiment succeed; but they are in no way thrust upon the experienced as novel observations.

In order to make reliable observations of electric resonance effects it is very desirable, if not necessary, to employ an alternator giving currents which rise and fall harmonically, as in working with make and break currents the observations are not always trustworthy, since many phenomena, which depend on the rate of change, may be produced with widely different frequencies. Even when making such observations with an alternator one is apt to be mistaken. When a circuit is connected to an alternator there are an indefinite number of values for capacity and self-induction which, in conjunction, will satisfy the condition of resonance. So there are in mechanics an infinite number of tuning forks which will respond to a note of a certain pitch, or loaded springs which have a definite period of vibration. But the resonance will be most perfectly attained in that case in which the motion is effected with the greatest freedom. Now in mechanics, considering the vibration in the common medium—that is, air—it is of comparatively little importance whether one tuning fork be somewhat larger than another, because the losses in the air are not very considerable. One may, of course, enclose a tuning fork in an exhausted vessel and by thus reducing the air resistance to a minimum obtain better resonant action. Still the difference would not be very great. But it would make a great difference if the tuning fork were immersed in mercury. In the electrical vibration it is of enormous importance to arrange the conditions so that the vibration is effected with the greatest freedom. The magnitude of the resonance effect depends, under otherwise equal conditions, on the quantity of electricity set in motion or on the strength of the current driven through the circuit. But the circuit opposes the passage of the currents by reason of its impedance and therefore, to secure the best action it is necessary to reduce the impedance to a minimum. It is impossible to overcome it entirely, but merely in part, for the ohmic resistance cannot be overcome. But when the frequency of the impulses is very great, the flow of the current is practically determined by self-induction. Now self-induction can be overcome by combining it with capacity. If the relation between these is such, that at the frequency used they annul each other, that is, have such values as to satisfy the condition of resonance, and the greatest quantity of electricity is made to flow through the external circuit, then the best result is obtained. It is simpler and safer to join the condenser in series with the self-induction. It is clear that in such

combinations there will be, for a given frequency, and considering
only the fundamental vibration, values which will give the best
result, with the condenser in shunt to the self-induction coil; of
course more such values than with the condenser in series. But
practical conditions determine the selection. In the latter case
in performing the experiments one may take a small self-induction
and a large capacity or a small capacity and a large self-induction, but the latter is preferable, because it is inconvenient to adjust a large capacity by small steps. By taking a coil with a very
large self-induction the critical capacity is reduced to a very small
value, and the capacity of the coil itself may be sufficient. It is
easy, especially by observing certain artifices, to wind a coil
through which the impedance will be reduced to the value of the
ohmic resistance only; and for any coil there is, of course, a frequency at which the maximum current will be made to pass
through the coil. The observation of the relation between self-

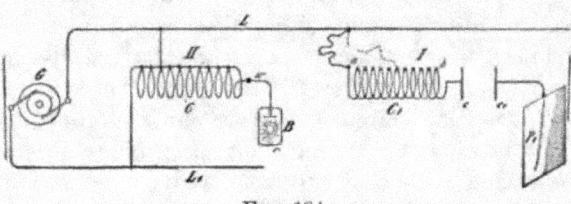

Fig. 184.

induction, capacity and frequency is becoming important in the
operation of alternate current apparatus, such as transformers or
motors, because by a judicious determination of the elements the
employment of an expensive condenser becomes unnecessary.
Thus it is possible to pass through the coils of an alternating
current motor under the normal working conditions the required
current with a low E. M. F. and do away entirely with the false
current, and the larger the motor, the easier such a plan becomes
practicable; but it is necessary for this to employ currents of very
high potential or high frequency.

In Fig. 184 I. is shown a plan which has been followed in the
study of the resonance effects by means of a high frequency alternator. c_1 is a coil of many turns, which is divided into small
separate sections for the purpose of adjustment. The final adjustment was made sometimes with a few thin iron wires (though
this is not always advisable) or with a closed secondary. The coil

c_1 is connected with one of its ends to the line L from the alternator G and with the other end to one of the plates c of a condenser c c_1, the plate (c_1) of the latter being connected to a much larger plate P_1. In this manner both capacity and self-induction were adjusted to suit the dynamo frequency.

As regards the rise of potential through resonant action, of course, theoretically, it may amount to anything since it depends on self-induction and resistance and since these may have any value. But in practice one is limited in the selection of these values and besides these, there are other limiting causes. One may start with, say, 1,000 volts and raise the E. M. F. to 50 times that value, but one cannot start with 100,000 and raise it to ten times that value because of the losses in the media which are great, especially if the frequency is high. It should be possible to start with, for instance, two volts from a high or low frequency circuit of a dynamo and raise the E. M. F. to many hundred times that value. Thus coils of the proper dimensions might be connected each with only one of its ends to the mains from a machine of low E. M. F., and though the circuit of the machine would not be closed in the ordinary acceptance of the term, yet the machine might be burned out if a proper resonance effect would be obtained. I have not been able to produce, nor have I observed with currents from a dynamo machine, such great rises of potential. It is possible, if not probable, that with currents obtained from apparatus containing iron the disturbing influence of the latter is the cause that these theoretical possibilities cannot be realized. But if such is the case I attribute it solely to the hysteresis and Foucault current losses in the core. Generally it was necessary to transform upward, when the E. M. F. was very low, and usually an ordinary form of induction coil was employed, but sometimes the arrangement illustrated in Fig. 184 II., has been found to be convenient. In this case a coil c is made in a great many sections, a few of these being used as a primary. In this manner both primary and secondary are adjustable. One end of the coil is connected to the line L_1 from the alternator, and the other line L is connected to the intermediate point of the coil. Such a coil with adjustable primary and secondary will be found also convenient in experiments with the disruptive discharge. When true resonance is obtained the top of the wave must of course be on the free end of the coil as, for instance, at the terminal of the phosphorescence bulb B. This is easily recognized by observing the potential of a point on the wire w near to the coil.

In connection with resonance effects and the problem of transmission of energy over a single conductor which was previously considered, I would say a few words on a subject which constantly fills my thoughts and which concerns the welfare of all. I mean the transmission of intelligible signals or perhaps even power to any distance without the use of wires. I am becoming daily more convinced of the practicability of the scheme; and though I know full well that the great majority of scientific men will not believe that such results can be practically and immediately realized, yet I think that all consider the developments in recent years by a number of workers to have been such as to encourage thought and experiment in this direction. My conviction has grown so strong, that I no longer look upon this plan of energy or intelligence transmission as a mere theoretical possibility, but as a serious problem in electrical engineering, which must be carried out some day. The idea of transmitting intelligence without wires is the natural outcome of the most recent results of electrical investigations. Some enthusiasts have expressed their belief that telephony to any distance by induction through the air is possible. I cannot stretch my imagination so far, but I do firmly believe that it is practicable to disturb by means of powerful machines the electrostatic condition of the earth and thus transmit intelligible signals and perhaps power. In fact, what is there against the carrying out of such a scheme? We now know that electric vibration may be transmitted through a single conductor. Why then not try to avail ourselves of the earth for this purpose? We need not be frightened by the idea of distance. To the weary wanderer counting the mile-posts the earth may appear very large, but to that happiest of all men, the astronomer, who gazes at the heavens and by their standard judges the magnitude of our globe, it appears very small. And so I think it must seem to the electrician, for when he considers the speed with which an electric disturbance is propagated through the earth all his ideas of distance must completely vanish.

A point of great importance would be first to know what is the capacity of the earth? and what charge does it contain if electrified? Though we have no positive evidence of a charged body existing in space without other oppositely electrified bodies being near, there is a fair probability that the earth is such a body, for by whatever process it was separated from other bodies—and this is the accepted view of its origin—it must have retained a charge, as occurs in all processes of mechanical separation. If it be a charged body insulated in space its capacity should be extremely small, less than one-thousandth of a farad. But the upper strata of the air are conducting, and so, perhaps, is the medium in free space beyond the atmosphere, and these may contain an opposite charge. Then the capacity might be incomparably greater. In any case it is of the greatest importance to get an idea of what quantity of electricity the earth contains. It is difficult to say whether we shall ever acquire this necessary knowledge, but there is hope that we may, and that is, by means of electrical resonance. If ever we can ascertain at what period the earth's charge, when disturbed, oscillates with respect to an oppositely electrified system or known circuit, we shall know a fact possibly of the greatest importance to the welfare of the human race. I propose to seek for the period by means of an electrical oscillator, or a source of alternating electric currents. One of the terminals of the source would be connected to earth as, for instance, to the city water mains, the other to an insulated body of large surface. It is possible that the outer conducting air strata, or free space, contain an opposite charge and that, together with the earth, they form a condenser of very large capacity. In such case the period of vibration may be very low and an alternating dynamo machine might serve for the purpose of the experiment. I would then transform the current to a potential as high as it would be found possible and connect the ends of the high tension secondary to the ground and to the insulated body. By varying the frequency of the currents and carefully observing the potential of the insulated body and watching for the disturbance at various neighboring points of the earth's surface resonance might be detected. Should, as the majority of scientific men in all probability believe, the period be extremely small, then a dynamo machine would not do and a proper electrical oscillator would have to be produced and perhaps it might not be possible to obtain such rapid vibrations. But whether this be possible or not, and whether the earth contains a charge or not, and whatever may be its period of vibration, it certainly is possible—for of this we have daily evidence—to produce some electrical disturbance sufficiently powerful to be perceptible by suitable instruments at any point of the earth's surface.

Assume that a source of alternating currentss be connected, as in Fig. 185, with one of its terminals to earth (conveniently to the water mains) and with the other to a body of large surface P.

When the electric oscillation is set up there will be a movement of electricity in and out of P, and alternating currents will pass through the earth, converging to, or diverging from, the point c where the ground connection is made. In this manner neighboring points on the earth's surface within a certain radius will be disturbed. But the disturbance will diminish with the distance, and the distance at which the effect will still be perceptible will depend on the quantity of electricity set in motion. Since the body P is insulated, in order to displace a considerable quantity, the potential of the source must be excessive, since there would be limitations as to the surface of P. The conditions might be adjusted so that the generator or source s will set up the same electrical movement as though its circuit were closed. Thus it is certainly practicable to impress an electric vibration at least of a certain low period upon the earth by means of proper machinery. At what distance such a vibration might be made perceptible can only be conjectured. I have on another occasion considered the question how the earth might behave to electric disturbances. There is no doubt that, since in such an experiment the electrical density at the surface could be but extremely small considering the size of the earth, the air would not act as a very disturbing factor, and there would be not much energy lost through the action of the air, which would be the case if the density were great. Theoretically, then, it could not require a great amount of energy to produce a disturbance perceptible at great distance, or even all over the surface of the globe. Now, it is quite certain that at any point within a certain radius of the source s a properly adjusted self-induction and capacity device can be set in action by resonance. But not only can this be done, but another source s₁, Fig. 185, similar to s, or any number of such sources, can be set

FIG. 185.

to work in synchronism with the latter, and the vibration thus intensified and spread over a large area, or a flow of electricity produced to or from the source s_1 if the same be of opposite phase to the source s. I think that beyond doubt it is possible to operate electrical devices in a city through the ground or pipe system by resonance from an electrical oscillator located at a central point. But the practical solution of this problem would be of incomparably smaller benefit to man than the realization of the scheme of transmitting intelligence, or perhaps power, to any distance through the earth or environing medium. If this is at all possible, distance does not mean anything. Proper apparatus must first be produced by means of which the problem can be attacked and I have devoted much thought to this subject. I am firmly convinced that it can be done and hope that we shall live to see it done.

ON THE LIGHT PHENOMENA PRODUCED BY HIGH-FREQUENCY CURRENTS OF HIGH POTENTIAL AND GENERAL REMARKS RELATING TO THE SUBJECT.

Returning now to the light effects which it has been the chief object to investigate, it is thought proper to divide these effects into four classes : 1. Incandescence of a solid. 2. Phosphorescence. 3. Incandescence or phosphorescence of a rarefied gas ; and 4. Luminosity produced in a gas at ordinary pressure. The first question is : How are these luminous effects produced ? In order to answer this question as satisfactorily as I am able to do in the light of accepted views and with the experience acquired, and to add some interest to this demonstration, I shall dwell here upon a feature which I consider of great importance, inasmuch as it promises, besides, to throw a better light upon the nature of most of the phenomena produced by high-frequency electric currents. I have on other occasions pointed out the great importance of the presence of the rarefied gas, or atomic medium in general, around the conductor through which alternate currents of high frequency are passed, as regards the heating of the conductor by the currents. My experiments, described some time ago, have shown that, the higher the frequency and potential difference of the currents, the more important becomes the rarefied gas in which the conductor is immersed, as a factor of the heating. The potential difference, however, is, as I then pointed out, a more important element than the frequency. When both of these are sufficiently high, the heating may be almost entirely due to the presence of the rarefied gas. The experiments to follow will show the importance of the rarefied gas, or, generally, of gas at ordinary or other pressure as regards the incandescence or other luminous effects produced by currents of this kind.

I take two ordinary 50-volt 16 c. p. lamps which are in every respect alike, with the exception, that one has been opened at the top and the air has filled the bulb, while the other is at the ordinary degree of exhaustion of commercial lamps. When I attach the lamp which is exhausted to the terminal of the secondary of the coil, which I have already used, as in experiments illustrated in Fig. 179a for instance, and turn on the current, the filament, as you have before seen, comes to high incandescence. When I attach the second lamp, which is filled with air, instead of the former, the filament still glows, but much less brightly. This experiment illustrates only in part the truth of the statements before made. The importance of the filament's being immersed in rarefied gas is plainly noticeable but not to such a degree as might be desirable. The reason is that the secondary of this coil is wound for low tension, having only 150 turns, and the potential difference at the terminals of the lamp is therefore small. Were I to take another coil with many more turns in the secondary, the effect would be increased, since it depends partially on the potential difference, as before remarked. But since the effect likewise depends on the frequency, it may be properly stated that it depends on the time rate of the variation of the potential difference. The greater this variation, the more important becomes the gas as an element of heating. I can produce a much greater rate of variation in another way, which, besides, has the advantage of doing away with the objections, which might be made in the experiment just shown, even if both the lamps were connected in series or multiple arc to the coil, namely, that in consequence of the reactions existing between the primary and secondary coil the conclusions are rendered uncertain. This result I secure by charging, from an ordinary transformer which is fed from the alternating current supply station, a battery of condensers, and discharging the latter directly through a circuit of small self-induction, as before illustrated in Figs. 183a, 183b, and 183c.

In Figs. 186a, 186b and 186c, the heavy copper bars BB_1, are connected to the opposite coatings of a battery of condensers, or generally in such way, that the high frequency or sudden discharges are made to traverse them. I connect first an ordinary 50-volt incandescent lamp to the bars by means of the clamps *c c*. The discharges being passed through the lamp, the filament is rendered incandescent, though the current through it is very small, and would not be nearly sufficient to produce a visible effect under the conditions of ordinary use of the lamp. Instead of this I now attach to the bars another lamp exactly like the first, but with the seal broken off, the bulb being therefore filled with air at ordinary pressure. When the discharges are directed through the filament, as before, it does not become incandescent. But the result might still be attributed to one of the many possible reactions. I therefore connect both the lamps in multiple arc as illustrated in Fig. 186*a*. Passing

FIG. 186a.   FIG. 186b.   FIG. 186c.

the discharges through both the lamps, again the filament in the exhausted lamp *l* glows very brightly while that in the non-exhausted lamp *l₁* remains dark, as previously. But it should not be thought that the latter lamp is taking only a small fraction of the energy supplied to both the lamps; on the contrary, it may consume a considerable portion of the energy and it may become even hotter than the one which burns brightly. In this experiment the potential difference at the terminals of the lamps varies in sign theoretically three to four million times a second. The ends of the filaments are correspondingly electrified, and the gas in the bulbs is violently agitated and a large portion of the supplied energy is thus converted into heat. In the non-exhausted bulb, there being a few million times more gas molecules than in the exhausted one, the bombardment, which is most violent at the ends of the filament, in the neck of the bulb, consumes a large portion of the energy without producing any visible effect. The reason is that, there being many molecules, the bombardment is quantitatively considerable, but the individual impacts are not very violent, as the speeds of the molecules are comparatively small owing to the small free path. In the exhausted bulb, on the contrary, the speeds are very great, and the individual impacts are violent and therefore better adapted to produce a visible effect. Besides, the convection of heat is greater in the former bulb. In both the bulbs the current traversing the filaments is very small, incomparably smaller than that which they require on an ordinary low-frequency circuit. The potential difference, however, at the ends of the filaments is very great and might be possibly 20,000 volts or more, if the filaments were straight and their ends far apart. In the ordinary lamp a spark generally occurs between the ends of the filament or between the platinum wires outside, before such a difference of potential can be reached.

It might be objected that in the experiment before shown the lamps, being in multiple arc, the exhausted lamp might take a much larger current and that the effect observed might not be exactly attributable to the action of the gas in the bulbs. Such objections will lose much weight if I connect the lamps in series, with the same result. When this is done and the discharges are directed through the filaments, it is again noted that the filament in the non-exhausted bulb l_1 remains dark, while that in the exhausted one (l) glows even more intensely than under its normal conditions of working, Fig. 186b. According to general ideas the current through the filaments should now be the same, were it not modified by the presence of the gas around the filaments.

At this juncture I may point out another interesting feature, which illustrates the effect of the rate of change of potential of the currents. I will leave the two lamps connected in series to the bars BB_1, as in the previous experiment, Fig. 186b, but will presently reduce considerably the frequency of the currents, which was excessive in the experiment just before shown. This I may do by inserting a self-induction coil in the path of the discharges, or by augmenting the capacity of the condensers. When I now pass these low-frequency discharges through the lamps, the exhausted lamp l again is as bright as before, but it is noted also that the non-exhausted lamp l_1 glows, though not quite as intensely as the other. Reducing the current through the lamps, I may bring the filament in the latter lamp to redness, and, though the filament in the exhausted lamp l is bright, Fig. 186c, the degree of its incandescence is much smaller than in Fig. 186b, when the currents were of a much higher frequency.

In these experiments the gas acts in two opposite ways in determining the degree of the incandescence of the filaments, that is, by convection and bombardment. The higher the frequency and potential of the currents, the more important becomes the bombardment. The convection on the contrary should be the smaller, the higher the frequency. When the currents are steady there is practically no bombardment, and convection may therefore with such currents also considerably modify the degree of incandescence and produce results similar to those just before shown. Thus, if two lamps exactly alike, one exhausted and one not exhausted, are connected in multiple arc or series to a direct-current machine, the filament in the non-exhausted lamp will require a considerably greater current to be rendered incandescent. This result is entirely due to convection, and the effect is the more prominent the thinner the filament. Professor Ayrton and Mr. Kilgour some time ago published quantitative results concerning the thermal emissivity by radiation and convection in which the effect with thin wires was clearly shown. This effect may be strikingly illustrated by preparing a number of small, short, glass tubes, each containing through its axis the thinnest obtainable platinum wire. If these tubes be highly exhausted, a number of them may be connected in multiple arc to a direct-current machine and all of the wires may be kept at incandescence with a smaller current than that required to render incandescent a single one of the wires if the tube be not exhausted. Could the tubes be so highly exhausted that convection would be nil, then the relative amounts of heat given off by convection and radiation could be determined without the difficulties attending thermal quantitative measurements. If a source of electric impulses of high frequency and very high potential is employed, a still greater number of the tubes may be taken and the wires rendered incandescent by a current not capable of warming perceptibly a wire of the same size immersed in air at ordinary pressure, and conveying the energy to all of them.

I may here describe a result which is still more interesting, and to which I have been led by the observation of these phenomena.   I noted that small differences in the density of the air produced a considerable difference in the degree of incandescence of the wires, and I thought that, since in a tube, through which a luminous discharge is passed, the gas is generally not of uniform density, a very thin wire contained in the tube might be rendered incandescent at certain places of smaller density of the gas, while it would remain dark at the places of greater density, where the convection would be greater and the bombardment less intense.   Accordingly a tube t was prepared, as illustrated in Fig. 187, which contained through the middle a very fine platinum wire w.   The tube was exhausted to a moderate degree and it was found that when it was attached to the terminal of a high-frequency coil the platinum wire w would indeed, become incandescent in patches, as illustrated in Fig. 187.   Later a number of these tubes with one or more wires were prepared, each showing this result.   The effect was best noted when the striated discharge occurred in the tube, but was also produced when the striæ were not visible, showing that, even then, the gas in the tube was not of uniform density.   The position of the striæ was generally such, that the rarefactions corresponded to the places of incandescence or greater brightness on the wire w.   But in a few instances it was noted, that the bright spots on the wire were covered by the dense parts of the striated discharge as indicated by l in Fig. 187, though the effect was barely perceptible.   This was explained in a plausible way by assuming that the convection was not widely different in the dense and rarefied places, and that the bombardment was greater on the dense places of the striated discharge.   It is, in fact, often observed in bulbs, that under certain conditions a thin wire is brought to higher incandescence when the air is not too highly rarefied.   This is the case when the potential of the coil is not high enough for the vacuum, but the result may be attributed to many different causes.   In all cases this curious phenomenon of incandescence disappears when the tube, or rather the wire, acquires throughout a uniform temperature.

Disregarding now the modifying effect of convection there are then two distinct causes which determine the incandescence of a wire or filament with varying currents, that is, conduction current and bombardment.   With steady currents we have to deal only with the former of these two causes, and the heating effect is a minimum, since the resistance is least to steady flow.   When the current is a varying one the resistance is greater, and hence the heating effect is increased. Thus if the rate of change of the current is very great, the resistance may increase to such an extent that the filament is brought to incandescence with inappreciable currents, and we are able to take a short and thick block of carbon or other material and bring it to bright incandescence with a current incomparably smaller than that required to bring to the same degree of incandescence an ordinary thin lamp filament with a steady or low frequency current. This result is important, and illustrates how rapidly our views on these subjects are changing, and how quickly our field of knowledge is ex-

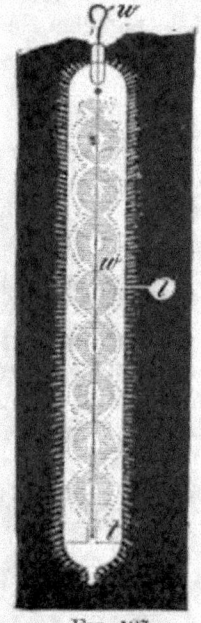

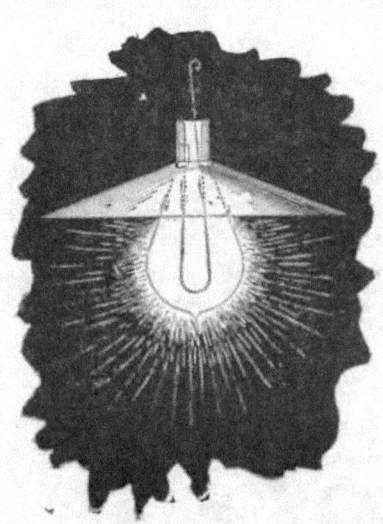

FIG. 187.   FIG. 188.

tending. In the art of incandescent lighting, to view this result in one aspect only, it has been commonly considered as an essential requirement for practical success, that the lamp filament should be thin and of high resistance. But now we know that the resistance of the filament to the steady flow does not mean anything; the filament might as well be short and thick; for if it be immersed in rarefied gas it will become incandescent by the passage of a small current. It all depends on the frequency and potential of the currents. We may conclude from this, that it would be of advantage, so far as the lamp is considered, to employ high frequencies for lighting, as they allow the use of short and thick filaments and smaller currents.

If a wire or filament be immersed in a homogeneous medium, all the heating is due to true conduction current, but if it be enclosed in an exhausted vessel the conditions are entirely different. Here the gas begins to act and the heating effect of the conduction current, as is shown in many experiments, may be very small compared with that of the bombardment. This is especially the case if the circuit is not closed and the potentials are of course very high. Suppose that a fine filament enclosed in an exhausted vessel be connected with one of its ends to the terminal of a high tension coil and with its other end to a large insulated plate. Though the circuit is not closed, the filament, as I have before shown, is brought to incandescence. If the frequency and potential be comparatively low, the filament is heated by the current passing *through it*. If the frequency and potential, and principally the latter, be increased, the insulated plate need be but very small, or may be done away with entirely; still the filament will become incandescent, practically all the heating being then due to the bombardment. A practical way of combining both the effects of conduction currents and bombardment is illustrated in Fig. 188, in which an ordinary lamp is shown provided with a very thin filament which has one of the ends of the latter connected to a shade serving the purpose of the insulated plate, and the other end to the terminal of a high tension source. It should not be thought that only rarefied gas is an important factor in the heating of a conductor by varying currents, but gas at ordinary pressure may become important, if the potential difference and frequency of the currents is excessive. On this subject I have already stated, that when a conductor is fused by a stroke of lightning, the current through it may be exceedingly small, not even sufficient to heat the conductor perceptibly, were the latter immersed in a homogeneous medium.

From the preceding it is clear that when a conductor of high resistance is connected to the terminals of a source of high frequency currents of high potential, there may occur considerable dissipation of energy, principally at the ends of the conductor, in consequence of the action of the gas surrounding the conductor. Owing to this, the current through a section of the conductor at a point midway between its ends may be much smaller than through a section near the ends. Furthermore, the current passes principally through the outer portions of the conductor, but this effect is to be distinguished from the skin effect as ordinarily interpreted, for the latter would, or should, occur also in a continuous incompressible medium. If a great many incandescent lamps are connected in series to a source of such currents, the lamps at the ends may burn brightly, whereas those in the middle may remain entirely dark. This is due principally to bombardment, as before stated. But even if the currents be steady, provided the difference of potential is very great, the lamps at the end will burn more brightly than those in the middle. In such case there is no rhythmical bombardment, and the result is produced entirely by leakage. This leakage or dissipation into space when the tension is high, is considerable when incandescent lamps are used, and still more considerable with arcs, for the latter act like flames. Generally, of course, the dissipation is much smaller with steady, than with varying, currents.

I have contrived an experiment which illustrates in an interesting manner the effect of lateral diffusion. If a very long tube is attached to the terminal of a high frequency coil, the luminosity is greatest near the terminal and falls off gradually towards the remote end. This is more marked if the tube is narrow.

A small tube about one-half inch in diameter and twelve inches long (Fig. 189), has one of its ends drawn out into a fine fibre f nearly three feet long. The tube is placed in a brass socket T which can be screwed on the terminal T_1 of the induction coil. The discharge passing through the tube first illuminates the bottom of the same, which is of comparatively large section; but through the long glass fibre the discharge cannot pass. But gradually the rarefied gas inside becomes warmed and more conducting and the discharge spreads into the glass fibre. This spreading is so slow, that it may take half a minute or more until the discharge has worked through up to the top of the glass fibre, then presenting the appearance of a strongly luminous thin thread. By adjusting the potential at the terminal the light may be made to travel upwards at any speed. Once, however, the glass fibre is heated, the discharge breaks through its entire length instantly. The interesting point to be noted is that, the higher the frequency of the currents, or in other words, the greater relatively the lateral dissipation, at a slower rate may the light be made to propagate through the fibre. This experiment

is best performed with a highly exhausted and freshly made tube.
When the tube has been used for some time the experiment
often fails. It is possible that the gradual and slow impairment
of the vacuum is the cause. This slow propagation of the discharge through a very narrow glass tube corresponds exactly to
the propagation of heat through a bar warmed at one end. The
quicker the heat is carried away laterally the longer time it will
take for the heat to warm the remote end. When the current
of a low frequency coil is passed through the fibre from end to
end, then the lateral dissipation is small and the discharge instantly breaks through almost without exception.

Fig. 189.   Fig. 190.

After these experiments and observations which have shown
the importance of the discontinuity or atomic structure of the
medium and which will serve to explain, in a measure at least,
the nature of the four kinds of light effects producible with
these currents, I may now give you an illustration of these
effects. For the sake of interest I may do this in a manner
which to many of you might be novel. You have seen before
that we may now convey the electric vibration to a body by
means of a single wire or conductor of any kind. Since the

human frame is conducting I may convey the vibration through my body.

First, as in some previous experiments, I connect my body with one of the terminals of a high-tension transformer and take in my hand an exhausted bulb which contains a small carbon button mounted upon a platinum wire leading to the outside of the bulb, and the button is rendered incandescent as soon as the transformer is set to work (Fig. 190). I may place a conducting shade on the bulb which serves to intensify the action, but is not necessary. Nor is it required that the button should be in conducting connection with the hand through a wire leading through the glass,

FIG. 191.

FIG. 192.

for sufficient energy may be transmitted through the glass itself by inductive action to render the button incandescent.

Next I take a highly exhausted bulb containing a strongly phosphorescent body, above which is mounted a small plate of aluminum on a platinum wire leading to the outside, and the currents flowing through my body excite intense phosphorescence in the bulb (Fig. 191). Next again I take in my hand a simple exhausted tube, and in the same manner the gas inside the tube is rendered highly incandescent or phosphorescent (Fig. 192). Finally, I may take in my hand a wire, bare or covered with thick insulation, it is quite immaterial; the electrical vibration is so intense as to cover the wire with a luminous film (Fig. 193).

A few words must now be devoted to each of these phenomena. In the first place, I will consider the incandescence of a button or of a solid in general, and dwell upon some facts which apply equally to all these phenomena. It was pointed out before that when a thin conductor, such as a lamp filament, for instance, is connected with one of its ends to the terminal of a transformer of high tension the filament is brought to incandescence partly by a conduction current and partly by bombardment. The shorter and thicker the filament the more important becomes the latter, and finally, reducing the filament to a mere button, all the heating must practically be attributed to the bombardment. So in the experiment before shown, the button is rendered incandescent by the rhythmical impact of freely movable small bodies in the bulb. These bodies may be the molecules of the residual gas, particles of dust or lumps torn from the electrode; whatever they are, it is certain that the heating of the button is essentially connected with the pressure of such freely movable particles, or of atomic matter in general in the bulb. The heating is the more intense the greater the number of impacts per second and the greater the energy of each impact. Yet the button would be heated also if it were connected to a source of a steady potential. In such a case electricity would be carried away from the button by the freely movable carriers or particles flying about, and the quantity of electricity thus carried away might be sufficient to bring the button to incandescence by its passage through the latter. But the bombardment could not be of great importance in such case. For this reason it would require a comparatively very great supply of energy to the button to maintain it at incandescence with a steady potential. The higher the frequency of the electric impulses the more economically can the button be maintained at incandescence. One of the chief reasons why this is so, is, I believe, that with impulses of very high frequency there is less exchange of the freely movable carriers around the electrode and this means, that in the bulb the heated matter is better confined to the neighborhood of the button. If a double bulb, as illustrated in Fig. 194 be made, comprising a large globe B and a small one b, each containing as usual a filament f mounted on a platinum wire w and w_1, it is found, that if the filaments $f f$ be exactly alike, it requires less energy to keep the filament in the globe b at a certain degree of incandescence, than that in the globe B. This is due to the confinement of the movable particles around the button. In this case it is also ascertained, that the filament in the small globe *b* is less deteriorated when maintained a certain length of time at incandescence. This is a necessary consequence of the fact that the gas in the small bulb becomes strongly heated and therefore a very good conductor, and less work is then performed on the button, since the bombardment becomes less intense as the conductivity of the gas increases. In this construction, of course, the small bulb becomes very hot and when it reaches an elevated temperature the convection and radiation on the outside increase. On another occasion I have shown bulbs in which this drawback was largely avoided. In these instances a very small bulb, containing a refractory button, was mounted in a large globe and the space be-

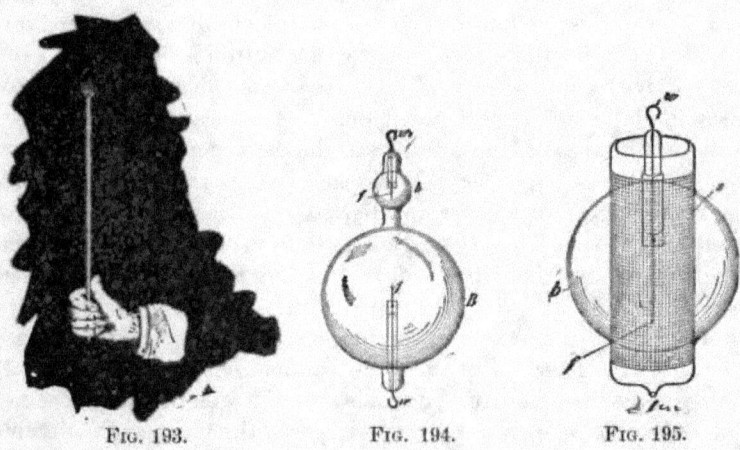

FIG. 193.   FIG. 194.   FIG. 195.

tween the walls of both was highly exhausted. The outer large globe remained comparatively cool in such constructions. When the large globe was on the pump and the vacuum between the walls maintained permanent by the continuous action of the pump, the outer globe would remain quite cold, while the button in the small bulb was kept at incandescence. But when the seal was made, and the button in the small bulb maintained incandescent some length of time, the large globe too would become warmed. From this I conjecture that if vacuous space (as Prof. Dewar finds) cannot convey heat, it is so merely in virtue of our rapid motion through space or, generally speaking, by the motin of the medium relatively to us, for a permanent condition could not be maintained without the medium being constantly renewed. A vacuum cannot, according to all evidence, be permanently maintained around a hot body.

In these constructions, before mentioned, the small bulb inside would, at least in the first stages, prevent all bombardment against the outer large globe. It occurred to me then to ascertain how a metal sieve would behave in this respect, and several bulbs, as illustrated in Fig. 195, were prepared for this purpose. In a globe b, was mounted a thin filament f (or button) upon a platinum wire w passing through a glass stem and leading to the outside of the globe. The filament f was surrounded by a metal sieve s. It was found in experiments with such bulbs that a sieve with wide meshes apparently did not in the slightest affect the bombardment against the globe b. When the vacuum was high, the shadow of the sieve was clearly projected against the globe and the latter would get hot in a short while. In some bulbs the sieve s was connected to a platinum wire sealed in the glass. When this wire was connected to the other terminal of the induction coil (the E. M. F. being kept low in this case), or to an insulated plate, the bombardment against the outer globe b was diminished. By taking a sieve with fine meshes the bombardment against the globe b was always diminished, but even then if the exhaustion was carried very far, and when the potential of the transformer was very high, the globe b would be bombarded and heated quickly, though no shadow of the sieve was visible, owing to the smallness of the meshes. But a glass tube or other continuous body mounted so as to surround the filament, did entirely cut off the bombardment and for a while the outer globe b would remain perfectly cold. Of course when the glass tube was sufficiently heated the bombardment against the outer globe could be noted at once. The experiments with these bulbs seemed to show that the speeds of the projected molecules or particles must be considerable (though quite insignificant when compared with that of light), otherwise it would be difficult to understand how they could traverse a fine metal sieve without being affected, unless it were found that such small particles or atoms cannot be acted upon directly at measurable distances. In regard to the speed of the projected atoms, Lord Kelvin has recently estimated it at about one kilometre a second or thereabouts in an ordinary Crookes bulb. As the potentials obtainable with a disruptive discharge coil are much higher than with ordinary coils, the speeds must, of course, be much greater when the bulbs are lighted from such a coil. Assuming the speed to be as high as five kilometres and uniform through the whole trajectory, as it should be in a very highly exhausted vessel, then if the alternate electrifications of the electrode would be of a frequency of five million, the greatest distance a particle could get away from the electrode would be one millimetre, and if it could be acted upon directly at that distance, the exchange of electrode matter or of the atoms would be very slow and there would be practically no bombardment against the bulb. This at least should be so, if the action of an electrode upon the atoms of the residual gas would be such as upon electrified bodies which we can perceive. A hot body enclosed in an exhausted bulb produces always atomic bombardment, but a hot body has no definite rhythm, for its molecules perform vibrations of all kinds.

If a bulb containing a button or filament be exhausted as high as is possible with the greatest care and by the use of the best artifices, it is often observed that the discharge cannot, at first, break through, but after some time, probably in consequence of some changes within the bulb, the discharge finally passes through and the button is rendered incandescent. In fact, it appears that the higher the degree of exhaustion the easier is the incandescence produced. There seem to be no other causes to which the incandescence might be attributed in such case except to the bombardment or similar action of the residual gas, or of particles of matter in general. But if the bulb be exhausted with the greatest care can these play an important part? Assume the vacuum in the bulb to be tolerably perfect, the great interest then centres in the question: Is the medium which pervades all space continuous or atomic? If atomic, then the heating of a conducting button or filament in an exhausted vessel might be due largely to ether bombardment, and then the heating of a conductor in general through which currents of high frequency or high potential are passed must be modified by the behavior of such medium; then also the skin effect, the apparent increase of the ohmic resistance, etc., admit, partially at least, of a different explanation.

It is certainly more in accordance with many phenomena observed with high-frequency currents to hold that all space is pervaded with free atoms, rather than to assume that it is devoid of these, and dark and cold, for so it must be, if filled with a continuous medium, since in such there can be neither heat nor light.

Is then energy transmitted by independent carriers or by the vibration of a continuous medium? This important question is by no means as yet positively answered. But most of the effects which are here considered, especially the light effects, incandescence, or phosphorescence, involve the presence of free atoms and would be impossible without these.

In regard to the incandescence of a refractory button (or filament) in an exhausted receiver, which has been one of the subjects of this investigation, the chief experiences, which may serve as a guide in constructing such bulbs, may be summed up as follows: 1. The button should be as small as possible, spherical, of a smooth or polished surface, and of refractory material which withstands evaporation best. 2. The support of the button should be very thin and screened by an aluminum and mica sheet, as I have described on another occasion. 3. The exhaustion of the bulb should be as high as possible. 4. The frequency of the currents should be as high as practicable. 5. The currents should be of a harmonic rise and fall, without sudden interruptions. 6. The heat should be confined to the button by inclosing the same in a small bulb or otherwise. 7. The space between the walls of the small bulb and the outer globe should be highly exhausted.

Most of the considerations which apply to the incandescence of a solid just considered may likewise be applied to phosphorescence. Indeed, in an exhausted vessel the phosphorescence is, as a rule, primarily excited by the powerful beating of the electrode stream of atoms against the phosphorescent body. Even in many cases, where there is no evidence of such a bombardment, I think that phosphorescence is excited by violent impacts of atoms, which are not necessarily thrown off from the electrode but are acted upon from the same inductively through the medium or through chains of other atoms. That mechanical shocks play an important part in exciting phosphorescence in a bulb may be seen from the following experiment. If a bulb, constructed as that illustrated in Fig. 174, be taken and exhausted with the greatest care so that the discharge cannot pass, the filament f acts by electrostatic induction upon the tube t and the latter is set in vibration. If the tube o be rather wide, about an inch or so, the filament may be so powerfully vibrated that whenever it hits the glass tube it excites phosphorescence. But the phosphorescence ceases when the filament comes to rest. The vibration can be arrested and again started by varying the frequency of the currents. Now the filament has its own period of vibration, and if the frequency of the currents is such that there is resonance, it is easily set vibrating, though the potential of the currents be small. I have often observed that the filament in the bulb is destroyed by such mechanical resonance. The filament vibrates as a rule so rapidly that it cannot be seen and the experimenter may at first be mystified. When such an experiment as the one described is carefully performed, the potential of the currents need be extremely small, and for this reason I infer that the phosphorescence is then due to the mechanical shock of the filament against the glass, just as it is produced by striking a loaf of sugar with a knife. The mechanical shock produced by the projected atoms is easily noted when a bulb containing a button is grasped in the hand and the current turned on suddenly. I believe that a bulb could be shattered by observing the conditions of resonance.

In the experiment before cited it is, of course, open to say, that the glass tube, upon coming in contact with the filament, retains a charge of a certain sign upon the point of contact. If now the filament again touches the glass at the same point while it is oppositely charged, the charges equalize under evolution of light. But nothing of importance would be gained by such an explanation. It is unquestionable that the initial charges given to the atoms or to the glass play some part in exciting phosphorescence. So, for instance, if a phosphorescent bulb be first excited by a high frequency coil by connecting it to one of the terminals of the latter and the degree of luminosity be noted, and then the bulb be highly charged from a Holtz machine by attaching it preferably to the positive terminal of the machine, it is found that when the bulb is again connected to the terminal of the high frequency coil, the phosphorescence is far more intense. On another occasion I have considered the possibility of some phosphorescent phenomena in bulbs being produced by the incandescence of an infinitesimal layer on the surface of the phosphorescent body. Certainly the impact of the atoms is powerful enough to produce intense incandescence by the collisions, since they bring quickly to a high temperature a body of considerable bulk. If any such effect exists, then the best appliance for producing phosphorescence in a bulb, which we know so far, is a disruptive discharge coil giving an enormous potential with but few fundamental discharges, say 25–30 per second, just enough to produce a continu-

ous impression upon the eye.   It is a fact that such a coil excites
phosphorescence under almost any condition and at all degrees
of exhaustion, and I have observed effects which appear to be due
to phosphorescence even at ordinary pressures of the atmosphere,
when the potentials are extremely high.   But if phosphorescent
light is produced by the equalization of charges of electrified
atoms (whatever this may mean ultimately), then the higher the
frequency of the impulses or alternate electrifications, the
more economical will be the light production.   It is a long
known and noteworthy fact that all the phosphorescent bodies
are poor conductors of electricity and heat, and that all bodies
cease to emit phosphorescent light when they are brought to a
certain temperature.   Conductors on the contrary do not possess
this quality.   There are but few exceptions to the rule.   Carbon
is one of them.   Becquerel noted that carbon phosphoresces at
at a certain elevated temperature preceding the dark red.   This
phenomenon may be easily observed in bulbs provided with a
rather large carbon electrode (say, a sphere of six millimetres diameter).   If the current is turned on after a few seconds, a snow
white film covers the electrode, just before it gets dark red.
Similar effects are noted with other conducting bodies, but many
scientific men will probably not attribute them to true phosphorescence.   Whether true incandescence has anything to do with
phosphorescence excited by atomic impact or mechanical shocks
still remains to be decided, but it is a fact that all conditions,
which tend to localize and increase the heating effect at the point
of impact, are almost invariably the most favorable for the production of phosphorescence.   So, if the electrode be very small,
which is equivalent to saying in general, that the electric density
is great; if the potential be high, and if the gas be highly rarefied, all of which things imply high speed of the projected atoms,
or matter, and consequently violent impacts—the phosphorescence is very intense.   If a bulb provided with a large and small
electrode be attached to the terminal of an induction coil, the
small electrode excites phosphorescence while the large one may
not do so, because of the smaller electric density and hence
smaller speed of the atoms.   A bulb provided with a large electrode may be grasped with the hand while the electrode is connected to the terminal of the coil and it may not phosphoresce;
but if instead of grasping the bulb with the hand, the same be
touched with a pointed wire, the phosphorescence at once spreads

through the bulb, because of the great density at the point of contact. With low frequencies it seems that gases of great atomic weight excite more intense phosphorescence than those of smaller weight, as for instance, hydrogen. With high frequencies the observations are not sufficiently reliable to draw a conclusion. Oxygen, as is well-known, produces exceptionally strong effects, which may be in part due to chemical action. A bulb with hydrogen residue seems to be most easily excited. Electrodes which are most easily deteriorated produce more intense phosphorescence in bulbs, but the condition is not permanent because of the impairment of the vacuum and the deposition of the electrode matter upon the phosphorescent surfaces. Some liquids, as oils, for instance, produce magnificent effects of phosphorescence (or fluorescence?), but they last only a few seconds. So if a bulb has a trace of oil on the walls and the current is turned on, the phosphorescence only persists for a few moments until the oil is carried away. Of all bodies so far tried, sulphide of zinc seems to be the most susceptible to phosphorescence. Some samples, obtained through the kindness of Prof. Henry in Paris, were employed in many of these bulbs. One of the defects of this sulphide is, that it loses its quality of emitting light when brought to a temperature which is by no means high. It can therefore, be used only for feeble intensities. An observation which might deserve notice is, that when violently bombarded from an aluminum electrode it assumes a black color, but singularly enough, it returns to the original condition when it cools down.

The most important fact arrived at in pursuing investigations in this direction is, that in all cases it is necessary, in order to excite phosphorescence with a minimum amount of energy, to observe certain conditions. Namely, there is always, no matter what the frequency of the currents, degree of exhaustion and character of the bodies in the bulb, a certain potential (assuming the bulb excited from one terminal) or potential difference (assuming the bulb to be excited with both terminals) which produces the most economical result. If the potential be increased, considerable energy may be wasted without producing any more light, and if it be diminished, then again the light production is not as economical. The exact condition under which the best result is obtained seems to depend on many things of a different nature, and it is to be yet investigated by other experimenters, but it will certainly have to be observed when such phosphorescent bulbs are operated, if the best results are to be obtained.

Coming now to the most interesting of these phenomena, the incandescence or phosphorescence of gases, at low pressures or at the ordinary pressure of the atmosphere, we must seek the explanation of these phenomena in the same primary causes, that is, in shocks or impacts of the atoms. Just as molecules or atoms beating upon a solid body excite phosphorescence in the same or render it incandescent, so when colliding among themselves they produce similar phenomena. But this is a very insufficient explanation and concerns only the crude mechanism. Light is produced by vibrations which go on at a rate almost inconceivable. If we compute, from the energy contained in the form of known radiations in a definite space the force which is necessary to set up such rapid vibrations, we find, that though the density of the ether be incomparably smaller than that of any body we know, even hydrogen, the force is something surpassing comprehension. What is this force, which in mechanical measure may amount to thousands of tons per square inch? It is electrostatic force in the light of modern views. It is impossible to conceive how a body of measurable dimensions could be charged to so high a potential that the force would be sufficient to produce these vibrations. Long before any such charge could be imparted to the body it would be shattered into atoms. The sun emits light and heat, and so does an ordinary flame or incandescent filament, but in neither of these can the force be accounted for if it be assumed that it is associated with the body as a whole. Only in one way may we account for it, namely, by identifying it with the atom. An atom is so small, that if it be charged by coming in contact with an electrified body and the charge be assumed to follow the same law as in the case of bodies of measurable dimensions, it must retain a quantity of electricity which is fully capable of accounting for these forces and tremendous rates of vibration. But the atom behaves singularly in this respect—it always takes the same " charge."

It is very likely that resonant vibration plays a most important part in all manifestations of energy in nature. Throughout space all matter is vibrating, and all rates of vibration are represented, from the lowest musical note to the highest pitch of the chemical rays, hence an atom, or complex of atoms, no matter what its period, must find a vibration with which it is in resonance.

When we consider the enormous rapidity of the light vibrations, we realize the impossibility of producing such vibrations directly with any apparatus of measurable dimensions, and we are driven to the only possible means of attaining the object of setting up waves of light by electrical means and economically, that is, to affect the molecules or atoms of a gas, to cause them to collide and vibrate. We then must ask ourselves—How can free molecules or atoms be affected?

It is a fact that they can be affected by electrostatic force, as is apparent in many of these experiments. By varying the electrostatic force we can agitate the atoms, and cause them to collide accompanied by evolution of heat and light. It is not demonstrated beyond doubt that we can affect them otherwise. If a luminous discharge is produced in a closed exhausted tube, do the atoms arrange themselves in obedience to any other but to electrostatic force acting in straight lines from atom to atom? Only recently I investigated the mutual action between two circuits with extreme rates of vibration. When a battery of a few jars (*c c c c*, Fig. 196) is discharged through a primary P of low resistance (the connections being as illustrated in Figs. 183*a*, 183*b* and 183*c*), and the frequency of vibration is many millions there are great differences of potential between points on the primary not more than a few inches apart. These differences may be 10,000 volts per inch, if not more, taking the maximum value of the E. M. F. The secondary *s* is therefore acted upon by electrostatic induction, which is in such extreme cases of much greater importance than the electro-dynamic. To such sudden impulses the primary as well as the secondary are poor conductors, and therefore great differences of potential may be produced by electrostatic induction between adjacent points on the secondary. Then sparks may jump between the wires and streamers become visible in the dark if the light of the discharge through the spark gap *d d* be carefully excluded. If now we substitute a closed vacuum tube for the metallic secondary *s*, the differences of potential produced in the tube by electrostatic induction from the primary are fully sufficient to excite portions of it; but as the points of certain differences of potential on the primary are not fixed, but are generally constantly changing in position, a luminous band is produced in the tube, apparently not touching the glass, as it should, if the points of maximum and minimum differences of potential were fixed on the primary. I do not exclude the possibility of such a tube being excited only by electro-dynamic induction, for very able physicists hold this view; but in my opinion, there is as yet no positive proof given that atoms of a gas in a closed tube may arrange themselves in chains under the action of an electromotive impulse produced by electro-dynamic induction in the tube. I have been unable so far to produce striæ in a tube, however long, and at whatever degree of exhaustion, that is, striæ at right angles to the supposed direction of the discharge or the axis of the tube; but I have distinctly observed in a large bulb, in which a wide luminous band was produced by passing a discharge of a battery through a wire surrounding the bulb, a circle of feeble luminosity between two luminous bands, one of which was more intense than the other. Furthermore, with my present experience I do not think that such a gas discharge in a closed tube can vibrate, that is, vibrate as a whole. I am convinced that no

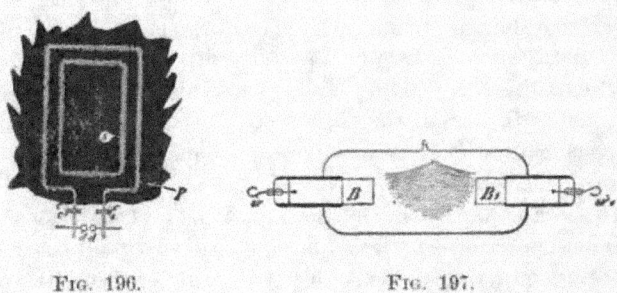

FIG. 196.   FIG. 197.

discharge through a gas can vibrate. The atoms of a gas behave very curiously in respect to sudden electric impulses. The gas does not seem to possess any appreciable inertia to such impulses, for it is a fact, that the higher the frequency of the impulses, with the greater freedom does the discharge pass through the gas. If the gas possesses no inertia then it cannot vibrate, for some inertia is necessary for the free vibration. I conclude from this that if a lightning discharge occurs between two clouds, there can be no oscillation, such as would be expected, considering the capacity of the clouds. But if the lightning discharge strike the earth, there is always vibration—in the earth, but not in the cloud. In a gas discharge each atom vibrates at its own rate, but there is no vibration of the conducting gaseous mass as a whole. This is an important consideration in the great problem of producing light economically, for it teaches us that to reach this result we must use impulses of very high frequency and necessarily also of high potential. It is a fact that oxygen produces a more intense light in a tube. Is it because oxygen atoms possess some inertia and the vibration does not die out instantly? But then nitrogen should be as good, and chlorine and vapors of many other bodies much better than oxygen, unless the magnetic properties of the latter enter prominently into play. Or, is the process in the tube of an electrolytic nature? Many observations certainly speak for it, the most important being that matter is always carried away from the electrodes and the vacuum in a bulb cannot be permanently maintained. If such process takes place in reality, then again must we take refuge in high frequencies, for, with such, electrolytic action should be reduced to a minimum, if not rendered entirely impossible. It is an undeniable fact that with very high frequencies, provided the impulses be of harmonic nature, like those obtained from an alternator, there is less deterioration and the vacua are more permanent. With disruptive discharge coils there are sudden rises of potential and the vacua are more quickly impaired, for the electrodes are deteriorated in a very short time. It was observed in some large tubes, which were provided with heavy carbon blocks B B₁, connected to platinum wires w w_1 (as illustrated in Fig. 197), and which were employed in experiments with the disruptive discharge instead of the ordinary air gap, that the carbon particles under the action of the powerful magnetic field in which the tube was placed, were deposited in regular fine lines in the middle of the tube, as illustrated. These lines were attributed to the deflection or distortion of the discharge by the magnetic field, but why the deposit occurred principally where the field was most intense did not appear quite clear. A fact of interest, likewise noted, was that the presence of a strong magnetic field increases the deterioration of the electrodes, probably by reason of the rapid interruptions it produces, whereby there is actually a higher E. M. F. maintained between the electrodes.

Much would remain to be said about the luminous effects produced in gases at low or ordinary pressures. With the present experiences before us we cannot say that the essential nature of these charming phenomena is sufficiently known. But investigations in this direction are being pushed with exceptional ardor. Every line of scientific pursuit has its fascinations, but electrical investigation appears to possess a peculiar attraction, for there is no experiment or observation of any kind in the domain of this wonderful science which would not forcibly appeal to us. Yet to me it seems, that of all the many marvelous things we observe, a vacuum tube, excited by an electric impulse from a distant source, bursting forth out of the darkness and illuminating the room with its beautiful light, is as lovely a phenomenon as can greet our eyes. More interesting still it appears when, reducing the fundamental discharges across the gap to a very small num-

FIG. 198.

ber and waving the tube about we produce all kinds of designs in luminous lines. So by way of amusement I take a straight long tube, or a square one, or a square attached to a straight tube, and by whirling them about in the hand, I imitate the spokes of a wheel, a Gramme winding, a drum winding, an alternate current motor winding, etc. (Fig. 198). Viewed from a distance the effect is weak and much of its beauty is lost, but being near or holding the tube in the hand, one cannot resist its charm.

In presenting these insignificant results I have not attempted to arrange and co-ordinate them, as would be proper in a strictly scientific investigation, in which every succeeding result should be a logical sequence of the preceding, so that it might be guessed in advance by the careful reader or attentive listener. I have preferred to concentrate my energies chiefly upon advancing novel facts or ideas which might serve as suggestions to others, and this may serve as an excuse for the lack of harmony. The explanations of the phenomena have been given in good faith and in the spirit of a student prepared to find that they admit of a better interpretation. There can be no great harm in a student taking an erroneous view, but when great minds err, the world must dearly pay for their mistakes.

CHAPTER XXIX.

TESLA ALTERNATING CURRENT GENERATORS FOR HIGH FREQUENCY, IN DETAIL.

It has become a common practice to operate arc lamps by alternating or pulsating, as distinguished from continuous, currents; but an objection which has been raised to such systems exists in the fact that the arcs emit a pronounced sound, varying with the rate of the alternations or pulsations of current. This noise is due to the rapidly alternating heating and cooling, and consequent expansion and contraction, of the gaseous matter forming the arc, which corresponds with the periods or impulses of the current. Another disadvantageous feature is found in the difficulty of maintaining an alternating current arc in consequence of the periodical increase in resistance corresponding to the periodical working of the current. This feature entails a further disadvantage, namely, that small arcs are impracticable.

Theoretical considerations have led Mr. Tesla to the belief that these disadvantageous features could be obviated by employing currents of a sufficiently high number of alternations, and his anticipations have been confirmed in practice. These rapidly alternating currents render it possible to maintain small arcs which, besides, possess the advantages of silence and persistency. The latter quality is due to the necessarily rapid alternations, in consequence of which the arc has no time to cool, and is always maintained at a high temperature and low resistance.

At the outset of his experiments Mr. Tesla encountered great difficulties in the construction of high frequency machines. A generator of this kind is described here, which, though constructed quite some time ago, is well worthy of a detailed description. It may be mentioned, in passing, that dynamos of this type have been used by Mr. Tesla in his lighting researches and experiments with currents of high potential and high frequency, and reference to them will be found in his lectures elsewhere printed in this volume.[1]

---

1. See pages 153-4 5.

In the accompaning engravings, Figs. 199 and 200 show the machine, respectively, in side elevation and vertical cross-section ; Figs. 201, 202 and 203 showing enlarged details of construction. As will be seen, A is an annular magnetic frame, the interior of which is provided with a large number of pole-pieces D.

Owing to the very large number and small size of the poles and the spaces between them, the field coils are applied by winding an insulated conductor F zigzag through the grooves, as shown in Fig. 203, carrying the wire around the annulus to form as many layers as is desired. In this way the pole-pieces D will be energized with alternately opposite polarity around the entire ring.

For the armature, Mr. Tesla employs a spider carrying a ring

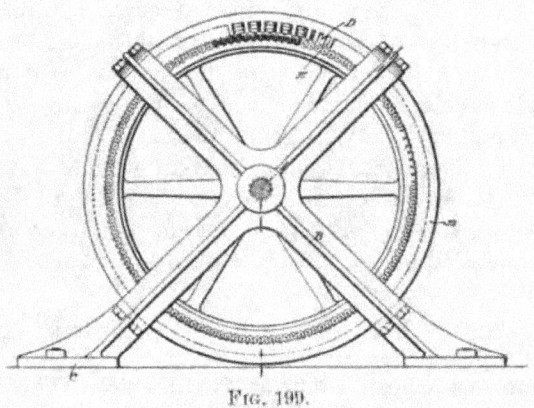

FIG. 199.

J, turned down, except at its edges, to form a trough-like receptacle for a mass of fine annealed iron wires K, which are wound in the groove to form the core proper for the armature-coils. Pins L are set in the sides of the ring J and the coils M are wound over the periphery of the armature-structure and around the pins. The coils M are connected together in series, and these terminals N carried through the hollow shaft H to contact-rings P P, from which the currents are taken off by brushes O.

In this way a machine with a very large number of poles may be constructed. It is easy, for instance, to obtain in this manner three hundred and seventy-five to four hundred poles in a machine that may be safely driven at a speed of fifteen hundred or sixteen hundred revolutions per minute, which will produce ten

thousand or eleven thousand alternations of current per second.
Arc lamps R R are shown in the diagram as connected up in series
with the machine in Fig. 200.   If such a current be applied to
running arc lamps, the sound produced by or in the arc becomes
practically inaudible, for, by increasing the rate of change in the
current, and consequently the number of vibrations per unit of
time of the gaseous material of the arc up to, or beyond, ten
thousand or eleven thousand per second, or to what is regarded
as the limit of audition, the sound due to such vibrations will not
be audible.   The exact number of changes or undulations necessary to produce this result will vary somewhat according to the
size of the arc—that is to say, the smaller the arc, the greater the

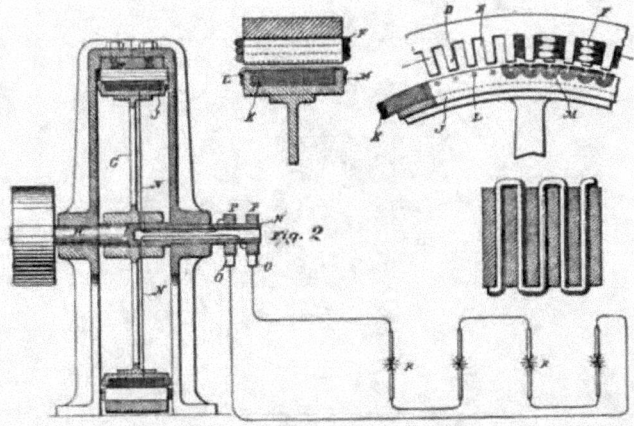

Figs. 200, 201, 202 and 203.

number of changes that will be required to render it inaudible
within certain limits. It should also be stated that the arc should
not exceed a certain length.

The difficulties encountered in the construction of these
machines are of a mechanical as well as an electrical nature.
The machines may be designed on two plans: the field may be
formed either of alternating poles, or of polar projections of the
same polarity.   Up to about 15,000 alternations per second in an
experimental machine, the former plan may be followed, but a
more efficient machine is obtained on the second plan.

In the machine above described, which was capable of running
two arcs of normal candle power, the field was composed of a

ring of wrought iron 32 inches outside diameter, and about 1 inch thick. The inside diameter was 30 inches. There were 384 polar projections. The wire was wound in zigzag form, but two wires were wound so as to completely envelop the projections. The distance between the projections is about $\frac{3}{16}$ inch, and they are a little over $\frac{1}{16}$ inch thick. The field magnet was made relatively small so as to adapt the machine for a constant current. There are 384 coils connected in two series. It was found impracticable to use any wire much thicker than No. 26 B. and S. gauge on account of the local effects. In such a machine the clearance should be as small as possible; for this reason the machine was made only $1\frac{1}{4}$ inch wide, so that the binding wires might be obviated. The armature wires must be wound with

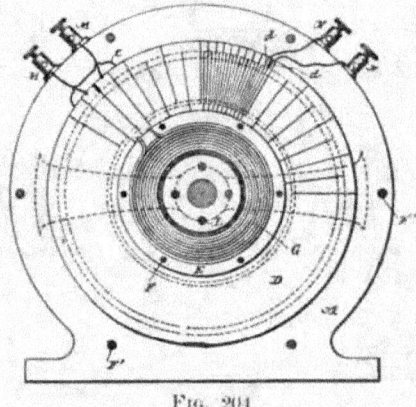

Fig. 204.

great care, as they are apt to fly off in consequence of the great peripheral speed. In various experiments this machine has been run as high as 3,000 revolutions per minute. Owing to the great speed it was possible to obtain as high as 10 amperes out of the machine. The electromotive force was regulated by means of an adjustable condenser within very wide limits, the limits being the greater, the greater the speed. This machine was frequently used to run Mr. Tesla's laboratory lights.

The machine above described was only one of many such types constructed. It serves well for an experimental machine, but if still higher alternations are required and higher efficiency is necessary, then a machine on a plan shown in Figs. 204 to

207, is preferable. The principal advantage of this type of machine is that there is not much magnetic leakage, and that a field may be produced, varying greatly in intensity in places not much distant from each other.

In these engravings, Figs. 204 and 205 illustrate a machine in which the armature conductor and field coils are stationary, while the field magnet core revolves. Fig. 206 shows a machine embodying the same plan of construction, but having a stationary field magnet and rotary armature.

The conductor in which the currents are induced may be arranged in various ways; but Mr. Tesla prefers the following method: He employs an annular plate of copper D, and by

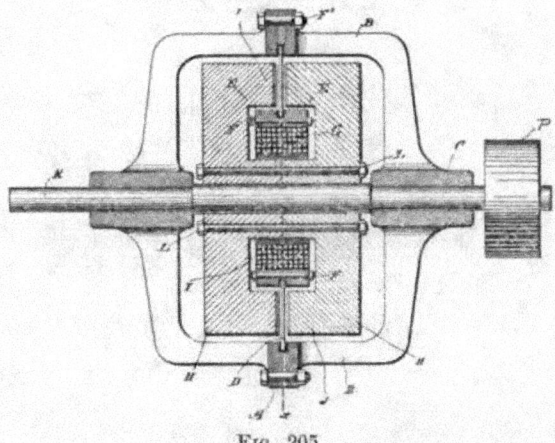

FIG. 205.

means of a saw cuts in it radial slots from one edge nearly through to the other, beginning alternately from opposite edges. In this way a continuous zigzag conductor is formed. When the polar projections are $\frac{1}{4}$ inch wide, the width of the conductor should not, under any circumstances, be more than $\frac{1}{32}$ inch wide; even then the eddy effect is considerable.

To the inner edge of this plate are secured two rings of non-magnetic metal E, which are insulated from the copper conductor, but held firmly thereto by means of the bolts F. Within the rings E is then placed an annular coil G, which is the energizing coil for the field magnet. The conductor D and the parts attached thereto are supported by means of the cylindrical shell or casting A A, the two parts of which are brought together and clamped to the outer edge of the conductor D.

The core for the field magnet is built up of two circular parts H H, formed with annular grooves I, which, when the two parts are brought together, form a space for the reception of the energizing coil G. The hubs of the cores are trued off, so as to fit closely against one another, while the outer portions or flanges which form the polar faces J J, are reduced somewhat in thickness to make room for the conductor D, and are serrated on their faces. The number of serrations in the polar faces is arbitrary:

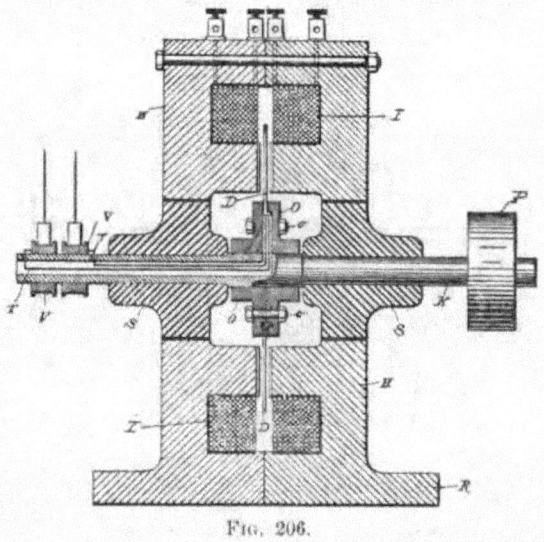

FIG. 206.

but there must exist between them and the radial portions of the conductor D certain relation, which will be understood by reference to Fig. 207 in which N N represent the projections or points on one face of the core of the field, and s s the points of the other face. The conductor D is shown in this figure in section a a' designating the radial portions of the conductor, and b the insulating divisions between them. The relative width of the parts a a' and the space between any two adjacent points N N or s s is such that when the radial portions a of the conductor are passing between the opposite points N s where the field is strongest, the intermediate radial portions a' are passing through the

widest spaces midway between such points and where the field is
weakest. Since the core on one side is of opposite polarity to
the part facing it, all the projections of one polar face will be of
opposite polarity to those of the other face. Hence, although
the space between any two adjacent points on the same face may
be extremely small, there will be no leakage of the magnetic
lines between any two points of the same name, but the lines of
force will pass across from one set of points to the other. The
construction followed obviates to a great degree the distortion of
the magnetic lines by the action of the current in the conductor
D, in which it will be observed the current is flowing at any given
time from the centre toward the periphery in one set of radial
parts a and in the opposite direction in the adjacent parts a'.

In order to connect the energizing coil G, Fig. 204, with a source
of continuous current, Mr. Tesla utilizes two adjacent radial portions of the conductor D for connecting the terminals of the coil
G with two binding posts M. For this purpose the plate D is cut

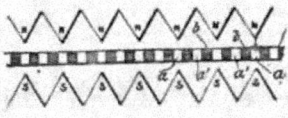

FIG. 207.

entirely through, as shown, and the break thus made is bridged
over by a short conductor c. The plate D is cut through to form
two terminals d, which are connected to binding posts N. The
core H H, when rotated by the driving pulley, generates in the conductors D an alternating current, which is taken off from the
binding posts N.

When it is desired to rotate the conductor between the faces
of a stationary field magnet, the construction shown in Fig.
206, is adopted. The conductor D in this case is or may be
made in substantially the same manner as above described by
slotting an annular conducting-plate and supporting it between
two heads o, held together by bolts o and fixed to the driving-shaft
K. The inner edge of the plate or conductor D is preferably
flanged to secure a firmer union between it and the heads o. It
is insulated from the head. The field-magnet in this case consists of two annular parts H H, provided with annular grooves I
for the reception of the coils. The flanges or faces surrounding

the annular groove are brought together, while the inner flanges are serrated, as in the previous case, and form the polar faces. The two parts H H are formed with a base R, upon which the machine rests. s s are non-magnetic bushings secured or set in the central opening of the cores. The conductor D is cut entirely through at one point to form terminals, from which insulated conductors T are led through the shaft to collecting-rings v.

In one type of machine of this kind constructed by Mr. Tesla, the field had 480 polar projections on each side, and from this machine it was possible to obtain 30,000 alternations per second. As the polar projections must necessarily be very narrow, very thin wires or sheets must be used to avoid the eddy current effects. Mr. Tesla has thus constructed machines with a stationary armature and rotating field, in which case also the field-coil was supported so that the revolving part consisted only of a wrought iron body devoid of any wire and also machines with a rotating armature and stationary field. The machines may be either drum or disc, but Mr. Tesla's experience shows the latter to be preferable.

In the course of a very interesting article contributed to the *Electrical World* in February, 1891, Mr. Tesla makes some suggestive remarks on these high frequency machines and his experiences with them, as well as with other parts of the high frequency apparatus. Part of it is quoted here and is as follows:—

The writer will incidentally mention that any one who attempts for the first time to construct such a machine will have a tale of woe to tell. He will first start out, as a matter of course, by making an armature with the required number of polar projections. He will then get the satisfaction of having produced an apparatus which is fit to accompany a thoroughly Wagnerian opera. It may besides possess the virtue of converting mechanical energy into heat in a nearly perfect manner. If there is a reversal in the polarity of the projections, he will get heat out of the machine; if there is no reversal, the heating will be less, but the output will be next to nothing. He will then abandon the iron in the armature, and he will get from the Scylla to the Charybdis. He will look for one difficulty and will find another, but, after a few trials, he may get nearly what he wanted.

Among the many experiments which may be performed with
such a machine, of not the least interest are those performed
with a high-tension induction coil. The character of the discharge is completely changed. The arc is established at much
greater distances, and it is so easily affected by the slightest current of air that it often wriggles around in the most singular
manner. It usually emits the rhythmical sound peculiar to the
alternate current arcs, but the curious point is that the sound
may be heard with a number of alternations far above ten thousand per second, which by many is considered to be about the
limit of audition. In many respects the coil behaves like a static
machine. Points impair considerably the sparking interval, electricity escaping from them freely, and from a wire attached to
one of the terminals streams of light issue, as though it were
connected to a pole of a powerful Toepler machine. All these
phenomena are, of course, mostly due to the enormous differences of potential obtained. As a consequence of the self-induction of the coil and the high frequency, the current is minute
while there is a corresponding rise of pressure. A current impulse of some strength started in such a coil should persist to
flow no less than four ten-thousandths of a second. As this time
is greater than half the period, it occurs that an opposing electromotive force begins to act while the current is still flowing. As
a consequence, the pressure rises as in a tube filled with liquid
and vibrated rapidly around its axis. The current is so small
that, in the opinion and involuntary experience of the writer, the
discharge of even a very large coil cannot produce seriously injurious effects, whereas, if the same coil were operated with a
current of lower frequency, though the electromotive force would
be much smaller, the discharge would be most certainly injurious. This result, however, is due in part to the high frequency.
The writer's experiences tend to show that the higher the frequency the greater the amount of electrical energy which may
be passed through the body without serious discomfort; whence
it seems certain that human tissues act as condensers.

One is not quite prepared for the behavior of the coil when
connected to a Leyden jar. One, of course, anticipates that since
the frequency is high the capacity of the jar should be small. He
therefore takes a very small jar, about the size of a small wine
glass, but he finds that even with this jar the coil is practically
short-circuited. He then reduces the capacity until he comes to

about the capacity of two spheres, say, ten centimetres in diameter and two to four centimetres apart. The discharge then assumes the form of a serrated band exactly like a succession of sparks viewed in a rapidly revolving mirror; the serrations, of course, corresponding to the condenser discharges. In this case one may observe a queer phenomenon. The discharge starts at the nearest points, works gradually up, breaks somewhere near the top of the spheres, begins again at the bottom, and so on. This goes on so fast that several serrated bands are seen at once. One may be puzzled for a few minutes, but the explanation is simple enough. The discharge begins at the nearest points, the air is heated and carries the arc upward until it breaks, when it is reestablished at the nearest points, etc. Since the current passes easily through a condenser of even small capacity, it will be found quite natural that connecting only one terminal to a body of the same size, no matter how well insulated, impairs considerably the striking distance of the arc.

Experiments with Geissler tubes are of special interest. An exhausted tube, devoid of electrodes of any kind, will light up at some distance from the coil. If a tube from a vacuum pump is near the coil the whole of the pump is brilliantly lighted. An incandescent lamp approached to the coil lights up and gets perceptibly hot. If a lamp have the terminals connected to one of the binding posts of the coil and the hand is approached to the bulb, a very curious and rather unpleasant discharge from the glass to the hand takes place, and the filament may become incandescent. The discharge resembles to some extent the stream issuing from the plates of a powerful Toepler machine, but is of incomparably greater quantity. The lamp in this case acts as a condenser, the rarefied gas being one coating, the operator's hand the other. By taking the globe of a lamp in the hand, and by bringing the metallic terminals near to or in contact with a conductor connected to the coil, the carbon is brought to bright incandescence and the glass is rapidly heated. With a 100-volt 10 c. p. lamp one may without great discomfort stand as much current as will bring the lamp to a considerable brilliancy; but it can be held in the hand only for a few minutes, as the glass is heated in an incredibly short time. When a tube is lighted by bringing it near to the coil it may be made to go out by interposing a metal plate on the hand between the coil and tube; but if the metal plate be fastened to a glass rod or otherwise insulated, the tube may remain lighted if the plate be interposed, or may even increase in luminosity. The effect depends on the position of the plate and tube relatively to the coil, and may be always easily foretold by *assuming* that conduction takes place from one terminal of the coil to the other. According to the position of the plate, it may either divert from or direct the current to the tube.

In another line of work the writer has in frequent experiments maintained incandescent lamps of 50 or 100 volts burning at any desired candle power with both the terminals of each lamp connected to a stout copper wire of no more than a few feet in length. These experiments seem interesting enough, but they are not more so than the queer experiment of Faraday, which has been revived and made much of by recent investigators, and in which a discharge is made to jump between two points of a bent copper wire. An experiment may be cited here which may seem equally interesting. If a Geissler tube, the terminals of which are joined by a copper wire, be approached to the coil, certainly no one would be prepared to see the tube light up. Curiously enough, it does light up, and, what is more, the wire does not seem to make much difference. Now one is apt to think in the first moment that the impedance of the wire might have something to do with the phenomenon. But this is of course immediately rejected, as for this an enormous frequency would be required. This result, however, seems puzzling only at first; for upon reflection it is quite clear that the wire can make but little difference. It may be explained in more than one way, but it agrees perhaps best with observation to assume that conduction takes place from the terminals of the coil through the space. On this assumption, if the tube with the wire be held in any position, the wire can divert little more than the current which passes through the space occupied by the wire and the metallic terminals of the tube; through the adjacent space the current passes practically undisturbed. For this reason, if the tube be held in any position at right angles to the line joining the binding posts of the coil, the wire makes hardly any difference, but in a position more or less parallel with that line it impairs to a certain extent the brilliancy of the tube and its facility to light up. Numerous other phenomena may be explained on the same assumption. For instance, if the ends of the tube be provided with washers of sufficient size and held in the line joining the terminals of the coil, it will not light up, and then nearly the whole of the current, which would otherwise pass uniformly through the space between the washers, is diverted through the wire. But if the tube be inclined sufficiently to that line, it will light up in spite of the washers. Also, if a metal plate be fastened upon a glass rod and held at right angles to the line joining the binding posts, and nearer to one of them, a tube held more or less parallel with the line will light up instantly when one of the terminals touches the plate, and will go out when separated from the plate. The greater the surface of the plate, up to a certain limit, the easier the tube will light up. When a tube is placed at right angles to the straight line joining the binding posts, and then rotated, its luminosity steadily increases until it is parallel with that line. The writer must state, however, that he does not favor the idea of a leakage or current through the space any more than as a suitable explanation, for he is convinced that all these experiments could not be performed with a static machine yielding a constant difference of potential, and that condenser action is largely concerned in these phenomena.

It is well to take certain precautions when operating a Ruhmkorff coil with very rapidly alternating currents. The primary current should not be turned on too long, else the core may get so hot as to melt the gutta-percha or paraffin, or otherwise injure the insulation, and this may occur in a surprisingly short time, considering the current's strength. The primary current being turned on, the fine wire terminals may be joined without great risk, the impedance being so great that it is difficult to force enough current through the fine wire so as to injure it, and in fact the coil may be on the whole much safer when the terminals of the fine wire are connected than when they are insulated; but special care should be taken when the terminals are connected to the coatings of a Leyden jar, for with anywhere near the critical capacity, which just counteracts the self-induction at the existing frequency, the coil might meet the fate of St. Polycarpus. If an expensive vacuum pump is lighted up by being near to the coil or touched with a wire connected to one of the terminals, the current should be left on no more than a few moments, else the glass will be cracked by the heating of the rarefied gas in one of the narrow passages—in the writer's own experience *quod erat demonstrandum.*[1]

---

1. It is thought necessary to remark that, although the induction coil may give quite a good result when operated with such rapidly alternating currents, yet its construction, quite irrespective of the iron core, makes it very unfit for such high frequencies, and to obtain the best results the construction should be greatly modified.

There are a good many other points of interest which may be observed in connection with such a machine. Experiments with the telephone, a conductor in a strong field or with a condenser or arc, seem to afford certain proof that sounds far above the usual accepted limit of hearing would be perceived. A telephone will emit notes of twelve to thirteen thousand vibrations per second; then the inability of the core to follow such rapid alternations begins to tell. If, however, the magnet and core be replaced by a condenser and the terminals connected to the high-tension secondary of a transformer, higher notes may still be heard. If the current be sent around a finely laminated core and a small piece of thin sheet iron be held gently against the core, a sound may be still heard with thirteen to fourteen thousand alternations per second, provided the current is sufficiently strong. A small coil, however, tightly packed between the poles of a powerful magnet, will emit a sound with the above number of alternations, and arcs may be audible with a still higher frequency. The limit of audition is variously estimated. In Sir William Thomson's writings it is stated somewhere that ten thousand per second, or nearly so, is the limit. Other, but less reliable, sources give it as high as twenty-four thousand per second. The above experiments have convinced the writer that notes of an incomparably higher number of vibrations per second would be perceived provided they could be produced with sufficient power. There is no reason why it should not be so. The condensations and rarefactions of the air would necessarily set the diaphragm in a corresponding vibration and some sensation would be produced, whatever—within certain limits—the velocity of transmission to their nerve centres, though it is probable that for want of exercise the ear would not be able to distinguish any such high note. With the eye it is different; if the sense of vision is based upon some resonance effect, as many believe, no amount of increase in the intensity of the ethereal vibration could extend our range of vision on either side of the visible spectrum.

The limit of audition of an arc depends on its size. The greater the surface by a given heating effect in the arc, the higher the limit of audition. The highest notes are emitted by the high-tension discharges of an induction coil in which the arc is, so to speak, all surface. If R be the resistance of an arc, and C the current, and the linear dimensions be n times increased, then the resistance is $\frac{R}{n}$, and with the same current density the current would be n^2C; hence the heating effect is n^3 times greater, while the surface is only n^2 times as great. For this reason very large arcs would not emit any rhythmical sound even with a very low frequency. It must be observed, however, that the sound emitted depends to some extent also on the composition of the carbon. If the carbon contain highly refractory material, this, when heated, tends to maintain the temperature of the arc uniform and the sound is lessened; for this reason it would seem that an alternating arc requires such carbons.

With currents of such high frequencies it is possible to obtain noiseless arcs, but the regulation of the lamp is rendered extremely difficult on account of the excessively small attractions or repulsions between conductors conveying these currents.

An interesting feature of the arc produced by these rapidly alternating currents is its persistency. There are two causes for it, one of which is always present, the other sometimes only. One is due to the character of the current and the other to a property of the machine. The first cause is the more important one, and is due directly to the rapidity of the alternations. When an arc is formed by a periodically undulating current, there is a corresponding undulation in the temperature of the gaseous column, and, therefore, a corresponding undulation in the resistance of the arc. But the resistance of the arc varies enormously with the temperature of the gaseous column, being practically infinite when the gas between the electrodes is cold. The persistence of the arc, therefore, depends on the inability of the column to cool. It is for this reason impossible to maintain an arc with the current alternating only a few times a second. On the other hand, with a practically continuous current, the arc is easily maintained, the column being constantly kept at a high temperature and low resistance. The higher the frequency the smaller the time interval during which the arc may cool and increase considerably in resistance. With a frequency of 10,000 per second or more in an arc of equal size excessively small variations of temperature are superimposed upon a steady temperature, like ripples on the surface of a deep sea. The heating effect is practically continuous and the arc behaves like one produced by a continuous current, with the exception, however, that it may not be quite as easily started, and that the electrodes are equally consumed; though the writer has observed some irregularities in this respect.

The second cause alluded to, which possibly may not be present, is due to the tendency of a machine of such high frequency to maintain a practically constant current. When the arc is lengthened, the electromotive force rises in proportion and the arc appears to be more persistent.

Such a machine is eminently adapted to maintain a constant current, but it is very unfit for a constant potential. As a matter of fact, in certain types of such machines a nearly constant current is an almost unavoidable result. As the number of poles or polar projections is greatly increased, the clearance becomes of great importance. One has really to do with a great number of very small machines. Then there is the impedance in the armature, enormously augmented by the high frequency. Then, again, the magnetic leakage is facilitated. If there are three or four hundred alternate poles, the leakage is so great that it is virtually the same as connecting, in a two-pole machine, the poles by a piece of iron. This disadvantage, it is true, may be obviated more or less by using a field throughout of the same polarity, but then one encounters difficulties of a different nature. All these things tend to maintain a constant current in the armature circuit.

In this connection it is interesting to notice that even to-day engineers are astonished at the performance of a constant current machine, just as, some years ago, they used to consider it an extraordinary performance if a machine was capable of maintaining a constant potential difference between the terminals. Yet one result is just as easily secured as the other. It must only be remembered that in an inductive apparatus of any kind, if constant potential is required, the inductive relation between the primary or exciting and secondary or armature circuit must be the closest possible; whereas, in an apparatus for constant current just the opposite is required. Furthermore, the opposition to the current's flow in the induced circuit must be as small as possible in the former and as great as possible in the latter case. But opposition to a current's flow may be caused in more than one way. It may be caused by ohmic resistance or self-induction. One may make the induced circuit of a dynamo machine or transformer of such high resistance that when operating devices of considerably smaller resistance within very wide limits a nearly constant current is maintained. But such high resistance involves a great loss in power, hence it is not practicable. Not so self-induction. Self-induction does not necessarily mean loss of power. The moral is, use self-induction instead of resistance. There is, however, a circumstance which favors the adoption of this plan, and this is, that a very high self-induction may be obtained cheaply by surrounding a comparatively small length of wire more or less completely with iron, and, furthermore, the effect may be exalted at will by causing a rapid undulation of the current. To sum up, the requirements for constant current are: Weak magnetic connection between the induced and inducing circuits, greatest possible self-induction with the least resistance, greatest practicable rate of change of the current. Constant potential, on the other hand, requires: Closest magnetic connection between the circuits, steady induced current, and, if possible, no reaction. If the latter conditions could be fully satisfied in a constant potential machine, its output would surpass many times that of a machine primarily designed to give constant current. Unfortunately, the type of machine in which these conditions may be satisfied is of little practical value, owing to the small electromotive force obtainable and the difficulties in taking off the current.

With their keen inventor's instinct, the now successful arc-light men have early recognized the desiderata of a constant current machine. Their arc light machines have weak fields, large armatures, with a great length of copper wire and few commutator segments to produce great variations in the current's strength and to bring self-induction into play. Such machines may maintain within considerable limits of variation in the resistance of the circuit a practically constant current. Their output is of course correspondingly diminished, and, perhaps with the object in view not to cut down the output too much, a simple device compensating exceptional variations is employed. The undulation of the current is almost essential to the commercial success of an arc-light system. It introduces in the circuit a steadying element taking the place of a large ohmic resistance, without involving a great loss in power, and, what is more important, it allows the use of simple clutch lamps, which with a current of a certain number of impulses per second, best suitable for each particular lamp, will, if properly attended to, regulate even better than the finest clock-work lamps. This discovery has been made by the writer—several years too late.

It has been asserted by competent English electricians that in a constant-current machine or transformer the regulation is effected by varying the phase of the secondary current. That this view is erroneous may be easily proved by using, instead of lamps, devices each possessing self-induction and capacity or self-induction and resistance—that is, retarding and accelerating components—in such proportions as to not affect materially the phase of the secondary current. Any number of such devices may be inserted or cut out, still it will be found that the regulation occurs, a constant current being maintained, while the electromotive force is varied with the number of the devices. The change of phase of the secondary current is simply a result following from the changes in resistance, and, though secondary reaction is always of more or less importance, yet the real cause of the regulation lies in the existence of the conditions above enumerated. It should be stated, however, that in the case of a machine the above remarks are to be restricted to the cases in which the machine is independently excited. If the excitation be effected by commutating the armature current, then the fixed position of the brushes makes any shifting of the neutral line of the utmost importance, and it may not be thought immodest of the writer to mention that, as far as records go, he seems to have been the first who has successfully regulated machines by providing a bridge connection between a point of the external circuit and the commutator by means of a third brush. The armature and field being properly proportioned and the brushes placed in their determined positions, a constant current or constant potential resulted from the shifting of the diameter of commutation by the varying loads.

In connection with machines of such high frequencies, the condenser affords an especially interesting study. It is easy to raise the electromotive force of such a machine to four or five times the value by simply connecting the condenser to the circuit, and the writer has continually used the condenser for the the purposes of regulation, as suggested by Blakesley in his book on alternate currents, in which he has treated the most frequently occurring condenser problems with exquisite simplicity and clearness. The high frequency allows the use of small capacities and renders investigation easy. But, although in most of the experiments the result may be foretold, some phenomena observed seem at first curious. One experiment performed three or four months ago with such a machine and a condenser may serve as an illustration. A machine was used giving about 20,000 alternations per second. Two bare wires about twenty feet long and two millimetres in diameter, in close proximity to each other, were connected to the terminals of the machine at the one end, and to a condenser at the other. A small transformer without an iron core, of course, was used to bring the reading within range of a Cardew voltmeter by connecting the voltmeter to the secondary. On the terminals of the condenser the electromotive force was about 120 volts, and from there inch by inch it gradually fell until at the terminals of the machine it was about 65 volts. It was virtually as though the condenser were a generator, and the line and armature circuit simply a resistance connected to it. The writer looked for a case of resonance, but he was unable to augment the effect by varying the capacity very carefully and gradually or by changing the speed of the machine. A case of pure resonance he was unable to obtain. When a condenser was connected to the terminals of the machine—the self-induction of the armature being first determined in the maximum and minimum position and the mean value taken —the capacity which gave the highest electromotive force corresponded most nearly to that which just counteracted the self-induction with the existing frequency. If the capacity was increased or diminished, the electromotive force fell as expected.

With frequencies as high as the above mentioned, the condenser effects are of enormous importance. The condenser becomes a highly efficient apparatus capable of transferring considerable energy.

In an appendix to this book will be found a description of the Tesla oscillator, which its inventor believes will among other great advantages give him the necessary high frequency conditions, while relieving him of the inconveniences that attach to generators of the type described at the beginning of this chapter.

CHAPTER XXX.

ALTERNATE CURRENT ELECTROSTATIC INDUCTION APPARATUS.[1]

ABOUT a year and a half ago while engaged in the study of
alternate currents of short period, it occurred to me that such
currents could be obtained by rotating charged surfaces in close
proximity to conductors. Accordingly I devised various forms

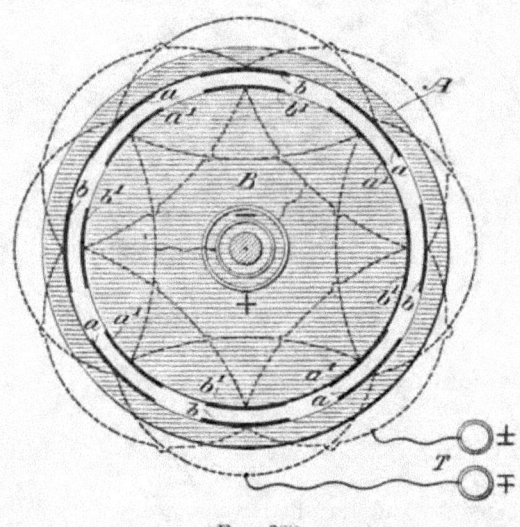

FIG. 208.

of experimental apparatus of which two are illustrated in the
accompanying engravings.

In the apparatus shown in Fig. 208, A is a ring of dry shellacked hard wood provided on its inside with two sets of tin-foil
coatings, a and b, all the a coatings and all the b coatings being
connected together, respectively, but independent from each
other. These two sets of coatings are connected to two termi-

---

1. Article by Mr. Tesla in *The Electrical Engineer*, N. Y., May 6, 1891.

nals, T. For the sake of clearness only a few coatings are shown. Inside of the ring A, and in close proximity to it there is arranged to rotate a cylinder B, likewise of dry, shellacked hard wood, and provided with two similar sets of coatings, a^1 and b^1, all the coatings a^1 being connected to one ring and all the others, b^1, to another marked $+$ and $-$. These two sets, a^1 and b^1 are charged to a high potential by a Holtz or Wimshurst machine, and may be connected to a jar of some capacity. The inside of ring A is coated with mica in order to increase the induction and also to allow higher potentials to be used.

When the cylinder B with the charged coatings is rotated, a

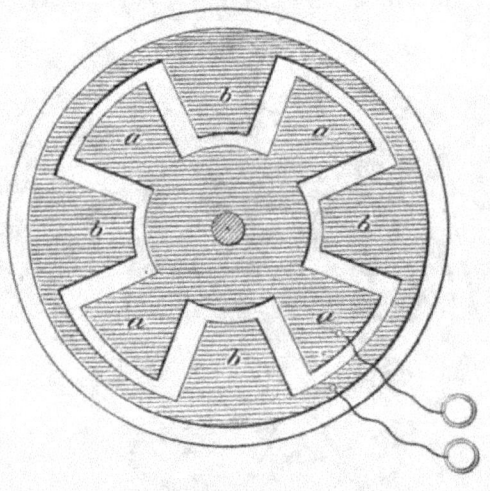

FIG. 209.

circuit connected to the terminals T is traversed by alternating currents. Another form of apparatus is illustrated in Fig. 209. In this apparatus the two sets of tin-foil coatings are glued on a plate of ebonite, and a similar plate which is rotated, and the coatings of which are charged as in Fig. 208, is provided.

The output of such an apparatus is very small, but some of the effects peculiar to alternating currents of short periods may be observed. The effects, however, cannot be compared with those obtainable with an induction coil which is operated by an alternate current machine of high frequency, some of which were described by me a short while ago.

CHAPTER XXXI.

"Massage" With Currents of High Frequency.[1]

I trust that the present brief communication will not be interpreted as an effort on my part to put myself on record as a "patent medicine" man, for a serious worker cannot despise anything more than the misuse and abuse of electricity which we have frequent occasion to witness. My remarks are elicited by the lively interest which prominent medical practitioners evince at every real advance in electrical investigation. The progress in recent years has been so great that every electrician and electrical engineer is confident that electricity will become the means of accomplishing many things that have been heretofore, with our existing knowledge, deemed impossible. No wonder then that progressive physicians also should expect to find in it a powerful tool and help in new curative processes. Since I had the honor to bring before the American Institute of Electrical Engineers some results in utilizing alternating currents of high tension, I have received many letters from noted physicians inquiring as to the physical effects of such currents of high frequency. It may be remembered that I then demonstrated that a body perfectly well insulated in air can be heated by simply connecting it with a source of rapidly alternating high potential. The heating in this case is due in all probability to the bombardment of the body by air, or possibly by some other medium, which is molecular or atomic in construction, and the presence of which has so far escaped our analysis—for according to my ideas, the true ether radiation with such frequencies as even a few millions per second must be very small. This body may be a good conductor or it may be a very poor conductor of electricity with little change in the result. The human body is, in such a case, a fine conductor, and if a person insulated in a room, or no matter where, is brought into contact with such a source of

---

1. Article by Mr. Tesla in *The Electrical Engineer* of Dec. 23d, 1891.

rapidly alternating high potential, the skin is heated by bombardment. It is a mere question of the dimensions and character of the apparatus to produce any degree of heating desired.

It has occurred to me whether, with such apparatus properly prepared, it would not be possible for a skilled physician to find in it a means for the effective treatment of various types of disease. The heating will, of course, be superficial, that is, on the skin, and would result, whether the person operated on were in bed or walking around a room, whether dressed in thick clothes or whether reduced to nakedness. In fact, to put it broadly, it is conceivable that a person entirely nude at the North Pole might keep himself comfortably warm in this manner.

Without vouching for all the results, which must, of course, be determined by experience and observation, I can at least warrant the fact that heating would occur by the use of this method of subjecting the human body to bombardment by alternating currents of high potential and frequency such as I have long worked with. It is only reasonable to expect that some of the novel effects will be wholly different from those obtainable with the old familiar therapeutic methods generally used. Whether they would all be beneficial or not remains to be proved.

CHAPTER XXXII.

ELECTRIC DISCHARGE IN VACUUM TUBES.[1]

In *The Electrical Engineer* of June 10 I have noted the description of some experiments of Prof. J. J. Thomson, on the "Electric Discharge in Vacuum Tubes," and in your issue of June 24 Prof. Elihu Thomson describes an experiment of the same kind. The fundamental idea in these experiments is to set up an electromotive force in a vacuum tube—preferably devoid of any electrodes—by means of electro-magnetic induction, and to excite the tube in this manner.

As I view the subject I should, think that to any experimenter who had carefully studied the problem confronting us and who attempted to find a solution of it, this idea must present itself as naturally as, for instance, the idea of replacing the tinfoil coatings of a Leyden jar by rarefied gas and exciting luminosity in the condenser thus obtained by repeatedly charging and discharging it. The idea being obvious, whatever merit there is in this line of investigation must depend upon the completeness of the study of the subject and the correctness of the observations. The following lines are not penned with any desire on my part to put myself on record as one who has performed similar experiments, but with a desire to assist other experimenters by pointing out certain peculiarities of the phenomena observed, which, to all appearances, have not been noted by Prof. J. J. Thomson, who, however, seems to have gone about systematically in his investigations, and who has been the first to make his results known. These peculiarities noted by me would seem to be at variance with the views of Prof. J. J. Thomson, and present the phenomena in a different light.

My investigations in this line occupied me principally during the winter and spring of the past year. During this time many different experiments were performed, and in my exchanges of ideas

---

1. Article by Mr. Tesla in *The Electrical Engineer*. N. Y., July 1, 1891.

on this subject with Mr. Alfred S. Brown, of the Western Union Telegraph Company, various different dispositions were suggested which were carried out by me in practice. Fig. 210 may serve as an example of one of the many forms of apparatus used. This consisted of a large glass tube sealed at one end and projecting into an ordinary incandescent lamp bulb. The primary, usually consisting of a few turns of thick, well-insulated copper sheet was inserted within the tube, the inside space of the bulb furnishing the secondary. This form of apparatus was arrived at after some experimenting, and was used principally with the view of enabling me to place a polished reflecting surface on the inside of the tube, and for this purpose the last turn of the primary was covered with a thin silver sheet. In all forms of apparatus used

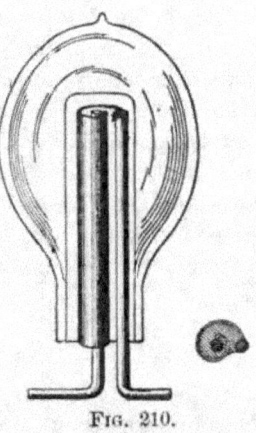

FIG. 210.

there was no special difficulty in exciting a luminous circle or cylinder in proximity to the primary.

As to the number of turns, I cannot quite understand why Prof. J. J. Thomson should think that a few turns were "quite sufficient," but lest I should impute to him an opinion he may not have, I will add that I have gained this impression from the reading of the published abstracts of his lecture. Clearly, the number of turns which gives the best result in any case, is dependent on the dimensions of the apparatus, and, were it not for various considerations, one turn would always give the best result.

I have found that it is preferable to use in these experiments an alternate current machine giving a moderate number of alternations per second to excite the induction coil for charging the Leyden jar which discharges through the primary—shown diagrammatically in Fig. 211,—as in such case, before the disruptive discharge takes place, the tube or bulb is slightly excited and the formation of the luminous circle is decidedly facilitated.

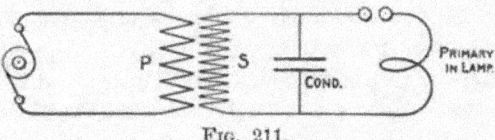

FIG. 211.

But I have also used a Wimshurst machine in some experiments.

Prof. J. J. Thomson's view of the phenomena under consideration seems to be that they are wholly due to electro-magnetic action. I was, at one time, of the same opinion, but upon carefully investigating the subject I was led to the conviction that they are more of an electrostatic nature. It must be remembered that in these experiments we have to deal with primary currents of an enormous frequency or rate of change and of high potential, and that the secondary conductor consists of a rarefied

FIG. 212.

gas, and that under such conditions electrostatic effects must play an important part.

In support of my view I will describe a few experiments made by me. To excite luminosity in the tube it is not absolutely necessary that the conductor should be closed. For instance, if an ordinary exhausted tube (preferably of large diameter) be surrounded by a spiral of thick copper wire serving as the primary; a feebly luminous spiral may be induced in the tube, roughly shown in Fig. 212. In one of these experiments a curious phenomenon was observed; namely, two intensely luminous circles, each of them close to a turn of the primary spiral, were formed inside of the tube, and I attributed this phenomenon to the existence of nodes on the primary. The circles were connected by a faint luminous spiral parallel to the primary and in close proximity to it. To produce this effect I have found it necessary to strain the jar to the utmost. The turns of the spiral tend to close and form circles, but this, of course, would be expected, and does not necessarily indicate an electro-magnetic effect; whereas the fact that a glow can be produced along the primary in the form of an open spiral argues for an electrostatic effect.

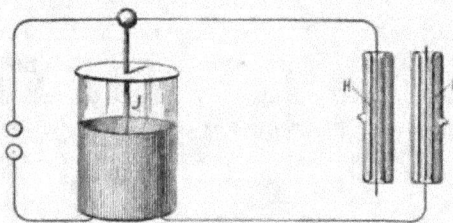

FIG. 213.

In using Dr. Lodge's recoil circuit, the electrostatic action is likewise apparent. The arrangement is illustrated in Fig. 213. In his experiment two hollow exhausted tubes H H were slipped over the wires of the recoil circuit and upon discharging the jar in the usual manner luminosity was excited in the tubes.

Another experiment performed is illustrated in Fig. 214. In this case an ordinary lamp-bulb was surrounded by one or two turns of thick copper wire P and the luminous circle L excited in the bulb by discharging the jar through the primary. The lamp-bulb was provided with a tinfoil coating on the side opposite to the primary and each time the tinfoil coating was connected to the ground or to a large object the luminosity of the circle was considerably increased. This was evidently due to electrostatic action.

In other experiments I have noted that when the primary touches the glass the luminous circle is easier produced and is

more sharply defined ; but I have not noted that, generally speaking, the circles induced were very sharply defined, as Prof. J. J.
Thomson has observed ; on the contrary, in my experiments they
were broad and often the whole of the bulb or tube was illuminated ; and in one case I have observed an intensely purplish

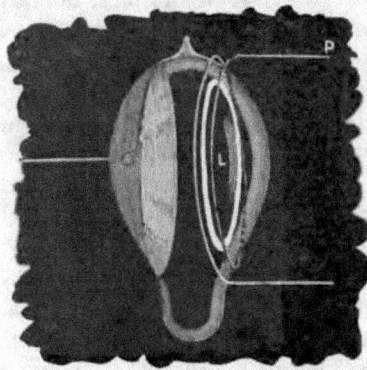

Fig. 214.

glow, to which Prof. J. J. Thomson refers.   But the circles were
always in close proximity to the primary and were considerably
easier produced when the latter was very close to the glass, much
more so than would be expected assuming the action to be elec-

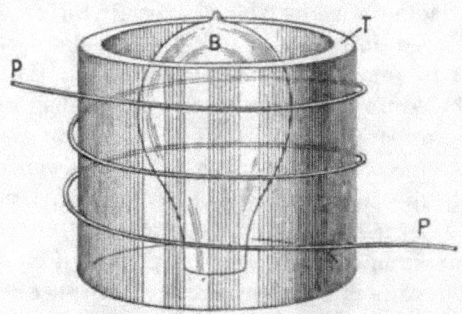

Fig. 215.

tromagnetic and considering the distance ; and these facts speak
for an electrostatic effect.
    Furthermore I have observed that there is a molecular bom-
bardment in the plane of the luminous circle at right angles to
the glass—supposing the circle to be in the plane of the primary

—this bombardment being evident from the rapid heating of the glass near the primary. Were the bombardment not at right angles to the glass the heating could not be so rapid. If there is a circumferential movement of the molecules constituting the luminous circle, I have thought that it might be rendered manifest by placing within the tube or bulb, radially to the circle, a thin plate of mica coated with some phosphorescent material and another such plate tangentially to the circle. If the molecules would move circumferentially, the former plate would be rendered more intensely phosphorescent. For want of time I have, however, not been able to perform the experiment.

Another observation made by me was that when the specific inductive capacity of the medium between the primary and secondary is increased, the inductive effect is augmented. This is roughly illustrated in Fig. 215. In this case luminosity was excited in an exhausted tube or bulb B and a glass tube T slipped between the primary and the bulb, when the effect pointed out was noted. Were the action wholly electromagnetic no change could possibly have been observed.

I have likewise noted that when a bulb is surrounded by a wire closed upon itself and in the plane of the primary, the formation of the luminous circle within the bulb is not prevented. But if instead of the wire a broad strip of tinfoil is glued upon the bulb, the formation of the luminous band was prevented, because then the action was distributed over a greater surface. The effect of the closed tinfoil was no doubt of an electrostatic nature, for it presented a much greater resistance than the closed wire and produced therefore a much smaller electromagnetic effect.

Some of the experiments of Prof. J. J. Thomson also would seem to show some electrostatic action. For instance, in the experiment with the bulb enclosed in a bell jar, I should think that when the latter is exhausted so far that the gas enclosed reaches the maximum conductivity, the formation of the circle in the bulb and jar is prevented because of the space surrounding the primary being highly conducting; when the jar is further exhausted, the conductivity of the space around the primary diminishes and the circles appear necessarily first in the bell jar, as the rarefied gas is nearer to the primary. But were the inductive effect very powerful, they would probably appear in the bulb also. If, however, the bell jar were exhausted to the highest degree they would very likely show themselves in the bulb only, that is, supposing the vacuous space to be non-conducting. On the assumption that in these phenomena electrostatic actions are concerned we find it easily explicable why the introduction of mercury or the heating of the bulb prevents the formation of the luminous band or shortens the after-glow; and also why in some cases a platinum wire may prevent the excitation of the tube. Nevertheless some of the experiments of Prof. J. J. Thomson would seem to indicate an electromagnetic effect. I may add that in one of my experiments in which a vacuum was produced by the Torricellian method, I was unable to produce the luminous band, but this may have been due to the weak exciting current employed.

My principal argument is the following: I have experimentally proved that if the same discharge which is barely sufficient to excite a luminous band in the bulb when passed through the primary circuit be so directed as to exalt the electrostatic inductive effect—namely, by converting upwards—an exhausted tube, devoid of electrodes, may be excited at a distance of several feet.

SOME EXPERIMENTS ON THE ELECTRIC DISCHARGE IN VACUUM TUBES.[1]

BY PROF. J. J. THOMSON, M.A., F.R.S.

The phenomena of vacuum discharges were, Prof. Thomson said, greatly simplified when their path was wholly gaseous, the complication of the dark space surrounding the negative electrode, and the stratifications so commonly observed in ordinary vacuum tubes, being absent. To produce discharges in

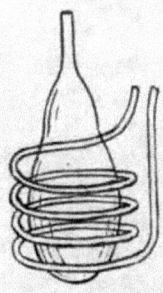

FIG. 216.   FIG. 217.

tubes devoid of electrodes was, however, not easy to accomplish, for the only available means of producing an electromotive force in the discharge circuit was by electro-magnetic induction. Ordinary methods of producing variable induction were valueless, and recourse was had to the oscillatory discharge of a

1. Abstract of a paper read before Physical Society of London.

Leyden jar, which combines the two essentials of a current whose maximum value is enormous, and whose rapidity of alternation is immensely great The discharge circuits, which may take the shape of bulbs, or of tubes bent in the form of coils, were placed in close proximity to glass tubes filled with mercury, which formed the path of the oscillatory discharge. The parts thus corresponded to the windings of an induction coil, the vacuum tubes being the secondary, and the tubes filled with mercury the primary. In such an apparatus the Leyden jar need not be large, and neither primary nor secondary need have many turns, for this would increase the self-induction of the former, and lengthen the discharge path in the latter. Increasing the self-induction of the primary reduces the E. M. F. induced in the secondary, whilst lengthening the secondary does not increase the E. M. F. per unit length. The two or three turns, as shown in Fig. 216, in each, were found to be quite sufficient, and, on discharging the Leyden jar between two highly polished knobs in the primary

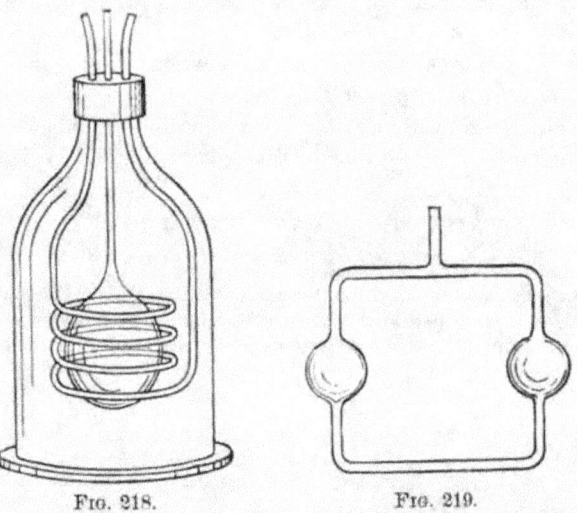

FIG. 218.   FIG. 219.

circuit, a plain uniform band of light was seen to pass round the secondary. An exhausted bulb, Fig. 217, containing traces of oxygen was placed within a primary spiral of three turns, and, on passing the jar discharge, a circle of light was seen within the bulb in close proximity to the primary circuit, accompanied by a purplish glow, which lasted for a second or more. On heating the bulb, the duration of the glow was greatly diminished, and it could be instantly extinguished by the presence of an electro-magnet. Another exhausted bulb, Fig. 218, surrounded by a primary spiral, was contained in a bell-jar, and when the pressure of air in the jar was about that of the atmosphere, the secondary discharge occurred in the bulb, as is ordinarily the case. On exhausting the jar, however, the luminous discharge grew fainter, and a point was reached at which no secondary discharge was visible. Further exhaustion of the jar caused the secondary discharge to appear outside of the bulb. The fact of obtaining no luminous discharge, either in the bulb or jar, the author could only explain on two suppositions, viz.: that under the conditions then existing the specific inductive capacity of the gas was very great, or that a discharge could pass without being luminous. The author had also observed that the conductivity of a vacuum tube without electrodes increased as the pressure diminished, until a certain point was reached, and afterwards diminished again, thus showing that the high resistance of a nearly perfect vacuum is in no way due to the presence of the electrodes. One peculiarity of the discharges was their local nature, the rings of light being much more sharply defined than was to be expected. They were also found to be most easily produced when the chain of molecules in the discharge were all of the same kind. For example, a discharge could be easily sent through a tube many feet long, but the introduction of a small pellet of mercury in the tube stopped the discharge, although the conductivity of the mercury was much greater than that of the vacuum. In some cases he had noticed that a very fine wire placed within a tube, on the side remote from the primary circuit, would prevent a luminous discharge in that tube.

Fig. 219 shows an exhausted secondary coil of one loop containing bulbs; the discharge passed along the inner side of the bulbs, the primary coils being placed within the secondary.

---

[1] In *The Electrical Engineer* of August 12, I find some remarks of Prof. J. J. Thomson, which appeared originally in the London *Electrician* and which have a bearing upon some experiments described by me in your issue of July 1.

I did not, as Prof. J. J. Thomson seems to believe, misunderstand his position in regard to the cause of the phenomena considered, but I thought that in his experiments, as well as in my own, electrostatic effects were of great importance. It did not appear, from the meagre description of his experiments, that all possible precautions had been taken to exclude these effects. I did not doubt that luminosity could be excited in a closed tube when electrostatic action is completely excluded. In fact, at the outset, I myself looked for a purely electrodynamic effect and believed that I had obtained it. But many experiments performed at that time proved to me that the electrostatic effects were generally of far greater importance, and admitted of a more satisfactory explanation of most of the phenomena observed.

In using the term *electrostatic* I had reference rather to the nature of the action than to a stationary condition, which is the usual acceptance of the term. To express myself more clearly, I will suppose that near a closed exhausted tube be placed a small sphere charged to a very high potential. The sphere would act inductively upon the tube, and by distributing electricity over

---

1. Article by Mr. Tesla in *The Electrical Engineer*, N. Y., August 26, 1891.

the same would undoubtedly produce luminosity (if the potential be sufficiently high), until a permanent condition would be reached. Assuming the tube to be perfectly well insulated, there would be only one instantaneous flash during the act of distribution. This would be due to the electrostatic action simply.

But now, suppose the charged sphere to be moved at short intervals with great speed along the exhausted tube. The tube would now be permanently excited, as the moving sphere would cause a constant redistribution of electricity and collisions of the molecules of the rarefied gas. We would still have to deal with an electrostatic effect, and in addition an electrodynamic effect would be observed. But if it were found that, for instance, the effect produced depended more on the specific inductive capacity than on the magnetic permeability of the medium—which would certainly be the case for speeds incomparably lower than that of light—then I believe I would be justified in saying that the effect produced was more of an electrostatic nature. I do not mean to say, however, that any similar condition prevails in the case of the discharge of a Leyden jar through the primary, but I think that such an action would be desirable.

It is in the spirit of the above example that I used the terms "more of an electrostatic nature," and have investigated the influence of bodies of high specific inductive capacity, and observed, for instance, the importance of the quality of glass of which the tube is made. I also endeavored to ascertain the influence of a medium of high permeability by using oxygen. It appeared from rough estimation that an oxygen tube when excited under similar conditions—that is, as far as could be determined—gives more light; but this, of course, may be due to many causes.

Without doubting in the least that, with the care and precautions taken by Prof. J. J. Thomson, the luminosity excited was due solely to electrodynamic action, I would say that in many experiments I have observed curious instances of the ineffectiveness of the screening, and I have also found that the electrification through the air is often of very great importance, and may, in some cases, determine the excitation of the tube.

In his original communication to the *Electrician*, Prof. J. J. Thomson refers to the fact that the luminosity in a tube near a wire through which a Leyden jar was discharged was noted by Hittorf. I think that the feeble luminous effect referred to has been noted by many experimenters, but in my experiments the effects were much more powerful than those usually noted.

The following is the communication[1] referred to :—

"Mr. Tesla seems to ascribe the effects he observed to electrostatic action, and I have no doubt, from the description he gives of his method of conducting his experiments, that in them electrostatic action plays a very important part. He seems, however, to have misunderstood my position with respect to the cause of these discharges, which is not, as he implies, that luminosity in tubes without electrodes cannot be produced by electrostatic action, but that it can also be produced when this action is excluded. As a matter of fact, it is very much easier to get the luminosity when these electrostatic effects are operative than when they are not. As an illustration of this I may mention that the first experiment I tried with the discharge of a Leyden jar produced luminosity in the tube, but it was not until after six weeks' continuous experimenting that I was able to get a discharge in the exhausted tube which I was satisfied was due to what is ordinarily called electrodynamic action. It is advisable to have a clear idea of what we mean by electrostatic action. If, previous to the discharge of the jar, the primary coil is raised to a high potential, it will induce over the glass of the tube a distribution of electricity. When the potential of the primary suddenly falls, this electrification will redistribute itself, and may pass through the rarefied gas and produce luminosity in doing so. Whilst the discharge of the jar is going on, it is difficult, and, from a theoretical point of view, undesirable, to separate the effect into parts, one of which is called electrostatic, the other electromagnetic ; what we can prove is that in this case the discharge is not such as would be produced by electromotive forces derived from a potential function. In my experiments the primary coil was connected to earth, and, as a further precaution, the primary was separated from the discharge tube by a screen of blotting paper, moistened with dilute sulphuric acid, and connected to earth. Wet blotting paper is a sufficiently good conductor to screen off a stationary electrostatic effect, though it is not a good enough one to stop waves of alternating electromotive intensity. When showing the experiments to the Physical Society I could not, of course, keep the tubes covered up, but, unless my memory deceives me, I stated the precautions which had been taken against the electrostatic effect. To correct misapprehension I may state that I did not read a formal paper to the Society, my object being to exhibit a few of the most typical experiments. The account of the experiments in the *Electrician* was from a reporter's note, and was not written, or even read, by me. I have now almost finished writing out, and hope very shortly to publish, an account of these and a large number of allied experiments, including some analogous to those mentioned by Mr. Tesla on the effect of conductors placed near the discharge tube, which I find, in some cases, to produce a diminution, in others an increase, in the brightness of the discharge, as well as some on the effect of the presence of substances of large specific inductive capacity. These seem to me to admit of a satisfactory explanation, for which, however, I must refer to my paper."

---

1.   Note by Prof. J. J. Thomson in the London *Electrician*, July 24, 1891.

PART III.

MISCELLANEOUS INVENTIONS AND WRITINGS.

CHAPTER XXXIII.

METHOD OF OBTAINING DIRECT FROM ALTERNATING CURRENTS.

THIS method consists in obtaining direct from alternating currents, or in directing the waves of an alternating current so as to produce direct or substantially direct currents by developing or producing in the branches of a circuit including a source of alternating currents, either permanently or periodically, and by electric, electro-magnetic, or magnetic agencies, manifestations of energy, or what may be termed active resistances of opposite electrical character, whereby the currents or current waves of opposite sign will be diverted through different circuits, those of one sign passing over one branch and those of opposite sign over the other.

We may consider herein only the case of a circuit divided into two paths, inasmuch as any further subdivision involves merely an extension of the general principle. Selecting, then, any circuit through which is flowing an alternating current, Mr. Tesla divides such circuit at any desired point into two branches or paths. In one of these paths he inserts some device to create an electromotive force counter to the waves or impulses of current of one sign and a similar device in the other branch which opposes the waves of opposite sign. Assume, for example, that these devices are batteries, primary or secondary, or continuous current dynamo machines. The waves or impulses of opposite direction composing the main current have a natural tendency to divide between the two branches; but by reason of the opposite electrical character or effect of the two branches, one will offer an easy passage to a current of a certain direction, while the other will offer a relatively high resistance to the passage of the same current. The result of this disposition is, that the waves of current of one sign will, partly or wholly, pass over one of the paths or branches, while those of the opposite sign pass over the other. There may thus be obtained from an alternating current two or more direct currents without the employment of any commutator

such as it has been heretofore regarded as necessary to use.   The
current in either branch may be used in the same way and for
the same purposes as any other direct current—that is, it may be
made to charge secondary batteries, energize electro-magnets, or
for any other analogous purpose.

Fig. 220 represents a plan of directing the alternating currents
by means of devices purely electrical in character. Figs. 221,
222, 223, 224, 225, and 226 are diagrams illustrative of other
ways of carrying out the invention.

In Fig. 220, A designates a generator of alternating currents,
and B B the main or line circuit therefrom.   At any given point
in this circuit at or near which it is desired to obtain direct currents, the circuit B is divided into two paths or branches c D.   In
each of these branches is placed an electrical generator, which
for the present we will assume produces direct or continuous cur-

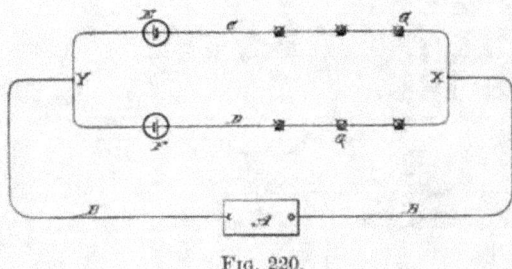

FIG. 220.

rents.   The direction of the current thus produced is opposite in
one branch to that of the current in the other branch, or, considering the two branches as forming a closed circuit, the generators E F are connected up in series therein, one generator in
each part or half of the circuit.   The electromotive force of the
current sources E and F may be equal to or higher or lower than
the electromotive forces in the branches c D, or between the points
x and y of the circuit B B.   If equal, it is evident that current
waves of one sign will be opposed in one branch and assisted in
the other to such an extent that all the waves of one sign will
pass over one branch and those of opposite sign over the other.
If, on the other hand, the electromotive force of the sources E F
be lower than that between x and y, the currents in both
branches will be alternating, but the waves of one sign will preponderate. One of the generators or sources of current E or F
may be dispensed with; but it is preferable to employ both, if

they offer an appreciable resistance, as the two branches will be thereby better balanced. The translating or other devices to be acted upon by the current are designated by the letters G, and they are inserted in the branches C D in any desired manner ; but in order to better preserve an even balance between the branches due regard should, of course, be had to the number and character of the devices.

Figs. 221, 222, 223, and 224 illustrate what may termed "electro-magnetic" devices for accomplishing a similar result—that is to say, instead of producing directly by a generator an electromotive force in each branch of the circuit, Mr. Tesla establishes a field or fields of force and leads the branches through the same in such manner that an active opposition of opposite effect or direction will be developed therein by the passage, or tendency to pass, of the alternations of current. In Fig. 221, for example, A is

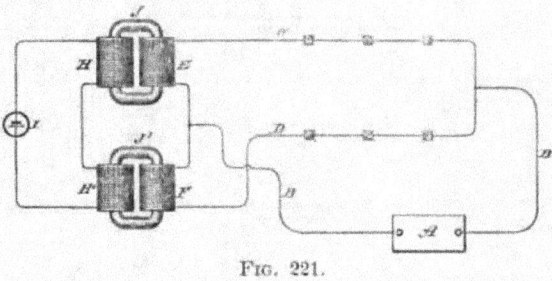

Fig. 221.

the generator of alternating currents, B B the line circuit, and C D the branches over which the alternating currents are directed. In each branch is included the secondary of a transformer or induction coil, which, since they correspond in their functions to the batteries of the previous figure, are designated by the letters E F. The primaries H H' of the induction coils or transformers are connected either in parallel or series with a source of direct or continuous currents I, and the number of convolutions is so calculated for the strength of the current from I that the cores J J' will be saturated. The connections are such that the conditions in the two transformers are of opposite character—that is to say, the arrangement is such that a current wave or impulse corresponding in direction with that of the direct current in one primary, as H, is of opposite direction to that in the other primary H'. It thus results that while one secondary offers a resistance or op-

position to the passage through it of a wave of one sign, the other
secondary similarly opposes a wave of opposite sign. In consequence, the waves of one sign will, to a greater or less extent, pass
by way of one branch, while those of opposite sign in like manner pass over the other branch.

In lieu of saturating the primaries by a source of continuous
current, we may include the primaries in the branches C D, respectively, and periodically short-circuit by any suitable mechanical devices—such as an ordinary revolving commutator—their
secondaries. It will be understood, of course, that the rotation
and action of the commutator must be in synchronism or in
proper accord with the periods of the alternations in order to
secure the desired results. Such a disposition is represented

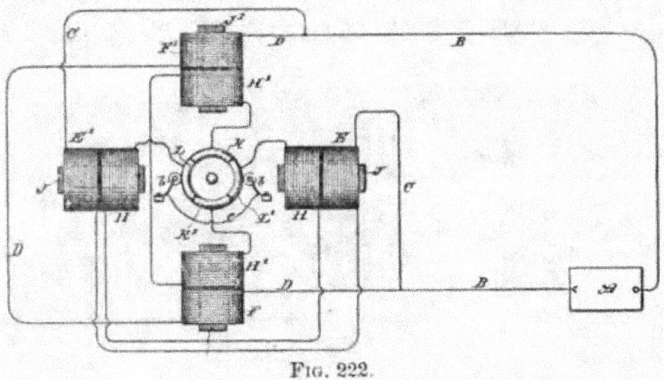

FIG. 222.

diagrammatically in Fig. 222. Corresponding to the previous
figures, A is the generator of alternating currents, B B the line,
and C D the two branches for the direct currents. In branch C
are included two primary coils E E', and in branch D are two
similar primaries F F' The corresponding secondaries for these
coils and which are on the same subdivided cores J or J', are in
circuits the terminals of which connect to opposite segments K
K', and L L', respectively, of a commutator. Brushes b b bear
upon the commutator and alternately short-circuit the plates K
and K', and L and L', through a connection c. It is obvious that
either the magnets and commutator, or the brushes, may revolve.

The operation will be understood from a consideration of the
effects of closing or short-circuiting the secondaries. For example, if at the instant when a given wave of current passes, one

set of secondaries be short-circuited, nearly all the current flows through the corresponding primaries; but the secondaries of the other branch being open-circuited, the self-induction in the primaries is highest, and hence little or no current will pass through that branch. If, as the current alternates, the secondaries of the two branches are alternately short-circuited, the result will be that the currents of one sign pass over one branch and those of the opposite sign over the other. The disadvantages of this arrangement, which would seem to result from the employment of sliding contacts, are in reality very slight, inasmuch as the electromotive force of the secondaries may be made exceedingly low, so that sparking at the brushes is avoided.

Fig. 223 is a diagram, partly in section, of another plan of carrying out the invention. The circuit B in this case is divided, as before, and each branch includes the coils of both the fields

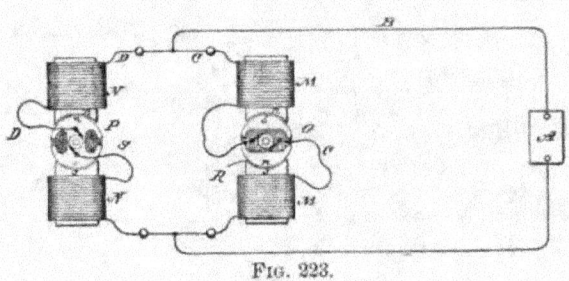

Fig. 223.

and revolving armatures of two induction devices. The armatures O P are preferably mounted on the same shaft, and are adjusted relatively to one another in such manner that when the self-induction in one branch, as C, is maximum, in the other branch D it is minimum. The armatures are rotated in synchronism with the alternations from the source A. The winding or position of the armature coils is such that a current in a given direction passed through both armatures would establish in one, poles similar to those in the adjacent poles of the field, and in the other, poles unlike the adjacent field poles, as indicated by *n n s s* in the diagram. If the like poles are presented, as shown in circuit D, the condition is that of a closed secondary upon a primary, or the position of least inductive resistance; hence a given alternation of current will pass mainly through D. A half revolution of the armatures produces an opposite effect and the succeeding current impulse passes through c. Using this figure as an illustration, it is evident that the fields N M may be permanent magnets or independently excited and the armatures O P driven, as in the present case, so as to produce alternate currents, which will set up alternately impulses of opposite direction in the two branches D C, which in such case would include the armature circuits and translating devices only.

In Fig. 224 a plan alternative with that shown in Fig. 222 is illustrated. In the previous case illustrated, each branch C and D contained one or more primary coils, the secondaries of which were periodically short circuited in synchronism with the alternations of current from the main source A, and for this purpose a commutator was employed. The latter may, however, be dispensed with and an armature with a closed coil substituted.

Referring to Fig. 224 in one of the branches, as C, are two coils

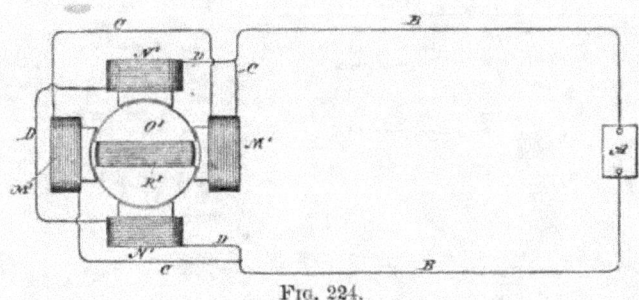

FIG. 224.

M′, wound on laminated cores, and in the other branches D are similar coils N′. A subdivided or laminated armature O′, carrying a closed coil R′, is rotatably supported between the coils M′ N′, as shown. In the position shown—that is, with the coil R′ parallel with the convolutions of the primaries N′ M′—practically the whole current will pass through branch D, because the self-induction in coils M′ M′ is maximum. If, therefore, the armature and coil be rotated at a proper speed relatively to the periods or alternations of the source A, the same results are obtained as in the case of Fig. 222.

Fig. 225 is an instance of what may be called, in distinction to the others, a "magnetic" means of securing the result. v and w are two strong permanent magnets provided with armatures v′ w′, respectively. The armatures are made of thin laminæ of soft iron or steel, and the amount of magnetic metal which they contain is so calculated that they will be fully or nearly saturated by the magnets. Around the armatures are coils E F, contained, respectively, in the circuits c and D. The connections and electrical conditions in this case are similar to those in Fig. 221, except that the current source of I, Fig. 221, is dispensed with and the saturation of the core of coils E F obtained from the permanent magnets.

The previous illustrations have all shown the two branches or paths containing the translating or induction devices as in derivation one to the other; but this is not always necessary. For example, in Fig. 226, A is an alternating-current generator; B B, the line wires or circuit. At any given point in the circuit let us form two paths, as D D′, and at another point two paths, as c c′. Either pair or group of paths is similar to the previous dis-

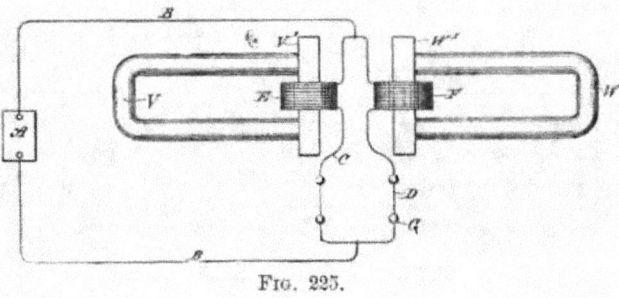

FIG. 225.

positions with the electrical source or induction device in one branch only, while the two groups taken together form the obvious equivalent of the cases in which an induction device or generator is included in both branches. In one of the paths, as D, are included the devices to be operated by the current. In the other branch, as D′, is an induction device that opposes the current impulses of one direction and directs them through the branch D. So, also, in branch c are translating devices G, and in branch c′ an induction device or its equivalent that diverts through c impulses of opposite direction to those diverted by the device in branch D′. The diagram shows a special form of induction device for this purpose. J J′ are the cores, formed with pole-pieces, upon which are wound the coils M N. Between these pole-pieces are mounted at right angles to one another the magnetic armatures O P, preferably mounted on the same shaft and designed to be rotated in synchronism with the alternations of current. When one of the armatures is in line with the poles or in the position occupied by armature P, the magnetic circuit of the induction device is practically closed; hence there will be the greatest opposition to the passage of a current through coils N N. The alternation will therefore pass by way of branch D. At the same time, the magnetic circuit of the other induction device being broken by the position of the armature o, there will be less opposition to the current in coils M, which will shunt the current from branch c. A reversal of the current being attended by a shifting of the armatures, the opposite effect is produced.

Other modifications of these methods are possible, but need not be pointed out. In all these plans, it will be observed, there

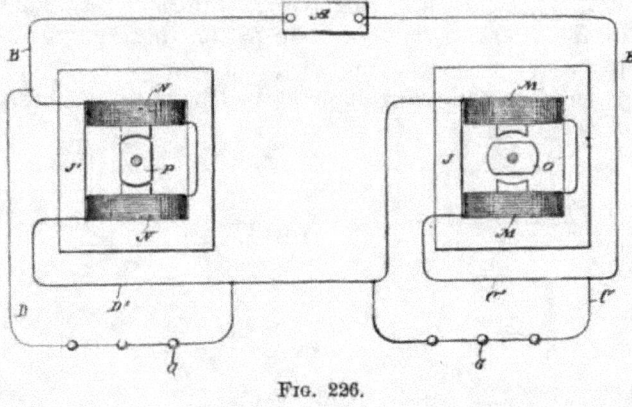

FIG. 226.

is developed in one or all of these branches of a circuit from a source of alternating currents, an active (as distinguished from a dead) resistance or opposition to the currents of one sign, for the purpose of diverting the currents of that sign through the other or another path, but permitting the currents of opposite sign to pass without substantial opposition.

Whether the division of the currents or waves of current of opposite sign be effected with absolute precision or not is immaterial, since it will be sufficient if the waves are only partially diverted or directed, for in such case the preponderating influence in each branch of the circuit of the waves of one sign secures the same practical results in many if not all respects as though the current were direct and continuous.

An alternating and a direct current have been combined so that the waves of one direction or sign were partially or wholly overcome by the direct current; but by this plan only one set of alternations are utilized, whereas by the system just described the entire current is rendered available. By obvious applications of this discovery Mr. Tesla is enabled to produce a self-exciting alternating dynamo, or to operate direct current meters on alternating-current circuits or to run various devices—such as arc lamps —by direct currents in the same circuit with incandescent lamps or other devices operated by alternating currents.

It will be observed that if an intermittent counter or opposing force be developed in the branches of the circuit and of higher electromotive force than that of the generator, an alternating current will result in each branch, with the waves of one sign preponderating, while a constantly or uniformly acting opposition in the branches of higher electromotive force than the generator would produce a pulsating current, which conditions would be, under some circumstances, the equivalent of those described.

CHAPTER XXXIV.

CONDENSERS WITH PLATES IN OIL.

IN experimenting with currents of high frequency and high potential, Mr. Tesla has found that insulating materials such as glass, mica, and in general those bodies which possess the highest specific inductive capacity, are inferior as insulators in such devices when currents of the kind described are employed compared with those possessing high insulating power, together with a smaller specific inductive capacity; and he has also found that it is very desirable to exclude all gaseous matter from the apparatus, or any ac-

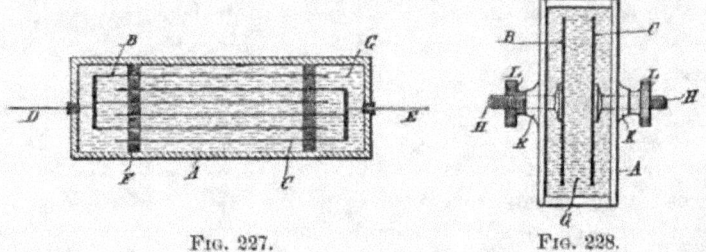

FIG. 227.   FIG. 228.

cess of the same to the electrified surfaces, in order to prevent heating by molecular bombardment and the loss or injury consequent thereon. He has therefore devised a method to accomplish these results and produce highly efficient and reliable condensers, by using oil as the dielectric[1]. The plan admits of a particular con-

---

1. Mr. Tesla's experiments, as the careful reader of his three lectures will perceive, have revealed a very important fact which is taken advantage of in this invention. Namely, he has shown that in a condenser a considerable amount of energy may be wasted, and the condenser may break down merely because gaseous matter is present between the surfaces. A number of experiments are described in the lectures, which bring out this fact forcibly and serve as a guide in the operation of high tension apparatus. But besides bearing upon this point, these experiments also throw a light upon investigations of a purely scientific nature and explain now the lack of harmony among the observations of various investigators. Mr. Tesla shows that in a fluid such as oil the losses are very small as compared with those incurred in a gas.

struction of condenser, in which the distance between the plates is adjustable, and of which he takes advantage.

In the accompanying illustrations, Fig. 227 is a section of a condenser constructed in accordance with this principle and having stationary plates; and Fig. 228 is a similar view of a condenser with adjustable plates.

Any suitable box or receptacle A may be used to contain the plates or armatures. These latter are designated by B and C and are connected, respectively, to terminals D and E, which pass out through the sides of the case. The plates ordinarily are separated by strips of porous insulating material F, which are used merely for the purpose of maintaining them in position. The space within the can is filled with oil G. Such a condenser will prove highly efficient and will not become heated or permanently injured.

In many cases it is desirable to vary or adjust the capacity of a condenser, and this is provided for by securing the plates to adjustable supports—as, for example, to rods H—passing through stuffing boxes K in the sides of case A and furnished with nuts L, the ends of the rods being threaded for engagement with the nuts.

It is well known that oils possess insulating properties, and it it has been a common practice to interpose a body of oil between two conductors for purposes of insulation; but Mr. Tesla believes he has discovered peculiar properties in oils which render them very valuable in this particular form of device.

CHAPTER XXXV.

Electrolytic Registering Meter.

An ingenious form of electrolytic meter attributable to Mr. Tesla is one in which a conductor is immersed in a solution, so arranged that metal may be deposited from the solution or taken away in such a manner that the electrical resistance of the conductor is varied in a definite proportion to the strength of the current the energy of which is to be computed, whereby this variation in resistance serves as a measure of the energy and also may actuate registering mechanism, whenever the resistance rises above or falls below certain limits.

In carrying out this idea Mr. Tesla employs an electrolytic cell, through which extend two conductors parallel and in close proximity to each other. These conductors he connects in series through a resistance, but in such manner that there is an equal difference of potential between them throughout their entire extent. The free ends or terminals of the conductors are connected either in series in the circuit supplying the current to the lamps or other devices, or in parallel to a resistance in the circuit and in series with the current consuming devices. Under such circumstances a current passing through the conductors establishes a difference of potential between them which is proportional to the strength of the current, in consequence of which there is a leakage of current from one conductor to the other across the solution. The strength of this leakage current is proportional to the difference of potential, and, therefore, in proportion to the strength of the current passing through the conductors. Moreover, as there is a constant difference of potential between the two conductors throughout the entire extent that is exposed to the solution, the current density through such solution is the same at all corresponding points, and hence the deposit is uniform along the whole of one of the conductors, while the metal is taken away uniformly from the other. The resistance of one conductor is by this means diminished, while that of the other is increased, both in proportion to the strength of the current passing through the conductors. From such variation in the resistance of either or both of the conductors forming the positive and negative electrodes of the cell, the current energy expended may be readily computed. Figs. 229 and 230 illustrate two forms of such a meter.

In Fig. 229 G designates a direct-current generator. L L are the conductors of the circuit extending therefrom. A is a tube of glass, the ends of which are sealed, as by means of insulating plugs or caps B B. c c' are two conductors extending through the tube A, their ends passing out through the plugs B to

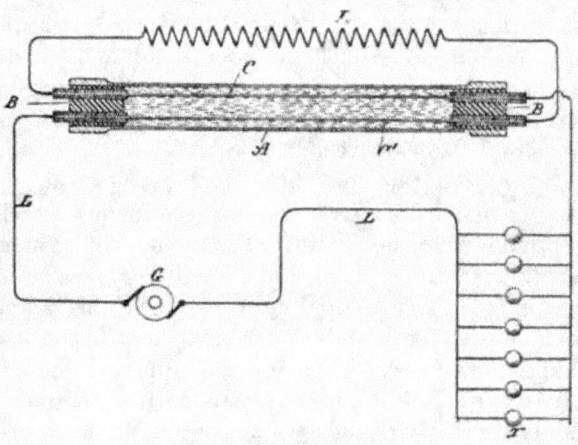

FIG. 229.

terminals thereon. These conductors may be corrugated or formed in other proper ways to offer the desired electrical resistance. R is a resistance connected in series with the two conductors c c', which by their free terminals are connected up in circuit with one of the conductors L.

The method of using this device and computing by means thereof the energy of the current will be readily understood. First, the resistances of the two conductors c c', respectively, are accurately measured and noted. Then a known current is passed through the instrument for a given time, and by a second measurement the increase and diminution of the resistances of the two conductors are respectively taken. From these data the constant is obtained—that is to say, for example, the increase of resistance of one conductor or the diminution of the resistance of the other per lamp hour. These two measurements evidently serve as a check, since the gain of one conductor should equal the loss of the other. A further check is afforded by measuring both wires in series with the resistance, in which case the resistance of the whole should remain constant.

In Fig. 230 the conductors c c' are connected in parallel, the current device at x passing in one branch first through a resistance R' and then through conductor c, while on the other branch it passes first through conductor c', and then through resistance

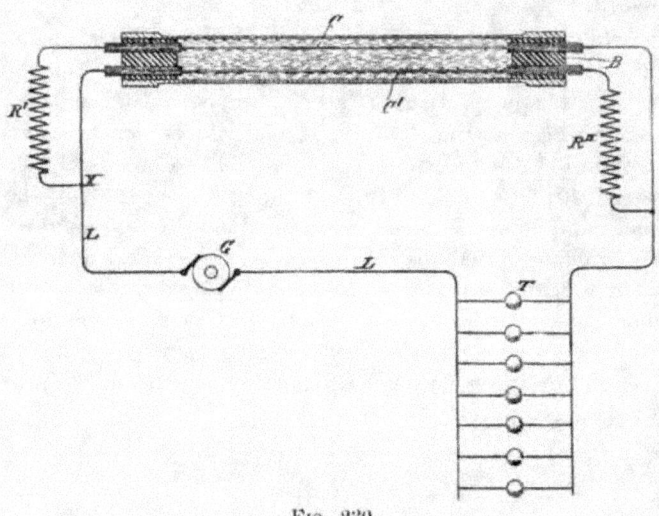

FIG. 230.

R''. The resistances R' R'' are equal, as also are the resistances of the conductors c c'. It is, moreover, preferable that the respective resistances of the conductors c c' should be a known and convenient fraction of the coils or resistances R' R''. It will be observed that in the arrangement shown in Fig. 230 there is a constant potential difference between the two conductors c c' throughout their entire length.

It will be seen that in both cases illustrated, the proportionality of the increase or decrease of resistance to the current strength will always be preserved, for what one conductor gains the other loses, and the resistances of the conductors c c' being small as compared with the resistances in series with them. It will be understood that after each measurement or registration of a given variation of resistance in one or both conductors, the direction of the current should be changed or the instrument reversed, so that the deposit will be taken from the conductor which has gained and added to that which has lost. This principle is capable of many modifications. For instance, since there is a section of the circuit—to wit, the conductor c or c'—that varies in resistance in proportion to the current strength, such variation may be utilized, as is done in many analogous cases, to effect the operation of various automatic devices, such as registers. It is better, however, for the sake of simplicity to compute the energy by measurements of resistance.

The chief advantages of this arrangement are, first, that it is possible to read off directly the amount of the energy expended by means of a properly constructed ohm-meter and without resorting to weighing the deposit; secondly it is not necessary to employ shunts, for the whole of the current to be measured may be passed through the instrument; third, the accuracy of the instrument and correctness of the indications are but slightly affected by changes in temperature. It is also said that such meters have the merit of superior economy and compactness, as well as of cheapness in construction. Electrolytic meters seem to need every auxiliary advantage to make them permanently popular and successful, no matter how much ingenuity may be shown in their design.

CHAPTER XXXVI.

THERMO-MAGNETIC MOTORS AND PYRO-MAGNETIC GENERATORS.

No electrical inventor of the present day dealing with the problems of light and power considers that he has done himself or his opportunities justice until he has attacked the subject of thermo-magnetism. As far back as the beginning of the seventeenth century it was shown by Dr. William Gilbert, the father of modern electricity, that a loadstone or iron bar when heated to redness loses its magnetism; and since that time the influence of heat on the magnetic metals has been investigated frequently, though not with any material or practical result.

For a man of Mr. Tesla's inventive ability, the problems in this field have naturally had no small fascination, and though he has but glanced at them, it is to be hoped he may find time to pursue the study deeper and further. For such as he, the investigation must undoubtedly bear fruit. Meanwhile he has worked out one or two operative devices worthy of note.[1] He obtains mechanical power by a reciprocating action resulting from the joint operations of heat, magnetism, and a spring or weight or other force—that is to say he subjects a body magnetized by induction or otherwise to the action of heat until the magnetism is sufficiently neutralized to allow a weight or spring to give motion to the body and lessen the action of the heat, so that the magnetism may be sufficiently restored to move the

---

1. It will, of course, be inferred from the nature of these devices that the vibration obtained in this manner is very slow owing to the inability of the iron to follow rapid changes in temperature. In an interview with Mr. Tesla on this subject, the compiler learned of an experiment which will interest students. A simple horseshoe magnet is taken and a piece of sheet iron bent in the form of an L is brought in contact with one of the poles and placed in such a position that it is kept in the attraction of the opposite pole delicately suspended. A spirit lamp is placed under the sheet iron piece and when the iron is heated to a certain temperature it is easily set in vibration oscillating as rapidly as 400 to 500 times a minute. The experiment is very easily performed and is interesting principally on account of the very rapid rate of vibration.

body in the opposite direction, and again subject the same to the demagnetizing power of the heat.

Use is made of either an electro-magnet or a permanent magnet, and the heat is directed against a body that is magnetized by induction, rather than directly against a permanent magnet, thereby avoiding the loss of magnetism that might result in the permanent magnet by the action of heat. Mr. Tesla also provides for lessening the volume of the heat or for intercepting the same during that portion of the reciprocation in which the cooling action takes place.

In the diagrams are shown some of the numerous arrangements that may be made use of in carrying out this idea. In all of these figures the magnet-poles are marked n s, the armature A, the Bunsen burner or other source of heat H, the axis of mo-

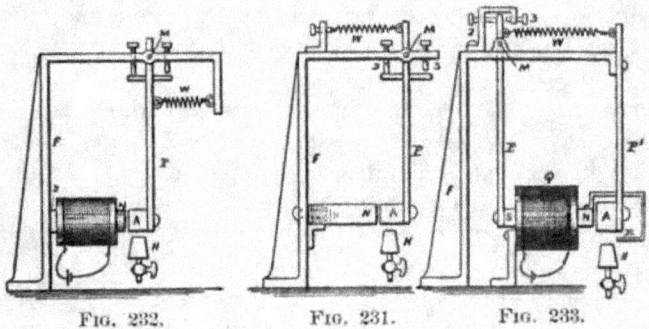

FIG. 232.   FIG. 231.   FIG. 233.

tion M, and the spring or the equivalent thereof—namely, a weight—is marked w.

In Fig. 231 the permanent magnet N is connected with a frame, F, supporting the axis M, from which the arm P hangs, and at the lower end of which the armature A is supported. The stops 2 and 3 limit the extent of motion, and the spring w tends to draw the armature A away from the magnet N. It will now be understood that the magnetism of N is sufficient to overcome the spring w and draw the armature A toward the magnet N. The heat acting upon the armature A neutralizes its induced magnetism sufficiently for the spring w to draw the armature A away from the magnet N and also from the heat at H. The armature now cools, and the attraction of the magnet N overcomes the spring w and draws the armature A back again above the burner

n, so that the same is again heated and the operations are repeated. The reciprocating movements thus obtained are employed as a source of mechanical power in any desired manner.
Usually a connecting-rod to a crank upon a fly-wheel shaft would
be made use of, as indicated in Fig. 240.

Fig. 232 represents the same parts as before described; but an

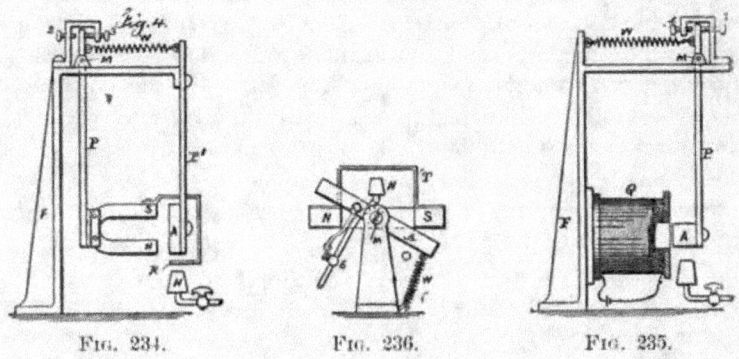

FIG. 234.   FIG. 236.   FIG. 235.

electro-magnet is illustrated in place of a permanent magnet.
The operations, however, are the same.

In Fig. 233 are shown the same parts as in Figs. 231 and 232,
but they are differently arranged. The armature A, instead of
swinging, is stationary and held by arm P', and the core N s of
the electro-magnet is made to swing within the helix Q, the
core being suspended by the arm P from the pivot M. A shield,
R, is connected with the magnet-core and swings with it, so
that after the heat has demagnetized the armature A to such an
extent that the spring w draws the core N s away from the armature A, the shield R comes between the flame H and armature A,
thereby intercepting the action of the heat and allowing the armature to cool, so that the magnetism, again preponderating,
causes the movement of the core N s toward the armature A and
the removal of the shield R from above the flame, so that the heat
again acts to lessen or neutralize the magnetism. A rotary or
other movement may be obtained from this reciprocation.

Fig. 234 corresponds in every respect with Fig. 233, except
that a permanent horseshoe-magnet, N s is represented as taking
the place of the electro-magnet in Fig. 233.

In Fig. 235 is shown a helix, Q, with an armature adapted to
swing toward or from the helix. In this case there may be a soft-

iron core in the helix, or the armature may assume the form of a
solenoid core, there being no permanent core within the helix.

Fig. 236 is an end view, and Fig. 237 a plan view, illustrating
the method as applied to a swinging armature, A, and a stationary
permanent magnet, N s.   In this instance Mr. Tesla applies the
heat to an auxiliary armature or keeper, T, which is adjacent to
and preferably in direct contact with the magnet.   This armature T, in the form of a plate of sheet-iron, extends across from
one pole to the other and is of sufficient section to practically
form a keeper for the magnet, so that when the armature T is
cool nearly all the lines of force pass over the same and very little
free magnetism is exhibited.   Then the armature A, which swings
freely on the pivots M in front of the poles N s, is very little attracted and the spring W pulls the same way from the poles into
the position indicated in the diagram.   The heat is directed upon
the iron plate T at some distance from the magnet, so as to allow
the magnet to keep comparatively cool.   This heat is applied beneath the plate by means of the burners n, and there is a connection from the armature A or its pivot to the gas-cock 6, or
other device for regulating the heat.   The heat acting upon the
middle portion of the plate T, the magnetic conductivity of the
heated portion is diminished or destroyed, and a great number of
the lines of force are deflected over the armature A, which is now

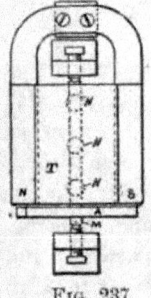

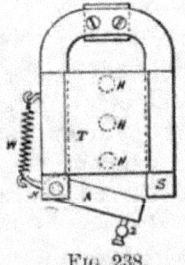

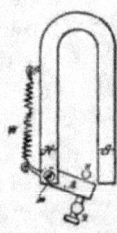

Fig. 237.   Fig. 238.   Fig. 239.

powerfully attracted and drawn into line, or nearly so, with the
poles N s.   In so doing the cock 6 is nearly closed and the plate
T cools, the lines of force are again deflected over the same, the
attraction exerted upon the armature A is diminished, and the
spring W pulls the same away from the magnet into the position
shown by full lines, and the operations are repeated.   The ar-

rangement shown in Fig. 236 has the advantages that the magnet and armature are kept cool and the strength of the permanent magnet is better preserved, as the magnetic circuit is constantly closed.

In the plan view, Fig. 238, is shown a permanent magnet and keeper plate, T, similar to those in Figs. 236 and 237, with the burners H for the gas beneath the same; but the armature is pivoted at one end to one pole of the magnet and the other end swings toward and from the other pole of the magnet. The spring W acts against a lever arm that projects from the armature, and the supply of heat has to be partly cut off by a connection to the swinging armature, so as to lessen the heat acting upon the keeper plate when the armature A has been attracted.

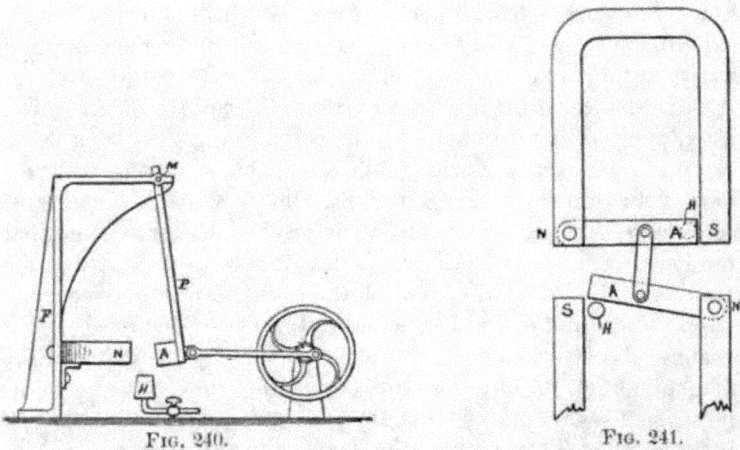

FIG. 240.   FIG. 241.

Fig. 239 is similar to Fig. 238, except that the keeper T is not made use of and the armature itself swings into and out of the range of the intense action of the heat from the burner H. Fig. 240 is a diagram similar to Fig. 231, except that in place of using a spring and stops, the armature is shown as connected by a link, to the crank of a fly-wheel, so that the fly-wheel will be revolved as rapidly as the armature can be heated and cooled to the necessary extent. A spring may be used in addition, as in Fig. 231. In Fig. 241 the armatures A A are connected by a link, so that one will be heating while the other is cooling, and the attraction exerted to move the cooled armature is availed of to draw away the heated armature instead of using a spring.

Mr. Tesla has also devoted his attention to the development of a pyromagnetic generator of electricity[1] based upon the following laws: First, that electricity or electrical energy is developed in any conducting body by subjecting such body to a varying magnetic influence; and second, that the magnetic properties of iron or other magnetic substance may be partially or entirely destroyed or caused to disappear by raising it to a certain temperature, but restored and caused to reappear by again lowering its temperature to a certain degree. These laws may be applied in the production of electrical currents in many ways, the principle of which is in all cases the same, viz., to subject a conductor to a varying magnetic influence, producing such variations by the application of heat, or, more strictly speaking, by the application or action of a varying temperature upon the source of the magnetism. This principle of operation may be illustrated by a simple experiment: Place end to end, and preferably in actual contact, a permanently magnetized steel bar and a strip or bar of soft iron. Around the end of the iron bar or plate wind a coil of insulated wire. Then apply to the iron between the coil and the steel bar a flame or other source of heat which will be capable of raising that portion of the iron to an orange red, or a temperature of about 600° centigrade. When this condition is reached, the iron somewhat suddenly loses its magnetic properties, if it be very thin, and the same effect is produced as though the iron had been moved away from the magnet or the heated section had been removed. This change of position, however, is accompanied by a shifting of the magnetic lines, or, in other words, by a variation in the magnetic influence to which the coil is exposed, and a current in the coil is the result. Then remove the flame or in any other way reduce the temperature of the iron. The lowering of its temperature is accompanied by a return of its magnetic properties, and another change of magnetic conditions occurs, accompanied by a current in an opposite direction in the coil. The same operation may be

1. The chief point to be noted is that Mr. Tesla attacked this problem in a way which was, from the standpoint of theory, and that of an engineer, far better than that from which some earlier trials in this direction started. The enlargement of these ideas will be found in Mr. Tesla's work on the pyromagnetic generator, treated in this chapter. The chief effort of the inventor was to economize the heat, which was accomplished by inclosing the iron in a source of heat well insulated, and by cooling the iron by means of steam, utilizing the steam over again. The construction also permits of more rapid magnetic changes per unit of time, meaning larger output.

repeated indefinitely, the effect upon the coil being similar to that which would follow from moving the magnetized bar to and from the end of the iron bar or plate.

The device illustrated below is a means of obtaining this result, the features of novelty in the invention being, first, the employment of an artificial cooling device, and, second, inclosing the source of heat and that portion of the magnetic circuit exposed to the heat and artificially cooling the heated part.

These improvements are applicable generally to the generators constructed on the plan above described—that is to say, we may use an artificial cooling device in conjunction with a variable or varied or uniform source of heat.

Fig. 242 is a central vertical longitudinal section of the com-

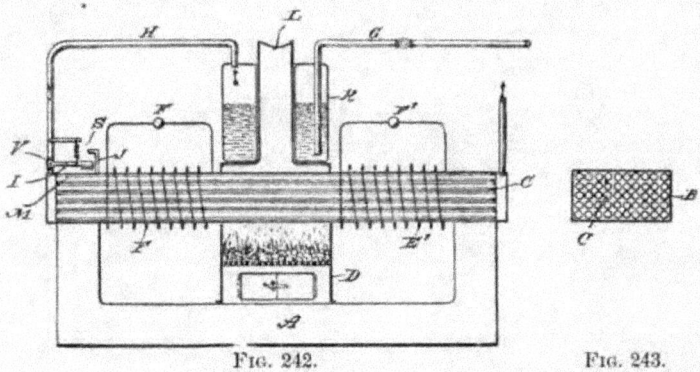

FIG. 242.   FIG. 243.

plete apparatus and Fig. 243 is a cross-section of the magnetic armature-core of the generator.

Let A represent a magnetized core or permanent magnet the poles of which are bridged by an armature-core composed of a casing or shell B inclosing a number of hollow iron tubes c. Around this core are wound the conductors E E′, to form the coils in which the currents are developed. In the circuits of these coils are current-consuming devices, as F F′.

D is a furnace or closed fire-box, through which the central portion of the core B extends. Above the fire is a boiler K, containing water. The flue L from the fire-box may extend up through the boiler.

G is a water-supply pipe, and H is the steam-exhaust pipe, which communicates with all the tubes c in the armature B, so that steam escaping from the boiler will pass through the tubes.

In the steam-exhaust pipe H is a valve V, to which is connected the lever I, by the movement of which the valve is opened or closed. In such a case as this the heat of the fire may be utilized for other purposes after as much of it as may be needed has been applied to heating the core B. There are special advantages in the employment of a cooling device, in that the metal of the core B is not so quickly oxidized. Moreover, the difference between the temperature of the applied heat and of the steam, air, or whatever gas or fluid be applied as the cooling medium, may be increased or decreased at will, whereby the rapidity of the magnetic changes or fluctuations may be regulated.

CHAPTER XXXVII.

ANTI-SPARKING DYNAMO BRUSH AND COMMUTATOR.

In direct current dynamos of great electromotive force—such, for instance, as those used for arc lighting—when one commutator bar or plate comes out of contact with the collecting-brush a spark is apt to appear on the commutator. This spark may be due to the break of the complete circuit, or to a shunt of low resistance formed by the brush between two or more commutator-bars. In the first case the spark is more apparent, as there is at the moment when the circuit is broken a discharge of the magnets through the field helices, producing a great spark or flash which causes an unsteady current, rapid wear of the commutator bars and brushes, and waste of power. The sparking may be reduced by various devices, such as providing a path for the current at the moment when the commutator segment or bar leaves the brush, by short-circuiting the field-helices, by increasing the number of the commutator-bars, or by other similar means; but all these devices are expensive or not fully available, and seldom attain the object desired.

To prevent this sparking in a simple manner, Mr. Tesla some years ago employed with the commutator-bars and intervening insulating material, mica, asbestos paper or other insulating and incombustible material, arranged to bear on the surface of the commutator, near to and behind the brush.

In the drawings, Fig. 244 is a section of a commutator with an asbestos insulating device; and Fig. 245 is a similar view, representing two plates of mica upon the back of the brush.

In Fig. 244, c represents the commutator and intervening insulating material; B B, the brushes. d d are sheets of asbestos paper or other suitable non-conducting material. f f are springs, the pressure of which may be adjusted by means of the screws g g.

In Fig. 245 a simple arrangement is shown with two plates of mica or other material. It will be seen that whenever one commutator segment passes out of contact with the brush, the formation of the arc will be prevented by the intervening insulating material coming in contact with the insulating material on the brush.

Asbestos paper or cloth impregnated with zinc-oxide, magnesia, zirconia, or other suitable material, may be used, as the

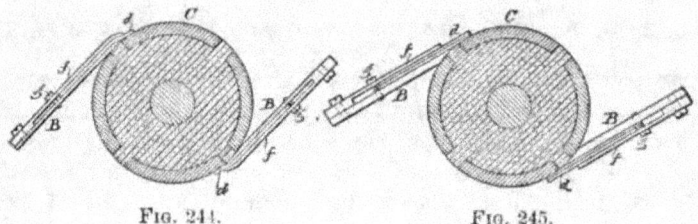

Fig. 244.   Fig. 245.

paper and cloth are soft, and serve at the same time to wipe and polish the commutator; but mica or any other suitable material can be employed, provided the material be an insulator or a bad conductor of electricity.

A few years later Mr. Tesla turned his attention again to the same subject, as, perhaps, was very natural in view of the fact that the commutator had always been prominent in his thoughts, and that so much of his work was even aimed at dispensing with it entirely as an objectionable and unnecessary part of dynamos and motors. In these later efforts to remedy commutator troubles, Mr. Tesla constructs a commutator and the collectors therefor in two parts mutually adapted to one another, and, so far as the essential features are concerned, alike in mechanical structure. Selecting as an illustration a commutator of two segments adapted for use with an armature the coils or coil of which have but two free ends, connected respectively to the segments, the bearing-surface is the face of a disc, and is formed of two metallic quadrant segments and two insulating segments of the same dimensions, and the face of the disc is smoothed off, so that the metal and insulating segments are flush. The part which takes the place of the usual brushes, or the "collector," is a disc of the same character as the commutator and has a surface similarly formed with two insulating and two metallic segments. These two parts are mounted with their faces in contact and in such manner that the rotation of the armature causes the commutator to turn upon the collector, whereby the currents induced in the coils are taken off by the collector segments and thence conveyed off by suitable conductors leading from the collector segments. This is the general plan of the construction adopted. Aside from certain adjuncts, the nature and functions of which are set forth later, this means of commutation will be seen to possess many important advantages. In the first place the short-circuiting and the breaking of the armature coil connected to the commutator-segments occur at the same instant, and from the nature of the construction this will be done with the greatest precision; secondly, the duration of both the break and of the short circuit will be reduced to a minimum. The first results in a reduction which amounts practically to a suppression of the spark, since the break and the short circuit produce opposite effects in the armature-coil. The second has the effect of diminishing the destructive effect of a spark, since this would be in a measure proportional to the duration of the spark; while lessening the duration of the short circuit obviously increases the efficiency of the machine.

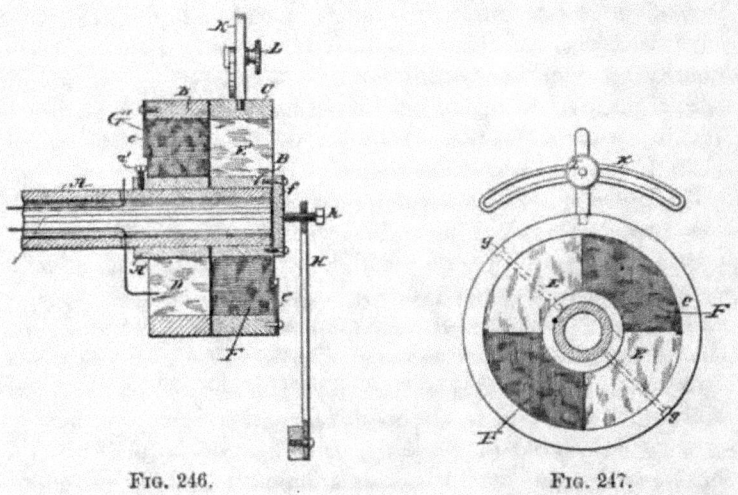

FIG. 246.   FIG. 247.

The mechanical advantages will be better understood by referring to the accompanying diagrams, in which Fig. 246 is a central longitudinal section of the end of a shaft with the improved commutator carried thereon. Fig. 247 is a view of the inner or bearing face of the collector. Fig. 248 is an end view from the armature side of a modified form of commutator. Figs.

249 and 250 are views of details of Fig. 248. Fig. 251 is a longitudinal central section of another modification, and Fig. 252 is a sectional view of the same. A is the end of the armature-shaft of a dynamo-electric machine or motor. A′ is a sleeve of insulating material around the shaft, secured in place by a screw, *a′*.

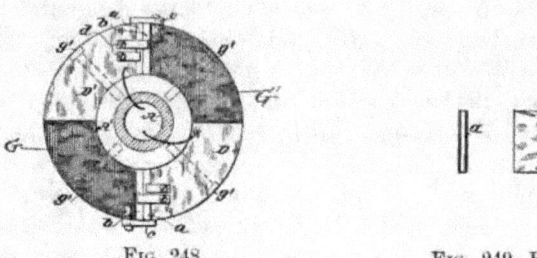

Fig. 248   Fig. 249. Fig. 250.

The commutator proper is in the form of a disc which is made up of four segments D D′ G G′, similar to those shown in Fig. 248. Two of these segments, as D D′, are of metal and are in electrical connection with the ends of the coils on the armature. The other two segments are of insulating material. The segments are held in place by a band, B, of insulating material. The disc is held in place by friction or by screws, *g′ g′*, Fig. 248, which secure the disc firmly to the sleeve A′.

The collector is made in the same form as the commutator. It is composed of the two metallic segments E E′ and the two insulating segments F F′, bound together by a band, C. The metallic segments E E′ are of the same or practically the same width or extent as the insulating segments or spaces of the commutator. The collector is secured to a sleeve, B′, by screws *g g*, and the sleeve is arranged to turn freely on the shaft A. The end of the sleeve B′ is closed by a plate, *f*, upon which presses a pivot-pointed screw, *h*, adjustable in a spring, H, which acts to maintain the collector in close contact with the commutator and to compensate for the play of the shaft. The collector is so fixed that it cannot turn with the shaft. For example, the diagram shows a slotted plate, K, which is designed to be attached to a stationary support, and an arm extending from the collector and carrying a clamping screw, L, by which the collector may be adjusted and set to the desired position.

Mr. Tesla prefers the form shown in Figs. 246 and 247 to fit

the insulating segments of both commutator and collector loosely
and to provide some means—as, for example, light springs, *e e*,
secured to the bands A' B', respectively, and bearing against the
segments—to exert a light pressure upon them and keep them in
close contact and to compensate for wear. The metal segments
of the commutator may be moved forward by loosening the
screw *a'*.

The line wires are fed from the metal segments of the collector,
being secured thereto in any convenient manner, the plan of connections being shown as applied to a modified form of the commutator in Fig. 251. The commutator and the collector in thus
presenting two flat and smooth bearing surfaces prevent most effectually by mechanical action the occurrence of sparks.

The insulating segments are made of some hard material capable of being polished and formed with sharp edges. Such materials as glass, marble, or soapstone may be advantageously used.
The metal segments are preferably of copper or brass; but they
may have a facing or edge of durable material—such as platinum
or the like—where the sparks are liable to occur.

In Fig. 248 a somewhat modified form of the invention is
shown, a form designed to facilitate the construction and replac-

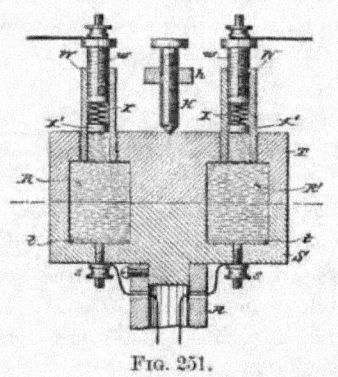

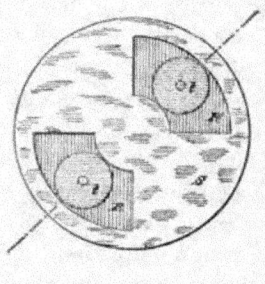

FIG. 251.   FIG. 252.

ing of the parts. In this modification the commutator and collector are made in substantially the same manner as previously
described, except that the bands B C are omitted. The four segments of each part, however, are secured to their respective sleeves
by screws *g' g'*, and one edge of each segment is cut away, so that
small plates *a b* may be slipped into the spaces thus formed. Of

these plates *a a* are of metal, and are in contact with the metal segments D D', respectively. The other two, *b b*, are of glass or marble, and they are all better square, as shown in Figs. 249 and 250, so that they may be turned to present new edges should any edge become worn by use. Light springs *d* bear upon these plates and press those in the commutator toward those in the collector, and insulating strips *c c* are secured to the periphery of the discs to prevent the blocks from being thrown out by centrifugal action. These plates are, of course, useful at those edges of the segments only where sparks are liable to occur, and, as they are easily replaced, they are of great advantage. It is considered best to coat them with platinum or silver.

In Figs. 251 and 252 is shown a construction where, instead of solid segments, a fluid is employed. In this case the commutator and collector are made of two insulating discs, s T, and in lieu of the metal segments a space is cut out of each part, as at R R', corresponding in shape and size to a metal segment. The two parts are fitted smoothly and the collector T held by the screw *h* and spring H against the commutator s. As in the other cases, the commutator revolves while the collector remains stationary. The ends of the coils are connected to binding-posts *s s*, which are in electrical connection with metal plates *t t* within the recesses in the two parts s T. These chambers or recesses are filled with mercury, and in the collector part are tubes w w, with screws *w w*, carrying springs x and pistons x', which compensate for the expansion and contraction of the mercury under varying temperatures, but which are sufficiently strong not to yield to the pressure of the fluid due to centrifugal action, and which serve as binding-posts.

In all the above cases the commutators are adapted for a single coil, and the device is particularly suited to such purposes. The number of segments may be increased, however, or more than one commutator used with a single armature. Although the bearing-surfaces are shown as planes at right angles to the shaft or axis, it is evident that in this particular the construction may be very greatly modified.

CHAPTER XXXVIII.

AUXILIARY BRUSH REGULATION OF DIRECT CURRENT DYNAMOS.

AN interesting method devised by Mr. Tesla for the regulation of direct current dynamos, is that which has come to be known as the "third brush" method. In machines of this type, devised by him as far back as 1885, he makes use of two main brushes to which the ends of the field magnet coils are connected, an auxiliary brush, and a branch or shunt connection from an intermediate point of the field wire to the auxiliary brush.[1]

The relative positions of the respective brushes are varied, either automatically or by hand, so that the shunt becomes inoperative when the auxiliary brush has a certain position upon the commutator; but when the auxiliary brush is moved in its relation to the main brushes, or the latter are moved in their relation to the auxiliary brush, the electric condition is disturbed and more or less of the current through the field-helices is diverted through the shunt or a current is passed over the shunt to the field-helices. By varying the relative position upon the commutator of the respective brushes automatically in proportion to the varying electrical conditions of the working-circuit, the current developed can be regulated in proportion to the demands in the working-circuit.

Fig. 253 is a diagram illustrating the invention, showing one core of the field-magnets with one helix wound in the same direction throughout. Figs. 254 and 255 are diagrams showing one core of the field-magnets with a portion of the helices wound in opposite directions. Figs. 256 and 257 are diagrams illustrating

1. The compiler has learned partially from statements made on several occasions in journals and partially by personal inquiry of Mr. Tesla, that a great deal of work in this interesting line is unpublished. In these inventions as will be seen, the brushes are automatically shifted, but in the broad method barely suggested here the regulation is effected without any change in the position of the brushes. This auxiliary brush invention, it will be remembered, was very much discussed a few years ago, and it may be of interest that this work of Mr. Tesla, then unknown in this field, is now brought to light.

the electric devices that may be employed for automatically adjusting the brushes, and Fig. 258 is a diagram illustrating the positions of the brushes when the machine is being energized at the start.

a and *b* are the positive and negative brushes of the main or working-circuit, and *c* the auxiliary brush. The working-circuit D extends from the brushes *a* and *b*, as usual, and contains electric lamps or other devices, D′, either in series or in multiple arc.

M M′ represent the field-helices, the ends of which are connected to the main brushes *a* and *b*. The branch or shunt wire *c′* extends from the auxiliary brush *c* to the circuit of the field-helices, and is connected to the same at an intermediate point, *x*.

H represents the commutator, with the plates of ordinary con-

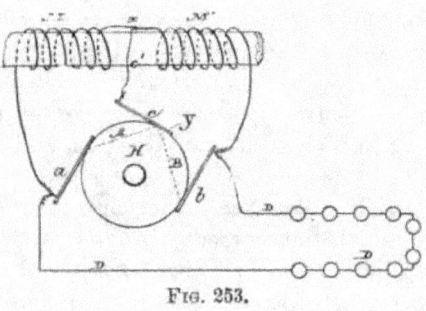

FIG. 253.

struction. When the auxiliary brush *c* occupies such a position upon the commutator that the electro-motive force between the brushes *a* and *c* is to the electro-motive force between the brushes *c* and *b* as the resistance of the circuit *a* M *c′* *c* A is to the resistance of the circuit *b* M′ *c′* *c* B, the potentials of the points *x* and Y will be equal, and no current will flow over the auxiliary brush; but when the brush *c* occupies a different position the potentials of the points *x* and Y will be different, and a current will flow over the auxiliary brush to and from the commutator, according to the relative position of the brushes. If, for instance, the commutator-space between the brushes *a* and *c*, when the latter is at the neutral point, is diminished, a current will flow from the point Y over the shunt *c* to the brush *b*, thus strengthening the current in the part M′, and partly neutralizing the current in part M; but if the space between the brushes *a* and *c* is increased, the current will flow over the auxiliary brush in an opposite direction, and the current in ᴍ will be strengthened, and in ᴍ' partly neutralized.

By combining with the brushes *a*, *b*, and *c* any usual automatic regulating mechanism, the current developed can be regulated in proportion to the demands in the working circuit. The parts ᴍ

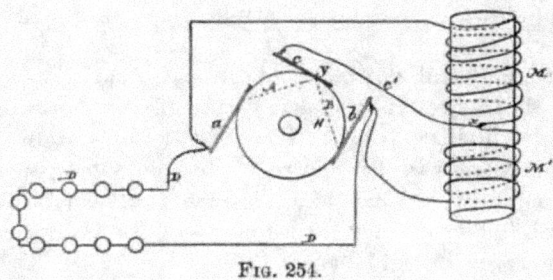

FIG. 254.

and ᴍ' of the field wire may be wound in the same direction. In this case they are arranged as shown in Fig. 253; or the part ᴍ may be wound in the opposite direction, as shown in Figs. 254 and 255.

It will be apparent that the respective cores of the field-magnets are subjected to neutralizing or intensifying effects of the current in the shunt through *c'*, and the magnetism of the cores will be partially neutralized, or the points of greatest magnetism shifted, so that it will be more or less remote from or approaching to the armature, and hence the aggregate energizing actions of the field magnets on the armature will be correspondingly varied.

In the form indicated in Fig. 253 the regulation is effected by shifting the point of greatest magnetism, and in Figs. 254 and 255 the same effect is produced by the action of the current in the shunt passing through the neutralizing helix.

The relative positions of the respective brushes may be varied by moving the auxiliary brush, or the brush *c* may remain stationary and the core ᴘ be connected to the main-brush holder ᴀ, so as to adjust the brushes *a b* in their relation to the brush *c*. If, however, an adjustment is applied to all the brushes, as seen in Fig. 257, the solenoid should be connected to both *a* and *c*, so as to move them toward or away from each other.

There are several known devices for giving motion in proportion to an electric current. In Figs. 256 and 257 the moving cores are shown as convenient devices for obtaining the required extent of motion with very slight changes in the current passing through the helices. It is understood that the adjustment of the main brushes causes variations in the strength of the current independently of the relative position of those brushes to the auxiliary brush. In all cases the adjustment should be such that no current flows over the auxiliary brush when the dynamo is running with its normal load.

In Figs. 256 and 257 A A indicate the main-brush holder, carrying the main brushes, and c the auxiliary-brush holder, carrying the auxiliary brush. These brush-holders are movable in arcs concentric with the centre of the commutator-shaft. An iron piston, P, of the solenoid s, Fig. 256, is attached to the auxiliary-brush holder c. The adjustment is effected by means of a spring and screw or tightener.

In Fig. 257 instead of a solenoid, an iron tube inclosing a coil is shown. The piston of the coil is attached to both brush-holders A A and c. When the brushes are moved directly by electrical devices, as shown in Figs. 256 and 257, these are so constructed that the force exerted for adjusting is practically uniform through the whole length of motion.

It is true that auxiliary brushes have been used in connection with the helices of the field-wire; but in these instances the

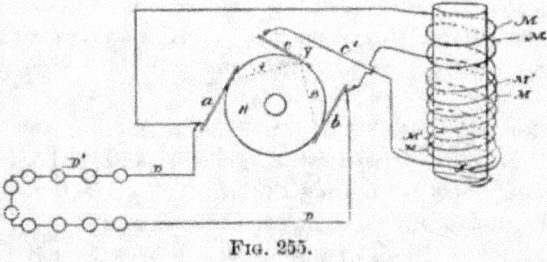

FIG. 255.

helices receive the entire current through the auxiliary brush or brushes, and these brushes could not be taken off without breaking the circuit through the field. These brushes cause, moreover, heavy sparking at the commutator. In the present case the auxiliary brush causes very little or no sparking, and can be taken off without breaking the circuit through the field-

helices. The arrangement has, besides, the advantage of facilitating
the self-excitation of the machine in all cases where the resistance of the field-wire is very great comparatively to the resistance of the main circuit at the start—for instance, on arc-light

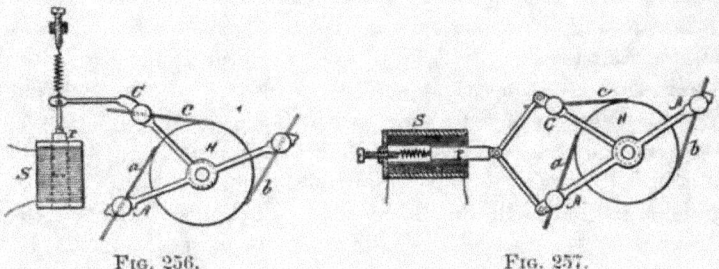

FIG. 256.   FIG. 257.

machines. In this case the auxiliary brush *c* is placed near to, or
better still in contact with, the brush *b*, as shown in Fig. 258.
In this manner the part M' is completely cut out, and as the part
M has a considerably smaller resistance than the whole length of
the field-wire the machine excites itself, whereupon the auxiliary
brush is shifted automatically to its normal position.

In a further method devised by Mr. Tesla, one or more auxiliary brushes are employed, by means of which a portion or the
whole of the field coils is shunted. According to the relative position upon the commutator of the respective brushes more or
less current is caused to pass through the helices of the field, and
the current developed by the machine can be varied at will by
varying the relative positions of the brushes.

In Fig. 259, *a* and *b* are the positive and negative brushes of
the main circuit, and *c* an auxiliary brush. The main circuit D

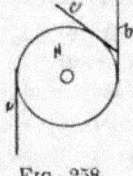

FIG. 258.

extends from the brushes *a* and *b*, as usual, and contains the
helices M of the field wire and the electric lamps or other working devices. The auxiliary brush *c* is connected to the point *x*
of the main circuit by means of the wire *c'*. H is a commutator

of ordinary construction. It will have been seen from what was said already that when the electro-motive force between the brushes *a* and *c* is to the electromotive force between the brushes *c* and *b* as the resistance of the circuit *a* M *c′ c* A is to the resistance of the circuit *b* c B *c c′* D, the potentials of the points *x* and *y* will be equal, and no current will pass over the auxiliary brush *c*; but if that brush occupies a different position relatively to the main brushes the electric condition is disturbed, and current will flow either from *y* to *x* or from *x* to *y*, according to the relative position of the brushes. In the first case the current through the field-helices will be partly neutralized and the magnetism of the field magnets will be diminished. In the second case the current will be increased and the magnets gain strength. By combining with the brushes *a b c* any automatic regulating mechanism, the current developed can be regulated automatically in proportion to the demands of the working circuit.

In Figs. 264 and 265 some of the automatic means are represented that may be used for moving the brushes. The core P, Fig. 264, of the solenoid-helix s is connected with the brush *c* to move the same, and in Fig. 265 the core P is shown as within the helix s, and connected with brushes *a* and *c*, so as to move the same toward or from each other, according to the strength of the current in the helix, the helix being within an iron tube, s′, that becomes magnetized and increases the action of the solenoid.

In practice it is sufficient to move only the auxiliary brush, as shown in Fig. 264, as the regulation is very sensitive to the slightest changes; but the relative position of the auxiliary brush to the main brushes may be varied by moving the main brushes, or both main and auxiliary brushes may be moved, as illustrated in Fig. 265. In the latter two cases, it will be understood, the motion of the main brushes relatively to the neutral line of the machine causes variations in the strength of the current independently of their relative position to the auxiliary brush. In all cases the adjustment may be such that when the machine is running with the ordinary load, no current flows over the auxiliary brush.

The field helices may be connected, as shown in Fig. 259, or a part of the field helices may be in the outgoing and the other part in the return circuit, and two auxiliary brushes may be employed as shown in Figs. 261 and 262. Instead of shunting the whole of the field helices, a portion only of such helices may be shunted, as shown in Figs. 260 and 262.

The arrangement shown in Fig. 262 is advantageous, as it diminishes the sparking upon the commutator, the main circuit being closed through the auxiliary brushes at the moment of the break of the circuit at the main brushes.

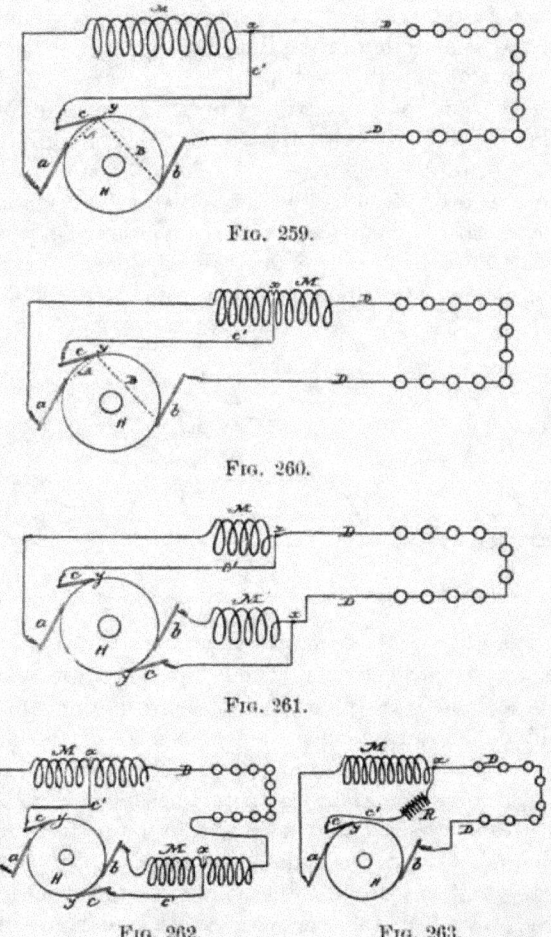

FIG. 259.

FIG. 260.

FIG. 261.

FIG. 262.   FIG. 263.

The field helices may be wound in the same direction, or a part may be wound in opposite directions.

The connection between the helices and the auxiliary brush or brushes may be made by a wire of small resistance, or a resistance may be interposed (R, Fig. 263,) between the point *x* and the auxiliary brush or brushes to divide the sensitiveness when the brushes are adjusted.

The accompanying sketches also illustrate improvements made by Mr. Tesla in the mechanical devices used to effect the shifting of the brushes, in the use of an auxiliary brush. Fig. 266 is an elevation of the regulator with the frame partly in section; and Fig. 267 is a section at the line xx, Fig. 266. c is the commutator; B and B', the brush-holders, B carrying the main brushes $a\ a'$, and B' the auxiliary or shunt brushes $b\ b$. The axis of the brush-holder B is supported by two pivot-screws, $p\ p$. The other brush-holder, B', has a sleeve, d, and is movable around the axis of the brush-holder B. In this way both brush-holders can turn very freely, the friction of the parts being reduced to a minimum. Over the brush-holders is mounted the solenoid s, which rests upon a forked column, c. This column

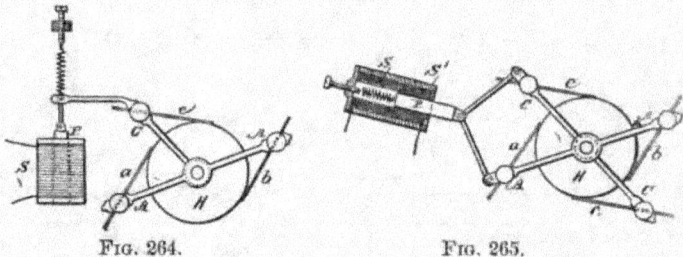

FIG. 264.   FIG. 265.

also affords a support for the pivots $p\ p$, and is fastened upon a solid bracket or projection, P, which extends from the base of the machine, and is cast in one piece with the same. The brush-holders B B' are connected by means of the links $e\ e$ and the cross-piece F to the iron core I, which slides freely in the tube T of the solenoid. The iron core I has a screw, s, by means of which it can be raised and adjusted in its position relatively to the solenoid, so that the pull exerted upon it by the solenoid is practically uniform through the whole length of motion which is required to effect the regulation. In order to effect the adjustment with greater precision, the core I is provided with a small iron screw, s'. The core being first brought very nearly in the required position relatively to the solenoid by means of the screw s, the small screw s' is then adjusted until the magnetic attraction upon the core is the same when the core is in any position. A convenient stop, t, serves to limit the upward movement of the iron core.

To check somewhat the movement of the core ɪ, a dash-pot, ᴋ, is used. The piston ʟ of the dash-pot is provided with a valve, v, which opens by a downward pressure and allows an easy downward movement of the iron core ɪ, but closes and checks the movement of the core when it is pulled up by the action of the solenoid.

To balance the opposing forces, the weight of the moving parts, and the pull exerted by the solenoid upon the iron core, the weights w w may be used. The adjustment is such that when the solenoid is traversed by the normal current it is just strong enough to balance the downward pull of the parts.

The electrical circuit-connections are substantially the same as

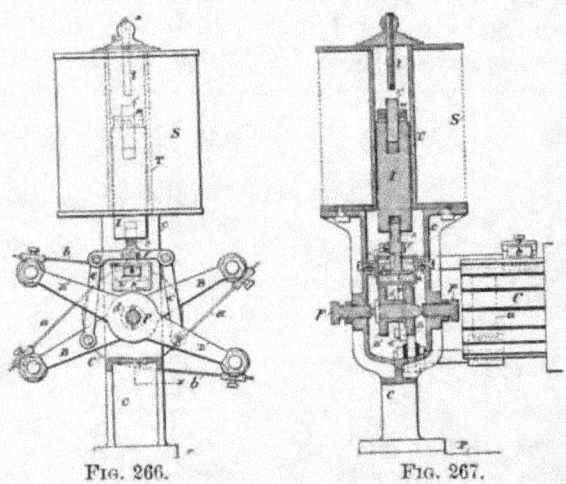

Fɪɢ. 266.   Fɪɢ. 267.

indicated in the previous diagrams, the solenoid being in series with the circuit when the translating devices are in series, and in shunt when the devices are in multiple arc. The operation of the device is as follows: When upon a decrease of the resistance of the circuit or for some other reason, the current is increased, the solenoid s gains in strength and pulls up the iron core ɪ, thus shifting the main brushes in the direction of rotation and the auxiliary brushes in the opposite way. This diminishes the strength of the current until the opposing forces are balanced and the solenoid is traversed by the normal current; but if from any cause the current in the circuit is diminished, then the weight of the moving parts overcomes the pull of the solenoid, the iron core 1 descends, thus shifting the brushes the opposite way and increasing the current to the normal strength. The dash-pot connected to the iron core 1 may be of ordinary construction; but it is better, especially in machines for arc lights, to provide the piston of the dash-pot with a valve, as indicated in the diagrams. This valve permits a comparatively easy downward movement of the iron core, but checks its movement when it is drawn up by the solenoid. Such an arrangement has the advantage that a great number of lights may be put on without diminishing the light-power of the lamps in the circuit, as the brushes assume at once the proper position. When lights are cut out, the dash-pot acts to retard the movement; but if the current is considerably increased the solenoid gets abnormally strong and the brushes are shifted instantly. The regulator being properly adjusted, lights or other devices may be put on or out with scarcely any perceptible difference. It is obvious that instead of the dash-pot any other retarding device may be used.

CHAPTER XXXIX.

Improvement in the Construction of Dynamos and Motors.

This invention of Mr. Tesla is an improvement in the construction of dynamo or magneto electric machines or motors, consisting in a novel form of frame and field magnet which renders the machine more solid and compact as a structure, which requires fewer parts, and which involves less trouble and expense in its manufacture. It is applicable to generators and motors generally, not only to those which have independent circuits adapted for use in the Tesla alternating current system, but to other continuous or alternating current machines of the ordinary type generally used.

Fig. 268 shows the machine in side elevation. Fig. 269 is a vertical sectional view of the field magnets and frame and an end view of the armature; and Fig. 270 is a plan view of one of the parts of the frame and the armature, a portion of the latter being cut away.

The field magnets and frame are cast in two parts. These parts are identical in size and shape, and each consists of the solid plates or ends A B, from which project inwardly the cores C D and the side bars or bridge pieces, E F. The precise shape of these parts is largely a matter of choice—that is to say, each casting, as shown, forms an approximately rectangular frame; but it might obviously be more or less oval, round, or square, without departure from the invention. It is also desirable to reduce the width of the side bars, E F, at the center and to so proportion the parts that when the frame is put together the spaces between the pole pieces will be practically equal to the arcs which the surfaces of the poles occupy.

The bearings G for the armature shaft are cast in the side bars E F. The field coils are either wound on the pole pieces or on a form and then slipped on over the ends of the pole pieces. The lower part or casting is secured to the base after being finished off. The armature K on its shaft is then mounted in the bearings of the lower casting and the other part of the frame placed in position, dowel pins L or any other means being used to secure the two parts in proper position.

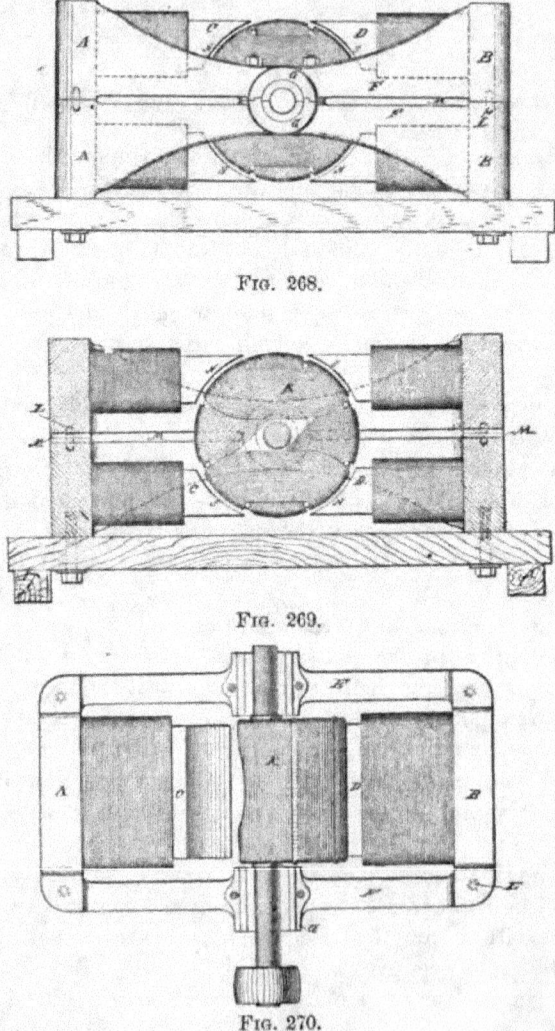

Fig. 268.

Fig. 269.

Fig. 270.

In order to secure an easier fit, the side bars E F, and end pieces, A B, are so cast that slots M are formed when the two parts are put together.

This machine possesses several advantages.   For example, if we magnetize the cores alternately, as indicated by the characters N s, it will be seen that the magnetic circuit between the poles of each part of a casting is completed through the solid iron side bars.   The bearings for the shaft are located at the neutral points of the field, so that the armature core is not affected by the magnetic condition of the field.

The improvement is not restricted to the use of four pole pieces, as it is evident that each pole piece could be divided or more than four formed by the shape of the casting.

CHAPTER XL.

TESLA DIRECT CURRENT ARC LIGHTING SYSTEM.

AT one time, soon after his arrival in America, Mr. Tesla was greatly interested in the subject of arc lighting, which then occupied public attention and readily enlisted the support of capital. He therefore worked out a system which was confided to a company formed for its exploitation, and then proceeded to devote his energies to the perfection of the details of his more celebrated "rotary field" motor system. The Tesla arc lighting apparatus appeared at a time when a great many other lamps and machines were in the market, but it commanded notice by its ingenuity. Its chief purpose was to lessen the manufacturing cost and simplify the processes of operation.

We will take up the dynamo first. Fig. 271 is a longitudinal section, and Fig. 272 a cross section of the machine. Fig. 273 is a top view, and Fig. 274 a side view of the magnetic frame. Fig. 275 is an end view of the commutator bars, and Fig. 276 is a section of the shaft and commutator bars. Fig. 277 is a diagram illustrating the coils of the armature and the connections to the commutator plates.

The cores $c\ c\ c\ c$ of the field-magnets are tapering in both directions, as shown, for the purposes of concentrating the magnetism upon the middle of the pole-pieces.

The connecting-frame F F of the field-magnets is in the form indicated in the side view, Fig. 274, the lower part being provided with the spreading curved cast legs $e\ e$, so that the machine will rest firmly upon two base-bars, $r\ r$.

To the lower pole, s, of the field-magnet M is fastened, by means of babbitt or other fusible diamagnetic material, the base B, which is provided with bearings b for the armature-shaft H. The base B has a projection, P, which supports the brush-holders and the regulating devices, which are of a special character devised by Mr. Tesla.

The armature is constructed with the view to reduce to a min-

inum the loss of power due to Foucault currents and to the
change of polarity, and also to shorten as much as possible the
length of the inactive wire wound upon the armature core.

It is well known that when the armature is revolved between
the poles of the field-magnets, currents are generated in the iron
body of the armature which develop heat, and consequently cause

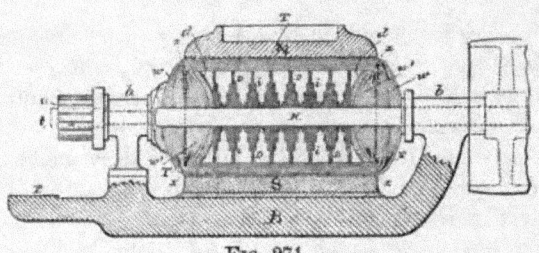

Fig. 271.

a waste of power. Owing to the mutual action of the lines of
force, the magnetic properties of iron, and the speed of the different portions of the armature core, these currents are generated
principally on and near the surface of the armature core, diminishing in strength gradually toward the centre of the core.
Their quantity is under some conditions proportional to the
length of the iron body in the direction in which these currents
are generated. By subdividing the iron core electrically in this
direction, the generation of these currents can be reduced to a
great extent. For instance, if the length of the armature-core is
twelve inches, and by a suitable construction it is subdivided
electrically, so that there are in the generating direction six inches
of iron and six inches of intervening air-spaces or insulating material, the waste currents will be reduced to fifty per cent.

As shown in the diagrams, the armature is constructed of thin
iron discs D D D, of various diameters, fastened upon the armature-shaft in a suitable manner and arranged according to their
sizes, so that a series of iron bodies, *i i i*, is formed, each of which
diminishes in thickness from the centre toward the periphery.
At both ends of the armature the inwardly curved discs *d d*, of
cast iron, are fastened to the armature shaft.

The armature core being constructed as shown, it will be easily
seen that on those portions of the armature that are the most
remote from the axis, and where the currents are principally developed, the length of iron in the generating direction is only a

small fraction of the total length of the armature core, and besides this the iron body is subdivided in the generating direction, and therefore the Foucault currents are greatly reduced. Another cause of heating is the shifting of the poles of the armature core. In consequence of the subdivision of the iron in the armature and the increased surface for radiation, the risk of heating is lessened.

The iron discs D D D are insulated or coated with some insulating-paint, a very careful insulation being unnecessary, as an electrical contact between several discs can only occur at places where the generated currents are comparatively weak. An armature core constructed in the manner described may be revolved between the poles of the field magnets without showing the slightest increase of temperature.

The end discs, *d d*, which are of sufficient thickness and, for the sake of cheapness, of cast-iron, are curved inwardly, as indicated in the drawings. The extent of the curve is dependent on the amount of wire to be wound upon the armatures. In this machine the wire is wound upon the armature in two superimposed parts, and the curve of the end discs, *d d*, is so calculated that the first part—that is, practically half of the wire—just fills

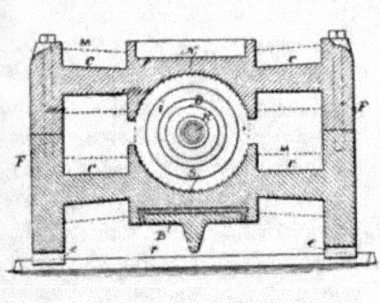

Fig. 272.

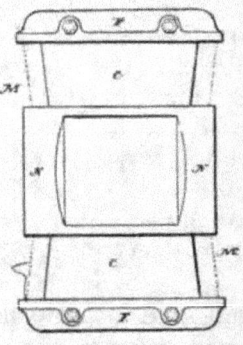

Fig. 273.

up the hollow space to the line *x x;* or, if the wire is wound in any other manner, the curve is such that when the whole of the wire is wound, the outside mass of wires, *w*, and the inside mass of wires, *w'*, are equal at each side of the plane *x x*. In this case the passive or electrically-inactive wires are of the smallest length practicable. The arrangement has further the advantage that the total lengths of the crossing wires at the two sides of the plane $x\,x$ are practically equal.

To equalize further the armature coils at both sides of the plates that are in contact with the brushes, the winding and connecting up is effected in the following manner: The whole wire is wound upon the armature-core in two superimposed parts,

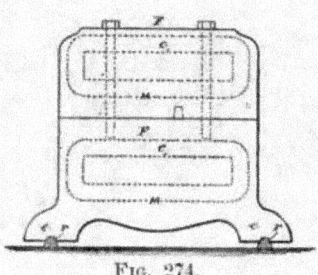

FIG. 274.

which are thoroughly insulated from each other. Each of these two parts is composed of three separated groups of coils. The first group of coils of the first part of wire being wound and connected to the commutator-bars in the usual manner, this group is insulated and the second group wound; but the coils of this second group, instead of being connected to the next following commutator bars, are connected to the directly opposite bars of the commutator. The second group is then insulated and the third group wound, the coils of this group being connected to those bars to which they would be connected in the usual way. The wires are then thoroughly insulated and the second part of wire is wound and connected in the same manner.

Suppose, for instance, that there are twenty-four coils—that is, twelve in each part—and consequently twenty-four commutator plates. There will be in each part three groups, each containing four coils, and the coils will be connected as follows:

	Groups.	*Commutator Bars.*
First part of wire	First	1— 5
	Second	17—21
	Third	9—13
Second part of wire	First	13—17
	Second	5— 9
	Third	21— 1

In constructing the armature core and winding and connecting the coils in the manner indicated, the passive or electrically inactive wire is reduced to a minimum, and the coils at each side of the plates that are in contact with the brushes are practically equal. In this way the electrical efficiency of the machine is increased.

The commutator plates *t* are shown as outside the bearing *b* of

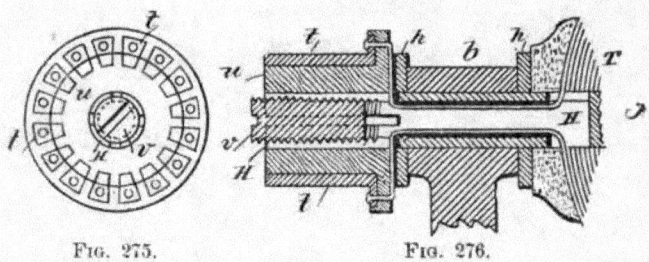

FIG. 275.   FIG. 276.

the armature shaft. The shaft *h* is tubular and split at the end portion, and the wires are carried through the same in the usual manner and connected to the respective commutator plates. The commutator plates are upon a cylinder, *u*, and insulated, and this cylinder is properly placed and then secured by expanding the split end of the shaft by a tapering screw plug, *r*.

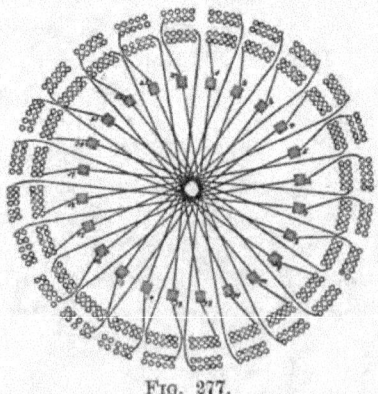

FIG. 277.

The arc lamps invented by Mr. Tesla for use on the circuits from the above described dynamo are those in which the separation and feed of the carbon electrodes or their equivalents is accomplished by means of electro-magnets or solenoids in connection with suitable clutch mechanism, and were designed for the purpose of remedying certain faults common to arc lamps.

He proposed to prevent the frequent vibrations of the movable carbon "point" and flickering of the light arising therefrom; to prevent the falling into contact of the carbons; to dispense with the dash pot, clock work, or gearing and similar devices; to render the lamp extremely sensitive, and to feed the carbon almost imperceptibly, and thereby obtain a very steady and uniform light.

In that class of lamps where the regulation of the arc is effected by forces acting in opposition on a free, movable rod or lever directly connected with the electrode, all or some of the forces being dependent on the strength of the current, any change in the electrical condition of the circuit causes a vibration and a corresponding flicker in the light. This difficulty is most apparent when there are only a few lamps in circuit. To lessen this difficulty lamps have been constructed in which the lever or armature, after the establishing of the arc, is kept in a fixed position and cannot vibrate during the feed operation, the feed mechanism acting independently; but in these lamps, when a clamp is employed, it frequently occurs that the carbons come into contact and the light is momentarily extinguished, and frequently parts of the circuit are injured. In both these classes of lamps it has been customary to use dash pot, clock work, or equivalent retarding devices; but these are often unreliable and objectionable, and increase the cost of construction.

Mr. Tesla combines two electro-magnets—one of low resistance in the main or lamp circuit, and the other of comparatively high resistance in a shunt around the arc—a movable armature lever, and a special feed mechanism, the parts being arranged so that in the normal working position of the armature lever the same is kept almost rigidly in one position, and is not affected even by considerable changes in the electric circuit; but if the carbons fall into contact the armature will be actuated by the magnets so as to move the lever and start the arc, and hold the carbons until the arc lengthens and the armature lever returns to the normal position. After this the carbon rod holder is released by the action of the feed mechanism, so as to feed the carbon and restore the arc to its normal length.

Fig. 278 is an elevation of the mechanism made use of in this arc lamp. Fig. 279 is a plan view. Fig. 280 is an elevation of the balancing lever and spring; Fig. 281 is a detached plan view of the pole pieces and armatures upon the friction clamp, and Fig. 282 is a section of the clamping tube.

M is a helix of coarse wire in a circuit from the lower carbon holder to the negative binding screw —. N is a helix of fine wire in a shunt between the positive binding screw + and the negative binding screw —. The upper carbon holder s is a parallel rod sliding through the plates s′ s² of the frame of the lamp, and hence the electric current passes from the positive binding

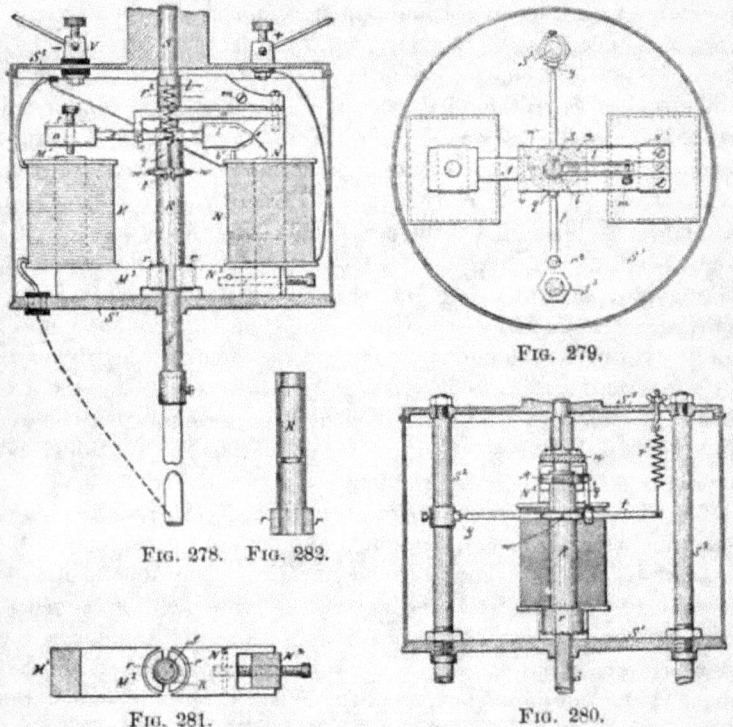

FIG. 279.

FIG. 278.   FIG. 282.

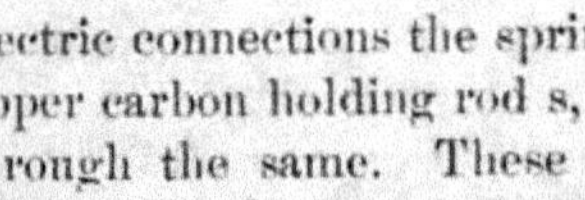

FIG. 281.

FIG. 280.

post + through the plate s², carbon holder s, and upper carbon to the lower carbon, and thence by the holder and a metallic connection to the helix M.

The carbon holders are of the usual character, and to insure electric connections the springs *l* are made use of to grasp the upper carbon holding rod s, but to allow the rod to slide freely through the same. These springs *l* may be adjusted in their pressure by the screw *m*, and the spring *l* may be sustained upon any suitable support. They are shown as connected with the upper end of the core of the magnet N.

Around the carbon-holding rod s, between the plates s' s², there is a tube, R, which forms a clamp. This tube is counter-bored, as seen in the section Fig. 282, so that it bears upon the rod s at its upper end and near the middle, and at the lower end of this tubular clamp R there are armature segments r of soft iron. A frame or arm, n, extending, preferably, from the core N², supports the lever A by a fulcrum-pin, o. This lever A has a hole, through which the upper end of the tubular clamp R passes freely, and from the lever A is a link, q, to the lever t, which lever is pivoted at y to a ring upon one of the columns s³. This lever t has an opening or bow surrounding the tubular clamp R, and there are pins or pivotal connections w between the lever t and this clamp R, and a spring, r², serves to support or suspend the weight of the parts and balance them, or nearly so. This spring is adjustable.

At one end of the lever A is a soft-iron armature block, a, over the core M' of the helix M, and there is a limiting screw, c, passing through this armature block a, and at the other end of the lever A is a soft iron armature block, b, with the end tapering or wedge-shaped, and the same comes close to and in line with the lateral projection e on the core N². The lower ends of the cores M' N² are made with laterally projecting pole-pieces M³ N³, respectively, and these pole-pieces are concave at their outer ends, and are at opposite sides of the armature segments r at the lower end of the tubular clamp R.

The operation of these devices is as follows: In the condition of inaction, the upper carbon rests upon the lower one, and when the electric current is turned on it passes freely, by the frame and spring l, through the rods and carbons to the coarse wire and helix M, and to the negative binding post V and the core M' thereby is energized. The pole piece M³ attracts the armature r, and by the lateral pressure causes the clamp R to grasp the rod s', and the lever A is simultaneously moved from the position shown by dotted lines, Fig. 278, to the normal position shown in full lines, and in so doing the link q and lever t are raised, lifting the clamp R and s, separating the carbons and forming the arc. The magnetism of the pole piece e tends to hold the lever A level, or nearly so, the core N² being energized by the current in the shunt which contains the helix N. In this position the lever A is not moved by any ordinary variation in the current, because the armature b is strongly attracted by the magnetism of c, and these parts are close to each other, and the magnetism of e acts at right angles to the magnetism of the core M'. If, now, the arc becomes too long, the current through the helix M is lessened, and the magnetism of the core N^3 is increased by the greater current passing through the shunt, and this core N^3, attracting the segmental armature r, lessens the hold of the clamp R upon the rod s, allowing the latter to slide and lessen the length of the arc, which instantly restores the magnetic equilibrium and causes the clamp R to hold the rod s. If it happens that the carbons fall into contact, then the magnetism of N^2 is lessened so much that the attraction of the magnet M will be sufficient to move the armature a and lever A so that the armature b passes above the normal position, so as to separate the carbons instantly; but when the carbons burn away, a greater amount of current will pass through the shunt until the attraction of the core N^2 will overcome the attraction of the core M' and bring the armature lever A again into the normal horizontal position, and this occurs before the feed can take place. The segmental armature pieces r are shown as nearly semicircular. They are square or of any other desired shape, the ends of the pole pieces M^3, N^3 being made to correspond in shape.

In a modification of this lamp, Mr. Tesla provided means for automatically withdrawing a lamp from the circuit, or cutting it out when, from a failure of the feed, the arc reached an abnormal length; and also means for automatically reinserting such lamp in the circuit when the rod drops and the carbons come into contact.

Fig. 283 is an elevation of the lamp with the case in section. Fig. 284 is a sectional plan at the line $x\,x$. Fig. 285 is an elevation, partly in section, of the lamp at right angles to Fig. 283. Fig. 286 is a sectional plan at the line $y\,y$ of Fig. 283. Fig. 287 is a section of the clamp in about full size. Fig. 288 is a detached section illustrating the connection of the spring to the lever that carries the pivots of the clamp, and Fig. 289 is a diagram showing the circuit-connections of the lamp.

In Fig. 283, M represents the main and N the shunt magnet, both securely fastened to the base A, which with its side columns, $s\,s$, are cast in one piece of brass or other diamagnetic material. To the magnets are soldered or otherwise fastened the brass washers or discs $a\,a\,a\,a$. Similar washers, $b\,b$, of fibre or other insulating material, serve to insulate the wires from the brass washers.

The magnets м and n are made very flat, so that their width exceeds three times their thickness, or even more. In this way a comparatively small number of convolutions is sufficient to produce the required magnetism, while a greater surface is offered for cooling off the wires.

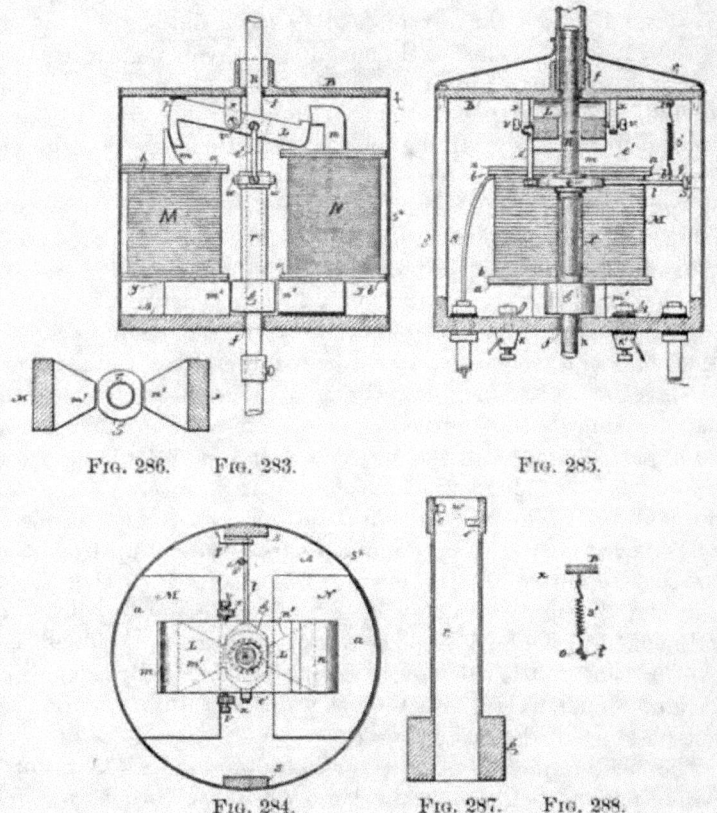

FIG. 286.   FIG. 283.   FIG. 285.

FIG. 284.   FIG. 287.   FIG. 288.

The upper pole pieces, *m n*, of the magnets are curved, as indicated in the drawings, Fig. 283. The lower pole pieces *m' n'*, are brought near together, tapering toward the armature *g*, as shown in Figs. 284 and 286. The object of this taper is to concentrate the greatest amount of the developed magnetism upon the armature, and also to allow the pull to be exerted always upon the middle of the armature *g*. This armature *g* is a piece of iron in the shape of a hollow cylinder, having on each side a segment cut away, the width of which is equal to the width of the pole pieces m' n'.

The armature is soldered or otherwise fastened to the clamp r, which is formed of a brass tube, provided with gripping-jaws e e, Fig. 287. These jaws are arcs of a circle of the diameter of the rod k, and are made of hardened German silver. The guides $f f$, through which the carbon-holding rod k slides, are made of the same material. This has the advantage of reducing greatly the wear and corrosion of the parts coming in frictional contact with the rod, which frequently causes trouble. The jaws e e are fastened to the inside of the tube r, so that one is a little lower than the other. The object of this is to provide a greater opening for the passage of the rod when the same is released by the clamp. The clamp r is supported on bearings w w, Figs. 283, 285 and 287, which are just in the middle between the jaws e e. The bearings w w are carried by a lever, t, one end of which rests upon an adjustable support, q, of the side columns, s, the other end being connected by means of the link e' to the armature-lever $\mathbf{L}$. The armature-lever $\mathbf{L}$ is a flat piece of iron in **N** shape, having its ends curved so as to correspond to the form of the upper pole-pieces of the magnets $\mathbf{M}$ and $\mathbf{N}$. It is hung upon the pivots v v, Fig. 284, which are in the jaw x of the top plate $\mathbf{B}$. This plate $\mathbf{B}$, with the jaw, is cast in one piece and screwed to the side columns, s s, that extend up from the base $\mathbf{A}$. To partly balance the overweight of the moving parts, a spring, s', Figs. 284 and 288, is fastened to the top plate, $\mathbf{B}$, and hooked to the lever t. The hook o is toward one side of the lever or bent a little sidewise, as seen in Fig. 288. By this means a slight tendency is given to swing the armature toward the pole-piece m' of the main magnet.

The binding-posts κ κ' are screwed to the base $\mathbf{A}$. A manual switch, for short-circuiting the lamp when the carbons are renewed, is also fastened to the base. This switch is of ordinary character, and is not shown in the drawings.

The rod k is electrically connected to the lamp-frame by means of a flexible conductor or otherwise. The lamp-case receives a removable cover, s^2, to inclose the parts.

The electrical connections are as indicated diagrammatically in Fig. 289. The wire in the main magnet consists of two parts, x' and p'. These two parts may be in two separated coils or in one single helix, as shown in the drawings. The part x' being normally in circuit, is, with the fine wire upon the shunt-magnet, wound and traversed by the current in the same direction, so as to tend to produce similar poles, N N or s s, on the corresponding pole-pieces of the magnets M and N. The part p' is only in circuit when the lamp is cut out, and then the current being in the opposite direction produces in the main magnet, magnetism of the opposite polarity.

The operation is as follows: At the start the carbons are to be in contact, and the current passes from the positive binding-post K to the lamp-frame, carbon-holder, upper and lower carbon, insulated return-wire in one of the side rods, and from there through the part x' of the wire on the main magnet to the nega-

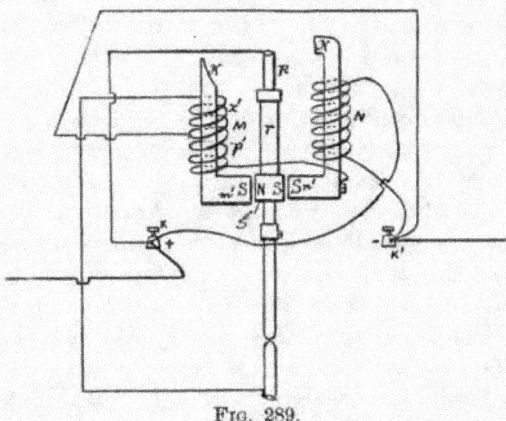

FIG. 289.

tive binding-post. Upon the passage of the current the main magnet is energized and attracts the clamping-armature g, swinging the clamp and gripping the rod by means of the gripping jaws e e. At the same time the armature lever L is pulled down and the carbons are separated. In pulling down the armature lever L the main magnet is assisted by the shunt-magnet N, the latter being magnetized by magnetic induction from the magnet M.

It will be seen that the armatures L and g are practically the keepers for the magnets M and N, and owing to this fact both magnets with either one of the armatures L and g may be considered as one horseshoe magnet, which we might term a "compound magnet." The whole of the soft-iron parts M, m', g, n', N and L form a compound magnet.

The carbons being separated, the fine wire receives a portion of the current. Now, the magnetic induction from the magnet M is such as to produce opposite poles on the corresponding ends of the magnet N; but the current traversing the helices tends to produce similar poles on the corresponding ends of both magnets, and therefore as soon as the fine wire is traversed by sufficient current the magnetism of the whole compound magnet is diminished.

With regard to the armature g and the operation of the lamp, the pole m' may be considered as the "clamping" and the pole n' as the "releasing" pole.

As the carbons burn away, the fine wire receives more current and the magnetism diminishes in proportion. This causes the armature lever L to swing and the armature g to descend gradually under the weight of the moving parts until the end p, Fig. 283, strikes a stop on the top plate, B. The adjustment is such that when this takes place the rod R is yet gripped securely by the jaws $e\,e$. The further downward movement of the armature lever being prevented, the arc becomes longer as the carbons are consumed, and the compound magnet is weakened more and more until the clamping armature g releases the hold of the gripping-jaws $e\,e$ upon the rod R, and the rod is allowed to drop a little, thus shortening the arc. The fine wire now receiving less current, the magnetism increases, and the rod is clamped again and slightly raised, if necessary. This clamping and releasing of the rod continues until the carbons are consumed. In practice the feed is so sensitive that for the greatest part of the time the movement of the rod cannot be detected without some actual measurement. During the normal operation of the lamp the armature lever L remains practically stationary, in the position shown in Fig. 283.

Should it happen that, owing to an imperfection in it, the rod and the carbons drop too far, so as to make the arc too short, or even bring the carbons in contact, a very small amount of current passes through the fine wire, and the compound magnet becomes sufficiently strong to act as at the start in pulling the armature lever L down and separating the carbons to a greater distance.

It occurs often in practical work that the rod sticks in the guides. In this case the arc reaches a great length, until it finally breaks. Then the light goes out, and frequently the fine wire is injured. To prevent such an accident Mr. Tesla provides this lamp with an automatic cut-out which operates as follows : When, upon a failure of the feed, the arc reaches a certain predetermined length, such an amount of current is diverted through the fine wire that the polarity of the compound magnet is reversed. The clamping armature g is now moved against the shunt magnet N until it strikes the releasing pole n'. As soon as the contact is established, the current passes from the positive binding post over the clamp r, armature g, insulated shunt magnet, and the helix p' upon the main magnet M to the negative binding post. In this case the current passes in the opposite direction and changes the polarity of the magnet M, at the same time maintaining by magnetic induction in the core of the shunt magnet the required magnetism without reversal of polarity, and the armature g remains against the shunt magnet pole n'. The lamp is thus cut out as long as the carbons are separated. The cut out may be used in this form without any further improvement ; but Mr. Tesla arranges it so that if the rod drops and the carbons come in contact the arc is started again. For this purpose he proportions the resistance of part p' and the number of the convolutions of the wire upon the main magnet so that when the carbons come in contact a sufficient amount of current is diverted through the carbons and the part x' to destroy or neutralize the magnetism of the compound magnet. Then the armature g, having a slight tendency to approach to the clamping pole m', comes out of contact with the releasing pole n'. As soon as this happens, the current through the part p' is interrupted, and the whole current passes through the part x. The magnet M is now strongly magnetized, the armature g is attracted, and the rod clamped. At the same time the armature lever L is pulled down out of its normal position and the arc started. In this way the lamp cuts itself out automatically when the arc gets too long, and reinserts itself automatically in the circuit if the carbons drop together.

CHAPTER XLI.

IMPROVEMENT IN "UNIPOLAR" GENERATORS.

ANOTHER interesting class of apparatus to which Mr. Tesla has directed his attention, is that of "unipolar" generators, in which a disc or a cylindrical conductor is mounted between magnetic poles adapted to produce an approximately uniform field. In the disc armature machines the currents induced in the rotating conductor flow from the centre to the periphery, or conversely, according to the direction of rotation or the lines of force as determined by the signs of the magnetic poles, and these currents are taken off usually by connections or brushes applied to the disc at points on its periphery and near its centre. In the case of the cylindrical armature machine, the currents developed in the cylinder are taken off by brushes applied to the sides of the cylinder at its ends.

In order to develop economically an electromotive force available for practicable purposes, it is necessary either to rotate the conductor at a very high rate of speed or to use a disc of large diameter or a cylinder of great length; but in either case it becomes difficult to secure and maintain a good electrical connection between the collecting brushes and the conductor, owing to the high peripheral speed.

It has been proposed to couple two or more discs together in series, with the object of obtaining a higher electro-motive force; but with the connections heretofore used and using other conditions of speed and dimension of disc necessary to securing good practicable results, this difficulty is still felt to be a serious obstacle to the use of this kind of generator. These objections Mr. Tesla has sought to avoid by constructing a machine with two fields, each having a rotary conductor mounted between its poles. The same principle is involved in the case of both forms of machine above described, but the description now given is confined to the disc type, which Mr. Tesla is inclined to favor for that machine. The discs are formed with flanges, after the manner of pulleys, and are connected together by flexible conducting bands or belts.

The machine is built in such manner that the direction of magnetism or order of the poles in one field of force is opposite to that in the other, so that rotation of the discs in the same direction develops a current in one from centre to circumference and in the other from circumference to centre. Contacts applied therefore to the shafts upon which the discs are mounted form the terminals of a circuit the electro-motive force in which is the sum of the electro-motive forces of the two discs.

It will be obvious that if the direction of magnetism in both

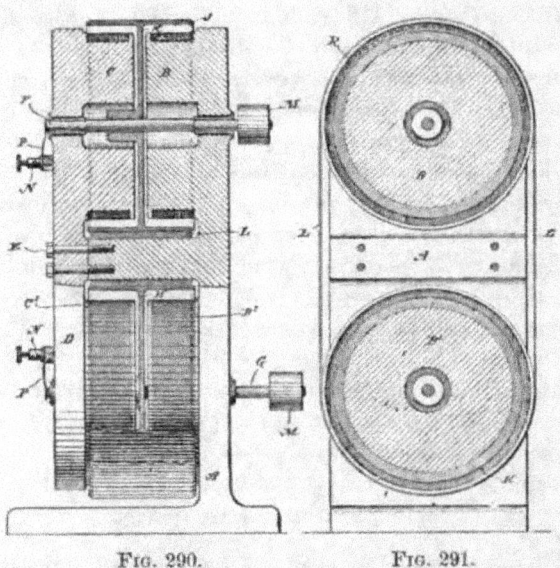

Fig. 290.　　　　　Fig. 291.

fields be the same, the same result as above will be obtained by driving the discs in opposite directions and crossing the connecting belts. In this way the difficulty of securing and maintaining good contact with the peripheries of the discs is avoided and a cheap and durable machine made which is useful for many purposes—such as for an exciter for alternating current generators, for a motor, and for any other purpose for which dynamo machines are used.

Fig. 290 is a side view, partly in section, of this machine. Fig. 291 is a vertical section of the same at right angles to the shafts.

In order to form a frame with two fields of force, a support, A, is cast with two pole pieces B B' integral with it. To this are joined by bolts E a casting D, with two similar and corresponding pole pieces c c'. The pole pieces B B' are wound and connected to produce a field of force of given polarity, and the pole pieces c c' are wound so as to produce a field of opposite polarity. The driving shafts F G pass through the poles and are journaled in insulating bearings in the casting A D, as shown.

H K are the discs or generating conductors. They are composed of copper, brass, or iron and are keyed or secured to their respective shafts. They are provided with broad peripheral flanges J. It is of course obvious that the discs may be insulated from their shafts, if so desired. A flexible metallic belt L is passed over the flanges of the two discs, and, if desired, may be used to drive one of the discs. It is better, however, to use this belt merely as a conductor, and for this purpose sheet steel, copper, or other suitable metal is used. Each shaft is provided with a driving pulley M, by which power is imparted from a driving shaft.

N N are the terminals. For the sake of clearness they are shown as provided with springs P, that bear upon the ends of the shafts. This machine, if self-exciting, would have copper bands around its poles; or conductors of any kind—such as wires shown in the drawings—may be used.

It is thought appropriate by the compiler to append here some notes on unipolar dynamos, written by Mr. Tesla, on a recent occasion.

NOTES ON A UNIPOLAR DYNAMO.[1]

It is characteristic of fundamental discoveries, of great achievements of intellect, that they retain an undiminished power upon the imagination of the thinker. The memorable experiment of Faraday with a disc rotating between the two poles of a magnet, which has borne such magnificent fruit, has long passed into every-day experience; yet there are certain features about this embryo of the present dynamos and motors which even to-day appear to us striking, and are worthy of the most careful study.

Consider, for instance, the case of a disc of iron or other metal

----

[1]. Article by Mr. Tesla, contributed to *The Electrical Engineer*, N. Y., Sept. 2, 1891.

revolving between the two opposite poles of a magnet, and the polar surfaces completely covering both sides of the disc, and assume the current to be taken off or conveyed to the same by contacts uniformly from all points of the periphery of the disc. Take first the case of a motor. In all ordinary motors the operation is dependent upon some shifting or change of the resultant of the magnetic attraction exerted upon the armature, this process being effected either by some mechanical contrivance on the motor or by the action of currents of the proper character. We may explain the operation of such a motor just as we can that of a water-wheel. But in the above example of the disc surrounded completely by the polar surfaces, there is no shifting of the magnetic action, no change whatever, as far as we know, and yet rotation ensues. Here, then, ordinary considerations do not apply : we cannot even give a superficial explanation, as in ordinary motors, and the operation will be clear to us only when we shall have recognized the very nature of the forces concerned, and fathomed the mystery of the invisible connecting mechanism.

Considered as a dynamo machine, the disc is an equally interesting object of study. In addition to its peculiarity of giving currents of one direction without the employment of commutating devices, such a machine differs from ordinary dynamos in that there is no reaction between armature and field. The armature current tends to set up a magnetization at right angles to that of the field current, but since the current is taken off uniformly from all points of the periphery, and since, to be exact, the external circuit may also be arranged perfectly symmetrical to the field magnet, no reaction can occur. This, however, is true only as long as the magnets are weakly energized, for when the magnets are more or less saturated, both magnetizations at right angles seemingly interfere with each other.

For the above reason alone it would appear that the output of such a machine should, for the same weight, be much greater than that of any other machine in which the armature current tends to demagnetize the field. The extraordinary output of the Forbes unipolar dynamo and the experience of the writer confirm this view.

Again, the facility with which such a machine may be made to excite itself is striking, but this may be due—besides to the absence of armature reaction—to the perfect smoothness of the current and non-existence of self-induction.

If the poles do not cover the disc completely on both sides, then, of course, unless the disc be properly subdivided, the machine will be very inefficient. Again, in this case there are points worthy of notice. If the disc be rotated and the field current interrupted, the current through the armature will continue to flow and the field magnets will lose their strength comparatively slowly. The reason for this will at once appear when we consider the direction of the currents set up in the disc.

Referring to the diagram Fig. 292, d represents the disc with the sliding contacts B B' on the shaft and periphery. N and s represent the two poles of a magnet. If the pole N be above, as indicated in the diagram, the disc being supposed to be in the

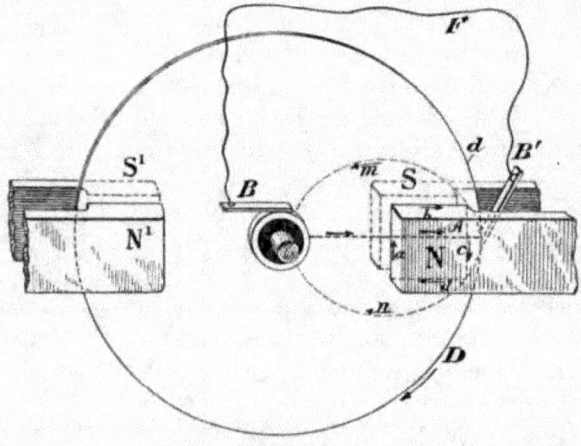

Fig. 292.

plane of the paper, and rotating in the direction of the arrow D, the current set up in the disc will flow from the centre to the periphery, as indicated by the arrow A. Since the magnetic action is more or less confined to the space between the poles N s, the other portions of the disc may be considered inactive. The current set up will therefore not wholly pass through the external circuit F, but will close through the disc itself, and generally, if the disposition be in any way similar to the one illustrated, by far the greater portion of the current generated will not appear externally, as the circuit F is practically short-circuited by the inactive portions of the disc. The direction of the resulting currents in the latter may be assumed to be as indicated by the dotted lines and arrows *m* and *n* ; and the direction of the energizing field current being indicated by the arrows *a b c d*, an inspection of the figure shows that one of the two branches of the eddy current; that is, A B' *m* B, will tend to demagnetize the field, while the other branch, that is, A B' *n* B, will have the opposite effect. Therefore, the branch A B' *m* B, that is, the one which is *approaching* the field, will repel the lines of the same, while branch A B' *n* B, that is, the one *leaving* the field, will gather the lines of force upon itself.

In consequence of this there will be a constant tendency to reduce the current flow in the path A B' *m* B, while on the other hand no such opposition will exist in path A B' *n* B, and the effect of the latter branch or path will be more or less preponderating over that of the former. The joint effect of both the assumed branch currents might be represented by that of one single current of the same direction as that energizing the field. In other words, the eddy currents circulating in the disc will energize the field magnet. This is a result quite contrary to what we might be led to suppose at first, for we would naturally expect that the resulting effect of the armature currents would be such as to oppose the field current, as generally occurs when a primary and secondary conductor are placed in inductive relations to each other. But it must be remembered that this results from the peculiar disposition in this case, namely, two paths being afforded to the current, and the latter selecting that path which offers the least opposition to its flow. From this we see that the eddy currents flowing in the disc partly energize the field, and for this reason when the field current is interrupted the currents in the disc will continue to flow, and the field magnet will lose its strength with comparative slowness and may even retain a certain strength as long as the rotation of the disc is continued.

The result will, of course, largely depend on the resistance and geometrical dimensions of the path of the resulting eddy current and on the speed of rotation; these elements, namely, determine the retardation of this current and its position relative to the field. For a certain speed there would be a maximum energizing action; then at higher speeds, it would gradually fall off to zero and finally reverse, that is, the resultant eddy current effect would be to weaken the field. The reaction would be best demonstrated experimentally by arranging the fields N s, N' s', freely movable on an axis concentric with the shaft of the disc. If the latter were rotated as before in the direction of the arrow D, the field would be dragged in the same direction with a torque, which, up to a certain point, would go on increasing with the speed of rotation, then fall off, and, passing through zero, finally become negative; that is, the field would begin to rotate in opposite direction to the disc. In experiments with alternate current motors in which the field was shifted by currents of differing phase, this interesting result was observed. For very low speeds of rotation of the field the motor would show a torque of 900 lbs. or more, measured on a pulley 12 inches in diameter. When the speed of rotation of the poles was increased, the torque would diminish, would finally go down to zero, become negative, and then the armature would begin to rotate in opposite direction to the field.

To return to the principal subject; assume the conditions to be such that the eddy currents generated by the rotation of the disc strengthen the field, and suppose the latter gradually removed while the disc is kept rotating at an increased rate. The current, once started, may then be sufficient to maintain itself and even increase in strength, and then we have the case of Sir William Thomson's "current accumulator." But from the above considerations it would seem that for the success of the experiment the employment of a disc *not subdivided*[1] would be essential, for if there should be a radial subdivision, the eddy currents could not form and the self-exciting action would cease. If such a radially subdivided disc were used it would be necessary to connect the spokes by a conducting rim or in any proper manner so as to form a symmetrical system of closed circuits.

The action of the eddy currents may be utilized to excite a machine of any construction. For instance, in Figs. 293 and 294 an arrangement is shown by which a machine with a disc armature might be excited. Here a number of magnets, N s, N s, are placed radially on each side of a metal disc D carrying on its rim a set of insulated coils, c c. The magnets form two separate fields, an internal and an external one, the solid disc rotating in the

---

1. Mr. Tesla here refers to an interesting article which appeared in July, 1865, in the *Phil. Magazine*, by Sir W. Thomson, in which Sir William, speaking of his "uniform electric current accumulator," assumes that for self-excitation it is desirable to subdivide the disc into an infinite number of infinitely thin spokes, in order to prevent diffusion of the current. Mr. Tesla shows that diffusion is absolutely necessary for the excitation and that when the disc is subdivided no excitation can occur.

field nearest the axis, and the coils in the field further from it. Assume the magnets slightly energized at the start ; they could be strengthened by the action of the eddy currents in the solid disc so as to afford a stronger field for the peripheral coils. Although there is no doubt that under proper conditions a machine might be excited in this or a similar manner, there being sufficient experimental evidence to warrant such an assertion, such a mode of excitation would be wasteful.

But a unipolar dynamo or motor, such as shown in Fig. 292, may be excited in an efficient manner by simply properly subdividing the disc or cylinder in which the currents are set up, and it is practicable to do away with the field coils which are usually employed. Such a plan is illustrated in Fig. 295. The disc or

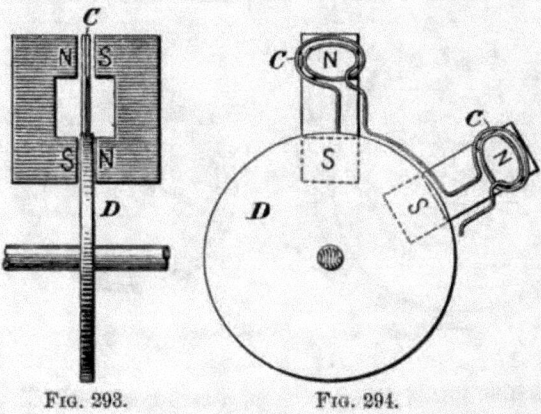

FIG. 293.   FIG. 294.

cylinder D is supposed to be arranged to rotate between the two poles N and S of a magnet, which completely cover it on both sides, the contours of the disc and poles being represented by the circles d and d^1 respectively, the upper pole being omitted for the sake of clearness. The cores of the magnet are supposed to be hollow, the shaft c of the disc passing through them. If the unmarked pole be below, and the disc be rotated screw fashion, the current will be, as before, from the centre to the periphery, and may be taken off by suitable sliding contacts, B B', on the shaft and periphery respectively. In this arrangement the current flowing through the disc and external circuit will have no appreciable effect on the field magnet.

But let us now suppose the disc to be subdivided spirally, as indicated by the full or dotted lines, Fig. 295. The difference of potential between a point on the shaft and a point on the periphery will remain unchanged, in sign as well as in amount. The only difference will be that the resistance of the disc will be augmented and that there will be a greater fall of potential from a point on the shaft to a point on the periphery when the same current is traversing the external circuit. But since the current is forced to follow the lines of subdivision, we see that it will tend either to energize or de-energize the field, and this will depend, other things being equal, upon the direction of the lines of subdivision. If the subdivision be as indicated by the full lines in Fig. 295, it is evident that if the current is of the same direction as before, that is, from centre to periphery, its effect will be to strengthen the field magnet; whereas, if the subdivision be as in-

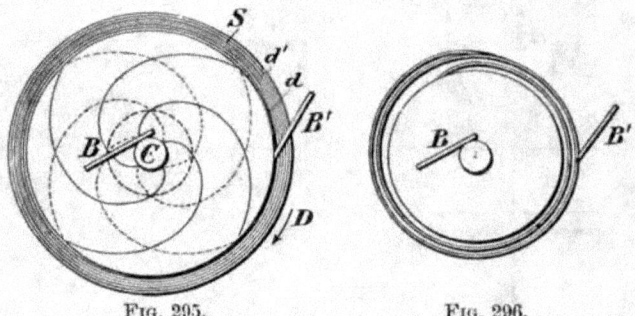

FIG. 295.   FIG. 296.

dicated by the dotted lines, the current generated will tend to weaken the magnet. In the former case the machine will be capable of exciting itself when the disc is rotated in the direction of arrow D; in the latter case the direction of rotation must be reversed. Two such discs may be combined, however, as indicated, the two discs rotating in opposite fields, and in the same or opposite direction.

Similar disposition may, of course, be made in a type of machine in which, instead of a disc, a cylinder is rotated. In such unipolar machines, in the manner indicated, the usual field coils and poles may be omitted and the machine may be made to consist only of a cylinder or of two discs enveloped by a metal casting.

Instead of subdividing the disc or cylinder spirally, as indicated in Fig. 295, it is more convenient to interpose one or more turns

between the disc and the contact ring on the periphery, as illustrated in Fig. 296.

A Forbes dynamo may, for instance, be excited in such a manner. In the experience of the writer it has been found that instead of taking the current from two such discs by sliding
contacts, as usual, a flexible conducting belt may be employed
to advantage. The discs are in such case provided with large
flanges, affording a very great contact surface. The belt should
be made to bear on the flanges with spring pressure to take up
the expansion. Several machines with belt contact were constructed by the writer two years ago, and worked satisfactorily;
but for want of time the work in that direction has been temporarily suspended. A number of features pointed out above have
also been used by the writer in connection with some types of
alternating current motors.

PART IV.

APPENDIX.—EARLY PHASE MOTORS AND THE TESLA MECHANICAL AND ELECTRICAL OSCILLATOR.

CHAPTER XLII.

MR. TESLA'S PERSONAL EXHIBIT AT THE WORLD'S FAIR.

WHILE the exhibits of firms engaged in the manufacture of electrical apparatus of every description at the Chicago World's Fair, afforded the visitor ample opportunity for gaining an excellent knowledge of the state of the art, there were also numbers of exhibits which brought out in strong relief the work of the individual inventor, which lies at the foundation of much, if not all, industrial or mechanical achievement. Prominent among such personal exhibits was that of Mr. Tesla, whose apparatus occupied part of the space of the Westinghouse Company, in Electricity Building.

This apparatus represented the results of work and thought covering a period of ten years. It embraced a large number of different alternating motors and Mr. Tesla's earlier high frequency apparatus. The motor exhibit consisted of a variety of fields and armatures for two, three and multiphase circuits, and gave a fair idea of the gradual evolution of the fundamental idea of the rotating magnetic field. The high frequency exhibit included Mr. Tesla's earlier machines and disruptive discharge coils and high frequency transformers, which he used in his investigations and some of which are referred to in his papers printed in this volume.

Fig. 297 shows a view of part of the exhibits containing the motor apparatus. Among these is shown at A a large ring intended to exhibit the phenomena of the rotating magnetic field. The field produced was very powerful and exhibited striking effects, revolving copper balls and eggs and bodies of various shapes at considerable distances and at great speeds. This ring was wound for two-phase circuits, and the winding was so distributed that a practically uniform field was obtained. This ring was prepared for Mr. Tesla's exhibit by Mr. C. F. Scott, electrician of the Westinghouse Electric and Manufacturing Company.

Fig. 297.

A smaller ring, shown at B, was arranged like the one exhibited at A but designed especially to exhibit the rotation of an armature in a rotating field. In connection with these two rings there was an interesting exhibit shown by Mr. Tesla which consisted of a magnet with a coil, the magnet being arranged to rotate in bearings. With this magnet he first demonstrated the identity between a rotating field and a rotating magnet; the latter, when rotating, exhibited the same phenomena as the rings when they were energized by currents of differing phase. Another prominent exhibit was a model illustrated at c which is a two-phase motor, as well as an induction motor and transformer. It consists of a large outer ring of laminated iron wound with two superimposed, separated windings which can be connected in a variety of ways. This is one of the first models used by Mr. Tesla as an induction motor and rotating transformer. The armature was either a steel or wrought iron disc with a closed coil. When the motor was operated from a two phase generator the windings were connected in two groups, as usual. When used as an induction motor, the current induced in one of the windings of the ring was passed through the other winding on the ring and so the motor operated with only two wires. When used as a transformer the outer winding served, for instance, as a secondary and the inner as a primary. The model shown at D is one of the earliest rotating field motors, consisting of a thin iron ring wound with two sets of coils and an armature consisting of a series of steel discs partly cut away and arranged on a small arbor.

At E is shown one of the first rotating field or induction motors used for the regulation of an arc lamp and for other purposes. It comprises a ring of discs with two sets of coils having different self-inductions, one set being of German silver and the other of copper wire. The armature is wound with two closed-circuited coils at right angles to each other. To the armature shaft are fastened levers and other devices to effect the regulation. At F is shown a model of a magnetic lag motor; this embodies a casting with pole projections protruding from two coils between which is arranged to rotate a smooth iron body. When an alternating current is sent through the two coils the pole projections of the field and armature within it are similarly magnetized, and upon the cessation or reversal of the current the armature and field repel each other and rotation is produced in this way.

Another interesting exhibit, shown at G, is an early model of a two field motor energized by currents of different phase. There are two independent fields of laminated iron joined by brass bolts; in each field is mounted an armature, both armatures being on the same shaft. The armatures were originally so arranged as to be placed in any position relatively to each other, and the fields also were arranged to be connected in a number of ways. The motor has served for the exhibition of a number of features; among other things, it has been used as a dynamo for the production of currents of any frequency between wide limits. In this case the field, instead of being energized by direct current, was energized by currents differing in phase, which

FIG. 298.

produced a rotation of the field; the armature was then rotated in the same or in opposite direction to the movement of the field; and so any number of alternations of the currents induced in the armature, from a small to a high number, determined by the frequency of the energizing field coils and the speed of the armature, was obtained.

The models H, I, J, represent a variety of rotating field, synchronous motors which are of special value in long distance transmission work. The principle embodied in these motors was enunciated by Mr. Tesla in his lecture before the American Institute of Electrical Engineers, in May, 1888[1]. It involves the production

1. See Part I, Chap. III, page 9.

of the rotating field in one of the elements of the motor by currents differing in phase and energizing the other element by direct currents. The armatures are of the two and three phase type. κ is a model of a motor shown in an enlarged view in Fig. 298. This machine, together with that shown in Fig. 299, was exhibited at the same lecture, in May, 1888. They were the first rotating field motors which were independently tested, having for that purpose been placed in the hands of Prof. Anthony in the winter of 1887–88. From these tests it was shown that the efficiency and output of these motors was quite satisfactory in every respect.

It was intended to exhibit the model shown in Fig. 299, but it was unavailable for that purpose owing to the fact that it was

FIG. 299.

some time ago handed over to the care of Prof. Ayrton in England. This model was originally provided with twelve independent coils; this number, as Mr. Tesla pointed out in his first lecture, being divisible by two and three, was selected in order to make various connections for two and three-phase operations, and during Mr. Tesla's experiments was used in many ways with from two to six phases. The model, Fig. 298, consists of a magnetic frame of laminated iron with four polar projections between which an armature is supported on brass bolts passing through the frame. A great variety of armatures was used in connection with these two and other fields. Some of the armatures are shown in front on the table, Fig. 297, and several are also shown enlarged in Figs. 300 to 310. An interesting exhibit is that shown at L, Fig. 297. This is an armature of hardened steel which was used in a demon-

stration before the Society of Arts in Boston, by Prof. Anthony.
Another curious exhibit is shown enlarged in Fig. 301. This
consists of thick discs of wrought iron placed lengthwise, with a
mass of copper cast around them. The discs were arranged
longitudinally to afford an easier starting by reason of the induced
current formed in the iron discs, which differed in phase from
those in the copper. This armature would start with a single circuit and run in synchronism, and represents one of the earliest
types of such an armature. Fig. 305 is another striking exhibit.

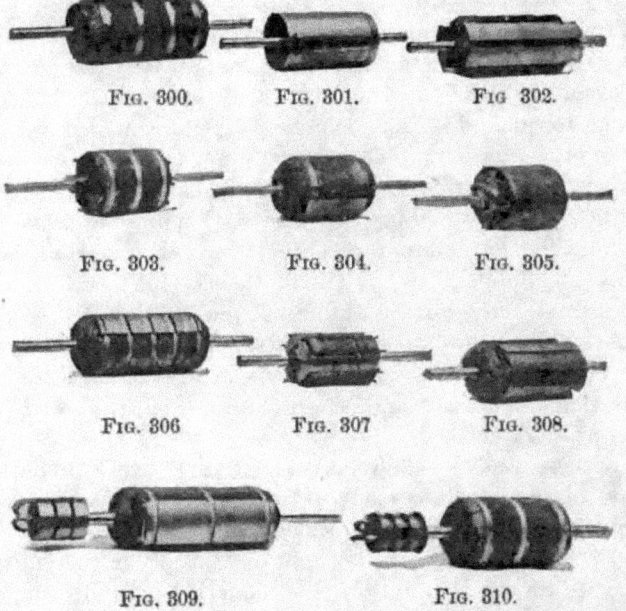

FIG. 300.   FIG. 301.   FIG 302.

FIG. 303.   FIG. 304.   FIG. 305.

FIG. 306   FIG. 307   FIG. 308.

FIG. 309.   FIG. 310.

This is one of the earliest types of an armature with holes beneath
the periphery, in which copper conductors are imbedded. The
armature has eight closed circuits and was used in many different
ways. Fig. 304 is a type of synchronous armature consisting of
a block of soft steel wound with a coil closed upon itself. This
armature was used in connection with the field shown in Fig. 298
and gave excellent results.

Fig. 302 represents a synchronous armature with a large coil
around a body of iron. There is another very small coil at right
angles to the first. This small coil was used for the purpose of

increasing the starting torque and was found very effective in this connection. Figs. 306 and 308 show a favorite construction of armature; the iron body is made up of two sets of discs cut away and placed at right angles to each other, the interstices being wound with coils. The one shown in Fig. 308 is provided with an additional groove on each of the projections formed by the discs, for the purpose of increasing the starting torque by a wire wound in these projections. Fig. 307 is a form of armature similarly constructed, but with four independent coils wound upon the four projections. This armature was used to reduce the speed of the motor with reference to that of the generator. Fig. 300 is still another armature with a great number of independent circuits closed upon themselves, so that all the dead points on the armature are done away with, and the armature has a large starting torque. Fig. 303 is another type of armature for a four-pole motor but with coils wound upon a smooth surface. A number of these armatures have hollow shafts, as they have been used in many ways. Figs. 309 and 310 represent armatures to which either alternating or direct current was conveyed by means of sliding rings. Fig. 309 consists of a soft iron body with a single coil wound around it, the ends of the coil being connected to two sliding rings to which, usually, direct current was conveyed. The armature shown in Fig. 310 has three insulated rings on a shaft and was used in connection with two or three phase circuits.

All these models shown represent early work, and the enlarged engravings are made from photographs taken early in 1888. There is a great number of other models which were exhibited, but which are not brought out sharply in the engraving, Fig. 297. For example at M is a model of a motor comprising an armature with a hollow shaft wound with two or three coils for two or three-phase circuits; the armature was arranged to be stationary and the generating circuits were connected directly to the generator. Around the armature is arranged to rotate on its shaft a casting forming six closed circuits. On the outside this casting was turned smooth and the belt was placed on it for driving with any desired appliance. This also is a very early model.

On the left side of the table there are seen a large variety of models, N, O, P, etc., with fields of various shapes. Each of these models involves some distinct idea and they all represent gradual development chiefly interesting as showing Mr. Tesla's efforts to adapt his system to the existing high frequencies.

On the right side of the table, at s, т, are shown, on separate supports, larger and more perfected armatures of commercial motors, and in the space around the table a variety of motors and generators supplying currents to them was exhibited.

The high frequency exhibit embraced Mr. Tesla's first original apparatus used in his investigations. There was exhibited a glass tube with one layer of silk-covered wire wound at the top and a copper ribbon on the inside. This was the first disruptive discharge coil constructed by him. At ʋ is shown the disruptive

FIG. 311.

discharge coil exhibited by him in his lecture before the American Institute of Electrical Engineers, in May, 1891.[1] At v and w are shown some of the first high frequency transformers. A number of various fields and armatures of small models of high frequency apparatus as shown at x and y, and others not visible in the picture, were exhibited. In the annexed space the dynamo then used by Mr. Tesla at Columbia College was exhibited; also another form of high frequency dynamo used.

In this space also was arranged a battery of Leyden jars and his large disruptive discharge coil which was used for exhibiting

1. See Part II, Chap. XXVI., page 145.

the light phenomena in the adjoining dark room. The coil was operated at only a small fraction of its capacity, as the necessary condensers and transformers could not be had and as Mr. Tesla's stay was limited to one week; notwithstanding, the phenomena were of a striking character. In the room were arranged two large plates placed at a distance of about eighteen feet from each other. Between them were placed two long tables with all sorts of phosphorescent bulbs and tubes; many of these were prepared with great care and marked legibly with the names which would shine with phosphorescent glow. Among them were some with the names of Helmholtz, Faraday, Maxwell, Henry, Franklin, etc. Mr. Tesla had also not forgotten the greatest living poet of his own country, Zmaj Jovan; two or three were prepared with inscriptions, like "Welcome, Electricians," and produced a beautiful effect. Each represented some phase of this work and stood for some individual experiment of importance. Outside the room was the small battery seen in Fig. 311, for the exhibition of some of the impedance and other phenomena of interest. Thus, for instance, a thick copper bar bent in arched form was provided with clamps for the attachment of lamps, and a number of lamps were kept at incandescence on the bar; there was also a little motor shown on the table operated by the disruptive discharge.

As will be remembered by those who visited the Exposition, the Westinghouse Company made a fine exhibit of the various commercial motors of the Tesla system, while the twelve generators in Machinery Hall were of the two-phase type constructed for distributing light and power. Mr. Tesla, also exhibited some models of his oscillators.

CHAPTER XLIII.

The Tesla Mechanical and Electrical Oscillators.

On the evening of Friday, August 25, 1893, Mr. Tesla delivered a lecture on his mechanical and electrical oscillators, before the members of the Electrical Congress, in the hall adjoining the Agricultural Building, at the World's Fair, Chicago. Besides the apparatus in the room, he employed an air compressor, which was driven by an electric motor.

Mr. Tesla was introduced by Dr. Elisha Gray, and began by stating that the problem he had set out to solve was to construct, first, a mechanism which would produce oscillations of a perfectly constant period independent of the pressure of steam or air applied, within the widest limits, and also independent of frictional losses and load. Secondly, to produce electric currents of a perfectly constant period independently of the working conditions, and to produce these currents with mechanism which should be reliable and positive in its action without resorting to spark gaps and breaks. This he successfully accomplished in his apparatus, and with this apparatus, now, scientific men will be provided with the necessaries for carrying on investigations with alternating currents with great precision. These two inventions Mr. Tesla called, quite appropriately, a mechanical and an electrical oscillator, respectively.

The former is substantially constructed in the following way. There is a piston in a cylinder made to reciprocate automatically by proper dispositions of parts, similar to a reciprocating tool. Mr. Tesla pointed out that he had done a great deal of work in perfecting his apparatus so that it would work efficiently at such high frequency of reciprocation as he contemplated, but he did not dwell on the many difficulties encountered. He exhibited, however, the pieces of a steel arbor which had been actually torn apart while vibrating against a minute air cushion.

With the piston above referred to there is associated in one of his models in an independent chamber an air spring, or dash pot, or else he obtains the spring within the chambers of the oscillator itself. To appreciate the beauty of this it is only necessary to say that in that disposition, as he showed it, no matter what the rigidity of the spring and no matter what the weight of the moving parts, in other words, no matter what the period of vibrations, the vibrations of the spring are always isochronous with the applied pressure. Owing to this, the results obtained with these vibrations are truly wonderful. Mr. Tesla provides for an air spring of tremendous rigidity, and he is enabled to vibrate big weights at an enormous rate, considering the inertia, owing to the recoil of the spring. Thus, for instance, in one of these experiments, he vibrates a weight of approximately 20 pounds at the rate of about 80 per second and with a stroke of about $\frac{7}{8}$ inch, but by shortening the stroke the weight could be vibrated many hundred times, and has been, in other experiments.

To start the vibrations, a powerful blow is struck, but the adjustment can be so made that only a minute effort is required to start, and, even without any special provision it will start by merely turning on the pressure suddenly. The vibration being, of course, isochronous, any change of pressure merely produces a shortening or lengthening of the stroke. Mr. Tesla showed a number of very clear drawings, illustrating the construction of the apparatus from which its working was plainly discernible. Special provisions are made so as to equalize the pressure within the dash pot and the outer atmosphere. For this purpose the inside chambers of the dash pot are arranged to communicate with the outer atmosphere so that no matter how the temperature of the enclosed air might vary, it still retains the same mean density as the outer atmosphere, and by this means a spring of constant rigidity is obtained. Now, of course, the pressure of the atmosphere may vary, and this would vary the rigidity of the spring, and consequently the period of vibration, and this feature constitutes one of the great beauties of the apparatus; for, as Mr. Tesla pointed out, this mechanical system acts exactly like a string tightly stretched between two points, and with fixed nodes, so that slight changes of the tension do not in the least alter the period of oscillation.

The applications of such an apparatus are, of course, numerous and obvious. The first is, of course, to produce electric currents, and by a number of models and apparatus on the lecture platform, Mr. Tesla showed how this could be carried out in practice by combining an electric generator with his oscillator. He pointed out what conditions must be observed in order that the period of vibration of the electrical system might not disturb the mechanical oscillation in such a way as to alter the periodicity, but merely to shorten the stroke. He combines a condenser with a self-induction, and gives to the electrical system the same period as that at which the machine itself oscillates, so that both together then fall in step and electrical and mechanical resonance is obtained, and maintained absolutely unvaried.

Next he showed a model of a motor with delicate wheelwork, which was driven by these currents at a constant speed, no matter what the air pressure applied was, so that this motor could be employed as a clock. He also showed a clock so constructed that it could be attached to one of the oscillators, and would keep absolutely correct time. Another curious and interesting feature which Mr. Tesla pointed out was that, instead of controlling the motion of the reciprocating piston by means of a spring, so as to obtain isochronous vibration, he was actually able to control the mechanical motion by the natural vibration of the electro-magnetic system, and he said that the case was a very simple one, and was quite analogous to that of a pendulum. Thus, supposing we had a pendulum of great weight, preferably, which would be maintained in vibration by force, periodically applied; now that force, no matter how it might vary, although it would oscillate the pendulum, would have no control over its period.

Mr. Tesla also described a very interesting phenomenon which he illustrated by an experiment. By means of this new apparatus, he is able to produce an alternating current in which the E. M. F. of the impulses in one direction preponderates over that of those in the other, so that there is produced the effect of a direct current. In fact he expressed the hope that these currents would be capable of application in many instances, serving as direct currents. The principle involved in this preponderating E. M. F. he explains in this way: Suppose a conductor is moved into the magnetic field and then suddenly withdrawn. If the current is not retarded, then the work performed will be a mere fractional one; but if the current is retarded, then the magnetic field acts as a spring. Imagine that the motion of the conductor is arrested by the current generated, and that at the instant when it stops to move into the field, there is still the maximum current flowing in the conductor; then this current will, according to Lenz's law, drive the conductor out of the field again, and if the conductor has no resistance, then it would leave the field with the velocity it entered it. Now it is clear that if, instead of simply depending on the current to drive the conductor out of the field, the mechanically applied force is so timed that it helps the conductor to get out of the field, then it might leave the field with higher velocity than it entered it, and thus one impulse is made to preponderate in E. M. F. over the other.

With a current of this nature, Mr. Tesla energized magnets strongly, and performed many interesting experiments bearing out the fact that one of the current impulses preponderates. Among them was one in which he attached to his oscillator a ring magnet with a small air gap between the poles. This magnet was oscillated up and down 80 times a second. A copper disc, when inserted within the air gap of the ring magnet, was brought into rapid rotation. Mr. Tesla remarked that this experiment also seemed to demonstrate that the lines of flow of current through a metallic mass are disturbed by the presence of a magnet in a manner quite independently of the so-called Hall effect. He showed also a very interesting method of making a connection with the oscillating magnet. This was accomplished by attaching to the magnet small insulated steel rods, and connecting to these rods the ends of the energizing coil. As the magnet was vibrated, stationary nodes were produced in the steel rods, and at these points the terminals of a direct current source were attached. Mr. Tesla also pointed out that one of the uses of currents, such as those produced in his apparatus, would be to select any given one of a number of devices connected to the same circuit by picking out the vibration by resonance. There is indeed little doubt that with Mr. Tesla's devices, harmonic and synchronous telegraphy will receive a fresh impetus, and vast possibilities are again opened up.

Mr. Tesla was very much elated over his latest achievements, and said that he hoped that in the hands of practical, as well as scientific men, the devices described by him would yield important results. He laid special stress on the facility now afforded for investigating the effect of mechanical vibration in all directions, and also showed that he had observed a number of facts in connection with iron cores.

The engraving, Fig. 312, shows, in perspective, one of the forms of apparatus used by Mr. Tesla in his earlier investigations in this field of work, and its interior construction is made plain by the sectional view shown in Fig. 313. It will be noted that the piston P is fitted into the hollow of a cylinder c which is provided with channel ports o o, and I, extending all around the inside surface. In this particular apparatus there are two channels o o

FIG. 312.

for the outlet of the working fluid and one, I, for the inlet. The piston P is provided with two slots s s' at a carefully determined distance, one from the other. The tubes T T which are screwed into the holes drilled into the piston, establish communication between the slots s s' and chambers on each side of the piston, each of these chambers connecting with the slot which is remote from it. The piston P is screwed tightly on a shaft A which passes through fitting boxes at the end of the cylinder c. The boxes project to a carefully determined distance into the hollow of the cylinder c, thus determining the length of the stroke.

Surrounding the whole is a jacket j. This jacket acts chiefly to diminish the sound produced by the oscillator and as a jacket when the oscillator is driven by steam, in which case a somewhat different arrangement of the magnets is employed. The apparatus here illustrated was intended for demonstration purposes, air being used as most convenient for this purpose.

A magnetic frame M M is fastened so as to closely surround the oscillator and is provided with energizing coils which establish

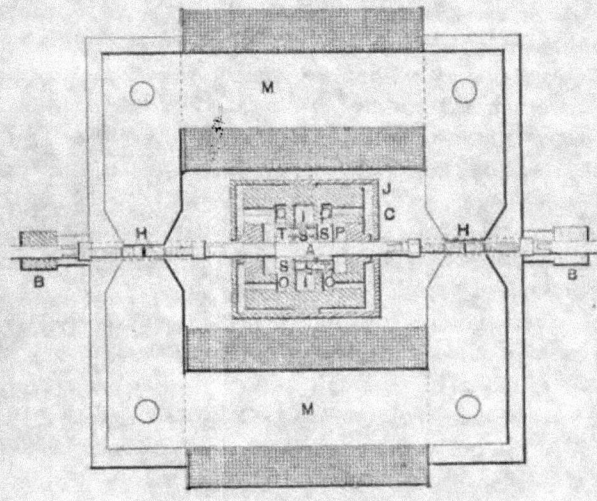

Fig. 313.

two strong magnetic fields on opposite sides. The magnetic frame is made up of thin sheet iron. In the intensely concentrated field thus produced, there are arranged two pairs of coils H H supported in metallic frames which are screwed on the shaft A of the piston and have additional bearings in the boxes B B on each side. The whole is mounted on a metallic base resting on two wooden blocks.

The operation of the device is as follows: The working fluid being admitted through an inlet pipe to the slot I and the piston being supposed to be in the position indicated, it is sufficient, though not necessary, to give a gentle tap on one of the shaft

ends protruding from the boxes B.   Assume that the motion imparted be such as to move the piston to the left (when looking at
the diagram) then the air rushes through the slot s' and tube T
into the chamber to the left.   The pressure now drives the piston towards the right and, owing to its inertia, it overshoots the
position of equilibrium and allows the air to rush through the
slot s and tube T into the chamber to the right, while the communication to the left hand chamber is cut off, the air of the
latter chamber escaping through the outlet o on the left.   On
the return stroke a similar operation takes place on the right
hand side.   This oscillation is maintained continuously and the
apparatus performs vibrations from a scarcely perceptible quiver
amounting to no more than ¹ of an inch, up to vibrations of a little
over ⅔ of an inch, according to the air pressure and load.   It is
indeed interesting to see how an incandescent lamp is kept burning with the apparatus showing a scarcely perceptible quiver.

   To perfect the mechanical part of the apparatus so that oscillations are maintained economically was one thing, and Mr. Tesla
hinted in his lecture at the great difficulties he had first encountered to accomplish this.   But to produce oscillations which would
be of constant period was another task of no mean proportions.
As already pointed out, Mr. Tesla obtains the constancy of period
in three distinct ways.   Thus, he provides properly calculated
chambers, as in the case illustrated, in the oscillator itself ; or he associates with the oscillator an air spring of constant resilience.   But
the most interesting of all, perhaps, is the maintenance of the constancy of oscillation by the reaction of the electromagnetic part of
the combination.   Mr. Tesla winds his coils, by preference, for high
tension and associates with them a condenser, making the natural
period of the combination fairly approximating to the average period
at which the piston would oscillate without any particular provision
being made for the constancy of period under varying pressure
and load.   As the piston with the coils is perfectly free to move,
it is extremely susceptible to the influence of the natural vibration set up in the circuits of the coils H H.   The mechanical efficiency of the apparatus is very high owing to the fact that friction
is reduced to a minimum and the weights which are moved are
small ; the output of the oscillator is therefore a very large one.

   Theoretically considered, when the various advantages which
Mr. Tesla holds out are examined, it is surprising, considering
the simplicity of the arrangement, that nothing was done in this

direction before. No doubt many inventors, at one time or other, have entertained the idea of generating currents by attaching a coil or a magnetic core to the piston of a steam engine, or generating currents by the vibrations of a tuning fork, or similar devices, but the disadvantages of such arrangements from an engineering standpoint must be obvious. Mr. Tesla, however, in the introductory remarks of his lecture, pointed out how by a series of conclusions he was driven to take up this new line of work by the necessity of producing currents of constant period and as a result of his endeavors to maintain electrical oscillation in the most simple and economical manner.

INDEX.

www.ingramcontent.com/pod-product-compliance
Lightning Source LLC
Chambersburg PA
CBHW081253170526
45165CB00011B/3299